WHAT EVERY CHEMICAL TECHNOLOGIST WANTS TO KNOW ABOUT...

Volume V

RESINS

Compiled by

Michael and Irene Ash

Chemical Publishing Co., Inc.
New York, N.Y.

Resins, Volume 5

ISBN: 978-0-8206-0058-1

Chemical Publishing Company:
www.chemical-publishing.com
www.chemicalpublishing.net

First Edition:

© **Chemical Publishing Company, Inc.** - New York 1990

Second Impression:

Chemical Publishing Company, Inc. - 2011

Printed in the United States of America

PREFACE

This reference book is the fifth volume in the set of books entitled WHAT EVERY CHEMICAL TECHNOLOGIST WANTS TO KNOW . . . SERIES. This compendium serves a unique function for those involved in the chemical industry—it provides the necessary information for making the decision as to which trademark chemical product is most suitable for a particular application.

The chemicals included in this fifth book of the series are resins, however, complete cross-referencing is provided for the multiple functions or classification of all the chemicals.

The first section which is the major portion of each volume contains the most common generic name of the chemicals as the main entry. All these generic entries are in alphabetical order. Synonyms for these chemicals are then listed. The CTFA name appears alongside the appropriate generic name. The structural and/or molecular formula of the chemical is listed whenever possible. The generic chemical is sold under various tradenames and these are listed here in alphabetical order for ease of reference along with their manufacturer in parentheses. *Modifications/Specialty Grades* lists those tradenames that contain or omit additives so that they are used for a variety of specialty purposes. The *Category* subheading classifies the resin by thermoplastic and/or thermoset types. The *Processing* category lists the tradenames that have been formulated to be used for specific or generalized plastics processing. Because of differences in form, activity, etc., individual tradenames of the generic chemical are used in particular applications more frequently. These are delineated in the *Applications* section. The differences in properties, toxicity/handling, storage/handling, and standard packaging are specified in the subsequent sections wherever distinguishing characteristics are known.

The second section of the volume TRADENAME PRODUCTS AND GENERIC EQUIVALENTS helps the user who only knows a chemical by one tradename to locate its main entry in section 1. The user can look up this tradename in this section of the book and be referred to the appropriate, main-entry, generic chemical name.

The third section GENERIC CHEMICAL SYNONYMS AND CROSS REFERENCES provides a way of locating the main entries by knowing only one of the synonyms. If the generic chemical is not in the volume, it will refer you to the volume in which it is contained.

The fourth section TRADENAME PRODUCT MANUFACTURERS lists the full addresses of the companies that manufacture or distribute the tradename products found in the first section.

The following is a list of the six volumes that comprise this series:

Volume I	Emulsifiers and Wetting Agents
Volume II	Dispersants, Solvents and Solubilizers
Volume III	Plasticizers, Stabilizers and Thickeners
Volume IV	Conditioners, Emollients and Lubricants
Volume V	Resins
Volume VI	Polymers and Plastics

This series has been made possible through long hours of research and compilation and the dedication and tireless efforts of Roberta Dakan who helped make this distinctive series possible. Our appreciation is extended to all the chemical manufacturers and distributors who supplied the technical information.

M. and I. Ash

NOTE

The information contained in this series is accurate to the best of our knowledge; however, no liability will be assumed by the publisher for the correctness or comprehensiveness of such information. The determination of the suitability of any of the products for prospective use is the responsibility of the user. It is herewith recommended that those who plan to use any of the products referenced seek the manufacturer's instructions for the handling of that particular chemical.

OTHER BOOKS BY MICHAEL AND IRENE ASH

A Formulary of Paints and Other Coatings, Volumes I and II
A Formulary of Detergents and Other Cleaning Agents
A Formulary of Adhesives and Sealants
A Formulary of Cosmetic Preparations
The Thesaurus of Chemical Products, Volumes I and II
Encyclopedia of Industrial Chemical Additives, Volumes I–IV
Encyclopedia of Surfactants, Volumes I–IV
Encyclopedia of Plastics, Polymers and Resins, Volumes I–IV
What Every Chemical Technologist Wants to Know About. . .
    Volume I—Emulsifiers and Wetting Agents
    Volume II—Dispersants, Solvents and Solubilizers
    Volume III—Plasticizers, Stabilizers and Thickeners
    Volume IV—Conditioners, Emollients, and Lubricants
Chemical Products Desk Reference

ABBREVIATIONS

@	at
anhyd.	anhydrous
APHA	American Public Health Association
approx.	approximately
aq.	aqueous
ASTM	American Society for Testing and Materials
avg.	average
B.P.	boiling point
Btu	British thermal unit
C	degrees Centigrade
CAS	Chemical Abstracts Service
cc	cubic centimeter(s)
CC	closed cup
cm	centimeter(s)
cm^3	cubic centimeter(s)
COC	Cleveland Open Cup
compd.	compound, compounded
conc.	concentrated, concentration
cP, cps	centipoise
cs, cSt	centistokes
CTFA	Cosmetic, Toiletry and Fragrance Association
DEA	diethanolamine
disp	dispersible, dispersion
dist	distilled
DOT	Department of Transportation
DW	distilled water
EO	ethylene oxide
equiv.	equivalent
F	degrees Fahrenheit
F.P.	freezing point
FDA	Food and Drug Administration
ft^3	cubic foot, cubic feet
g	gram(s)
gal	gallon(s)
HLB	hydrophile-lipophile balance
insol.	insoluble
IPA	isopropyl alcohol
kg	kilogram(s)
l, L	liter(s)
lb	pound(s)
M.P.	melting point
M.W.	molecular weight
max	maximum
MEA	monoethanolamine
MEK	methyl ethyl ketone
mfg.	manufacture
MIBK	methyl isobutyl ketone
min	minute(s)
min.	mineral, minimum
MIPA	monoisopropanolamine

misc.	miscible
ml	milliliter(s)
mm	millimeter(s)
NF	National Formulary
no.	number
o/w	oil-in-water
OC	open crucible
PEG	polyethylene glycol
pH	hydrogen-ion concentration
pkgs	packages
PMCC	Pensky Marten closed cup
POE	polyoxyethylene, polyoxyethylated
POP	polyoxypropylene
PPG	polypropylene glycol
pt.	point
R&B	Ring & Ball
RD	Recognized Disclosure
ref.	refractive
rpm	revolutions per minute
R.T.	room temperature
s	second(s)
sol.	soluble, solubility
sol'n.	solution
sp.gr.	specific gravity
SS	stainless steel
std.	standard
SUS	Saybolt Universal seconds
TCC	Taggart closed cup
TEA	triethanolamine
tech.	technical
temp.	temperature
theoret.	theoretical
TLV	threshold limit value
TOC	Taggart open cup
UL	Underwriter's Laboratory
USP	United States Pharmacopoeia
uv, UV	ultraviolet
veg	vegetable
visc.	viscosity, viscous
w/o	water-in-oil
wt	weight
$\approx$	approximately equal to
$<$	less than
$>$	greater than
$\leq$	less than or equal to
$\geq$	greater than or equal to

TABLE OF CONTENTS

ABS resin

SYNONYMS:
Acrylonitrile-butadiene-styrene resin

TRADENAME EQUIVALENTS:
ABS 124ESG, 124ESGE, 125M, 224M, 236F, 236MA, 236R, 301K, 323MA, 400P, 410ESG, 414MA, 420A, 424ESM, 500FR-1, 504ESG, 506ESG, 510EC, 522ESM, 550ES, 606ED, 707H, 707K, 808K, 910K [Mobil]

Abson 010, 042, 110, 120, 130, 135, 140, 161, 171 [Mobay]

Bapolan 8445, 8635, 8640 [Bamberger]

Cycolac AR, CG, CGA, CTB, DFA-R, DH, EP-3510, EPB-3570, ETS-3540, FBK, GSE, GSM, GT, HM, KGA, KJB, KJM, KJT, KJU, KJW, L, LS, LXB, SF, T, TA, X-11, X-15, X-17, X-37, Z36, Z48, Z86, ZA2, ZA4, ZA7, ZB2, ZB6, ZC7 [Borg-Warner]

Dow ABS 213, 350, 500, PG912, PG940 Resin [Dow]

Lucky ABS HF-350, HF-380, HI-100, HI-121, HI-151, HR-420, HR-450, MP-211, RS-600, SH-610 [Standard Polymers]

Lustran ABS 248, 252, 448, 452, 456, 545, 648, 743, 750-10802, 752, 780-10802, 782-10846, 860, 865, 1152, -HR850, -HR851, PG-298, PG-299, PG-300, FR ABS911, FR ABS914HM, FR ABS921UV, Ultra ABS HX, MX [Monsanto]

Magnum ABS 213, 341, 343, 545, 941, 4500UV, 9010, 9020, 9030, 9408, 9450P, CLR95, FG960 [Dow]

Novodur HGV, P2T, P2T-AT, PH-AT, PHE, PK, PK-AT, PKT, PL-AT, PLT-AT, PM, PM3C, PM5C, PM-AT, PME, PMT, PMT-AT, PTE, PX [Bayer AG]

PDX-84357 [LNP]

Carbon-Reinforced:
Stat-Kon AC-1003 (15% carbon fiber) [LNP]

Glass-Reinforced:
ABS-G1FG-2 (20% glass) [Washington Penn]

1

ABS resin *(cont'd.)*

AS-10GF (10% glass fiber), -20GF (20% glass fiber), -30GF (30% glass fiber), -40GF (40% glass fiber) [Compounding Technology]

Cycolac KGA [Borg-Warner]

Novodur PHGV (17% glass fiber), PHGV-AT (17% glass fiber) [Bayer AG]

RTP 601 (10% glass fiber), 603 (20% glass fiber), 605 (30% glass fiber), 607 (40% glass fiber) [Fiberite]

Thermocomp AF-1004 (20% glass fiber), AF-1006 (30% glass fiber), AF-1008 (40% glass fiber) [LNP]

ABS-stainless steel composite:

PDX-84357 [LNP]

MODIFICATIONS/SPECIALTY GRADES:

Flame Retardant:

ABS 500 FR-1; Abson 010, 042; Cycolac FBK, KGA, KJB, KJM, KJT, KJU, KJW, Z86, ZC7; Lustran ABS 860, 865, FR ABS911, FR ABS914HM, FR ABS921UV; Magnum ABS 4500UV; PDX-84357

Heat Resistant:

Bapolan 8445; Cycolac X-11, X-15, X-17, X-37, Z36, Z48; Lucky ABS HR-420, HR-450; Lustran ABS-HR850, HR851; Magnum ABS 4500UV, ABS 9048

High Flow:

ABS 125M, 707H; Abson 135; Cycolac DFA-R, EP-3510, FBK, GT, T, TA; Dow ABS 213; Lucky ABS HF-350, HF-380; Novodur PL-AT, PLT-AT

High Impact:

ABS 236F, 323MA, 301K, 420A, 500FR-1, 504ESG, 506ESG, 522ESM, 550ES, 606ED, 808K; Abson 135, 161, 171; Bapolan 8445, 8635, 8640; Cycolac CG, CGA, EP-3510, ETS-3540, GSE, GSM, HM, L, LS, LXB, T, ZA2, ZB2, ZB6; Dow ABS PG940 Resin; Lucky ABS HI-100, HI-121, HI-151, MP-211, SH-610; Lustran ABS 448, 648, 743, 750-10802, 752, 780-10802, 782-10846, 115, PG-298, PG-300, Ultra ABS HX; Magnum ABS 545, 941, 4500UV, 9020, 9030

High Gloss:

ABS 414MA, 504ESG, 506ESG, 808K; Abson 042, 110, 120, 130, 135, 140, 161; Cycolac GSE, GT, LS, T; Dow ABS 350 Resin; Lucky ABS HI-151; Lustran ABS 248, 252, 448, 452, 545, 648, 752, -HR850, -HR851, Ultra ABS HX, MX; Magnum ABS 941, 9010, 9020, 9030

Semigloss:

ABS 124ESGE, 400P, 410ESG, 707K

Medium Gloss:

Lustran ABS 1152

Satin Gloss:

Abson 010

Low Gloss:

ABS 224M; Cycolac CG, CGA, ZA4, ZA7, ZB2; Magnum ABS 213, 341, 343, 545

Matte Finish:

ABS 420A, 424ESM, 522ESM; Dow ABS 213; Lustran ABS 456

ABS resin *(cont'd.)*

Plateable Grade:

ABS 910K; Cycolac EP-3510, EPB-3570, ETS-3540; Lucky ABS MP-211; Lustran ABS PG-298, PG-299, PG-300; Magnum ABS 9450P; Novodur PM3C, PM5C

CATEGORY:

Thermoplastic resin

PROCESSING:

Extrusion:

ABS 124ESGE, 236MA, 410ESG, 420A, 424ESM, 504ESG, 506ESG, 510EC, 522ESM, 550ES, 606ED; Abson 171; Bapolan 8635; Cycolac CG, CGA, GSE, LS, SF, Z-36; Dow ABS 213, 350, 500, PG912, PG940 Resin; Lucky ABS HI-100, HI-121, RS-600, SH-610; Lustran ABS 252, 452, 456, 545, 752, 780-10802, 782-10846, 860, 865, 1152; Magnum ABS FG960; Novodur HGV, PHE, PK, PME, PTE

Injection Molding:

ABS 124ESG, 125M, 224M, 236F, 236R, 301K, 323MA, 400P, 414MA, 500FR-1, 707H, 707K, 808K; ABS-G1FG-2; Abson 010, 042, 110, 120, 130, 135, 140, 161; AS-10GF, -20GF, -30GF, -40GF; Bapolan 8445, 8635, 8640; Cycolac AR, CTB, DFA-R, DH, FBK, GSM, GT, HM, KGA, KJB, KJM, KJT, KJU, KJW, L, LXB, T, TA, X-11, X-15, X-17, X-37, Z48, Z86, ZA2, ZA4, ZA7, , ZB6, ZC7; Dow ABS 213, 350, 500, PG912, PG940 Resin; Lucky ABS HF-350, HF-380, HI-100, HI-121, HI-151, HR-420, HR-450; Lustran ABS 248, 448, 648, 743, 750-10802, –HR850, -HR851, FR ABS911, FR ABS914HM, FR ABS921UV, Ultra ABS HX, MX; Magnum ABS 213, 341, 343, 545, 941, 4500UV, 9010, 9020, 9030, 9408, CLR95; Novodur P2T, P2T-AT, PH-AT, PHGV, PHGV-AT, PK, PK-AT, PKT, PL-AT, PLT-AT, PM, PM3C, PM5C, PM-AT, PMT, PMT-AT, PTE, PX; PDX-84357; RTP 601, 603, 605, 607; Stat-Kon AC-1003; Thermocomp AF-1004, AF-1006, AF-1008

Compression Molding:

Dow ABS 213, 350, 500, PG912, PG940 Resin

Structural Foam Processing:

Cycolac FBK, ZC7

Thermoforming:

Cycolac CGA; Lustran ABS 860

APPLICATIONS:

Automotive applications: (ABS 236MA, 236MA, 323MA, 424ESM, 500FR-1; ABS-G1FG-2; AS-10GF, -20GF, -30GF, -40GF; Cycolac ETS-3540, HM, X-15, X-17, ZA2, ZA4, ZA7; Dow ABS 213, 350, 500; Lucky ABS HI-121, SH-610; Lustran ABS-HR850, -HR851; Novodur Series, PMT; RTP 601, 603, 605, 607; Thermocomp AF-1004, AF-1006, AF-1008); ATV (ABS 522ESM); bumpers (Cycolac LXB); consoles (ABS 236MA, TA); exterior parts (ABS 910K; Cycolac AR, EP-3510, EPB-3570; Lustran ABS 252, 452, 752); heating/air conditioner ducts (Cycolac AR; Lucky ABS HR-420, HR-450; Lustran ABS-HR850, -HR851); instrument panels (ABS 414MA; Lucky ABS HR-420, HR-450; Lustran ABS 252,

ABS resin (cont'd.)

452, 456, 752, -HR850, -HR851); interior parts (ABS 236F; Cycolac AR, EP-3510, TA, X-11, Z48; Lustran ABS-HR850, -HR851); light housings (Lustran ABS-HR850, -HR851); motorcycle components (Lustran ABS 252, 452, 752); RV exteriors (Lustran ABS 252, 452, 752); RV interiors (ABS 420A); snowmobile components (Abson 171; Cycolac L; Lustran ABS 252, 452, 752)

Aviation industry: (ABS 500FR-1); avionics housings (PDX-84357)

Consumer products: (Lucky ABS HF-350, HF-380, HI-121, HI-151, HR-420, HR-450); carrying cases (ABS 224M); cosmetic cases (Cycolac EP-3510; Lucky ABS HI-151); furniture (ABS 500FR-1; Cycolac DFA-R, T; Lucky ABS HF-350, HF-380, HI-121; Lustran ABS 456; Novodur Series); hardware (Cycolac DFA-R); housewares (ABS 707K; Cycolac DFA-R; Lustran ABS 248); lawn and garden equipment (ABS 707H; Cycolac GSM; Lustran ABS 448, 648, 743; Lustran Ultra ABS HX, MX); luggage (ABS 424ESM, 522ESM; Dow ABS 213, 350, 500; Lucky ABS HI-121, SH-610; Lustran ABS 252, 452, 456, 752); musical instruments (Lucky ABS HF-350, HF-380, HI-121); photographic equipment (Cycolac X-37; Lustran ABS 248; Novodur Series; Thermocomp AF-1004, AF-1006, AF-1008); plumbing fixtures (Cycolac EPB-3570; Lustran ABS 456); recreational equipment (ABS 424ESM, 504ESG, 506ESG, 707K; Lustran ABS 252, 452, 752; Novodur Series); safety helmets (Abson 161; Cycolac L; Lucky ABS HI-100); shoes (Lucky ABS HI-100; Lustran ABS 448); smoke detector housings (ABS 500FR-1); sporting goods (ABS 224M, 808K; Cycolac GSM; Dow ABS 213, 350, 500; Lucky ABS HI-121; Lustran ABS 448, 648; Lustran Ultra ABS HX); toys (ABS 125M; Abson 135; Bapolan 8445; Cycolac CTB, DFA-R, T; Dow ABS 213, 350, 500; Lucky ABS HI-121; Lustran ABS 248, 448, 648; Lustran Ultra ABS HX, MX; Novodur Series); vacuum cleaner components (Cycolac GSM; Lustran ABS-HR850, -HR851)

Electrical/electronics industry: (Cycolac KJU, Z86; Novodur Series; PDX-84357); casette housings (Cycolac CTB, L; Lustran Ultra ABS HX, MX); computers (Bapolan 8640; Cycolac KJB, KJU; Stat-Kon AC-1003); electrical appliances (Cycolac KJB, KJT); electrical components (Cycolac L; Stat-Kon AC-1003); electrical enclosures/packaging (Lustran ABS 865; Stat-Kon AC-1003); electronic instruments (Bapolan 8640); lighting equipment (Lucky ABS HI-121, HR-420, HR-450); power tools (Cycolac KJB, KJT, X-17, X-37; Lustran ABS 648, 743; Lustran Ultra ABS HX, MX); speaker cabinets (Lucky ABS HI-151); telecommunications (ABS 808K; Abson 135; Bapolan 8445, 8640; Cycolac FBK, GT, KGA, T, X-37, ZB2, ZB6, ZC7; Lucky ABS HF-350, HF-380, HI-151; Lustran ABS 448; Lustran Ultra ABS HX, MX); TV cabinets (Abson 010, 042, Z86; Lucky ABS HF-350, HF-380, HR-420, HR-450); TV components (Cycolac FBK, KJT)

Industrial applications: (Lucky ABS RS-600); appliances (AS-10GF, -20GF, -30GF, -40GF; Bapolan 8445, 8640; Cycolac DFA-R, EPB-3570; Cycolac FBK, L, X-17, X-37; Dow ABS 213, 350, 500; Lucky ABS HF-350, HF-380; Lucky ABS HI-121, HI-151; Lustran ABS 248, 448, 648, 743, -HR850, -HR851; Lustran Ultra ABS HX, MX; Novodur Series; Thermocomp AF-1004, AF-1006, AF-1008); appliance housings (ABS 400P, 506ESG; Cycolac EP-3510, T; Lucky ABS HI-121); appliance parts (ABS 124ESGE, 236MA, 410ESG, 500FR-1, 707H, 808K; Abson 135; Cycolac EP-3510); business machines and office equipment (ABS 808K; Abson 010, 042; AS-10GF, -20GF, -30GF, -40GF; Bapolan 8445; Cycolac FBK, KJB, KJT, X-37; Lucky ABS HF-350, HF-380, HI-151; Lustran ABS 248, 448, 648, 865; Lustran Ultra ABS HX, MX; Novodur Series; PDX-84357; Thermocomp AF-1004, AF-1006, AF-1008); closures (ABS 125M); construction applications (AS-10GF, -20GF, -30GF, -40GF; Cycolac X-11); core laminates (ABS 550ES); fittings (Lucky ABS HI-100; Lustran ABS 750-10802); housings (ABS 506ESG, 707H, 707K; Abson 042; AS-10GF, -20GF, -30GF, -40GF; Bapolan 8635; Cycolac FBK, GSM, KJT, ZB2; Dow ABS 213, 350, 500; Lucky ABS HF-350, HF-380, HI-121, HI-151, HR-420, HR-450; Lustran ABS 648, 743, -HR850, -HR851; Lustran Ultra ABS HX, MX; PDX-84357); intricate moldings (ABS 125M; Cycolac EP-3510, TA); knobs (Lucky ABS MP-211); large parts (Cycolac CG, LS; Lucky ABS HF-350, HF-380); molded parts (ABS 124ESG, 125M; Lucky ABS HI-100); packaging/displays (Cycolac CTB; Lustran ABS 248); pipe (ABS 301K, 510EC, 606ED; Dow ABS 213, 350, 500, PG912, PG940; Lucky ABS HI-100, HI-121; Lustran ABS 750-10802, 780-10802, 782-10846; Magnum ABS FG960); profile and sheet (ABS 124ESGE, 410ESG, 504ESG, 506ESG, 522ESM, 550ES; Cycolac SF); refrigerator liners/parts (ABS 504ESG; Bapolan 8445; Cycolac CTB; Dow ABS 213, 350, 500; Lucky ABS HI-121, MP-211, RS-600; Lustran ABS 252, 452, 545, 752); sheet (Bapolan 8635; Lucky ABS HI-121, RS-600, SH-610; Lustran ABS 860); swimming pool valves (ABS 707H); textile applications (Novodur Series, HGV); thick sections (Cycolac L); thin sections (ABS 125M; Cycolac TA); tools (Cycolac FBK, GSM; Dow ABS 213, 350, 500; Lustran ABS 248, 448)

Marine equipment: (Cycolac EPB-3570); boats (Lucky ABS HI-121; Lustran ABS 252, 452, 752)

PROPERTIES:
Form:

Solid (ABS 124ESG, 124ESGE, 125M, 224M, 236F, 236MA, 236MA, 301K, 323MA, 400P, 410ESG, 414MA, 420A, 424ESM, 500FR-1, 504ESG, 506ESG, 510EC, 522ESM, 550ES, 606ED, 707H, 707K, 808K, 910K; Dow ABS 213, 350, 500, PG912, PG940)

Pellets (Cycolac FBK)

Color:

Transparent (Cycolac CTB; Magnum ABS CLR95)

Natural (ABS 124ESG, 124ESGE. 125M, 224M, 236F, 236MA, 236MA, 323MA, 400P, 410ESG, 414MA, 420A, 424ESM, 500FR-1, 504ESG, 506ESG, 522ESM, 550ES, 707H, 707K, 808K; Cycolac FBK, KJB, X-37; Lustran ABS 248, 448, 456, 648, 743, 860, -HR850, -HR851; Lustran FR ABS911, FR ABS914HM, FR ABS921UV; Lustran Ultra ABS HX, MX; Magnum ABS 213, 341, 343, 545, 941, 4500UV, 9010, 9020, 9030, 9408, 9450P, FG960; Novodur Series; RTP 601, 603, 605, 607)

ABS resin (cont'd.)

White (Lustran ABS 545)

Light (Dow ABS 213, 350, 500)

Colors (ABS 236F, 236MA, 400P; Lustran ABS 456, 545, 860, -HR850, -HR851)

Opaque, colorable (Abson 010, 042, 110, 120, 130, 135, 140, 161, 171)

Opaque colors (Novodur Series)

Gray (ABS 910K; Cycolac EP-3510, EPB-3570; Lustran ABS PG-298, PG-299, PG-300)

Black (ABS 301K, 510EC, 606ED; Lustran ABS 780-10802, 782-10846; RTP 601, 603, 605, 607)

GENERAL PROPERTIES:

Melt Flow:

0.9 g/10 min (Lustran ABS 865)

1.5–2.5 g/10 min (ABS-G1FG-2)

1.8 g/10 min (Lustran ABS 1152)

2.0 g/10 min (Bapolan 8635; Magnum ABS 941)

2.6 g/10 min (Magnum ABS FG960)

3.8 g/10 min (Magnum ABS 9408)

4.0 g/10 min (Bapolan 8445, 8640)

5.0 g/10 min (Magnum ABS 341, 9030)

5.5 g/10 min (Magnum ABS 213, 545, 9020)

6.0 g/10 min (Magnum ABS CLR95)

6.5 g/10 min (Magnum ABS 4500UV)

8.0 g/10 min (Magnum ABS 343, 9010, 9450P)

11 cc/s (Lucky ABS HI-100)

15 cc/s (Lucky ABS HI-121, HR-420, MP-211)

19 cc/s (Lucky ABS HI-151)

33 cc/s (Lucky ABS HR-450)

43 cc/s (Lucky ABS HF-380)

54 cc/s (Lucky ABS HF-350)

Sp.gr.: 0.97 (Cycolac SF)

1.02 (Cycolac CG, L, LS, LXB)

1.03 (ABS 506ESG, 808K; Cycolac CGA, HM; Lucky ABS HI-100; Lustran ABS 1152)

1.04 (ABS 504ESG, 522ESM, 606ED, 707H, 707K; Bapolan 8445; Cycolac DFA-R, GSE, GSM, T, TA, X-11, ZA2, ZA4; Lucky ABS HF-350, HI-121, MP-211; Lustran ABS 456, 648, 743, 752, 780-10802; Lustran Ultra ABS HX; Magnum ABS 545, 9010, 9020, 9030, 9450P)

1.05 (ABS 124ESG, 124ESGE, 125M, 224M, 236F, 236MA, 236R, 301K, 323MA, 400P, 410ESG, 414MA, 420A, 424ESM, 510EC, 550ES; Cycolac AR, DH, ETS-3540, X-11, X-15, X-17, ZA7; Dow ABS 213, 500; Lucky ABS HF-380, HI-151, HR-420, HR-450; Lustran ABS 448, 750-10802, 782-10846, -HR850; Lustran Ultra ABS MX; Magnum ABS 213, 341, 343, 941, 9408, FG960)

1.06 (ABS 910K; Cycolac EP-3510, EPB-3570, X-37, Z-36, Z48; Dow ABS 350;

ABS resin *(cont'd.)*

Lustran ABS 248, 452, -HR851, PG-298, PG-299, PG-300)
1.07 (Cycolac CTB; Lustran ABS 252)
1.09 (Magnum ABS CLR95)
1.10 (AS-10GF; Lustran ABS 545)
1.11 (RTP 601; Stat-Kon AC-1003)
1.18 (RTP 603)
1.20 (AS-20GF; Cycolac KJM, KJT, Z86; Lustran ABS 860, 865; Lustran FR ABS914HM, FR ABS921UV; Thermocomp AF-1004)
1.21 (Lustran FR ABS911; Magnum ABS 4500UV)
1.22 (Cycolac KJB, KJU, KJW)
1.28 (RTP 605; Thermocomp AF-1006)
1.29 (AS-30GF)
1.30 (PDX-84357)
1.33 (ABS 500FR-1)
1.38 (AS-40GF; RTP 607; Thermocomp AF-1008)

Sp. Vol.:
20.1 in.^{3}/lb (Thermocomp AF-1008)
21.6 in.^{3}/lb (Thermocomp AF-1006)
23.1 in.^{3}/lb (Thermocomp AF-1004)

Density:
1.03 g/cc (Novodur PK, PK-AT)
1.04 g/cc (Novodur PM, PM3C, PM-AT, PME)
1.05 g/cc (Novodur PH-AT, PHE, PKT, PMT, PMT-AT, PX, X)
1.06 g/cc (Novodur P2T, P2T-AT, PM5C)
1.09 g/cc (Novodur HGV)
1.15 g/cc (Novodur PHGV)
1.16 g/cc (Novodur PHGV-AT)
1.17 g/cc (ABS-G1FG-2)
1.02 mg/m^3 (Abson 161, 171)
1.04 mg/m^3 (Abson 135)
1.06 mg/m^3 (Abson 110)
1.17 mg/m^3 (Abson 010)
1.22 mg/m^3 (Abson 042)
43 lb/ft^3 (bulk) (Magnum ABS 213, 341, 343, 545, 941, 4500UV, 9010, 9020, 9030, 9408, 9450P, CLR95, FG960)

Visc.:
3.9 kPa•s (450 F, apparent) (Lustran ABS 545)
4.14 kPa•s (232 C, apparent) (Lustran ABS 456)
4.48 kPa•s (232 C, apparent) (Lustran ABS 252, 452, 752)
7.24 kPa•s (190 C, apparent) (Lustran ABS 860)
75 poise (450 F, melt) (Magnum ABS 213)
90 poise (450 F, melt) (Magnum ABS 4500UV)
120 poise (450 F, melt) (Magnum ABS 343)

7

ABS resin *(cont'd.)*

130 poise (450 F, melt) (Magnum ABS CLR95)
150 poise (450 F, melt) (Magnum ABS 545)
200 poise (450 F, melt) (Magnum ABS 9010, 9450P)
240 poise (450 F, melt) (Magnum ABS 9020)
260 poise (450 F, melt) (Magnum ABS 341)
280 poise (450 F, melt) (Magnum ABS 9030, 9408)
438 poise (450 F, melt) (Magnum ABS 941)

M.P.:

460–500 F (AS-10GF)

Stability:

Resistant to aq. acids, alkalis, and salts; conc. oxidizing acids produce disintegration; aromatic or chlorinated hydrocarbons and high KB solvents cause marked swelling (Cycolac KJM, KJU, Z86)

Excellent thermal stability; excellent chemical resistance to organic acids, aliphatic amines, bases, polyglycols, pharmaceuticals; good resistance to weak and strong inorganic acids, alcohols (Dow ABS 213, 350, 500, PG912, PG940)

Highly resistant to chemicals; unaffected by alkalis, dilute organic and inorganic acids, aliphatic hydrocarbons, and most oils and fats; swelling or superficial dissolution may be caused by aromatics, ketones, ethers, esters, and chlorinated hydrocarbons; sensitive to oxygen and uv radiation, resulting in yellowing and other aging effects (Novodur Series)

Excellent resistance to bases; good resistance to acids; poor resistance to solvents (Thermocomp AF-1008)

Ref. Index:

1.536 (Cycolac CTB)

60° Gloss:

10–50% (Magnum ABS 9408)
10–70% (Magnum ABS 213, FG960)
25–80% (Magnum ABS 343, 545, 941)
40–50% (Lustran ABS 865)
60% (Magnum ABS 4500UV)
60–90% (Magnum ABS 341)
70–80% (Lustran ABS 1152)
95% (Magnum ABS 9010, 9020, 9030)

MECHANICAL PROPERTIES:

Tens. Str.:

13–27 MPa (Cycolac FBK)
13.1 MPa (yield) (Lustran ABS 456)
16 MPa (Cycolac SF)
28 MPa (break) (Abson 010)
29 MPa (Cycolac CGA)
30 MPa (Cycolac CG); (break) (Abson 161)
31 MPa (break) (Abson 171); (yield) (Lustran ABS 1152)

33 MPa (Cycolac HM)
34 MPa (Cycolac ETS-3540, L, LXB); (yield) (Lustran ABS 743)
35 MPa (yield) (Lustran ABS 752)
36 MPa (Cycolac LS); (yield) (Novodur PK, PK-AT)
38 MPa (Cycolac EP-3510)
39 MPa (break) (Abson 135)
40 MPa (Cycolac KJB, KJU, ZA4); (yield) (Lustran ABS 452, 865)
41 MPa (Cycolac KJW, T, TA, X-11); (break) (Abson 042); (yield) (Lustran ABS 545, 648; Novodur PM3C)
42 MPa (yield) (Lustran ABS 448; Lustran Ultra ABS HX)
43 MPa (Cycolac DFA-R, GSM, KJT, ZA2); (yield) (Lustran ABS 252)
44 MPa (Cycolac EPB-3570); (yield) (Novodur HGV)
45 MPa (Cycolac GSE, X-11, Z-36, ZA7); (yield) (Novodur PHE)
46 MPa (Cycolac AR, Z86); (yield) (Lustran Ultra ABS MX; Novodur PM, PM-AT)
47 MPa (yield) (Lustran ABS 248, -HR850; Novodur PKT, PLT-AT)
47.6 MPa (yield) (Lustran ABS 860)
48 MPa (Cycolac X-37, Z48); (yield) (Novodur PHGV-AT)
50 MPa (Cycolac KJM, X-17); (yield) (Novodur PH-AT, PL-AT, PTE)
51 MPa (break) (Abson 110)
52 MPa (Cycolac DH); (yield) (Lustran ABS-HR851; Novodur PMT-AT)
54 MPa (yield) (Novodur P2T-AT, PMT, PX)
55 MPa (yield) (Novodur PM5C)
56 MPa (yield) (Novodur PME)
57 MPa (yield) (Novodur PHGV)
58 MPa (yield) (Novodur P2T)
350 kgf/cm² (break) (Dow ABS 213)
365 kgf/cm² (break) (Dow ABS 500)
380 kg/cm² (Lucky ABS HI-100)
410 kgf/cm² (break) (Dow ABS 350)
420 kg/cm² (Lucky ABS HF-350)
460 kg/cm² (, SH-610)
470 kg/cm² (Lucky ABS HI-121, MP-211)
480 kg/cm² (Lucky ABS HF-380)
500 kg/cm² (Lucky ABS HI-151)
510 kg/cm² (Lucky ABS HR-450)
520 kg/cm² (Lucky ABS RS-600)
550 kg/cm² (Lucky ABS HR-420)
3400 psi (ABS 522ESM)
3500 psi (ABS 420A)
3800 psi (break) (Magnum ABS 4500UV)
4000 psi (ABS 424ESM)
4400 psi (ABS 506ESG)
4500 psi (break) (Magnum ABS 9030)

ABS resin *(cont'd.)*

4600 psi (ABS 500FR-1); (break) (Dow ABS PG912)
4650 psi (Lustran ABS 780-10802)
4700 psi (break) (Magnum ABS 941)
4800 psi (break) (Lustran ABS PG-300; Magnum ABS 9020, FG960)
5000 psi (ABS 808K; Lustran ABS 782-10846); (break) (Magnum ABS 213, 343, 545)
5100 psi (ABS 410ESG)
5200 psi (ABS 606ED); (break) (Dow ABS PG940; Lustran ABS PG-299; Magnum
    ABS 9010, 9450P)
5300 psi (break) (Magnum ABS 9408)
5400 psi (ABS 125M); (break) (Magnum ABS CLR95)
5500 psi (ABS 301K, 400P, 504ESG, 707H)
5600 psi (ABS 236F, 323MA, 550ES, 707K; Lustran ABS 750-10802)
5700 psi (ABS 236MA)
5800 psi (yield) (Lustran FR ABS921UV)
6000 psi (ABS 124ESGE; PDX-84357); (break) (Magnum ABS 341); (yield) (Lustran
    FR ABS911)
6100 psi (break) (Lustran ABS PG-298)
6200 psi (ABS 910K)
6300 psi (Cycolac CTB)
6500 psi (ABS 224M, 510EC)
7100 psi (ABS 236MA; Bapolan 8445)
7200 psi (ABS 414MA)
7400 psi (yield) (Lustran FR ABS914HM)
7700 psi (ABS 124ESG)
9000 psi (ABS-G1FG-2; RTP 601)
9800 psi (AS-10GF)
12,500 psi (AS-20GF)
13,000 psi (RTP 603)
13,500 psi (Thermocomp AF-1004)
13,800 psi (AS-30GF)
14,500 psi (Thermocomp AF-1006)
15,500 psi (AS-40GF; Stat-Kon AC-1003)
16,000 psi (RTP 605; Thermocomp AF-1008)
18,000 psi (RTP 607)

Tens. Elong.:

1–3% (AS-40GF)
1.5% (RTP 607)
1.8% (RTP 605)
2% (Novodur PHGV, PHGV-AT; RTP 601, 603; Stat-Kon AC-1003)
2–3% (Thermocomp AF-1008)
2–4% (AS-20GF, -30GF)
2.4% (yield) (Lustran ABS 545)
2.5% (Novodur PME); (break) (ABS-G1FG-2)

3% (Novodur P2T, P2T-AT, PH-AT, PHE, PK, PK-AT, PKT, PL-AT, PLT-AT, PM, PM3C, PM5C, PM-AT, PMT, PMT-AT, PX)
3–4% (Thermocomp AF-1004, AF-1006)
3–5% (AS-10GF)
3.5% (Novodur PTE)
4% (Novodur HGV; PDX-84357)
10% (ABS 124ESG, 707K); (break) (Dow ABS 500; Magnum ABS 341)
11% (ABS 236MA, 808K; Dow ABS PG940)
15% (ABS 414MA, 510EC, 550ES, 707H, 910K); (break) (Abson 110, 135)
18% (Lucky ABS RS-600); (break) (Lustran ABS PG-298, PG-299); (yield) (Bapolan 8445)
20% (ABS 500FR-1, 504ESG; Lucky ABS HI-151, SH-610)
23% (break) (Lustran ABS PG-300)
25% (ABS 124ESGE, 606ED; Lucky ABS HF-350, HF-380, HI-121, HR-420, HR-450, MP-211); (break) (Dow ABS 350; Magnum ABS 545, 941, 9030)
30% (ABS 125M, 224M, 301K, 400P; Lucky ABS HI-100)
35% (ABS 236MA, 410ESG)
40% (ABS 323MA, 506ESG); (break) (Magnum ABS 9010, 9408, 9450P, FG960)
50% (ABS 236F; Dow ABS PG912); (break) (Dow ABS 213; Lustran ABS 545; Magnum ABS 213, 4500UV, 9020, CLR95)
60% (ABS 424ESM); (break) (Abson 010; Magnum ABS 343)
70% (ABS 420A, 522ESM); (break) (Abson 161)
110% (break) (Abson 171)
Tens. Mod.:
0.83 GPa (Cycolac SF)
1.0 GPa (Lustran ABS 456)
1.10–1.72 GPa (Cycolac FBK)
1.5 GPa (Novodur PK, PK-AT)
1.6 GPa (yield) (Lustran ABS 1152)
1.7 GPa (Cycolac CG, CGA, LS, LXB)
1.8 GPa (Cycolac HM, L; Lustran ABS 743)
2.08 GPa (Novodur PM-AT)
2.1 GPa (Cycolac ETS-3540, X-1, Z86; Lustran ABS 860; Novodur PM3C)
2.12 GPa (Novodur PM)
2.2 GPa (Cycolac GSE, GSM, KJB, KJU, KJW, TA, X-11, ZA2; Lustran ABS 752); (yield) (Lustran ABS 865)
2.3 GPa (Cycolac DFA-R, EP-3510, KJT, T, Z-36, Z48, ZA4; Novodur PL-AT, PTE, PX)
2.32 GPa (Novodur PKT)
2.34 GPa (Novodur PHE)
2.38 GPa (Novodur PH-AT)
2.4 GPa (Cycolac AR, X-17, X-37, ZA7; Lustran ABS 648; Lustran Ultra ABS HX, MX)

ABS resin *(cont'd.)*

2.5 GPa (Cycolac KJM; Lustran ABS 448, 452, 545, -HR850; Novodur PLT-AT, PMT-AT)

2.51 GPa (Novodur HGV)

2.53 GPa (Novodur PME)

2.57 GPa (Novodur PMT)

2.59 GPa (Novodur P2T-AT)

2.6 GPa (Cycolac DH, EPB-3570; Lustran ABS 248, 252, -HR851)

2.65 GPa (Novodur PM5C)

2.75 GPa (Novodur P2T)

4.77 GPa (Novodur PHGV-AT)

5.00 GPa (Novodur PHGV)

23,900 kgf/cm^2 (Dow ABS 500)

25,300 kgf/cm^2 (Dow ABS 213)

28,100 kgf/cm^2 (Dow ABS 350)

200,000 psi (ABS 522ESM)

225,000 psi (ABS 420A, 506ESG; Lustran ABS 780-10802)

235,000 psi (ABS 424ESM, 808K; Lustran ABS 782-10846)

250,000 psi (Lustran ABS 750-10802)

255,000 psi (ABS 500FR-1)

265,000 psi (ABS 410ESG)

270,000 psi (ABS 236F)

272,000 psi (ABS 550ES)

275,000 psi (ABS 323MA)

280,000 psi (ABS 301K, 504ESG, 606ED; Magnum ABS 4500UV, 9030)

285,000 psi (ABS 236MA)

290,000 psi (ABS 400P; Dow ABS PG912; Magnum ABS 941, FG960)

300,000 psi (ABS 510EC, 707H, 707K; Lustran ABS PG-300; Magnum ABS 9408)

305,000 psi (ABS 224M)

310,000 psi (ABS 124ESGE, 125M, 236MA; Magnum ABS 545)

320,000 psi (ABS 910K; Dow ABS PG940; Lustran FR ABS911; Magnum ABS 343, 9010, 9450P)

330,000 psi (ABS 414MA; Cycolac CTB)

335,000 psi (Magnum ABS 9020)

340,000 psi (ABS 124ESG; Lustran ABS PG-299)

360,000 psi (Lustran FR ABS921UV; Magnum ABS 213, 341)

380,000 psi (Lustran FR ABS914HM; Magnum ABS CLR95)

415,000 psi (Lustran ABS PG-298)

600,000 psi (RTP 601)

675,000 psi (AS-10GF)

880,000 psi (RTP 603)

950,000 psi (AS-20GF)

1,100,000 psi (AS-30GF)

1,200,000 psi (AS-40GF; RTP 605)

1,500,000 psi (RTP 607)

Flex. Str.:
26 MPa (yield) (Cycolac SF)
28.5 MPa (yield) (Lustran ABS 456)
29–54 MPa (yield) (Cycolac FBK)
48 MPa (yield) (Lustran ABS 1152)
50 MPa (yield) (Lustran ABS 743)
51 MPA (Novodur PK, PK-AT); (break) (Abson 010, 161)
52 MPa (break) (Abson 171)
54 MPa (yield) (Cycolac CG, CGA)
55 MPa (yield) (Cycolac HM)
59 MPa (yield) (Cycolac ETS-3540, L, LS, LXB)
61 MPa (yield) (Lustran ABS 752)
63 MPa (yield) (Lustran ABS 865)
65 MPa (Novodur PHE)
66 MPa (Novodur HGV)
67 MPa (Lustran Ultra ABS HX; Novodur PM3C, PM-AT)
69 MPa (Novodur PM); (yield) (Cycolac EP-3510, KJB, KJT, KJU, KJW, X-11; Lustran ABS 648)
70 MPa (Novodur PH-AT, PKT, PLT-AT, PTE); (break) (Abson 042)
72 MPa (break) (Abson 135); (yield) (Cycolac T, TA, ZA4; Lustran ABS 448, 452, 545)
73 MPa (yield) (Cycolac Z86)
74 MPa (Novodur PL-AT); (yield) (Cycolac GSM, ZA2; Lustran ABS 248)
75.8 MPa (yield) (Lustran ABS 860)
76 MPa (Novodur PMT-AT, PX); (yield) (Cycolac GSE, Z-36; Lustran ABS 252)
77 MPa (Novodur PME)
78 MPa (Lustran Ultra ABS MX); (yield) (Cycolac X-11)
79 MPa (yield) (Cycolac DFA-R)
80 MPa (Novodur PHGV-AT, PMT); (yield) (Cycolac EPB-3570; Lustran ABS-HR850)
81 MPa (Novodur P2T-AT, PM5C); (yield) (Cycolac ZA7)
83 MPa (yield) (Cycolac AR, X-37, Z48)
85 MPa (Novodur P2T)
86 MPa (Lustran ABS-HR851; Novodur PHGV); (yield) (Cycolac KJM, X-17)
90 MPa (yield) (Cycolac DH)
100 MPa (break) (Abson 110)
630 kg/cm² (Lucky ABS HI-100)
665 kgf/cm² (Dow ABS 213)
700 kg/cm² (Lucky ABS HF-350)
760 kgf/cm² (Dow ABS 500)
800 kg/cm² (Lucky ABS HF-380, HI-121, MP-211)
820 kg/cm² (Lucky ABS SH-610)

ABS resin *(cont'd.)*

850 kg/cm² (Lucky ABS HI-151, HR-450)
900 kg/cm² (Lucky ABS HR-420)
905 kgf/cm² (Dow ABS 350)
920 kg/cm² (Lucky ABS RS-600)
5300 psi (ABS 522ESM)
5500 psi (ABS 420A)
6100 psi (ABS 424ESM, 500FR-1)
6700 psi (ABS 506ESG)
7400 psi (yield) (Magnum ABS 4500UV)
7700 psi (ABS 808K)
7900 psi (ABS 606ED)
8200 psi (ABS 236F)
8300 psi (ABS 323MA, 410ESG)
8400 psi (yield) (Magnum ABS 941)
8500 psi (ABS 504ESG)
9000 psi (ABS 124ESGE, 125M, 301K, 550ES, 707H); (yield) (Magnum ABS 9030)
9300 psi (ABS 707K); (yield) (Magnum ABS FG960)
9500 psi (ABS 236MA, 400P); (yield) (Lustran ABS PG-300; Magnum ABS 213, 9020)
9800 psi (yield) (Magnum ABS 9408)
9900 psi (yield) (Magnum ABS 343)
10,000 psi (ABS 224M, 510EC, 910K); (yield) (Lustran ABS PG-299; Lustran FR ABS911, FR ABS921UV)
10,500 psi (ABS 414MA); (yield) (Cycolac CTB; Magnum ABS 545)
11,000 psi (PDX-84357); (yield) (Magnum ABS 9010, 9450P)
11,100 psi (ABS 236MA)
11,370 psi (Bapolan 8445)
11,800 psi (yield) (Magnum ABS CLR95)
12,000 psi (ABS 124ESG)
12,500 psi (yield) (Lustran ABS PG-298, FR ABS914HM; Magnum ABS 341)
13,000 psi (ABS-G1FG-2)
14,000 psi (RTP 601)
14,900 psi (AS-10GF)
16,500 psi (AS-20GF)
17,000 psi (RTP 603)
17,500 psi (Thermocomp AF-1004)
18,000 psi (AS-30GF)
18,500 psi (Thermocomp AF-1006)
19,000 psi (RTP 605)
20,000 psi (Thermocomp AF-1008)
20,500 psi (AS-40GF)
21,000 psi (RTP 607)
22,000 psi (Stat-Kon AC-1003)

Flex. Mod.:
0.9 GPa (Cycolac SF)
1.2 GPa (Lustran ABS 456)
1.45–1.93 GPa (Cycolac FBK)
1.5 GPA (Novodur PK-AT)
1.55 GPa (Abson 161)
1.58 GPa (Abson 171)
1.6 GPA (Novodur PK)
1.7 GPa (Cycolac CG, CGA, LXB)
1.8 GPa (Cycolac HM, LS; Lustran ABS 1152; Novodur PM-AT)
1.85 GPa (Novodur PM, PM3C)
1.86 GPa (Novodur HGV)
1.9 GPa (Cycolac L; Lustran ABS 743)
1.98 GPa (Novodur PHE)
2.00 GPa (Novodur PMT-AT)
2.07 GPa (Abson 010)
2.1 GPa (Cycolac ETS-3540, X-11; Lustran ABS 752)
2.13 GPa (Novodur PH-AT)
2.19 GPa (Novodur PX)
2.2 GPa (Lustran ABS 865; Novodur PM5C, PMT)
2.21 GPa (Novodur PME)
2.22 GPa (Novodur P2T-AT)
2.3 GPa (Abson 135; Cycolac GSE, GSM, KJB, KJT, KJU, KJW, T, X-11, ZA2, ZA4)
2.32 GPa (Novodur P2T)
2.4 GPa (Cycolac TA, X-37, Z-36, Z86, ZA7)
2.41 GPa (Abson 042)
2.5 GPa (Cycolac AR, DFA-R, Z48; Lustran ABS 452, 648; Lustran Ultra ABS HX)
2.6 GPa (Cycolac X-17; Lustran ABS 448, 860, -HR850; Lustran Ultra ABS MX)
2.7 GPa (Cycolac EPB-3570; Lustran ABS-HR851)
2.8 GPa (Cycolac DH, KJM)
2.9 GPa (Lustran ABS 545)
3.0 GPa (Lustran ABS 252)
3.10 GPa (Abson 110)
3.15 GPa (Novodur PHGV-AT)
3.66 GPa (Novodur PHGV)
18,000 kg/cm^2 (Lucky ABS HI-100)
23,000 kg/cm^2 (Lucky ABS HI-121, MP-211)
24,000 kg/cm^2 (Lucky ABS HF-350)
24,500 kg/cm^2 (Lucky ABS SH-610)
25,000 kg/cm^2 (Lucky ABS HR-450)
26,000 kgf/cm^2 (Dow ABS 500)
26,700 kgf/cm^2 (Dow ABS 213)
27,000 kg/cm^2 (Lucky ABS HF-380, HI-151, HR-420)

ABS resin (cont'd.)

29,500 kg/cm^2 (Lucky ABS RS-600)
30,900 kgf/cm^2 (Dow ABS 350)
194,000 psi (ABS 522ESM)
214,000 psi (ABS 420A)
230,000 psi (ABS 424ESM)
235,000 psi (ABS 506ESG)
270,000 psi (ABS 500FR-1, 808K)
280,000 psi (ABS 606ED)
290,000 psi (Magnum ABS 4500UV)
295,000 psi (ABS 504ESG)
300,000 psi (Magnum ABS 941, 9030, FG960)
310,000 psi (Magnum ABS 9408)
320,000 psi (ABS 236F, 414MA; Magnum ABS 343)
321,000 psi (ABS 550ES)
330,000 psi (ABS 124ESG, 410ESG; Lustran ABS PG-300; Lustran FR ABS911;
   Magnum ABS 9020)
340,000 psi (ABS 124ESGE)
348,000 psi (ABS 323MA)
350,000 psi (ABS 707H, 707K, 910K; Cycolac CTB, GT)
355,000 psi (ABS 125M, 400P)
360,000 psi (ABS 224M, 236MA, 301K, 510EC)
370,000 psi (ABS 236MA; Lustran ABS PG-299)
375,000 psi (Lustran FR ABS921UV; Magnum ABS 545)
380,000 psi (Magnum ABS 213, 9010, 9450P)
390,000 psi (Cycolac ZC7; Magnum ABS 341)
400,000 psi (Cycolac ZB6; Lustran FR ABS914HM)
420,000 psi (Magnum ABS CLR95)
425,000 psi (Lustran ABS PG-298)
430,000 psi (PDX-84357)
500,000 psi (RTP 601)
650,000 psi (ABS-G1FG-2)
680,000 psi (AS-10GF)
800,000 psi (RTP 603)
850,000 psi (Thermocomp AF-1004)
900,000 psi (AS-20GF)
940,000 psi (Cycolac KGA)
1,000,000 psi (RTP 605)
1,100,000 psi (Stat-Kon AC-1003; Thermocomp AF-1006)
1,150,000 psi (AS-30GF)
1,250,000 psi (AS-40GF)
1,300,000 psi (RTP 607)
1,400,000 psi (Thermocomp AF-1008)

Compr. Str.:
5.5 MPa (Cycolac T)
30 MPa (Cycolac FBK)
40 MPa (Cycolac L)
53 MPa (Cycolac KJB)
59 MPa (Cycolac X-37)
9200 psi (RTP 601)
12,300 psi (AS-10GF)
13,500 psi (RTP 603; Thermocomp AF-1004)
14,000 psi (AS-20GF)
14,500 psi (Thermocomp AF-1006)
15,000 psi (AS-30GF)
16,000 psi (AS-40GF; Thermocomp AF-1008)
17,000 psi (RTP 605)
19,000 psi (RTP 607)

Compr. Mod.:
1.2 GPa (Cycolac L)
1.3 GPa (Cycolac T)
1.4 GPa (Cycolac KJB)
2.3 GPa (Cycolac X-37)

Shear Str.:
6500 psi (AS-20GF)
7000 psi (AS-30GF; Thermocomp AF-1004)
7300 psi (AS-40GF)
7500 psi (Thermocomp AF-1006)
8000 psi (Thermocomp AF-1008)

Impact Str. (Dart):
14 J (Cycolac FBK)

Impact Str. (Izod):
8 kJ/m^2 notched (Novodur P2T, P2T-AT)
9 kJ/m^2 notched (Novodur PL-AT, PM5C)
10 kJ/m^2 notched (Novodur PH-AT, PLT-AT, PMT, PMT-AT, PX)
11 kJ/m^2 notched (Novodur HGV)
12 kJ/m^2 notched (Novodur PHE, PM, PM-AT)
13 kJ/m^2 notched (Novodur PTE)
14 kJ/m^2 notched (Novodur PHGV, PM3C)
15 kJ/m^2 notched (Novodur PME, PHGV-AT, PKT)
20 kJ/m^2 notched (Novodur PK, PK-AT)
85 J/m (Lustran ABS 252)
107 J/m notched (Abson 110; Lustran ABS 545)
112 J/m notched (Lustran ABS 865)
123 J/m notched (Cycolac X-17)
134 J/m notched (Abson 010)

150 J/m notched (Lustran ABS-HR851)

160 J/m notched (Cycolac X-37)

185 J/m notched (Lustran Ultra ABS MX)

187 J/m (Cycolac Z48; Lustran ABS 456)

192 J/m notched (Cycolac SF)

198 J/m notched (Cycolac KJT)

203 J/m (Lustran ABS 860)

214 J/m (Cycolac KJU; Lustran ABS 452); notched (Cycolac DFA-R, KJB, Z-36; Lustran ABS 248)

215 J/m notched (Cycolac KJW)

230 J/m notched (Cycolac X-11)

235 J/m notched (Cycolac DH)

240 J/m notched (Abson 042; Cycolac Z86)

267 J/m (Cycolac ZA7); notched (Cycolac EPB-3570; Lustran ABS-HR850)

283 J/m (Cycolac TA)

320 J/m (Cycolac AR, X-11, ZA4)

321 J/m notched (Abson 135)

331 J/m notched (Lustran ABS 448)

336 J/m (Lustran ABS 752)

347 J/m notched (Cycolac T; Lustran ABS 1152)

350 J/m notched (Lustran Ultra ABS HX)

358 J/m notched (Lustran ABS 648)

374 J/m notched (Cycolac CGA, EP-3510, GSE, GSM, HM, ZA2; Lustran ABS 743)

400 J/m notched (Cycolac L)

401 J/m notched (Cycolac CG)

405 J/m (Cycolac ETS-3540)

427 J/m notched (Cycolac LS, LXB)

480 J/m notched (Abson 171)

481 J/m notched (Abson 161)

9.8 cm kgf/cm notched (Dow ABS 213)

15 kg cm/cm (Lucky ABS RS-600)

17 kg cm/cm² (Lucky ABS HR-420)

17.5 cm kgf/cm notched (Dow ABS 350)

20 kg cm/cm² (Lucky ABS HF-380, HI-151, HR-450)

21.8 cm kgf/cm notched (Dow ABS 500)

25 kg cm/cm² (Lucky ABS HF-350)

30 kg cm/cm² (Lucky ABS HI-121, MP-211, SH-610)

40 kg cm/cm² (Lucky ABS HI-100)

0.6 ft lb/in. notched (PDX-84357)

1.0 ft lb/in. notched (ABS 414MA; ABS-G1FG-2; AS-40GF; Stat-Kon AC-1003)

1.1 ft lb/in. notched (AS-30GF)

1.2 ft lb/in. notched (RTP 607)

1.25 ft lb/in. notched (AS-20GF)

1.3 ft lb/in. notched (AS-10GF; RTP 605; Thermocomp AF-1008)

1.4 ft lb/in. notched (RTP 603; Thermocomp AF-1006)

1.5 ft lb/in. notched (ABS 125M; RTP 601; Thermocomp AF-1004)

2.0 ft lb/in. notched (ABS 124ESGE; Magnum ABS 213, CLR95)

2.3 ft lb/in. notched (Cycolac ZC7)

2.5 ft lb/in. notched (ABS 124ESG, 224M, 236MA, 510EC; Cycolac CTB; Dow ABS PG912)

2.8 ft lb/in. notched (Magnum ABS 341)

3.0 ft lb/in. notched (ABS 424ESM, 910K; Magnum ABS 343)

3.2 ft lb/in. unnotched (Cycolac KGA)

4.0 ft lb/in. notched (ABS 400P, 707H, 707K; Cycolac GT; Lustran ABS PG-299, FR ABS914HM; Magnum ABS 9010, 9408, 9450P)

4.2 ft lb/in. notched (Magnum ABS FG960)

4.5 ft lb/in. notched (ABS 236MA, 410ESG; Lustran FR ABS911, FR ABS921UV)

4.6 ft lb/in. (Bapolan 8445)

5.0 ft lb/in. notched (ABS 420A, 500FR-1; Magnum ABS 545)

5.2 ft lb/in. notched (Lustran ABS PG-298; Magnum ABS 4500UV)

5.5 ft lb/in. notched (Lustran ABS PG-300)

5.8 ft lb/in. notched (Lustran ABS 750-10802)

5.9 ft lb/in. notched (ABS 236F; Cycolac ZB6; Dow ABS PG940)

6.0 ft lb/in. notched (ABS 301K, 323MA, 808K; Cycolac ZB2; Magnum ABS 9020)

6.5 ft lb/in. notched (ABS 504ESG)

7.0 ft lb/in. notched (ABS 522ESM)

7.5 ft lb/in. notched (ABS 506ESG; Magnum ABS 9030)

8.0 ft lb/in. notched (Lustran ABS 782-10846)

8.5 ft lb/in. notched (ABS 606ED)

9.0 ft lb/in. notched (ABS 550ES; Lustran ABS 780-10802)

10.5 ft lb/in. notched (Magnum ABS 941)

Hardness:

Ball Indentation 61 MPa (Novodur PK-AT)

Ball Indentation 63 (Novodur PK)

Ball Indentation 76 MPa (Novodur PKT)

Ball Indentation 78 MPa (Novodur HGV)

Ball Indentation 79 MPa (Novodur PHE, PM-AT)

Ball Indentation 80 MPa (Novodur PM3C)

Ball Indentation 81 MPa (Novodur PM)

Ball Indentation 82 MPa (Novodur PL-AT)

Ball Indentation 84 MPa (Novodur PTE)

Ball Indentation 86 MPa (Novodur PLT-AT)

Ball Indentation 90 MPa (Novodur PH-AT)

Ball Indentation 91 MPa (Novodur PX)

Ball Indentation 92 MPa (Novodur PMT-AT)

Ball Indentation 93 MPa (Novodur PMT)

ABS resin *(cont'd.)*

Ball Indentation 95 MPa (Novodur PM5C)
Ball Indentation 98 MPa (Novodur P2T-AT)
Ball Indentation 101 MPa (Novodur P2T)
Ball Indentation 103 MPa (Novodur PME)
Ball Indentation 120 MPa (Novodur PHGV-AT)
Ball Indentation 125 (Novodur PHGV)
Rockwell M97 (Thermocomp AF-1004)
Rockwell M99 (Thermocomp AF-1006)
Rockwell M102 (Thermocomp AF-1008)
Rockwell M105 (Bapolan 8445)
Rockwell R60–70 (Cycolac FBK)
Rockwell R68 (Lustran ABS 456)
Rockwell R70 (ABS 500FR-1)
Rockwell R81 (Cycolac CGA)
Rockwell R82 (Cycolac CG)
Rockwell R85 (ABS 506ESG, 522ESM)
Rockwell R88 (Abson 161)
Rockwell R89 (Cycolac HM)
Rockwell R90 (ABS 420A; Abson 010; Cycolac L; Lucky ABS HI-100)
Rockwell R91 (Cycolac LS)
Rockwell R92 (ABS 424ESM, 606ED; Abson 171; Cycolac LXB)
Rockwell R93 (ABS 808K; Lustran ABS 1152)
Rockwell R97 (Abson 042; Cycolac KJB, KJU)
Rockwell R98 (Lustran ABS 780-10802)
Rockwell R99 (Cycolac KJW)
Rockwell R100 (ABS 504ESG, 550ES; Cycolac X-11; Lustran ABS 782-10846; Lustran FR ABS911, FR ABS921UV; Magnum ABS 941)
Rockwell R101 (Magnum ABS 9030)
Rockwell R102 (Cycolac GSM, KJT, TA, Z86, ZA2; Lucky ABS HF-350, HI-121, HR-450, MP-211; Lustran ABS 752)
Rockwell R103 (ABS 236F, 301K, 323MA; Cycolac EP-3510, GSE, T, ZA4)
Rockwell R104 (ABS 410ESG; Cycolac ETS-3540; Dow ABS 500; Lucky ABS SH-610; Magnum ABS 343, 545, 9408, FG960; RTP 601)
Rockwell R105 (ABS 707K; Cycolac CTB; Dow ABS 213; Lucky ABS HI-151, HR-420; Lustran ABS 648, 743, PG-300; Magnum ABS 213)
Rockwell R106 (ABS 236MA, 400P; Abson 135; Cycolac Z-36, ZA7; Lustran ABS 865)
Rockwell R106–109 (Lustran ABS 545)
Rockwell R107 (ABS 707H; Cycolac X-11; Lustran Ultra ABS HX; Magnum ABS 9010, 9020, 9450P; RTP 603)
Rockwell R108 (Cycolac AR, DFA-R, Z48; Lucky ABS RS-600; Lustran ABS PG-299)
Rockwell R109 (ABS 125M, 910K; Cycolac EPB-3570, X-37; Lustran ABS 448,

–HR850)
Rockwell R110 (ABS 124ESGE, 224M, 510EC; Lustran ABS 860; Lustran Ultra ABS
MX; Magnum ABS 341, CLR95; RTP 605, 607)
Rockwell R111 (Cycolac DH, KJM, X-17; Lustran ABS PG-298, FR ABS914HM)
Rockwell R112 (ABS 124ESG, 414MA; Dow ABS 350; Lustran ABS 248)
Rockwell R113 (Lustran ABS 452, -HR851)
Rockwell R114 (ABS 236MA)
Rockwell R115 (Abson 110)
Rockwell R118 (Lustran ABS 252)
Rockwell R122 (AS-20GF)
Rockwell R123 (AS-30GF, -40GF)
Shore D 65 (Cycolac SF)
Shore D 81 (ABS-G1FG-2)

Mold Shrinkage:
0.004–0.006 cm/cm (Lustran Ultra ABS HX, MX)
0.005–0.007 cm/cm (Cycolac ETS-3540, Z48)
0.005–0.008 cm/cm (Cycolac KJM, KJU, Z86)
0.006–0.008 cm/cm (Cycolac AR, TA, ZA4, ZA7)
0.007–0.008 cm/cm (Cycolac HM)
0.007–0.009 cm/cm (Cycolac LXB, ZA2)
0.000–0.002 in./in. (ABS-G1FG-2)
0.002 in./in. (Stat-Kon AC-1003)
0.003–0.005 in./in. (Magnum ABS CLR95)
0.004 in./in. (PDX-84357)
0.004–0.006 in./in. (Lustran FR ABS911, FR ABS914HM, FR ABS921UV; Magnum
ABS 213, 341, 343, 545, 941, 9408, FG960)
0.004–0.007 in./in. (Magnum ABS 4500UV, 9010, 9020, 9030, 9450P)

Water Absorp.:
0.02% (ABS-G1FG-2)
0.18% (PDX-84357; Stat-Kon AC-1003)

THERMAL PROPERTIES:

Soften. Pt. [Vicat]:
87 C (Novodur PK-AT)
88 C (Novodur PK)
90 C (Novodur PH-AT, PM-AT)
91 C (Novodur PL-AT)
93 C (Novodur PM)
94 C (Novodur PX)
96 C (Novodur PHE, PLT-AT, PM3C)
97 C (Novodur HGV)
98 C (Novodur PHGV-AT)
99 C (Novodur PME, PHGV)
102 C (Dow ABS 213, 350, 500; Novodur PMT, PMT-AT, PTE)

ABS resin *(cont'd.)*

    105 C (Novodur P2T-AT, PKT)
    106 C (Novodur PM5C)
    108 C (Novodur P2T)
    198 F (ABS 500FR-1)
    205 F (Magnum ABS CLR95)
    208 F (ABS 707K)
    211 F (Magnum ABS 4500UV)
    212 F (ABS 301K, 323MA, 707H; Bapolan 8445)
    215 F (ABS 410ESG, 522ESM)
    216 F (ABS 506ESG)
    217 F (ABS 420A, 550ES)
    218 F (ABS 125M, 236F, 510EC, 808K)
    219 F (ABS 424ESM)
    220 F (ABS 124ESGE, 224M, 236MA, 236R, 400P, 504ESG, 606ED, 910K; Dow
        ABS PG912; Magnum ABS 213, 343, 941, FG960)
    221 F (AS-10GF; Magnum ABS 341, 545)
    222 F (compr. molded, Dow ABS PG940)
    224 F (ABS 414MA)
    225 F (ABS 124ESG; Magnum ABS 9010, 9030, 9450P)
    226 F (Magnum ABS 9020)
Conduct.:
    0.17 W/Km (Novodur PKT, PMT, PX)
    0.18 W/Km (Novodur P2T, PH-AT, PK-AT, PM-AT, PM5C)
    0.19 W/Km (Novodur PHE, PL-AT, PLT-AT, PM, PM3C, PMT-AT)
    0.20 W/Km (Novodur PHGV, PK)
    0.21 W/Km (Novodur P2T-AT, PTE)
    0.22 W/Km (Novodur HGV, PHGV-AT)
    1.25 Btu/h/ft^2/F/in. (RTP 601)
    1.4 Btu/h/ft^2/F/in. (AS-20GF; RTP 603)
    1.5 Btu/h/ft^2/F/in. (AS-30GF; RTP 605)
    1.6 Btu/h/ft^2/F/in. (AS-40GF; RTP 607)
    3.2 Btu/h/ft^2/F/in. (Stat-Kon AC-1003)
Distort. Temp.:
    64 C (1.82 MPa) (Abson 010)
    66 C (1.82 MPa) (Cycolac SF)
    67–78 C (1.82 MPa) (Cycolac FBK)
    72 C (1.82 MPa) (Lustran ABS 860)
    78 C (1.82 MPa) (Lustran ABS 865)
    80 C (1.82 MPa) (Lustran ABS 456)
    81 C (1.82 MPa) (Lustran Ultra ABS MX)
    82 C (1.82 MPa) (Abson 042; Cycolac Z86; Lustran ABS 648, 743, 1152)
    83 C (1.82 MPa) (Cycolac TA)
    84 C (1.82 MPa) (Cycolac DFA-R; Lustran ABS 752; Novodur PH-AT, PK-AT, PM-

AT)

85 C (Lucky ABS SH-610); (1.82 MPa) (Cycolac HM; Lustran ABS 448, 452; Lustran Ultra ABS HX)

86 C (Lucky ABS RS-600; Lustran ABS 545); (1.82 MPa) (Lustran ABS 248; Novodur PX)

87 C (1.82 MPa) (Abson 135; Cycolac L, T; Lustran ABS 252; Novodur PK, PL-AT, PM); (264 psi) (Lustran ABS 780-10802)

88 C (1.82 MPa) (Cycolac CGA, ETS-3540, KJB, KJT, KJU, KJW); (18.5 kg/cm²) (Lucky ABS HF-350)

89 C (1.82 MPa) (Abson 161; Cycolac EP-3510, GSM, LS, ZA2; Novodur HGV, PHE, PHGV, PKT, PME); (264 psi) (Lustran ABS 782-10846)

90 C (18.5 kg/cm²) (Lucky ABS HF-380, HI-100)

91 C (1.82 MPa) (Abson 110, 171; Cycolac CG, LXB)

92 C (1.82 MPa) (Cycolac AR, EPB-3570, GSE, ZA4; Novodur PHGV-AT); (18.5 kg/cm²) (Lucky ABS HI-121, MP-211)

93 C (1.82 MPa) (Novodur P2T-AT); (18.5 kg/cm²) (Lucky ABS HI-151)

94 C (1.82 MPa) (Cycolac DH, KJM; Novodur PMT-AT)

96 C (1.82 MPa) (Cycolac X-11; Novodur PMT, PTE)

98 C (1.82 MPa) (Lustran ABS-HR850)

99 C (1.82 MPa) (Cycolac ZA7; Novodur P2T, PM5C)

102 C (1.82 MPa) (Cycolac Z-36; Lustran ABS-HR851); (18.5 kg/cm²) (Lucky ABS HR-450)

104 C (1.82 MPa) (Cycolac X-17)

105 C (1.82 MPa) (Cycolac X-11)

110 C (1.82 MPa) (Cycolac X-37, Z48)

112 C (18.5 kg/cm²) (Lucky ABS HR-420)

92 F (18.6 kgf/cm²) (Dow ABS 500)

93 F (18.6 kgf/cm²) (Dow ABS 213)

96 F (18.6 kgf/cm²) (Dow ABS 350)

161 F (264 psi) (Magnum ABS 4500UV, CLR95)

165 F (264 psi) (Magnum ABS 213, 9030)

167 F (264 psi) (Magnum ABS 941)

168 F (264 psi) (Magnum ABS 341, 9020)

170 F (264 psi) (Cycolac CTB; Magnum ABS 9010, 9450P, FG960)

175 F (264 psi) (Magnum ABS 343, 545)

180 F (264 psi) (Lustran ABS PG-300); (264 psi, compr. molded) (Dow ABS PG912)

183 F (264 psi) (Lustran FR ABS921UV)

185 F (264 psi) (Bapolan 8445; Lustran ABS PG-299; Lustran FR ABS911)

187 F (264 psi) (ABS 500FR-1; Lustran FR ABS914HM; Magnum ABS 9408)

190 F (264 psi) (Cycolac GT); (264 psi, compr. molded) (Dow ABS PG940)

192 F (264 psi) (ABS 707K)

194 F (264 psi) (ABS-G1FG-2)

200 F (264 psi) (Cycolac ZB2, ZB6, ZC7)

ABS resin *(cont'd.)*

202 F (264 psi) (ABS 707H)
203 F (264 psi) (ABS 236F, 410ESG, 510EC, 522ESM, 808K)
204 F (264 psi) (ABS 400P, 414MA, 506ESG)
205 F (264 psi) (ABS 124ESGE, 125M, 224M, 301K, 323MA, 420A, 504ESG; AS-10GF)
206 F (264 psi) (ABS 550ES)
207 F (264 psi) (ABS 424ESM)
208 F (264 psi) (ABS 236MA, 910K)
209 F (264 psi) (ABS 236R, 606ED; Lustran ABS PG-298)
210 F (264 psi) (AS-20GF; Cycolac KGA; PDX-84357)
215 F (264 psi) (ABS 124ESG; AS-30GF; RTP 601; Stat-Kon AC-1003; Thermocomp AF-1004)
220 F (264 psi) (AS-40GF; RTP 603; Thermocomp AF-1006)
225 F (264 psi) (Thermocomp AF-1008)
230 F (264 psi) (RTP 605)
240 F (264 psi) (RTP 607)

Coeff. of Linear Exp.:

39×10^{-6} K^{-1} (Novodur PHGV-AT)
47×10^{-6} K^{-1} (Novodur PHGV)
80×10^{-6} K^{-1} (Novodur HGV)
82×10^{-6} K^{-1} (Novodur P2T, PM5C)
86×10^{-6} K^{-1} (Novodur PMT-AT)
87×10^{-6} K^{-1} (Novodur P2T-AT, PME)
88×10^{-6} K^{-1} (Novodur PH-AT)
89×10^{-6} K^{-1} (Novodur PMT)
93×10^{-6} K^{-1} (Novodur PM3C, PX)
94×10^{-6} K^{-1} (Novodur PM)
95×10^{-6} K^{-1} (Novodur PTE)
96×10^{-6} K^{-1} (Novodur PL-AT, PLT-AT)
102×10^{-6} K^{-1} (Novodur PHE)
104×10^{-6} K^{-1} (Novodur PK-AT)
106×10^{-6} K^{-1} (Novodur PM-AT)
114×10^{-6} K^{-1} (Novodur PK)
4.30×10^{-5} mm/mm/C (Abson 110)
5.00×10^{-5} mm/mm/C (Abson 135)
5.80×10^{-5} mm/mm/C (Abson 161, 171)
8.57×10^{-5} mm/mm/C (Abson 010)
6.5×10^{-5} cm/cm/C (Cycolac EPB-3570)
6.7×10^{-5} cm/cm/C (Cycolac X-17)
7.0×10^{-5} cm/cm/C (Cycolac EP-3510)
7.2×10^{-5} cm/cm/C (Lustran ABS-HR851)
7.4×10^{-5} cm/cm/C (Cycolac ETS-3540)
7.5×10^{-5} cm/cm/C (Lustran ABS 252)

7.9 × 10^{-5} cm/cm/C (Cycolac X-37, Z48; Lustran ABS 545)

8.1 × 10^{-5} cm/cm/C (Lustran ABS 248)

8.3 × 10^{-5} cm/cm/C (Cycolac KJT)

8.4 × 10^{-5} cm/cm/C (Lustran Ultra ABS MX)

8.5 × 10^{-5} cm/cm/C (Cycolac X-11, ZA7; Lustran ABS 860, -HR850)

8.6 × 10^{-5} cm/cm/C (Cycolac Z-36)

8.8 × 10^{-5} cm/cm/C (Cycolac DFA-R)

9.0 × 10^{-5} cm/cm/C (Lustran ABS 448)

9.2 × 10^{-5} cm/cm/C (Cycolac X-11, ZA4; Lustran ABS 452, 648)

9.4 × 10^{-5} cm/cm/C (Lustran ABS 752)

9.5 × 10^{-5} cm/cm/C (Cycolac GSE, GSM, T, ZA2; Lustran ABS 456, 743; Lustran Ultra ABS HX)

10.1 × 10^{-5} cm/cm/C (Cycolac HM)

10.3 × 10^{-5} cm/cm/C (Cycolac CG)

10.8 × 10^{-5} cm/cm/C (Cycolac LS)

11.0 × 10^{-5} cm/cm/C (Cycolac CGA, L)

11.7 × 10^{-5} cm/cm/C (Cycolac LXB)

13.9 × 10^{-5} cm/cm/C (Cycolac SF)

8.9 × 10^{-5} in./in./C (Cycolac CTB)

1.4 × 10^{-5} in./in./F (AS-40GF)

1.5 × 10^{-5} in./in./F (Stat-Kon AC-1003)

1.8 × 10^{-5} in./in./F (AS-30GF)

2.1 × 10^{-5} in./in./F (AS-20GF)

2.4 × 10^{-5} in./in./F (AS-10GF)

4.4 × 10^{-5} in./in./F (Magnum ABS 9010, 9450P)

4.5 × 10^{-5} in./in./F (Lustran ABS PG-298)

4.9 × 10^{-5} in./in./F (Magnum ABS 9020)

5.0 × 10^{-5} in./in./F (Lustran ABS PG-299)

5.1 × 10^{-5} in./in./F (Magnum ABS 9030)

5.3 × 10^{-5} in./in./F (Lustran ABS PG-300; Magnum ABS 343, CLR95)

5.6 × 10^{-5} in./in./F (Magnum ABS 213, 341, FG960)

5.7 × 10^{-5} in./in./F (Magnum ABS 545)

6.9 × 10^{-5} in./in./F (Magnum ABS 9408)

7.0 × 10^{-5} in./in./F (Magnum ABS 4500UV)

7.7 × 10^{-5} in./in./F (Magnum ABS 941)

1.2 × 10^5 in./in./F (Thermocomp AF-1008)

1.6 × 10^5 in./in./F (Thermocomp AF-1006)

2.0 × 10^5 in./in./F (Thermocomp AF-1004)

1.3 in./in./F (RTP 607)

1.7 in./in./F (RTP 605)

2.0 in./in./F (RTP 603)

3.2 in./in./F (RTP 601)

ABS resin *(cont'd.)*

Flamm.:
3.1 cm/min (Cycolac CG)
3.3 cm/min (Cycolac ZA4)
3.6 cm/min (Cycolac CGA, LXB, X-11, Z48)
4.3 cm/min (Cycolac SF)
4.6 cm/min (Cycolac Z-36)
B2 (DIN4102) (Novodur HGV, P2T, P2T-AT, PH-AT, PHE, PK, PK-AT, PKT, PL-AT, PLT-AT, PM, PM3C, PM5C, PM-AT, PME, PMT, PMT-AT, PTE, PX)
V-0/5V (Cycolac FBK; Cycolac KGA, KJB, KJW; Lustran FR ABS911; Magnum ABS 4500UV)
V-0 (ABS 500FR-1; Abson 010, 042; Cycolac KJM, KJT, KJU, Z86, ZC7; Lustran FR ABS921UV; PDX-84357)
V-0/V-1 (Lustran ABS 860, 865)
V-0/V-2 (Lustran FR ABS914HM)
HB (ABS 400P; Abson 110, 135, 161, 171; AS-20GF, -30GF, -40GF; Cycolac DFA-R, DH, EP-3510, EPB-3570, GSE, GSM, GT, L, LS, LXB, T, X-11, X-17, X-37, ZB2, ZB6; Lustran ABS 252, 452, 456, 545, 752, 1152; Lustran Ultra ABS HX, MX; Magnum ABS 213, 341, 343, 545, 941, 9010, 9020, 9030, 9408, 9450P, CLR95, FG960; RTP 601, 603, 605, 607; Stat-Kon AC-1003; Thermocomp AF-1004, AF-1006, AF-1008)

ELECTRICAL PROPERTIES:

Dissip. Factor:
0.004 (50 Hz) (Novodur PHGV)
0.005 (1 kHz) (Dow ABS 213); (1 MHz) (AS-40GF)
0.006 (50 Hz) (Novodur PX)
0.0062 (1 kHz) (Dow ABS 500)
0.007 (50 Hz) (Novodur HGV, PKT)
0.0071 (1 kHz) (Dow ABS 350)
0.008 (50 Hz) (Novodur P2T-AT, PHE, PHGV-AT, PK-AT, PM-AT, PME)
0.009 (50 Hz) (Novodur P2T, PK, PL-AT, PLT-AT, PM, PM5C, PMT, PMT-AT, PTE); (1 MHz) (AS-20GF, -30GF; RTP 601, 603, 605, 607)
0.010 (50 Hz) (Novodur PM3C)
0.012 (50 Hz) (Novodur PH-AT)

Dielec. Str.:
15 kV/mm (Cycolac DH)
16 kV/mm (Cycolac KJB, KJU, T)
17 kV/mm (Cycolac EP-3510, GSM)
21 kV/mm (Cycolac KJM)
23 kV/mm (50 Hz) (Novodur HGV, PHGV-AT, PME)
24 kV/mm (50 Hz) (Novodur PH-AT, PX)
25 kV/mm (Cycolac KJT); (50 Hz) (Novodur P2T, P2T-AT, PL-AT, PLT-AT, PM5C)
26 kV/mm (50 Hz) (Novodur PHE, PK-AT, PMT-AT)
27 kV/mm (50 Hz) (Novodur PTE)

28 kV/mm (50 Hz) (Novodur PK)
30 kV/mm (50 Hz) (Novodur PMT)
31 kV/mm (50 Hz) (Novodur PM, PM-AT)
31.8 kV/mm (Cycolac LXB)
32 kV/mm (Cycolac DFA-R, L, Z86); (50 Hz) (Novodur PHGV)
33 kV/mm (Cycolac X-11); (50 Hz) (Novodur PKT)
36 kV/mm (Cycolac X-17)
37 kV/mm (Cycolac EPB-3570, X-37); (50 Hz) (Novodur PM3C)
112 V/mil (Magnum ABS 9408)
400 V/mil (AS-10GF, -20GF, -30GF, -40GF; RTP 601, 603, 605, 607)
784 V/mil (Magnum ABS 9030)
800 V/mil (Magnum ABS 9020)
843 V/mil (Magnum ABS 9010. 9450P)
900 V/mil (Magnum ABS 4500UV)

Dielec. Const.:

2.6 (50 Hz) (Novodur PMT-AT)
2.7 (50 Hz) (Novodur P2T, PH-AT, PK, PL-AT, PLT-AT, PM, PM5C, PM-AT, PMT, PTE)
2.8 (50 Hz) (Novodur P2T-AT, PHE, PK-AT, PKT, PME)
2.83 (1000 Hz) (Dow ABS 213)
2.85 (1000 Hz) (Dow ABS 500)
2.9 (50 Hz) (Novodur PHGV, PHGV-AT)
2.92 (1000 Hz) (Dow ABS 350)
3.0 (50 Hz) (Novodur PM3C, PX)
3.2 (1 MC) (RTP 601)
3.3 (10^6 Hz) (AS-20GF)
3.4 (10^6 Hz) (AS-30GF); (1 MC) (RTP 603)
3.5 (10^6 Hz) (AS-40GF); (1 MC) (RTP 605)
3.6 (1 MHz) (RTP 607)
4.3 (50 Hz) (Novodur HGV)

Vol. Resist.:

9.3×10^7 megohm-cm (Cycolac DH)
1.1×10^8 megohm-cm (Cycolac T)
1.42×10^8 megohm-cm (Cycolac KJB, KJU)
3.04×10^8 megohm-cm (Cycolac KJM)
1.1×10^9 megohm-cm (Cycolac KJT)
1.8×10^9 megohm-cm (Cycolac L, LXB)
2.15×10^9 megohm-cm (Cycolac EP-3510)
2.7×10^9 megohm-cm (Cycolac DFA-R)
3.2×10^9 megohm-cm (Cycolac EPB-3570)
3.7×10^9 megohm-cm (Cycolac GSM, X-17)
7.0×10^9 megohm-cm (Cycolac X-37)
8.8×10^9 megohm-cm (Cycolac X-11)

ABS resin *(cont'd.)*

1.08×10^{10} megohm-cm (Cycolac Z86)
1.7×10^{14} megohm-cm (Magnum ABS 4500UV)
3.1×10^{15} megohm-cm (Magnum ABS 9030)
5.7×10^{15} megohm-cm (Magnum ABS 9020)
9.8×10^{15} megohm-cm (Magnum ABS 9010, 9450P)
100 ohm-cm (PDX-84357)
10,000 ohm-cm (Stat-Kon AC-1003)
5×10^{13} ohm-cm (Novodur HGV)
7×10^{14} ohm-cm (Novodur PK-AT)
9×10^{14} ohm-cm (Novodur PKT)
1×10^{15} ohm-cm (AS-10GF, -20GF, -30GF, -40GF; Novodur PM-AT, PMT-AT; RTP 601, 603, 605, 607)
2×10^{15} ohm-cm (Novodur P2T, P2T-AT, PH-AT, PHE, PHGV-AT, PL-AT, PLT-AT, PM, PM5C, PME, PMT, PTE)
3×10^{15} ohm-cm (Novodur PHGV, PK)
7×10^{15} ohm-cm (Novodur PX)
2×10^{16} ohm-cm (Novodur PM3C)

Surf. Resist.:

100 ohm/sq. (PDX-84357)
10,000 ohm/sq. (Stat-Kon AC-1003)
8×10^{12} ohm (Novodur HGV)
1×10^{14} ohm (Novodur PM3C)
2×10^{14} ohm (Novodur PK-AT)
3×10^{14} ohm (Novodur PHGV-AT, PL-AT, PLT-AT)
4×10^{14} ohm (Novodur P2T-AT, PH-AT, PM-AT, PTE)
5×10^{14} ohm (Novodur P2T, PKT, PM5C, PME, PMT-AT)
6×10^{14} ohm (Novodur PK, PMT)
7×10^{14} ohm (Novodur PHGV)
8×10^{14} ohm (Novodur PHE, PM, PX)

Arc Resist.:

16 s (Cycolac KJM)
19 s (Cycolac KJB)
24 s (Abson 010)
27 s (Cycolac Z86)
28 s (Cycolac KJT)
50 s (Abson 110)
53 s (Cycolac EPB-3570)
66 s (Cycolac X-11)
71 s (Cycolac KJU)
74 s (Abson 161, 171)
75 s (RTP 601, 603, 605, 607)
83 s (Cycolac DFA-R)
84 s (Magnum ABS 9020)

85 s (Cycolac L, LXB, X-17)
86 s (Cycolac DH)
87 s (Magnum ABS 9010, 9450P)
89 s (Cycolac T)
93 s (Cycolac X-37; Magnum ABS 9030)
103 s (Cycolac EP-3510)
114 s (Abson 135)
117 s (Magnum ABS 343, FG960)
120 s (Magnum ABS 341)
123 s (Magnum ABS 941)
128 s (Magnum ABS 545)
130 s (Cycolac GSM)
131 s (Magnum ABS 213)
145 s (Magnum ABS CLR95)

TOXICITY/HANDLING:

Inhalation of dusts can be hazardous (Dow ABS 213, 350, 500)

Vapors can be hazardous; products of combustion can be toxic (Dow ABS PG912, PG940)

Low toxicity under normal conditions; inhalation of dusts and vapors can be irritating (Magnum ABS 213, 341, 343, 545, 941, 4500UV, 9010, 9020, 9030, 9408, 9450P, CLR95, FG960)

Nontoxic; Novodur AT types not suitable for food packaging (Novodur Series)

STORAGE/HANDLING:

Resin will burn (Dow ABS 213, 350, 500, PG912, PG940)

Combustible (Magnum ABS 213, 341, 343, 545, 941, 4500UV, 9010, 9020, 9030, 9408, 9450P, CLR95, FG960)

Acetal resin

SYNONYMS:

Polyacetal

TRADENAME EQUIVALENTS:

Acetal homopolymer:

Delrin 100, 100 AF, 500, 500 AF, 500 CL, 500 F, 570, 900, 900 F, 907 F [DuPont]
Fulton 441D [LNP]
Polypenco Acetal [Polymer Corp.]

Acetal copolymer:

Celcon AS270, AS450, C-400, C-401, EP90, LW90, LW90-S2, LW90-SC, M25, M25-01, M25-04, M50, M90, M90-04, M90-08, M140, M270, M270-04, M450, U10, U10-11, U V25, UV90, WR25 Black, WR90 Black [Hoechst-Celanese]

Acetal resin *(cont'd.)*

Fulton 441 [LNP]

Hostaform C 2521, C 2541, C 9021, C 9021 M, C 9021 TF, C 13021, C 13031, C 27021, C 32021, C 32021 AST, C 52021, S 9063, S 9064, S 27063, S 27064, S 27073, S 27076, T 1020 [Hoechst-Celanese]

Polypenco Acetal [Polymer Corp.]

Ultraform H 2320, N 2211 PVX, N 2320, N 2320 Black 11001 UV, N 2320 Black 11005 MO, S 2320, W 2320 [Badische]

Carbon-reinforced acetal copolymer:

Stat-Kon KC-1002 (10% carbon fiber), KCL-4022 (10% carbon fiber) [LNP]

Thermocomp KC-1004 (20% PAN carbon fiber) [LNP]

Glass-reinforced acetal copolymer:

AT-20GF (20% glass fiber), -30GF (30% glass fiber), -40GF (40% glass fiber) [Compounding Technology]

Celcon GC-25A (25% glass fiber), LWGC-S2 (25% glass-coupled) [Hoechst-Celanese]

Hostaform C 9021 GV1/30 (26% glass fiber), C 9021 GV1/40 (40% glass fiber), C 9021 GV3/10 (10% glass bead), C 9021 GV3/20 (20% glass bead), C 9021 GV3/30 (30% glass bead) [Hoechst-Celanese]

RTP 800TFE20, 801 (10% glass fiber), 803 (20% glass fiber), 805 (30% glass fiber), 805TFE15, 807 (40% glass fiber) [Fiberite]

Thermocomp KB-1008 (40% glass bead), KF-1006 (30% glass fiber), KFX-1002 (10% glass fiber), KFX-1006 (30% glass fiber), KFX-1008 (40% glass fiber), KFX-1008MG (40% mixed glass) [LNP]

Ultraform® N 2200 G5 [Badische]

Mineral-filled acetal copolymer:

Celcon MC90, MC90-HM, MC270, MC270-HM [Hoechst-Celanese]

Hostaform C 9021 K [Hoechst-Celanese]

MODIFICATIONS/SPECIALTY GRADES:

Plateable grade:

Celcon EP90

UV-stabilized:

Celcon M90-08, UV25, UV90, WR25 Black, WR90 Black; Ultraform N 2320 Black 11001 UV

Weather-resistant:

Celcon WR25 Black, WR90 Black

Low friction/low wear:

Celcon AS270, AS450, C-400, C-401, LW90, LW90-S2, LW90-SC, LWGC-S2, M25; Delrin 100 AF, 500 AF, 500 CL; Hostaform C 9021 K, C 9021 M, C 9021 TF; Ultraform N 2211 PVX

High flow:

Celcon M90-04, M140, M270, M270-04, M450; Delrin 900; Hostaform C 13021, C 13031, C 27021, C 32021, C 32021 AST, C 52021, S 27063, S 27064, S 27073; Ultraform S 2320, W 2320

High impact:
Hostaform S 27076
Impact-modified:
Celcon C-400, C-401; Hostaform S 9063, S 9064, S 27063, S 27064, S 27073
Antistat-modified:
Hostaform C 32021 AST
Chemically coupled:
Thermocomp KFX-1002, KFX-1006, KFX-1008, KFX-1008MG
Lubricated:
Celcon M25-04 (internally lubricated), M90-04 (internally lubricated), M270-04 (internally lubricated); Delrin 500 CL (chemically lubricated); Ultraform N 2211 PVX
Molybdenum disulfide-lubricated:
Hostaform C 9021 M; Ultraform N 2320 Black 11005 MO
TFE/PTFE-lubricated:
Delrin 100 AF, 500 AF; Hostaform C 9021 TF; RTP 800TFE20, 805TFE15; Stat-Kon KCL-4022 (10% PTFE)
Silicone-lubricated:
Celcon LW90-S2 (2% silicone), LWGC-S2 (2% silicone), LW90-SC (20% silicone); Fulton 441, 441D

CATEGORY:
Thermoplastic resin
PROCESSING:
Extrusion:
Celcon LW90, LW90-S2, LW90-SC, LWGC-S2, M25, M25-01, M50, U10, U10-11; Hostaform C 2521, C 2541, C 9021; Polypenco Acetal; Ultraform H 2320
Injection molding:
AT-20GF, -30GF, -40GF; Celcon AS270, AS450, C-400, C-401, GC-25A, LW90, LW90-S2, LW90-SC, LWGC-S2, M25, M25-04, M50, M90, M90-04, M90-08, M140, M270, M270-04, M450, MC90, MC90-HM, MC270, MC270-HM, UV25, UV90, WR25 Black, WR90 Black; Delrin 100, 100 AF, 500, 500 AF, 500 CL, 500 F, 570, 900, 900 F, 907 F; Fulton 441, 441D; Hostaform C 2521, C 9021, C 9021 GV1/30, C 9021 GV1/40, C 9021 GV3/10, C 9021 GV3/20, C 9021 GV3/30, C 9021 K, C 9021 M, C 9021 TF, C 13021, C 13031, C 27021, C 32021, C 32021 AST, C 52021, S 9063, S 9064, S 27063, S 27064, S 27073, S 27076, T 1020; Polypenco Acetal; RTP 800TFE20, 801, 803, 805, 805TFE15, 807; Stat-Kon KC-1002, KCL-4022; Thermocomp KFX-1002, KFX-1006, KFX-1008, KFX-1008MG; Ultraform H 2320, N 2200 G5, N 2211 PVX, N 2320, N 2320 Black 11001 UV, N 2320 Black 11005 MO, S 2320, W 2320
Blow molding:
Celcon U10, U10-11; Hostaform T 1020
Compression/transfer molding:
Celcon U10

Acetal resin *(cont'd.)*

APPLICATIONS:

Agriculture industry: (Celcon M25; Delrin 100, 100 AF, 500, 500 AF, 500 CL, 500 F, 570, 900, 900 F, 907 F); irrigation applications (Celcon M25, WR25 Black, WR90 Black)

Automotive applications: (AT-20GF, -30GF, -40GF; Celcon C-400, C-401. EP90, M25, WR25 Black, WR90 Black; Thermocomp KFX-1006, KFX-1008); interior parts (Celcon UV25, UV90; Delrin 100, 100 AF, 500, 500 AF, 500 CL, 500 F, 570, 900, 900 F, 907 F); windshield wiper pivots (Celcon GC-25A)

Consumer products: (Celcon EP90, M25); clock mechanisms (Delrin 100, 100 AF, 500, 500 AF, 500 CL, 500 F, 570, 900, 900 F, 907 F); hardware (AT-20GF, -30GF, -40GF; Celcon M25); outdoor applications (Celcon M90-08, UV25, UV90; Ultraform N 2320 Black 11001 UV); plumbing fixtures (AT-20GF, -30GF, -40GF; Celcon EP90, M25); recreational equipment (Celcon UV25, UV90, WR25 Black, WR90 Black); smoke detectors (Stat-Kon KC-1002, KCL-4022); toys (Celcon UV25, UV90)

Electrical/electronic industry: (Celcon AS270, AS450, M25); audio/video cassettes (Celcon AS270, AS450, M450; Delrin 100, 100 AF, 500, 500 AF, 500 CL, 500 F, 570, 900, 900 F, 907 F); computers (Stat-Kon KC-1002, KCL-4022); electrical components (Polypenco Acetal; Stat-Kon KC-1002, KCL-4022); electrical enclosures/packaging (Stat-Kon KC-1002, KCL-4022); insulators (Polypenco Acetal)

Industrial applications: (Celcon C-400, C-401, M25, U10-11; Thermocomp KFX-1006, KFX-1008); aerosol containers (Celcon U10, U10-11); appliances (Celcon EP90, M25; Thermocomp KFX-1006, KFX-1008); bearing applications (Fulton 441, 441D); bearings and bushings (Celcon LW90, LW90-S2, LW90-SC, LWGC-S2; Polypenco Acetal; RTP 801, 803, 805, 807; Ultraform N 2320 Black 11005 MO); cams (Celcon M90, MC90, MC90-HM, MC270, MC270-HM, UV25, UV90; RTP 801, 803, 805, 807); containers (Celcon U10-11); conveyor plates (Celcon LW90, LW90-S2, LW90-SC, LWGC-S2); gear pump covers (Celcon MC90, MC90-HM, MC270, MC270-HM); gears (Celcon GC-25A, M90, MC90, MC90-HM, MC270, MC270-HM; Polypenco Acetal; RTP 801, 803, 805, 807); housings (Celcon M270); intricate/hard-to-fill moldings (Celcon M140, M270, M270-04); large parts (Hostaform S 27076); levers (Celcon UV25, UV90); machinery (Delrin 100, 100 AF, 500, 500 AF, 500 CL, 500 F, 570, 900, 900 F, 907 F); material handling equipment (Celcon M25); mechanical parts (Polypenco Acetal); micro gears (Celcon M450); molded parts (Celcon M25, M25-04, M50; Delrin 100, 100 AF, 500, 500 AF, 500 CL, 500 F, 570, 900, 900 F, 907 F); plated parts (Celcon EP90); precision parts (Hostaform C 13021, C 27021, C 32021, C 32021 AST, C 52021); profiles (Celcon U10, U10-11); pulleys (Celcon GC-25A); rods (Celcon M25, M25-01, M50, U10, U10-11; Polypenco Acetal); sheet (Celcon M25, M25-01, M50); slab (Celcon M25, M25-01, M50, U10, U10-11); springs (Celcon M90); tanks and floats (Celcon U10); thick sections (Hostaform C 2521, C 2541, T 1020; Ultraform H 2320); thick stock for machining (Ultraform H 2320); thin sections (Celcon M270, M450; Hostaform C 13021, C 27021, C 32021, C 32021 AST, C

52021; Ultraform S 2320); thin tubes and panels (Ultraform H 2320); tubing (Celcon M25, M25-01, M50, U10, U10-11; Ultraform H 2320); valves (Celcon M90; Delrin 100, 100 AF, 500, 500 AF, 500 CL, 500 F, 570, 900, 900 F, 907 F); wear plates (Celcon LW90, LW90-S2, LW90-SC, LWGC-S2); wire coatings (Celcon M25, M25-01)

Medical applications: (Celcon AS270, AS450)

PROPERTIES:

Form:

Free-flowing pellets (Celcon AS270, AS450, C-400, C-401, EP90, GC-25A, LW90, LW90-S2, LW90-SC, LWGC-S2, M25, M25-01, M25-04, M50, M90, M90-04, M90-08, M140, M270, M270-04, M450, MC90, MC90-HM, MC270, MC270-HM, U10, U10-11, UV25, UV90, WR25 Black, WR90 Black)

Avail. in rod, plate, bar, film, etc. (Polypenco Acetal)

Color:

Translucent white (Celcon GC-25A, M25, M25-01, M25-04, M50, M90-04, M90-08, M140, M270-04, M450, U10, U10-11)

Translucent white and colors (Celcon M90, M270)

Natural and custom colors (Celcon UV25, UV90)

Natural, black (RTP 800TFE20, 801, 805TFE15, 807)

Black (Ultraform N 2200 G5)

GENERAL PROPERTIES:

Melt Flow:

1.0 g/10 min (Celcon U10; Hostaform T 1020)

2.5 g/10 min (Celcon M25; Hostaform C 2521, C 2541; Ultraform H 2320)

4.0 g/10 min (Hostaform C 9021 GV1/40)

5.0 g/10 min (Celcon M50; Hostaform C 9021 GV1/30; Ultraform N 2200 G5)

7.0 g/10 min (Hostaform S 9064)

8.0 g/10 min (Hostaform C 9021 TF, S 9063)

8.5 g/10 min (Hostaform C 9021 GV3/30)

9.0 g/10 min (Celcon M90; Hostaform C 9021, C 9021 K, C 9021 M; Ultraform N 2211 PVX, N 2320, N 2320 Black 11001 UV, N 2320 Black 11005 MO)

10.0 g/10 min (Hostaform C 9021 GV3/20, S 27076)

11.0 g/10 min (Hostaform C 9021 GV3/10)

13.0 g/10 min (Hostaform C 13021, C 13031; Ultraform S 2320)

14.0 g/10 min (Celcon M140)

18.0 g/10 min (Hostaform S 27064)

20.0 g/10 min (Hostaform S 27063, S 27073)

23.0 g/10 min (Ultraform W 2320)

27.0 g/10 min (Celcon M270; Hostaform C 27021)

32.0 g/10 min (Hostaform C 32021, C 32021 AST)

45.0 g/10 min (Celcon M450)

50.0 g/10 min (Hostaform C 52021)

Acetal resin *(cont'd.)*

Sp. Gr.:
    1.29 (Hostaform S 27076)
    1.37 (Hostaform S 9064, S 27064)
    1.39 (Hostaform S 9063, S 27063, S 27073)
    1.40 (Fulton 441)
    1.41 (Celcon M25, M25-01, M25-04, M90, M90-04, M90-08, M270, M270-04;
        Hostaform C 2521, C 2541, C 9021, C 13021, C 13031, C 27021, C 32021, C 32021
        AST, C 52021, T 1020)
    1.41–1.42 (Polypenco Acetal)
    1.42 (Delrin 100, 500, 500 CL, 500 F, 900, 900 F, 907 F; Fulton 441D; Hostaform C
        9021 M)
    1.44 (Hostaform C 9021 K; Stat-Kon KC-1002)
    1.46 (Thermocomp KC-1004)
    1.47 (Hostaform C 9021 GV3/10; Thermocomp KFX-1002)
    1.48 (RTP 801; Stat-Kon KCL-4022)
    1.51 (RTP 800TFE20)
    1.52 (Hostaform C 9021 TF)
    1.53 (Hostaform C 9021 GV3/20)
    1.54 (Delrin 100 AF, 500 AF)
    1.55 (AT-20GF; RTP 803)
    1.56 (Delrin 570)
    1.59 (Celcon GC-25A; Hostaform C 9021 GV3/30)
    1.61 (Hostaform C 9021 GV1/30)
    1.63 (AT-30GF; RTP 805; Thermocomp KF-1006, KFX-1006)
    1.71 (Thermocomp KB-1008, KFX-1008, KFX-1008MG)
    1.72 (Hostaform C 9021 GV1/40)
    1.74 (AT-40GF; RTP 807)
    1.76 (RTP 805TFE15)

Sp. Vol.:
    0.709 cm³/g (Celcon M25, M90, M270)
    16.2 in.³/lb (Thermocomp KB-1008, KFX-1008MG)
    17.0 in.³/lb (Thermocomp KF-1006, KFX-1006)
    17.54 in.³/lb (Celcon GC-25A)
    18.9 in.³/lb (Thermocomp KFX-1002)
    19.6 in.³/lb (Fulton 441, 441D)
    19.7 in.³/lb (Celcon M25-01, M25-04, M90-04, M90-08, M270-04)

Density:
    1.41 g/cm³ (Celcon M25, M90; Ultraform H 2320, N 2211 PVX, N 2320, N 2320 Black
        11001 UV, S 2320, W 2320)
    1.42 g/cm³ (Ultraform N 2320 Black 11005 MO)
    1.58 g/cm³ (Ultraform N 2200 G5)
    0.0507 lb/in.³ (Celcon M25-01, M25-04, M90-04, M90-08, M270-04)
    0.057 lb/in.³ (Celcon GC-25A)

Acetal resin (cont'd.)

M.P.:

164–168 C (Ultraform H 2320, N 2200 G5, N 2211 PVX, N 2320, N 2320 Black 11001 UV, N 2320 Black 11005 MO, S 2320, W 2320)

165 C (Celcon M25, M90, M270)

175 C (Delrin 100, 100 AF, 500, 500 AF, 500 CL, 500 F, 570, 900, 900 F, 907 F)

329 F (Celcon M25-01, M25-04, M90-04, M90-08, M270-04)

329–347 F (Polypenco Acetal)

331 F (Celcon GC-25A)

370 F (Hostaform C 52021)

385 F (Hostaform C 2521, C 2541, C 32021, C 32021 AST, S 9063, S 9064, S 27063, S 27064, S 27073, S 27076, T 1020)

390 F (AT-20GF, AT-30GF, AT-40GF; Hostaform C 27021)

395 F (Hostaform C 13021, C 13031)

400 F (Hostaform C 9021, C 9021 GV1/30, C 9021 GV1/40, C 9021 GV3/10, C 9021 GV3/20, C 9021 GV3/30, C 9021 K, C 9021 M, C 9021 TF)

Stability:

Excellent chemical resistance; unaffected by common solvents (acetone, ethyl alcohol, and lacquer solvents @ R.T.); little affected by oils (mineral oil, motor oil, and brake fluids); exceptionally resistant to strong alkalis; attacked by strong mineral acids (sulfuric, nitric, and hydrochloric) (Celcon GC-25A, M25-01, M25-04, M90-04, M90-08, M270-04, U10-11)

Resists weak acids (pH 4) and weak bases (pH 9); not recommended in strong acids or alkalis; excellent resistance to a wide variety of solvents, ethers, oil, greases, gasoline, and other petroleum hydrocarbons (Delrin 100, 100 AF, 500, 500 AF, 500 CL, 500 F, 570, 900, 900 F, 907 F)

Good abrasion and impact resistance; good resistance to solvents and other chemicals; attacked by strong acids and bases (Polypenco Acetal)

Excellent resistance to solvents; good resistance to bases; fair resistance to acids (Thermocomp KC-1004, KFX-1008)

MECHANICAL PROPERTIES:

Tens. Str.:

47.6 MPa (Delrin 500 AF)

52.4 MPa (Delrin 100 AF)

58.6 MPa (Delrin 570)

60.7 MPa (yield) (Celcon M25, M90, M270)

64 MPa (yield) (Ultraform N 2320 Black 11005 MO)

65.5 MPa (Delrin 500 CL)

66.9 MPa (Delrin 500 F, 900 F, 907 F)

68 MPa (yield) (Ultraform N 2211 PVX, N 2320 Black 11001 UV, S 2320, W 2320)

68.9 MPa (Delrin 100, 500, 900)

70 MPa (yield) (Ultraform H 2320, N 2320)

140 MPa (yield) (Ultraform N 2200 G5)

2900 psi (yield) (Hostaform S 27076)

Acetal resin (cont'd.)

5900 psi (yield) (Hostaform C 9021 GV3/30)
6000 psi (RTP 800TFE20; Thermocomp KB-1008)
6100 psi (yield) (Hostaform S 9064, S 27064)
6800 psi (yield) (Hostaform C 9021 GV3/20)
7100 psi (yield) (Hostaform C 9021 TF)
7300 psi (yield) (Hostaform S 27063, S 27073)
7700 psi (yield) (Hostaform S 9063)
7800 psi (yield) (Hostaform C 9021 GV3/10)
8000 psi (Fulton 441)
8800 psi (Celcon M25-01, M25-04, M90-04, M90-08, M270-04)
8800–12,000 psi (Polypenco Acetal)
9000 psi (yield) (Hostaform C 2521, C 2541, C 9021 K)
9300 psi (yield) (Hostaform C 9021, C 9021 M, C 13021, C 27021, C 32021, C 32021
    AST, C 52021, T 1020)
9500 psi (Fulton 441D)
9900 psi (RTP 801)
10,400 psi (yield) (Hostaform C 13031)
11,000 psi (Stat-Kon KCL-4022)
11,800 psi (Thermocomp KC-1004)
12,000 psi (RTP 803, 805TFE15; Stat-Kon KC-1002)
12,500 psi (RTP 805; Thermocomp KFX-1002)
13,000 psi (RTP 807; Thermocomp KF-1006)
13,500 psi (AT-20GF; Thermocomp KFX-1008MG)
14,000 psi (AT-40GF)
15,500 psi (AT-30GF)
16,000 psi (Celcon GC-25A)
18,000 psi (yield) (Hostaform C 9021 GV1/30, C 9021 GV1/40)
19,500 psi (Thermocomp KFX-1006)
21,500 psi (yield) (Thermocomp KFX-1008)
Tens. Elong.:
1.2% (RTP 805TFE15)
1.5% (RTP 807)
1.6% (RTP 805)
2% (AT-20GF, AT-30GF, AT-40GF); (yield) (Thermocomp KF-1006)
2–3% (Celcon GC-25A)
2–4% (break) (Ultraform N 2200 G5)
2.2% (RTP 803)
2.5% (Stat-Kon KCL-4022)
3% (Stat-Kon KC-1002); (break) (Hostaform C 9021 GV1/30, C 9021 GV1/40);
    (yield) (Thermocomp KFX-1008MG)
3–4% (yield) (Thermocomp KFX-1006)
4% (RTP 801)
5–6% (yield) (Thermocomp KB-1008)

36

6% (yield) (Thermocomp KFX-1002)

8% (RTP 800TFE20); (yield) (Ultraform N 2211 PVX, N 2320, N 2320 Black 11001 UV, N 2320 Black 11005 MO, S 2320, W 2320)

8–10% (yield) (Ultraform H 2320)

10% (break) (Hostaform C 9021 GV3/30)

11% (break) (Hostaform C 9021 GV3/20)

12% (Delrin 570)

12–75% (Polypenco Acetal)

14% (break) (Hostaform C 9021 GV3/10)

15% (Celcon M270-04; Delrin 500 AF); (break) (Hostaform C 9021 TF, C 52021)

15–20% (break) (Ultraform W 2320)

20% (Celcon M90-04, M90-08); (break) (Hostaform C 9021 K, C 27021, C 32021, C 32021 AST; Ultraform N 2320 Black 11001 UV, N 2320 Black 11005 MO, S 2320)

22% (Delrin 100 AF)

25% (Delrin 500 F, 900); (break) (Hostaform C 9021 M, C 13021, C 13031; Ultraform N 2211 PVX, N 2320)

30% (Delrin 900 F, 907 F); (break) (Hostaform C 9021, T 1020)

35% (Fulton 441); (break) (Hostaform C 2521, C 2541)

40% (Celcon M270; Delrin 500, 500 CL; Fulton 441D); (break) (Ultraform H 2320)

60% (Celcon M90); (break) (Hostaform S 9063, S 27063, S 27073)

75% (Celcon M25, M25-01, M25-04; Delrin 100)

80% (break) (Hostaform S 27064)

90% (break) (Hostaform S 9064)

> 150% (break) (Hostaform S 27076)

Tens. Mod.:

2829 MPa (Celcon M25, M90, M270)

3000 MPa (Ultraform H 2320, N 2320 Black 11005 MO)

3100 MPa (Ultraform S 2320, W 2320)

3200 MPa (Ultraform N 2211 PVX, N 2320, N 2320 Black 11001 UV)

9100 MPa (Ultraform N 2200 G5)

320,000 psi (RTP 800TFE20)

410,000 psi (Celcon M25-01, M25-04, M90-04, M90-08, M270-04)

410,000–520,000 psi (Polypenco Acetal)

700,000 psi (RTP 801)

1.2×10^6 psi (Celcon GC-25A; RTP 803)

1.3×10^6 psi (AT-20GF)

1.4×10^6 psi (AT-30GF; RTP 805, 805TFE15)

1.6×10^6 psi (AT-40GF; RTP 807)

Flex. Str.:

89.7 MPa (Celcon M25, M90, M270)

10,000 psi (Fulton 441)

10,250 psi (RTP 800TFE20)

11,500 psi (Thermocomp KB-1008)

Acetal resin *(cont'd.)*

13,000 psi (Celcon M25-01, M25-04, M90-04, M90-08, M270-04)
13,000–15,500 psi (Polypenco Acetal)
13,700 psi (Thermocomp KC-1004)
14,000 psi (Fulton 441D; Stat-Kon KCL-4022)
15,000 psi (RTP 805TFE15; Stat-Kon KC-1002)
15,400 psi (RTP 801)
16,500 psi (AT-20GF; RTP 803)
16,800 psi (RTP 805)
17,000 psi (RTP 807)
17,500 psi (AT-30GF; Thermocomp KF-1006)
18,000 psi (AT-40GF)
18,500 psi (Thermocomp KFX-1002)
21,000 psi (Thermocomp KFX-1008MG)
29,000 psi (Thermocomp KFX-1006)

Flex. Mod.:
2340 MPa (Delrin 100 AF)
2410 MPa (Delrin 500 AF)
2588 MPa (Celcon M25, M90, M270)
2620 MPa (Delrin 100)
2760 MPa (Delrin 500 CL, 500 F, 900 F, 907 F)
2830 MPa (Delrin 500)
2960 MPa (Delrin 900)
5030 MPa (Delrin 570)
120,000 psi (Hostaform S 27076)
220,000 psi (Hostaform S 9064)
260,000 psi (Hostaform S 27064)
290,000 psi (Hostaform S 9063)
300,000 psi (Fulton 441; Hostaform S 27063, S 27073; RTP 800TFE20)
350,000 psi (Hostaform C 9021 TF)
370,000 psi (Hostaform C 2521, C 2541)
375,000 psi (Celcon M90-04, M90-08, M270-04)
375,000–550,000 psi (Polypenco Acetal)
380,000 psi (Hostaform T 1020)
390,000 psi (Hostaform C 9021 M)
400,000 psi (Hostaform C 9021)
410,000 psi (Fulton 441D; Hostaform C 9021 K, C 13021, C 27021, C 32021, C 32021
    AST, C 52021)
430,000 psi (Hostaform C 9021 GV3/10)
450,000 psi (Hostaform C 13031)
470,000 psi (Hostaform C 9021 GV3/20)
510,000 psi (Hostaform C 9021 GV3/30)
550,000 psi (Thermocomp KFX-1002)
600,000 psi (Thermocomp KB-1008)

630,000 psi (RTP 801)
900,000 psi (AT-20GF; RTP 803)
1.0×10^6 psi (RTP 805TFE15; Stat-Kon KCL-4022)
1.05×10^6 psi (Celcon GC-25A)
1.1×10^6 psi (AT-30GF; Hostaform C 9021 GV1/30; RTP 805; Stat-Kon KC-1002;
    Thermocomp KFX-1008MG)
1.3×10^6 psi (AT-40GF; RTP 807; Thermocomp KF-1006)
1.35×10^6 psi (Thermocomp KC-1004)
1.4×10^6 psi (Hostaform C 9021 GV1/40; Thermocomp KFX-1006)
1.8×10^6 psi (Thermocomp KFX-1008)

Compr. Str.:
31.0 MPa (Celcon M25, M90, M270; Delrin 100 AF, 500 AF, 500 CL, 500 F, 900 F,
    907 F)
34.5 MPa (Delrin 900)
35.9 MPa (Delrin 100, 500, 570)
4500 psi (Celcon M25-01, M25-04, M90-04, M90-08, M270-04)
8600 psi (RTP 805TFE15)
9500 psi (RTP 801)
11,000 psi (AT-40GF; RTP 807)
11,500 psi (AT-30GF; RTP 805)
12,000 psi (AT-20GF; RTP 803)

Shear Str.:
53.1 MPa (Celcon M25, M90, M270)
55.2 MPa (Delrin 100 AF, 500 AF)
65.5 MPa (Delrin 100, 500)
66.5 MPa (Delrin 500 CL, 570, 900 F, 907 F)
68.9 MPa (Delrin 500 F, 900)
6000 psi (Thermocomp KB-1008)
7700 psi (Celcon M25-01, M25-04, M90-04, M90-08, M270-04)
7700–9500 psi (Polypenco Acetal)
8000 psi (Thermocomp KFX-1002)
8200 psi (Thermocomp KC-1004, KF-1006, KFX-1008MG)
9000 psi (Thermocomp KFX-1006)
13,000 psi (Celcon GC-25A)

Impact Str. (Izod):
37.4 J/m notched (Delrin 500 AF)
42.7 J/m notched (Delrin 570)
53.4 J/m notched (Celcon M270)
55 J/m notched (Ultraform W 2320)
60 J/m notched (Ultraform N 2320 Black 11001 UV, N 2320 Black 11005 MO, S 2320)
64.1 J/m notched (Delrin 100 AF)
65 J/m notched (Ultraform N 2200 G5, N 2211 PVX)
69.4 J/m notched (Celcon M90; Delrin 900, 900 F, 907 F)

70 J/m notched (Ultraform N 2320)

74.7 J/m notched (Delrin 500, 500 CL, 500 F)

80.0 J/m notched (Celcon M25; Ultraform H 2320)

123 J/m notched (Delrin 100)

0.5 ft lb/in. notched (−40 F) (Hostaform C 9021 GV3/30, C 52021)

0.6 ft lb/in. notched (RTP 805TFE15); (−40 F) (Hostaform C 9021 GV3/10, C 9021 GV3/20, C 9021 K, C 9021 TF, C 32021, C 32021 AST)

0.7 ft lb/in. notched (RTP 800TFE20; Stat-Kon KCL-4022; Thermocomp KB-1008)

0.8 ft lb/in. notched (AT-30GF; Thermocomp KF-1006); (−40 F) (Hostaform C 9021 M, C 27021, S 9063, S 27063, S 27064, S 27073)

0.9 ft lb/in. notched (AT-40GF; RTP 803, 805, 807; Stat-Kon KC-1002); (−40 F) (Hostaform C 9021 GV1/40, C 13021, S 9064)

1.0 ft lb/in. notched (AT-20GF; Celcon M270-04; Thermocomp KC-1004); (−40 F) (Hostaform C 9021, C 13031, T 1020)

1.1 ft lb/in. notched (Celcon GC-25A; Thermocomp KFX-1008MG); (−40 F) (Hostaform C 9021 GV1/30)

1.2 ft lb/in. notched (Fulton 441D); (−40 F) (Hostaform C 2521, C 2541)

1.3 ft lb/in. notched (Celcon M90-04, M90-08; RTP 801)

1.4 ft lb/in. notched (Thermocomp KFX-1002, KFX-1008)

1.5 ft lb/in. notched (Celcon M25-01, M25-04)

1.6 ft lb/in. notched (Fulton 441)

1.7 ft lb/in. notched (−40 F) (Hostaform S 27076)

1.8 ft lb/in. notched (Thermocomp KFX-1006)

Tens. Impact:

126 kJ/m^3 (Celcon M270)

147 kJ/m^3 (Celcon M90)

189 kJ/m^3 (Celcon M25)

40–90 ft lb/in.^{2} (Polypenco Acetal)

Hardness:

Rockwell M40 (Hostaform S 9064)

Rockwell M53 (Hostaform S 27064)

Rockwell M64 (Hostaform C 9021 GV3/30, S 9063)

Rockwell M65 (Hostaform C 9021 TF)

Rockwell M68 (Hostaform C 9021 GV3/20)

Rockwell M70 (Hostaform S 27063, S 27073)

Rockwell M75 (Hostaform C 9021 GV3/10)

Rockwell M78 (Celcon M25; Delrin 100 AF, 500 AF)

Rockwell M80 (Celcon M25-01, M25-04, M90, M90-04, M90-08, M270, M270-04; Ultraform H 2320)

Rockwell M82 (Hostaform C 2521, C 2541, C 9021 M, T 1020; Thermocomp KFX-1002)

Rockwell M84 (Thermocomp KB-1008, KFX-1008MG)

Rockwell M85 (Hostaform C 9021, C 9021 K, C 13021, C 27021, C 32021, C 32021

AST, C 52021; Ultraform N 2211 PVX, N 2320, N 2320 Black 11001 UV, N 2320
  Black 11005 MO, S 2320, W 2320)
Rockwell M86 (Thermocomp KF-1006, KFX-1006)
Rockwell M87 (Hostaform C 9021 GV1/30; Thermocomp KFX-1008)
Rockwell M90 (Delrin 500 CL, 570; Hostaform C 9021 GV1/40, C 13031; Ultraform
  N 2200 G5)
Rockwell M92 (Delrin 500 F, 900 F, 907 F)
Rockwell M94 (Delrin 100, 500, 900)
Rockwell R58 (Hostaform S 27076)
Rockwell R107 (RTP 801, 803)
Rockwell R110 (RTP 800TFE20)
Rockwell R112 (RTP 805)
Rockwell R113 (RTP 805TFE15)
Rockwell R118 (RTP 807)
Rockwell R119–122 (Polypenco Acetal)

Mold Shrinkage:
0.018 cm/cm (Celcon M270)
0.022 cm/cm (Celcon M25, M90)
0.004–0.018 in./in. (Hostaform C 9021 GV1/30, C 9021 GV1/40)
0.008–0.010 in./in. (Stat-Kon KC-1002, KCL-4022)
0.010–09.015 in./in. (Hostaform S 27076)
0.012–0.015 in./in. (Hostaform C 9021 GV3/30)
0.014–0.017 in./in. (Hostaform C 9021 GV3/20)
0.014–0.018 in./in. (Hostaform C 9021 GV3/10)
0.018–0.020 in./in. (Hostaform C 9021, C 9021 K, C 9021 M, C 9021 TF, C 13021, C
  13031, C 27021, C 32021, C 32021 AST, C 52021, S 9063, S 9064, S 27063, S
  27064, S 27073)
0.020–0.022 in./in. (Hostaform C 2521, C 2541)
0.023–0.025 in./in. (Hostaform T 1020)

Water Absorp.:
0.15% (Hostaform C 9021 TF)
0.20% (Hostaform C 2521, C 2541, C 9021, C 9021 K, C 9021 M, C 13021, C 13031,
  C 27021, C 32021, C 32021 AST, C 52021, T 1020; Ultraform H 2320, N 2211 PVX,
  N 2320, N 2320 Black 11001 UV, N 2320 Black 11005 MO, S 2320, W 2320)
0.22% (Celcon M25, M90, M270; Hostaform C 9021 GV3/10; Ultraform N 2200 G5)
0.25% (Hostaform C 9021 GV3/20; Stat-Kon KCL-4022)
0.26% (Stat-Kon KC-1002)
0.27% (Hostaform C 9021 GV1/30, C 9021 GV3/30)
0.31% (Hostaform S 9063, S 27063, S 27073)
0.40% (Hostaform C 9021 GV1/40, S 27076)
0.41% (Hostaform S 9064, S 27064)

Acetal resin (cont'd.)

THERMAL PROPERTIES:
Soften. Pt. (Vicat):
  162 C (Celcon M25, M90, M270)
  150 F (Hostaform S 27076)
  263 F (Hostaform S 9064, S 27064)
  284 F (Hostaform S 9063, S 27063, S 27073)
  305 F (Hostaform C 9021 TF)
  323 F (Hostaform C 9021 M)
  324 F (Celcon GC-25A, M25-01, M25-04, M90-04, M90-08, M270-04)
  325 F (Hostaform C 2521, C 2541, C 9021, C 9021 GV1/30, C 9021 GV3/10, C 9021
      GV3/20, C 9021 GV3/30, C 9021 K, C 13021, C 27021, C 32021, C 32021 AST,
      C 52021, T 1020)
  335 F (Hostaform C 9021 GV1/40, C 13031)

Conduct.:
  0.00552 cal/s/cm²/C/cm (Celcon M25, M270)
  0.37 W/m•K (Delrin 100, 500, 900)
  1.6 Btu/h/ft²/F/in. (Celcon M25-01, M25-04, M90-04, M90-08, M270-04; Fulton 441)
  1.7 Btu/h/ft²/F/in. (Fulton 441D)
  1.8 Btu/h/ft²/F/in. (RTP 800TFE20, 801)
  2.0 Btu/h/ft²/F/in. (RTP 803)
  2.1 Btu/h/ft²/F/in. (AT-20GF)
  2.2 Btu/h/ft²/F/in. (AT-30GF; RTP 805)
  2.3 Btu/h/ft²/F/in. (AT-40GF; RTP 805TFE15, 807)
  4.0 Btu/h/ft²/F/in. (Stat-Kon KCL-4022)
  4.2 Btu/h/ft²/F/in. (Stat-Kon KC-1002)
  4.6 Btu/h/ft²/F/in. (Thermocomp KC-1004)

Distort. Temp.:
  100 C (1.8 MPa) (Delrin 100 AF, 500 AF)
  110 C (264 psi) (Celcon M25, M90, M270; Ultraform H 2320, N 2211 PVX, N 2320,
      N 2320 Black 11001 UV, N 2320 Black 11005 MO, S 2320, W 2320)
  121 C (1.8 MPa) (Delrin 500 CL, 500 F, 900 F, 907 F)
  124 C (1.8 MPa) (Delrin 100, 500, 900)
  157 C (1.8 MPa) (Delrin 570)
  162 C (264 psi) (Ultraform N 2200 G5)
  132 F (264 psi) (Hostaform S 27076)
  183 F (264 psi) (Hostaform S 9064, S 27064)
  195 F (264 psi) (Hostaform S 9063, S 27063, S 27073)
  210 F (264 psi) (Hostaform C 9021 K, C 9021 M, C 9021 TF, T 1020)
  226 F (264 psi) (Hostaform C 9021 GV3/10)
  230 F (264 psi) (Celcon M25-01, M25-04, M90-04, M90-08, M270-04; Fulton 441;
      Hostaform C 2521, C 2541, C 9021, C 9021 GV3/20, C 13021, C 27021, C 32021,
      C 32021 AST, C 52021)
  230–255 F (264 psi) (Polypenco Acetal)

Acetal resin (cont'd.)

233 F (264 psi) (Hostaform C 9021 GV3/30)
240 F (264 psi) (Fulton 441D)
250 F (264 psi) (Hostaform C 13031; Thermocomp KB-1008)
295 F (264 psi) (RTP 801)
300 F (264 psi) (Thermocomp KFX-1002)
310 F (264 psi) (Thermocomp KFX-1008MG)
320 F (264 psi) (Hostaform C 9021 GV1/30, C 9021 GV1/40; Stat-Kon KC-1002, KCL-4022; Thermocomp KC-1004)
322 F (264 psi) (Celcon GC-25A)
325 F (264 psi) (AT-20GF, AT-30GF; RTP 803, 805, 805TFE15; Thermocomp KF-1006)
328 F (264 psi) (RTP 807)
330 F (264 psi) (Thermocomp KFX-1006); (66 psi) (AT-40GF)
335 F (264 psi) (Thermocomp KFX-1008)

Coeff. of Linear Exp.:
4.7×10^{-5} cm/cm/C (Celcon M25, M90, M270)
$3.6–8.1 \times 10^{-5}$ m/m/C (Delrin 570)
12.2×10^{-5} m/m/C (Delrin 100, 100 AF, 500, 500 AF, 500 CL, 500 F, 900, 900 F, 907 F)
$3–4 \times 10^{-5}$ 1/K (Ultraform N 2200 G5)
11×10^{-5} 1/K (Ultraform H 2320, N 2211 PVX, N 2320, N 2320 Black 11001 UV, N 2320 Black 11005 MO, S 2320, W 2320)
0.2×10^{-4} in./in./F (Hostaform C 9021 GV1/30, C 9021 GV1/40)
0.5×10^{-4} in./in./F (Hostaform C 9021 GV3/30)
0.6×10^{-4} in/in./F (Hostaform C 2521, C 2541, C 9021, C 9021 GV3/10, C 9021 GV3/20, C 9021 K, C 9021 M, C 9021 TF, C 13021, C 13031, C 27021, C 32021, C 32021 AST, C 52021, T 1020)
1.8×10^{-5} in./in.F (Thermocomp KFX-1008)
1.9×10^{-5} in./in./F (AT-40GF)
2.2×10^{-5} in./in./F (Thermocomp KFX-1006)
2.4×10^{-5} in./in./F (AT-30GF; Thermocomp KF-1006, KFX-1008MG)
2.6×10^{-5} in./in./F (AT-20GF; Stat-Kon KC-1002; Thermocomp KFX-1002)
2.7×10^{-5} in./in./F (Stat-Kon KCL-4022)
3.0×10^{-5} in./in./F (Thermocomp KB-1008)
$4.2–4.7 \times 10^{-5}$ in./in./F (Polypenco Acetal)
5.0×10^{-5} in./in./F (Fulton 441)
5.4×10^{-5} in./in./F (Fulton 441D)

Sp. Heat:
1.5 kJ/ (kg•K) (Ultraform H 2320, N 2200 G5, N 2211 PVX, N 2320, N 2320 Black 11001 UV, N 2320 Black 11005 MO, S 2320, W 2320)
0.35 cal/g/C (Celcon M25, M90, M270)

Flamm.:
HB (Celcon M25, M90, M270; Delrin 100, 100 AF, 500, 500 AF, 500 CL, 500 F, 570,

Acetal resin *(cont'd.)*

900, 900 F, 907 F; Fulton 441, 441D; Hostaform C 2521, C 2541, C 9021, C 9021 GV1/30, C 9021 GV1/40, C 9021 GV3/10, C 9021 GV3/20, C 9021 GV3/30, C 9021 K, C 9021 M, C 9021 TF, C 13021, C 13031, C 27021, C 32021, C 32021 AST, C 52021, S 9063, S 9064, S 27063, S 27064, S 27073, S 27076, T 1020; RTP 800TFE20, 801, 803, 805, 805TFE15, 807; Stat-Kon KC-1002, KCL-4022; Thermocomp KB-1008, KC-1004, KF-1006, KFX-1002, KFX-1006, KFX-1008, KFX-1008MG; Ultraform H 2320, N 2200 G5, N 2211 PVX, N 2320, N 2320 Black 11001 UV, N 2320 Black 11005 MO, S 2320, W 2320)

ELECTRICAL PROPERTIES:

Dissip. Factor:

0.001 (100 Hz) (Celcon M25, M90, M270)

0.0024 (Ultraform H 2320, N 2211 PVX, N 2320, N 2320 Black 11005 MO, S 2320, W 2320)

0.0025 (Ultraform N 2200 G5)

0.003 (100 Hz) (Celcon GC-25A)

0.0035 (Ultraform N 2320 Black 11001 UV)

0.005 (1 MHz) (Delrin 100, 100 AF, 500, 500 AF, 500 CL, 500 F, 570, 900, 900 F, 907 F; Hostaform C 2521, C 2541, C 9021, C 9021 GV1/30, C 9021 GV1/40, C 9021 GV3/10, C 9021 GV3/20, C 9021 GV3/30, C 9021 K, C 9021 M, C 9021 TF, C 13021, C 27021, C 32021, C 32021 AST, C 52021, T 1020)

0.0055 (1 MHz) (RTP 801, 803, 805, 807)

0.006 (1 MHz) (AT-20GF, AT-30GF, AT-40GF)

0.007 (1 MHz) (Hostaform C 13031)

0.05 (1 MHz) (RTP 800TFE20, 805TFE15)

Dielec. Str.:

15.7 kV/mm (Delrin 100 AF, 500 AF, 500 CL)

19.3 kV/mm (Delrin 570)

19.7 kV/mm (Delrin 100, 500, 500 F, 900, 900 F, 907 F)

> 55 kV/mm (Ultraform H 2320, N 2200 G5, N 2211 PVX, N 2320, N 2320 Black 11001 UV, N 2320 Black 11005 MO, S 2320, W 2320)

380–500 V/mil (Polypenco Acetal)

400 V/mil (Hostaform C 9021 TF)

460 V/mil (Hostaform C 9021 GV1/40)

465 V/mil (RTP 800TFE20)

480 V/mil (Hostaform C 9021 GV1/30, C 9021 GV3/10, C 9021 GV3/20, C 9021 GV3/30, C 9021 K; RTP 805TFE15)

500 V/mil (AT-20GF, AT-30GF, AT-40GF; Hostaform C 2521, C 2541, C 9021, C 9021 M, C 13021, C 13031, C 27021, C 32021, C 32021 AST, C 52021, T 1020; RTP 801, 803, 805, 807)

2100 V/mil (Celcon M25, M90, M270)

600 V/0.001 in. (Celcon GC-25A)

Dielec. Const.:

3.1 (10^2–10^6 Hz) (Delrin 100 AF, 500 AF)

3.5 (10^2–10^6 Hz) (Delrin 500 CL)

3.6 (1 MHz) (Hostaform C 9021 TF)

3.65 (1 MHz) (RTP 800TFE20)

3.7 (60 Hz) (Polypenco Acetal); (100 Hz) (Celcon M25, M90, M270); (10^2–10^6 Hz) (Delrin 100, 500, 500 F, 900, 900 F, 907 F); (1 MHz) (Hostaform C 13021, C 13031, C 27021, C 32021, C 32021 AST, C 52021)

3.8 (10^5 Hz) (Ultraform H 2320, N 2211 PVX, N 2320, N 2320 Black 11005 MO, S 2320, W 2320); (1 MHz) (Hostaform C 9021, C 9021 GV3/10; RTP 801)

3.9 (50–100 Hz) (AT-20GF, AT-30GF, AT-40GF); (10^2–10^6 Hz) (Delrin 570); (1 MHz) (Hostaform C 2521, C 2541, T 1020; RTP 805TFE15)

4.0 (10^5 Hz) (Ultraform N 2200 G5, N 2320 Black 11001 UV); (1 MHz) (Hostaform C 9021 GV3/20, C 9021 K, C 9021 M; RTP 803, 805, 807)

4.12 (100 Hz) (Celcon GC-25A)

4.5 (1 MHz) (Hostaform C 9021 GV3/30)

4.8 (1 MHz) (Hostaform C 9021 GV1/30)

Vol. Resist.:

8 ohm-cm (Celcon GC-25A)

10,000 ohm-cm (Stat-Kon KC-1002, KCL-4022)

10^{13} ohm-cm (Hostaform S 27076)

10^{14} ohm-cm (AT-20GF, AT-30GF; Celcon M25, M90, M270; Hostaform S 9063, S 9064, S 27063, S 27064, S 27073; RTP 800TFE20, 801, 803, 805, 805TFE15, 807; Ultraform N 2200 G5, N 2211 PVX, N 2320 Black 11001 UV)

10^{14}–10^{15} ohm-cm (Polypenco Acetal)

5×10^{14} ohm-cm (Delrin 500 CL, 570)

10^{15} ohm-cm (Delrin 100, 500, 500 F, 900, 900 F, 907 F; Hostaform C 2521, C 2541, C 9021, C 9021 GV1/30, C 9021 GV1/40, C 9021 GV3/10, C 9021 GV3/20, C 9021 GV3/30, C 9021 K, C 9021 M, C 9021 TF, C 13021, C 13031, C 27021, C 32021, C 32021 AST, C 52021, T 1020; Ultraform H 2320, N 2320, N 2320 Black 11005 MO, S 2320, W 2320)

3×10^{16} ohm-cm (Delrin 100 AF, 500 AF)

Surf. Resist.:

2000 ohm/sq. (Thermocomp KC-1004)

10,000 ohm/sq. (Stat-Kon KC-1002, KCL-4022)

1.3×10^{16} ohm (Celcon M25, M90, M270)

2.0×10^{16} ohm (Celcon GC-25A)

Arc Resist.:

90 s (RTP 805TFE15)

120 s (Hostaform C 2521, C 2541, C 9021, C 9021 K, C 9021 M, C 9021 TF, C 13021, C 13031, C 27021, C 32021, C 32021 AST, C 52021, T 1020)

128 s (Hostaform C 9021 GV1/30)

142 s (Celcon GC-25A)

168 s (Delrin 570)

181 s (RTP 807)

Acetal resin *(cont'd.)*

182 s (RTP 805)
183 s (Delrin 100 AF, 500 AF)
184 s (RTP 803)
186 s (RTP 801)
200 s (Delrin 500 F, 900 F, 907 F)
220 s (Delrin 100, 500, 900)
240 s (burns) (Celcon M25, M90, M270)
465 s (RTP 800TFE20)

TOXICITY/HANDLING:

Process in areas of adequate ventilation (Celcon AS270, AS450, C-400, C-401, EP90, LW90, LW90-S2, LW90-SC, LWGC-S2, M25, M50, M90, M140, M270, M450, MC90, MC90-HM, MC270, MC270-HM, U10, UV25, UV90, WR25 Black, WR90 Black)

Releases formaldehyde when heated > melt temp.; use adequate ventilation for heating and processing; keep dry (Celcon GC-25A, M25-04, M90-04, M90-08, M270-04, U10-11)

STORAGE/HANDLING:

Keep dry; do not heat above 240 C or store in heated chamber for long periods above 193 C; acetal copolymer and PVC are mutually destructive—do not allow to mix even in trace quantities (Celcon AS270, AS450, C-400, C-401, EP90, LW90, LW90-S2, LW90-SC, LWGC-S2, M25, M50, M90, M140, M270, M450, MC90, MC90-HM, MC270, MC270-HM, U10, UV25, UV90, WR25 Black, WR90 Black)

STD. PKGS.:

25-kg multwall bags, 500-kg boxes (Celcon AS270, AS450, C-400, C-401, EP90, LW90, LW90-S2, LW90-SC, LWGC-S2, M25, M50, M90, M140, M270, M450, MC90, MC90-HM, MC270, MC270-HM, U10, UV25, UV90, WR25 Black, WR90 Black)

50 lb multiwall bags; 1200 lb boxes (Celcon GC-25A, M25-04, M90-04, M90-08, M270-04, U10-11)

Acrylic resin

STRUCTURE:

(Basic acrylic formula)

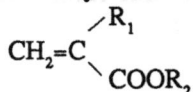

$$CH_2 = C \Big\langle \begin{matrix} R_1 \\ COOR_2 \end{matrix}$$

where R_1 or R_2 or both can be replaced by hydrogen or an aliphatic group, e.g., methyl (CH_3) or ethyl (C_2H_5)

Acrylic resin (cont'd.)

TRADENAME EQUIVALENTS:
Thermoplastic resins:
Asbestos-free acrylic:
Nobestos D-7700 [Rogers Corp.]
100% liquid resin:
Acrylic/acrylate copolymer: [CAS #25133-97-5; RD #977062-96-6]
Carboset 515 [B.F. Goodrich]
Solid acrylics:
Joncryl 67, 678 [S.C. Johnson]
Polymethacrylate resins:
NeoCryl B-700, B-705, B-723, B-725, B-726, B-728, B-734, B-750, B-1019, B-1041, B-1042, B-1054, EX-450, EX-482, EX-484, EX-485, EX-487 [Polyvinyl Chem. Industries]
Acryloid A-11, A-30 [Rohm & Haas]
Acrylic/acrylate copolymer: [CAS #25133-97-5; RD #977062-96-6]
Carboset 525 [BF Goodrich]
Acrylic molding pellets:
CP-51, CP-61, CP-80, CP-81, CP-82 [Continental Polymers]
Plexiglas DR, DR G, DR M, HFI-7, HFI-10, MI-7, MI-7G, V-044, V-045, V-052, V-811, V-920, VM, VS [Rohm & Haas]
Acrylic solutions:
Acryloid NAD-10 [Rohm & Haas]
Carboset XL-11, XL-19 [BF Goodrich]
Crystic LS7 [Scott Bader]
Joncryl 61LV, 74, 77, 78, 85, 130 [SC Johnson]
NeoCryl A-550, A-601, A-603A, A-610, A-630, A-1031, A-1044, A-1045, BT-7, CC-6, CL-300, SR-270, SR-272, SR-274, SR-276, S-1004 [Polyvinyl Chem. Industries]
Polymethacrylate solutions:
Acryloid A-10, A-21, A-21LV, A-101, B-44, B-48N, B-50, B-82, B-99, C-10LV [Rohm & Haas]
Acrylic/acrylate copolymer solution: [CAS #25133-97-5; RD #977062-96-6]
Carboset 514A, 514H [B.F. Goodrich]
Acrylic emulsions:
Daratak 74L [W.R. Grace]
Joncryl 142 [S.C. Johnson]
NeoCryl A-602, A-604, A-620, A-621, A-624, P-920A Glaze [Polyvinyl Chem. Industries]
Parcryl 200, 250, 300, 311, 400, 450, 475, 500, 777, 900, 966 [Thibaut & Walker]
Rhoplex AC-22, AC-34, AC-35, AC-61, AC-73, AC-388, AC-490, AC-507, AC-707, MV-1 [Rohm & Haas]
Texicryl 13-003, 13-010, 13-011, 13-104, 13-203 (self-crosslinking), 13-205 (self-crosslinking), 13-439, 13-442, 13-206 (self-crosslinking), 13-210 (self-crosslink-

Acrylic resin *(cont'd.)*

ing) [Scott Bader]

Ucar Acrylic 503, 505, 515, 516, 518 [Union Carbide]

Ucar Latex 123, 153, 154, 163, 173 (crosslinkable), 174 (crosslinkable), 175 (crosslinkable), 803 (self-crosslinking), 812 (self-crosslinking), 874 (self-crosslinking) [Union Carbide]

Ucar Vehicle 407, 4630, 4358, 4414, 4431, 4550 (crosslinking) [Union Carbide]

Versaflex 1, 2, 4, 5 [W.R. Grace]

Carboxyl-type acrylic emulsion:

Texicryl 13-100, 13-101, 13-300, 13-302 [Scott Bader]

Acrylic cast rod:

Polypenco Cast Acrylic Rod [Polymer Corp.]

Acrylic cast sheet:

Acrylite GP, MS-2 Sheet [Cyro Industries]

XT Polymer 250, 375 [Cyro Industries]

Polysilicate resin-modified cast sheet:

Lucite SAR [DuPont]

Methyl methacrylate-based cast sheet:

Lucite L [DuPont]

Acrylic film:

Korad 2, A, Klear [Georgia-Pacific]

Presto Adhesive 2125, 2125M88, 2125U88, 2127+, 2127M6 [Presto Mfg.]

Thermoset resins:

Acrylic solution:

Carboset 531 [B.F. Goodrich]

NeoCryl A-1038, BT-8, BT-20, BT-24, BT-175 [Polyvinyl Chem. Industries]

Hydroxyl-type solution:

Acryloid AT-50, AT-51, AT-56, AT-63, AT-64, WR-97 (crosslinkable) [Rohm & Haas]

Carboxyl-type solution:

Acryloid AT-70, AT-71, AT-75 [Rohm & Haas]

Acrylic emulsion:

Ucar Vehicle 4620 [Union Carbide]

Carboxyl-type emulsion:

Acrysol I-94 (crosslinkable), WS-12, WS-24, WS-32, WS-50 [Rohm & Haas]

Acrylic film:

Hysol PC20 [Hysol/Dexter Corp.]

Acrylic elastomer:

Cyanacryl C, L, R [Amer. Cyanamid]

MODIFICATIONS/SPECIALTY GRADES:

High impact:

Plexiglas DR G, DR M, HFI-10

High flow:

CP-51; Plexiglas VS

UV-stabilized:
Crystic LS7
Gamma radiation-resistant:
Plexiglas DR G, MI-7G
Optical quality:
Plexiglas V-044, V-045, V-052, V-811, V-920, VM; Polypenco Cast Acrylic Rod
Polyester carrier-reinforced:
Presto Adhesive 2125M88, 2127M6 [Presto Mfg.]
CATEGORY:
Thermoplastic resin; thermoset resin
PROCESSING:
Extrusion:
CP-51, CP-61, CP-80, CP-81, CP-82; Cyanacryl C, L; Plexiglas DR, DR G, DR M, MI-7, MI-7G, V-044, V-045, V-052, V-811, V-920, VM
Injection molding:
CP-51, CP-61, CP-82; Cyanacryl C, L, R; Plexiglas DR, DR G, DR M, HFI-7, HFI-10, MI-7, MI-7G, V-045, V-052, V-811, V-920, VM, VS
Compression molding:
Cyanacryl C
Transfer molding:
Cyanacryl C
Thermoforming:
XT Polymer 250, 375
APPLICATIONS:
Agriculture industry: farm equipment finishes (NeoCryl B-734)
Architectural applications: door panels (Acrylite GP, MS-2 Sheet); glazing for buildings (Lucite L); lighting (Acrylite GP, MS-2 Sheet); skylights (Acrylite GP, MS-2 Sheet; Lucite L); tub/shower enclosures (Acrylite GP, MS-2 Sheet); window glazing (Acrylite GP, MS-2 Sheet)
Automotive applications: (Nobestos D-7700; Ucar Latex 173); engines (Nobestos D-7700); finishes (Acryloid A-21, A-21LV, AT-56, B-44, B-48N, B-50, B-99; Korad Klear; NeoCryl B-734); parts (Cyanacryl L)
Aviation industry: (Nobestos D-7700); aircraft finishes (Acryloid A-21, A-21LV, B-44, B-82); glazing (Lucite L)
Consumer products: ; appliance finishes (Acryloid AT-50, AT-64, AT-70, B-44, B-48N, B-50, B-82, WR-97; Korad A, Klear); carpet shampoos (NeoCryl A-550); furniture (Polypenco Cast Acrylic Rod); furniture finishes (Korad A; NeoCryl A-604, A-624); metal furniture finishes (Acryloid B-44, B-48N, B-50, B-82); outdoor applications (Korad 2, A, Klear; Plexiglas DR G, DR M, HFI-7, HFI-10, MI-7, MI-7G)
Electrical/electronic industry: (Polypenco Cast Acrylic Rod); lighting equipment (Korad Klear; Lucite L)

Acrylic resin (cont'd.)

FDA-approved applications: (Plexiglas DR, DR G, MI-7, MI-7G)

Food-contact applications: (Plexiglas DR, DR G, MI-7, MI-7G); food packaging (XT Polymer 250, 375)

Functional additives: antisoilant (NeoCryl A-550); binder (NeoCryl A-1031, A-1038, BT-8; Texicryl 13-003, 13-011, 13-100, 13-101, 13-203, 13-205, 13-206, 13-210; Ucar Latex 874); dispersant (Acryloid WR-97); modifier (NeoCryl A-610; Rhoplex AC-73; Ucar Vehicle 4431); thickener (Texicryl 13-300, 13-302); vehicle (NeoCryl BT-175)

Industrial applications: adhesives (Acryloid C-10LV; Cyanacryl R; Daratak 74L; NeoCryl A-1031, A-1038, B-728; Presto Adhesive 2125, 2125M88, 2125U88, 2127+, 2127M6; Texicryl 13-101, 13-439, 13-442, 13-210; Ucar Latex 173, 174, 175); aerosols (NeoCryl B-723, EX-482); barrier coatings (NeoCryl B-728; Ucar Latex 123, 163); caulks, sealants, mastics (Parcryl 250, 777; Texicryl 13-003; Ucar Latex 153, 154, 163; Versaflex 1); cement (NeoCryl A-1044; Ucar Vehicle 407); coatings (Acryloid A-10, A-11, A-21, A-21LV, A-30, A-101, C-10LV, NAD-10; Acrysol I-94, WS-24, WS-32, WS-50; Daratak 74L; Joncryl 67, 74, 77, 130; NeoCryl A-602, A-603A, A-604, A-1031, A-1038, BT-8, BT-175, EX-487, S-1004; Ucar Vehicle 4414, 4550, 4620, 4630; Versaflex 4); coil coating (Acryloid AT-56, AT-63, AT-71, WR-97; NeoCryl B-723; Ucar Vehicle 4630); conveyor belts (Cyanacryl R); displays (Lucite L; Polypenco Cast Acrylic Rod); extruded goods (Cyanacryl C); films (Korad A; Rhoplex AC-73; Texicryl 13-003, 13-010, 13-100, 13-101, 13-439, 13-442, 13-206); floor sealer/paints (NeoCryl A-1045, CC-6; Rhoplex AC-61); foam compounds (Ucar Latex 812); gaskets (Cyanacryl C, L, R; Nobestos D-7700); glaze (NeoCryl P-920A Glaze); hose (Cyanacryl C, R); inks (Acryloid A-10, A-101, B-82, B-99, NAD-10; Acrysol I-94; Joncryl 61LV, 67, 74, 77, 78, 85, 130, 142, 678; NeoCryl B-723, B-728, B-750, B-1019, B1042, B-1054, BT-8, BT-20, BT-175, EX-482, , EX-484, EX-485, EX-487, S-1004; Texicryl 13-011); lacquers (NeoCryl EX-484); laminating (Ucar Latex 874); lenses (Polypenco Cast Acrylic Rod); masonry/concrete coatings (Acryloid B-44, B-82; NeoCryl A-601, B-700, B-725, B-726, B-1019, CC-6, EX-450, EX-482, EX-485; Parcryl 200, 250, 300, 400; Rhoplex AC-61, AC-388; Versaflex 1, 4, 5); mechanical goods (Cyanacryl C); metal coatings (Acryloid B-44, B-48N, B-50; Acrysol WS-12; Korad 2, A; NeoCryl A-601, A-610, A-620, A-621, A-630, B-700, B-723, B-725, B-726, B-728, B-1019, B-1042, B-1054, BT-20, BT-24, EX-450, EX-482, EX-485; Parcryl 900; Rhoplex MV-1; Ucar Vehicle 4358); metal decorating (Acryloid AT-71, AT-75); nonwoven fabrics (NeoCryl A-1031; Texicryl 13-011, 13-203, 13-205, 13-206, 13-210; Ucar Latex 803, 812, 874); o-rings (Cyanacryl C, L, R); packaging (XT Polymer 250, 375); paints (Parcryl 200, 300, 400, 450, 475, 500, 966; Rhoplex AC-22, AC-388, AC-490, AC-507, AC-707; Texicryl 13-011, 13-300, 13-302; Ucar Acrylic 503, 505, 515, 516, 518; Ucar Vehicle 5348; Versaflex 1, 4, 5); paper coating (Joncryl 61LV, 85; NeoCryl A-610, A-630, B-728, BT-175; Texicryl 13-003, 13-100, 13-101, 13-104; Ucar Latex 173; Versaflex 2, 5); pigments (NeoCryl A-1031; Texicryl 13-003); plastics coatings (Acryloid B-44, B-

48N, B-50, B-82; Acrysol WS-12; Korad 2, A; NeoCryl A-601, A-610, A-620, A-630, B-700, B-723, B-725, B-726, B-1019, B-1041, B-1042, B-1054, BT-20, BT-24, EX-450, EX-482, EX-485; Ucar Latex 173); plastisol primers (Acryloid A-21, A-21LV); polishes (NeoCryl B-705, B-1054, CC-6, CL-300, S-1004, SR-270, SR-272, SR-274, SR-276; Texicryl 13-010); printed circuit coating (Hysol PC20); product finishes (Acryloid AT-50, AT-51, AT-56, AT-63, AT-64, AT-70, B-44, B-48N, B-50, B-82, B-99, WR-97; Ucar Vehicle 4358, 4431); roof and tennis court coatings (Parcryl 200, 300, 311; Versaflex 4); rubber latexes (Texicryl 13-300, 13-302); seals (Cyanacryl C, L, R); sheet (Crystic LS7); signs (Korad Klear; Lucite L; Polypenco Cast Acrylic Rod); textiles (Acryloid B-82; NeoCryl B-725, B-726, B-728, B-1019, BT-7, BT-8, EX-450. EX-485; Texicryl 13-003, 13-101, 13-205, 13-210; Ucar Latex 803; Versaflex 5); wood coatings (Korad 2, A; NeoCryl A-601, A-604, A-610, A-620, A-624, B-700, B-705, B-725, B-726, B-728, B-1019, B-1054, EX-450; Parcryl 900; Rhoplex AC-34, AC-35, AC-61, AC-388; Versaflex 1, 2, 4, 5)

Marine applications: finishes (Acryloid B-48N, B-50, B-99); glazing (Lucite L)

Medical applications: (Plexiglas DR, DR G, MI-7G); hospital and lab equipment finishes (Acryloid AT-64)

Transportation industry: glazing (Lucite L); rail car finishes (Acryloid A-21, A-21LV, B-48N, B-50)

PROPERTIES:

Form:

Clear liquid (Carboset 514A, 515; Joncryl 61LV, 78)

Liquid (Acryloid A-10, A-21, A-21LV, A-101, AT-50, AT-51, AT-56, AT-63, AT-64, AT-70, AT-71, AT-75, B-44, B-48N, B-50, B-82, B-99, C-10LV, NAD-10, WR-97; Acrysol I-94, WS-12, WS-24, WS-32, WS-50; Carboset 514A, 515; Crystic LS7; Joncryl 74, 77, 85, 130; NeoCryl A-550, A-601, A-603A, A-610, A-630, A-1031, A-1044, A-1045, BT-7, CC-6, CL-300, SR-270, SR-272, SR-274, SR-276, S-1004)

Translucent solution (Carboset 531)

Dispersion (Carboset 514H, XL-11, XL-19)

Emulsion (NeoCryl A-602, A-604, A-620, A-621, A-624; Parcryl 200, 250, 300, 311, 400, 450, 475, 500, 777, 900, 966; Rhoplex AC-22, AC-34, AC-35, AC-61, AC-73, AC-388, AC-490, AC-507, AC-707, MV-1)

Emulsion; particle size 0.1 μ (Versaflex 2, 4, 5)

Emulsion; particle size 0.15 μ (Texicryl 13-010; Ucar Latex 173, 174, 175; Ucar Vehicle 4358, 4414)

Emulsion; particle size 0.2 μ (Texicryl 13-011, 13-100, 13-101, 13-203, 13-205, 13-210, 13-300, 13-302; Ucar Latex 803, 812; Ucar Vehicle 4414; Versaflex 1)

Emulsion; particle size 0.25 μ (Texicryl 13-003, 13-104, 13-206, 13-439, 13-442; Ucar Vehicle 4630)

Emulsion; particle size 0.3 μ (Ucar Vehicle 407, 4550, 4620)

Emulsion; particle size 0.4 μ (Ucar Acrylic 503; Ucar Latex 154, 163, 874)

Acrylic resin *(cont'd.)*

  Emulsion; particle size 0.45 μ (Ucar Latex 153)
  Emulsion; particle size 0.5 μ (Ucar Acrylic 515, 516, 518; Ucar Latex 123)
  Emulsion; particle size 0.6 μ (Ucar Acrylic 505)
  Emulsion; particle size 1.0 μ (Daratak 74L)
  Opaque emulsion (Joncryl 142)
  Solid (Acryloid A-11; NeoCryl B-1054)
  Slab (Cyanacryl C)
  Granular solid (Carboset 525)
  Flakes (Joncryl 67, 678)
  Powder (Acryloid A-30)
  Pellets (CP-51, CP-61, CP-80, CP-81, CP-82; Plexiglas DR, DR G, DR M, HFI-7, HFI-10, MI-7, MI-7G, V-044, V-045, V-052, V-811, V-920, VM, VS)
  Beads (NeoCryl B-700, B-705, B-723, B-725, B-726, B-728, B-734, B-750, B-1019, B-1041, B-1042, EX-450, EX-482, EX-484, EX-485, EX-487)
  Film (Korad Klear)
  Film (2–6 mil) (Korad A)
  Film (2, 3- and 6-mil) (Korad 2)
  Film 0.0017 in. adhesive thickness (Presto Adhesive 2125)
  Film 0.002 in. adhesive thickness; 1/2 mil carrier thickness (Presto Adhesive 2127+)
  Film 0.006 in. adhesive thickness (Presto Adhesive 2125U88)
  Film 0.006 in. adhesive thickness; 1/2 mil carrier thickness (Presto Adhesive 2125M88, 2127M6)
  Sheet (Lucite SAR; XT Polymer 250, 375)
  Sheet avail. in thicknesses from 0.030–4.25 in. (Acrylite GP, MS-2 Sheet)
  Sheet avail. in thicknesses from 0.125–0.375 in. (Lucite L)
  Rod (Polypenco Cast Acrylic Rod)
Color:
  Colorless (Carboset 514A, 515)
  Water-white (Joncryl 67)
  Clear (Korad Klear)
  Clear, translucent white, many colors (Acrylite GP, MS-2 Sheet)
  Clear, white, black, and colors (Korad A)
  Transparent solar tints, white, and custom colors (Lucite L)
  Transparent, translucent, and opaque colors (Plexiglas V-044, V-045, V-052, V-811, V-920, VM, VS)
  White (Carboset 525, 531; Daratak 74L; Joncryl 142; Versaflex 1, 2, 4, 5)
  Milky white (Carboset 514H, XL-11, XL-19; NeoCryl A-602, A-604, A-620, A-621, A-624)
  Milky (Joncryl 74, 77)
  Off-white (Cyanacryl C)
  Pale yellow (Crystic LS7)
  Straw (Joncryl 78)
  Very pale straw (XT Polymer 375)

Amber (Joncryl 61LV)
Wide range of colors (Korad 2)
Gardner 1 max. (50% in ethanol) (Joncryl 678)
Odor:
None (Acrylite GP, MS-2 Sheet; Lucite L)
Low (Joncryl 67, 77; NeoCryl EX-487, P-920A Glaze)
Slight, characteristic (Daratak 74L; Versaflex 1, 2, 4, 5)
Taste:
None (Acrylite GP, MS-2 Sheet; Lucite L)
Composition:
0.8–1.1% chlorine content (Cyanacryl L, R)
18% solids (Crystic LS7)
25% in ammonia water (Carboset 531)
30% in ammonia water (Carboset XL-11)
30% in Cellosolve acetate (Acryloid A-10)
30% solids in toluene/butanol (90/10) (Acryloid A-21)
30% solids in toluene/MEK/butanol (50/40/10) (Acryloid A-21LV)
30% nonvolatiles in water (NeoCryl A-1045)
30% nonvolatiles in water-ethanol (Joncryl 85)
30 ± 1% nonvolatiles in water (Joncryl 78)
32% nonvolatiles in water (NeoCryl A-601, P-920A Glaze)
32 ± 1% nonvolatiles in water (NeoCryl A-602, A-604)
35% solids (Joncryl 61LV)
35 ± 1% solids (Texicryl 13-010, 13-302)
36% nonvolatiles (NeoCryl A-603A)
37 ± 1% nonvolatiles in water (Joncryl 130)
38% nonvolatiles in water (NeoCryl SR-272)
39 ± 1% nonvolatiles (Joncryl 142)
40% nonvolatiles (NeoCryl B-1054)
40% nonvolatiles in water (NeoCryl A-550, BT-7, BT-8, BT-20, BT-175, CC-6, CL-
    300, SR-270, SR-274, SR-276)
40% solids in ammonia water (Carboset 514H, XL-19)
40% solids in MEK (Acryloid A-101)
40% solids in toluene (Acryloid B-48N, C-10LV)
40% solids in toluene/methyl Cellosolve (95/5) (Acryloid B-44)
40% solids in VM&P naphtha (Acryloid NAD-10)
40–42% solids (Versaflex 2)
40 ± 1% nonvolatiles in water (NeoCryl A-620, A-621, A-624)
41 ± 1% solids (Texicryl 13-011, 13-203, 13-300)
42% nonvolatiles (NeoCryl A-610)
43% solids (Ucar Vehicle 4431)
43 ± 0.5% nonvolatiles (Parcryl 900, 966)
44% nonvolatiles in water (NeoCryl A-630)

Acrylic resin *(cont'd.)*

44–45% solids (Rhoplex AC-22)
45% nonvolatiles (Joncryl 77)
45% nonvolatiles in water (NeoCryl BT-24)
45% solids (Ucar Vehicle 407, 4358, 4414, 4550, 4620)
45% solids in toluene (Acryloid B-50)
45 ± 1% solids (Texicryl 13-210)
45–47% solids (Versaflex 4, 5)
45.5–46.5% solids (Rhoplex MV-1)
46% nonvolatiles in water (NeoCryl A-1031, A-1038)
46–47% solids (Rhoplex AC-34, AC-35, AC-61, AC-73, AC-490, AC-507)
46 ± 1% solids (Texicryl 13-100, 13-101, 13-205)
46.5 ± 0.5% nonvolatiles (Parcryl 200, 250, 300, 311, 450, 500)
47% nonvolatiles in water (NeoCryl A-1044)
48% solids (Ucar Vehicle 4630)
48 ± 0.5% nonvolatiles (Parcryl 475)
49% solids (Ucar Latex 175)
49.5–50.5% solids (Rhoplex AC-388)
50% nonvolatiles in Isopar G (NeoCryl S-1004)
50% solids (Ucar Acrylic 518; Ucar Latex 173, 174)
50% solids in Solvesso 150/Cellosolve acetate (75/25) (Acryloid AT-71, AT-75)
50% solids in toluene (Acryloid B-82, B-99)
50% solids in xylene (Acryloid AT-63)
50% solids in xylene/butanol/methyl Cellosolve (60/22/18) (Acryloid AT-50)
50% solids in xylene/butanol (78/22) (Acryloid AT-51)
50% solids in xylene/butanol (90/10) (Acryloid AT-56, AT-64)
50% solids in xylene/Cellosolve acetate (75/25) (Acryloid AT-70)
50 ± 0.5% nonvolatiles (Parcryl 400)
50 ± 1% solids (Texicryl 13-104, 13-206, 13-439, 13-442)
51% solids (Ucar Latex 803)
53% solids (Ucar Acrylic 515, 516)
54–56% solids (Daratak 74L; Versaflex 1)
55% solids (Ucar Acrylic 505; Ucar Latex 153, 812)
55 ± 0.5% nonvolatiles (Parcryl 777)
55 ± 1% solids (Texicryl 13-003)
58% solids (Ucar Acrylic 503; Ucar Latex 163)
59% nonvolatiles (Joncryl 74)
60% solids (Ucar Latex 123, 154, 874)
64.5–65.5% solids (Rhoplex AC-707)
70% in isopropanol (Carboset 514A)
70% solids in isopropanol/butyl Cellosolve (83/17) (Acryloid WR-97)
98% nonvolatiles (NeoCryl B-700, B-705, B-725, B-726, B-728, B-734, B-1019, B-1041, B-1042, EX-450, EX-482, EX-484, EX-485, EX-487)
99% nonvolatiles (NeoCryl B-750)

100% conc. (Carboset 515, 525)
100% solids (Acryloid A-11, A-30; Korad A)

GENERAL PROPERTIES:

Solubility:
Sol. in acetone (Acryloid A-11)
Sol. in alcohol (Acryloid A-11); dilutable with alcohols (Joncryl 78, 142)
Sol. in alkalies (NeoCryl BT-7, BT-8, BT-20, BT-175, CL-300)
Sol. in most chlorinated hydrocarbons (Acryloid A-11)
Sol. in diacetone (Acryloid A-11)
Sol. in diethylene glycol (Joncryl 67)
Sol. in diethylene glycol monoethyl ether (Joncryl 67)
Sol. in dioxane (Acryloid A-11)
Sol. in most esters (Acryloid A-11)
Sol. in ethanol (3A) (Joncryl 67)
Sol. in ethylene glycol monoethyl ether (Joncryl 67)
Dilutable with glycol ethers (Joncryl 78)
Dilutable with glycols (Joncryl 78)
Sol. in isopropanol (Joncryl 67)
Sol. in MEK (Acryloid A-11; Joncryl 67)
Sol. in min. spirits to 0 C (NeoCryl B-705)
Sol. in n-propanol (Joncryl 67)
Sol. in propylene glycol (Joncryl 67)
Able to combine with other resins incl. casein, shellac, and acrylic polymers in all proportions (Joncryl 61LV)
Sol. in toluene (Acryloid A-11)
Sol. in triethylene glcyol (Joncryl 67)
Sol. in water (Joncryl 85); dilutable with water (Joncryl 78); dissolves readily in alkaline water (Joncryl 678)

M.W.:
7000 (Carboset 515)
30,000 (Carboset 514A, 514H, XL-19)
40,000 (Carboset XL-11)
260,000 (Carboset 525)
10^6 (Carboset 531, cured)

Melt Flow:
1.0 g/10 min (Plexiglas DR, DR G, DR M)
1.7 g/10 min (Plexiglas V-044, V-045)
2.0 g/10 min (Plexiglas V-052)
2.8 g/10 min (CP-80)
2.9 g/10 min (CP-81)
3.0 g/10 min (Plexiglas HFI-10, MI-7, MI-7G)
3.2 g/10 min (CP-82)
5.0 g/10 min (Plexiglas V-811)

Acrylic resin *(cont'd.)*

    7.1 g/10 min (Plexiglas V-920)
    10.0 g/10 min (CP-61)
    11.0 g/10 min (Plexiglas HFI-7)
    15.0 g/10 min (Plexiglas VM)
    21.5 g/10 min (CP-51)
    24 g/10 min (Plexiglas VS)

Sp. Gr.:
    0.89 (Crystic LS7)
    1.00 (Texicryl 13-442; Ucar Latex 175)
    1.01 (Texicryl 13-439)
    1.02 (NeoCryl S-1004)
    1.03 (Texicryl 13-210)
    1.04 (NeoCryl B-700, B-705; Texicryl 13-010, 13-104, 13-203)
    1.05 (NeoCryl EX-485)
    1.06 (Texicryl 13-206)
    1.07 (Joncryl 61LV; Texicryl 13-003, 13-011, 13-100, 13-101)
    1.08 (NeoCryl B-1042; Texicryl 13-205, 13-300, 13-302)
    1.09 (Ucar Vehicle 4358, 4414)
    1.10 (Cyanacryl R; NeoCryl B-734; Ucar Latex 174; XT Polymer 375)
    1.11 (NeoCryl B-725, B-726, B-750, B-1019, EX-450; Ucar Latex 173)
    1.12 (Ucar Latex 153, 163, 812; Ucar Vehicle 4431, 4620, 4630; XT Polymer 250)
    1.13 (Cyancryl C; Ucar Latex 123, 803; Ucar Vehicle 407, 4550)
    1.14 (Ucar Latex 154, 874)
    1.15 (NeoCryl EX-482; Plexiglas DR, DR G, DR M, HFI-10)
    1.17 (NeoCryl B-723, EX-487; Plexiglas HFI-7, MI-7, MI-7G)
    1.18 (CP-51, CP-61; Plexiglas VM, VS; Ucar Acrylic 503, 505, 515, 516, 518)
    1.18–1.20 (Polypenco Cast Acrylic Rod)
    1.19 (Acrylite GP, MS-2 Sheet, 0.25 in.; CP-80, CP-81, CP-82; Lucite L, SAR; NeoCryl B-728, EX-484; Plexiglas V-044, V-045, V-052, V-811, V-920)
    1.20 (NeoCryl B-1041)
    1.26 (Korad A)

Density:
    1030 kg/m³ (Carboset 514A)
    1040 kg/m³ (Carboset 531, XL-11)
    1052 kg/m³ (Carboset 514H, XL-19)
    1100 kg/m³ (Carboset 515)
    7.1 lb/gal (Acryloid NAD-10)
    7.8 lb/gal (Acryloid A-21, A-21LV)
    7.9 lb/gal (Acryloid A-101)
    8.1 lb/gal (Acryloid B-44, B-48N, B-50)
    8.2 lb/gal (Acryloid C-10LV)
    8.3 lb/gal (Joncryl 85)
    8.4 lb/gal (Acryloid B-82, B-99; NeoCryl A-601, A-603A, P-920A Glaze; Ucar Latex

175)
8.48 lb/gal (NeoCryl A-621)
8.5 lb/gal (NeoCryl A-604, A-620, A-624)
8.6 lb/gal (Acryloid A-10; NeoCryl A-602, A-610, A-630; Ucar Latex 173, 174)
8.7 lb/gal (Joncryl 77; NeoCryl A-1031, A-1045, SR-272; Ucar Vehicle 4358, 4414, 4431)
8.7 ± 0.1 lb/gal (Joncryl 61LV, 130)
8.8 lb/gal (Joncryl 74; NeoCryl A-1044, B-1054, BT-7, BT-8, BT-20, BT-24, BT-175, CC-6; Parcryl 200, 250, 300, 311, 400, 450, 475, 500, 900, 966; Rhoplex AC-35, AC-388, AC-507; Ucar Vehicle 407, 4620, 4630)
8.8 ± 0.1 lb/gal (Joncryl 78, 142)
8.85 lb/gal (Rhoplex AC-61, AC-490)
8.9 lb/gal (NeoCryl A-1038, CL-300, SR-270, SR-274, SR-276; Parcryl 777; Rhoplex AC-22, AC-34, AC-73, MV-1; Ucar Latex 123, 153, 163, 803, 812; Ucar Vehicle 4550)
8.97 lb/gal (Rhoplex AC-707)
9.0 lb/gal (NeoCryl A-550; Ucar Acrylic 518; Ucar Latex 154, 874)
9.1 lb/gal (Ucar Acrylic 505, 515, 516; Versaflex 2)
9.2 lb/gal (Ucar Acrylic 503)
9.25 lb/gal (Daratak 74L)
9.3 lb/gal (Versaflex 1)
9.7 lb/gal (Versaflex 4, 5)
Visc.:
0.05–0.30 poise (Texicryl 13-010, 13-011, 13-300, 13-302)
0.1–0.4 poise (Texicryl 13-203)
0.15–0.75 poise (Texicryl 13-100, 13-101)
0.2–0.4 poise (Texicryl 13-104, 13-210)
0.2–0.8 poise (Texicryl 13-205)
0.5–1.25 poise (Texicryl 13-439)
0.5–1.5 poise (Texicryl 13-206)
0.6–1.25 poise (Texicryl 13-442)
0.75–1.75 poise (Texicryl 13-003)
35 MPa•s (Carboset XL-19)
50 MPa•s (Carboset XL-11)
350 MPa•s (Carboset 514H)
2000 MPa•s (Carboset 531)
10,000–15,000 MPa•s (Carboset 514A)
2.0×10^6 MPa•s (Carboset 515)
15 cps (NeoCryl B-1054, BT-7, BT08, CC-6)
18 cps (Crystic LS7)
20 cps (NeoCryl BT-20)
20–60 cps (Versaflex 2)
20–200 cps (Versaflex 4)

Acrylic resin *(cont'd.)*

25 cps (Joncryl 142; NeoCryl BT-24)

30 cps (NeoCryl A-1044)

38–82 cps (Acryloid C-10LV)

50 cps (NeoCryl SR-274); (30% toluene) (NeoCryl EX-450)

50–500 cps (Parcryl 200, 250, 300, 311, 900)

75 cps (Ucar Vehicle 4414); (10% toluene) (NeoCryl B-1041); (30% MEK/ethyl acetate 1/1) (NeoCryl B-1042)

90 cps (Ucar Vehicle 4358)

100 cps (NeoCryl A-550, A-601, A-602, BT-175, SR-270; Ucar Vehicle 407, 4630)

100–300 cps (Versaflex 5)

125 cps (NeoCryl A-1045)

150 cps (NeoCryl A-603A, CL-300, SR-276; Ucar Vehicle 4550)

150–300 cps (NeoCryl A-604)

200 cps (Ucar Latex 123, 153); (50% in ethanol) (Joncryl 67)

200–800 cps (Versaflex 1)

210–280 cps (Acryloid A-21LV)

235–365 cps (Acryloid A-21)

250 cps (NeoCryl A-620, A-624; Ucar Acrylic 516; Ucar Vehicle 4620)

290–600 cps (Acryloid NAD-10)

300 cps (NeoCryl A-1031, A-1038); (40% toluene) (NeoCryl B-1019, EX-482); (50% in ethanol) (Joncryl 678); (Ucar Acrylic 515 latex)

350 cps (Ucar Latex 154); (40% toluene) (NeoCryl B-725)

360 cps (40% toluene) (NeoCryl EX-485)

400 cps (Ucar Latex 163, 874); (40% MEK) (NeoCryl EX-484)

450 cps (Joncryl 74, 77); (30% toluene) (NeoCryl B-723); (40% toluene) (NeoCryl B-750)

450–700 cps (Acryloid AT-56)

500 cps (NeoCryl P-920A Glaze, SR-272; Ucar Acrylic 518 latex; Ucar Latex 173, 812; Ucar Vehicle 4431)

500–2500 cps (Parcryl 777)

500 ± 200 cps (Joncryl 130)

550 cps (40% toluene) (NeoCryl B-734); (40% Isopar G) (NeoCryl S-1004)

600 cps (NeoCryl A-621, A-630; Ucar Latex 174, 803); (40% toluene) (NeoCryl 487)

625–1755 cps (Acryloid AT-50)

700–1400 cps (@ 35% solids) (Acryloid A-101)

750 ± 200 cps (Joncryl 78)

800 cps (Ucar Acrylic 505, latex)

800–1200 cps (Acryloid A-10)

855–1700 cps (Acryloid B-44)

1000 cps (Ucar Acrylic 503, latex)

1000 cps max. (NeoCryl A-610)

1000–3000 cps (Acryloid AT-63; Parcryl 966)

1070–2265 cps (Acryloid AT-51)

1200 cps (35% MEK) (NeoCryl B-728)
1200–2500 cps (Acryloid AT-70)
1500 cps (Ucar Latex 175); (40% xylene) (NeoCryl B-726)
1500–3300 cps (Acryloid AT-64)
1500–4000 cps (Parcryl 400, 450, 475, 500)
2000–4000 cps (Daratak 74L)
2500 cps (Joncryl 61LV)
2700 cps (40% Varsol 3) (NeoCryl B-705)
2700–3600 cps (Acryloid AT-71, AT-75)
3000 cps (40% Varsol 3) (NeoCryl B-700)
4000–7000 cps (Acryloid B-82, B-99)
5000–9000 cps (Acryloid B-50)
6000–10,000 cps (Acryloid B-48N)
14,000–17,000 cps (Acryloid WR-97)
22,000 cps (Joncryl 85)
Mooney 30–40 (212 F, ML-4) (Cyanacryl C)
Mooney 40–52 (Cyanacryl L, R)

Soften. Pt.:

137 C (R&B) (NeoCryl EX-487)
155 C (R&B) (NeoCryl B-1042)
160 C (R&B) (NeoCryl B-725, B-750, B-1019, EX-450)
163 C (R&B) (NeoCryl B-734)
170 C (R&B) (NeoCryl B-705)
172 C (R&B) (NeoCryl B-700)
181 C (R&B) (NeoCryl B-726)
194 C (R&B) (NeoCryl B-723)
> 200 C (R&B) (NeoCryl B-728, B-1041)
290 F (R&B) (Joncryl 67, 678)

Flash Pt.:

−1 C (Abel CC) (Crystic LS7)
28 F (TOC) (Acryloid A-101)
32 F (TOC) (Acryloid a-21LV)
38 F (Acryloid B-44)
40 F (TOC) (Acryloid C-10LV)
50 F (TOC) (Acryloid A-21)
64 F (TOC) (Acryloid NAD-10)
66 F (TOC) (Acryloid B-48N)
67 F (PMCC) (Acryloid WR-97)
68 F (TOC) (Acryloid B-82, B-99)
72 F (TOC) (Acryloid B-50)
74 F (PMCC) (Acryloid AT-64)
76 F (PMCC) (Acryloid AT-51, AT-63)
78 F (PMCC) (Acryloid AT-50, AT-56)

Acrylic resin (cont'd.)

82 F (PMCC) (Acryloid AT-70)
115 F (Joncryl 61LV)
116 F (PMCC) (Acryloid AT-71)
118 F (PMCC) (Acryloid AT-75)
120 F (Joncryl 142)
128 F (TOC) (Acryloid A-10)
140 F (PM) (NeoCryl P-920A Glaze)

Acid No.:

29–35 (Carboset XL-19)
60–70 (Carboset 514A, 514H, 531)
60–65 (Carboset 515)
70 (Joncryl 61LV)
70–80 (Carboset XL-11)
76–85 (Carboset 525)
200 (Joncryl 67, 678)

Stability:

Good resistance to many chemicals incl. solutions of inorganic alkalies, e.g., ammonia, dilute acids, e.g., sulfuric acid, aliphatic hydrocarbons, e.g., hexane and VM&P naphtha; good weather, breakage, and heat resistance (Acrylite GP, MS-2 Sheet)

Good high and low temp. resistance; excellent oil and compression set resistance; acceptable corrosion resistance by the GMC test procedure (Cyanacryl C)

Good resistance to oxygen, ozone, and weathering; good oil resistance @ 350 F (Cyanacryl R)

Emulsions exhibit 20+ min mechanical stability; films exhibit excellent light stability and aging characteristics (Daratak 74L)

Good grease, water, and scuff resistance (Joncryl 67, 678)

Good heat, wet and dry rub resistance, and press and fountain stability (Joncryl 142)

Good water and grease resistance (Joncryl 74)

Outstanding water and grease resistance (Joncryl 77)

High resistance to the effects of uv radiation, excellent retention of flexibility on exterior aging, and outstanding resistance to yellowing under both heat and light exposure (Korad A)

Excellent resistance to outdoor exposure (no noticeable chalking, fading, loss of adhesion, or other coating deterioration after 16 yrs), humidity, abrasion and marring, and chemicals (unaffected by most acids, alkalies, and aliphatic hydrocarbons) (Korad 2)

Excellent resistance to physical abuse and environmental attack (Korad Klear)

Resistant to many chemicals which attack other commonly used transparent plastic sheet, incl. acids, bases, hydrocarbons, esters, and alcohols (Lucite SAR)

Excellent mechanical stability; its films have excellent resistance to water, detergents, abrasion, and uv light (NeoCryl A-602)

Good mechanical stability; films offer good print, mar, chemical, and uv resistance (NeoCryl A-604)

Excellent mechanical stability and heat stability to 52 C; good water and alcohol resistance (NeoCryl A-620)

Excellent mechanical stability and heat stability to 52 C; films possess mar, water, alcohol, and cold check resistance (NeoCryl A-624)

Excellent mechanical stability and heat stability (1 mo. @ 52 C); good water, alcohol, and corrosion resistance (NeoCryl A-621)

Resistant to detergents, heat, aging (NeoCryl B-725)

Resistant to abrasion, chemicals, blocking (NeoCryl B-728)

Resistant to heat and aging (NeoCryl B-750)

High alcohol tolerance; moisture and smudge resistant (NeoCryl BT-8)

Films have good gasoline resistance (NeoCryl EX-482)

Good tolerance for alcohols (NeoCryl EX-485)

Resistant to aromatic fuels, high swelling oils, and water (Nobestos D-7700)

Excellent mechanical stability and water resistance (Parcryl 200, 250, 300, 311, 400, 450, 475, 500, 777, 900, 966)

Outstanding weatherability; resists sunlight, aging, and maintains good stability under variable conditions of heat, cold, moisture, etc.; high resistance to heat distortion (Polypenco Cast Acrylic Rod)

Emulsions exhibit good mechanical stability (20+ min); films exhibit good water resistance (Versaflex 1, 2, 4, 5)

Excellent resistance to oils and fats; resistant to water; does not discolor or become brittle over long periods of time or extremes of temp. (XT Polymer 250, 375)

Storage Stability:

Several years storage life @ 20 C (Crystic LS7)

Excellent freeze/thaw stability (Joncryl 74, 77, 85; NeoCryl A-603A)

Good freeze/thaw stability (Joncryl 61LV, 142)

Freeze-thaw stable (Rhoplex AC-22, AC-34, AC-35, AC-61, AC-73, AC-388, AC-490, AC-507, AC-707)

Marginal freeze-thaw stability (Rhoplex MV-1)

Freeze-thaw unstable (Texicryl 13-003, 13-010, 13-011, 13-100, 13-101, 13-104, 13-203, 13-205, 13-206, 13-210, 13-300, 13-302, 13-439, 13-442)

Ref. Index:

1.43 (Lucite SAR)

1.49 (Acrylite GP, MS-2 Sheet, 0.25 in.; CP-51, CP-61, CP-80, CP-81, CP-82; Lucite L; Plexiglas DR, DR G, DR M, HFI-7, HFI-10, MI-7, MI-7G, V-044, V-045, V-052, V-811, V-920, VM, VS)

pH:

2.0–4.0 (Texicryl 13-011, 13-300, 13-302, 13-439)

4.0 (NeoCryl BT-175)

4.0–5.0 (Texicryl 13-104)

4.0–6.0 (Texicryl 13-203)

4.3 (Ucar Latex 163)

4.5 (NeoCryl A-1031; Ucar Latex 154 , 874)

Acrylic resin *(cont'd.)*

4.5–5.5 (Daratak 74L)

4.7 (Ucar Vehicle 4550)

4.8 (Ucar Vehicle 407)

5.0 (NeoCryl A-550, A-1038, BT-7, BT-8, BT-20, BT-24; Ucar Latex 175, 803, 812)

5.0–6.0 (Parcryl 777; Versaflex 1)

5.0–7.0 (Texicryl 13-010)

5.5 (NeoCryl P-920A Glaze; Ucar Latex 153, 173)

5.8 (Ucar Latex 123)

6.0 (Ucar Acrylic 503, 505; Ucar Latex 174)

6.0 ± 0.1 (Joncryl 142)

6.0–7.0 (Texicryl 13-100, 13-101, 13-205, 13-206, 13-210)

6.4–7.0 (Carboset XL-11)

6.5–7.0 (Versaflex 2)

6.5–7.5 (Versaflex 5)

6.8–7.0 (Joncryl 85)

7.0–8.0 (NeoCryl A-610)

7.2–7.8 (Carboset XL-19)

7.3 (NeoCryl B-1054, CC-6)

7.4–7.8 (NeoCryl A-603A)

7.5 (NeoCryl CL-300, SR-272)

7.5–8.0 (NeoCryl A-601, A-630)

7.5–8.5 (NeoCryl A-602, A-620)

7.7 (Ucar Vehicle 4431)

8.0 (NeoCryl SR-270, SR-274; Ucar Vehicle 4630)

8.0 ± 0.2 (Joncryl 78)

8.0–8.5 (NeoCryl A-624)

8.0–9.0 (Parcryl 966; Versaflex 4)

8.0–9.5 (Texicryl 13-442)

8.3 (Joncryl 74, 77)

8.3–8.6 (Joncryl 61LV)

8.5 (NeoCryl A-1045, SR-276; Ucar Acrylic 515, 518; Ucar Vehicle 4358)

8.5 ± 0.2 (Joncryl 130)

8.5–9.0 (NeoCryl A-621)

8.7 (Ucar Vehicle 4414)

9.0–9.5 (Texicryl 13-003)

9.0–9.7 (Rhoplex AC-707)

9.0–10.0 (Parcryl 200, 250, 300, 311, 400, 450, 475, 500, 900; Rhoplex AC-35, AC-73, MV-1)

9.2–9.5 (Rhoplex AC-490)

9.4–9.7 (Rhoplex AC-388)

9.5 (NeoCryl A-1044; Ucar Acrylic 516)

9.5–10.0 (Rhoplex AC-22, AC-34, AC-61, AC-507)

Transmittance:
  90% (Plexiglas DR, DR G, DR M, HFI-10)
  92% (Plexiglas HFI-7, MI-7, MI-7G, V-044, V-045, V-052, V-811, V-920, VM, VS)

MECHANICAL PROPERTIES:

Tens. Str.:
  760 kg/cm² (break) (Lucite L)
  10.3 MPa (Nobestos D-7700)
  11.0 MPa (Carboset 514A, 514H)
  26.2 MPa (Carboset 525)
  1250 psi min. (post-cured) (Cyanacryl L)
  1550 psi min. (post-cured) (Cyanacryl R)
  4400 psi (Korad Klear)
  5100 psi (yield) (Korad A)
  5500 psi (Plexiglas DR, DR G, DR M, HIF-10)
  7000 psi (Plexiglas HFI-7, MI-7, MI-7G; XT Polymer 375)
  8000 psi (XT Polymer 250)
  8700 psi (Plexiglas VS)
  9000 psi (Polypenco Cast Acrylic Rod)
  9600 psi (Plexiglas VM)
  10,000 psi (Acrylite GP, MS-2 Sheet, 0.25 in.; CP-51, CP-61)
  10,500 psi (CP-80, CP-81, CP-82; Lucite SAR; Plexiglas V-044, V-045, V-052, V-811, V-920)

Tens. Elong.:
  2.0% (Polypenco Cast Acrylic Rod)
  3.6% (yield) (XT Polymer 250, 375)
  4.2% (Acrylite GP, MS-2 Sheet, 0.25 in.)
  4.5% (break) (Lucite L)
  15% (break) (XT Polymer 250)
  28% (break) (XT Polymer 375)
  50% (Carboset 514A, 514H)
  75% (Korad A)
  130% (Korad Klear)
  150% min (post-cured) (Cyanacryl L, R)
  165% (Carboset 525)

Tens. Mod.:
  30,000 kg/cm² (Lucite L)
  350,000–500,000 psi (Polypenco Cast Acrylic Rod)
  370,000 psi (XT Polymer 375)
  400,000 psi (Acrylite GP, MS-2 Sheet, 0.25 in.)
  430,000 psi (CP-51, CP-61, CP-80, CP-81, CP-82; XT Polymer 250)

Flex. Str.:
  1050 kg/cm² (break) (Lucite L)
  8000 psi (Plexiglas DR, DR G, DR M)

Acrylic resin (cont'd.)

9000 psi (Plexiglas HFI-10)
10,500 psi (Plexiglas HFI-7)
11,000 psi (XT Polymer 375)
12,000 psi (Plexiglas MI-7, MI-7G)
13,000 psi (CP-51; XT Polymer 250)
14,000 psi (CP-61; Plexiglas VM, VS; Polypenco Cast Acrylic Rod)
15,000 psi (Plexiglas V-044, V-045, V-052, V-811, V-920)
16,000 psi (CP-80, CP-81, CP-82; Lucite SAR)
16,500 psi (Acrylite GP, MS-2 Sheet, 0.25 in.)

Flex. Mod.:

30,000 kg/cm² (Lucite L)
240,000 psi (Plexiglas HFI-10)
250,000 psi (Plexiglas DR, DR G, SR M)
330,000 psi (Plexiglas HFI-7)
350,000 psi (Plexiglas MI-7, MI-7G; XT Polymer 375)
350,000–500,000 psi (Polypenco Cast Acrylic Rod)
400,000 psi (XT Polymer 250)
420,000 psi (Plexiglas VS)
440,000 psi (Plexiglas VM)
450,000 psi (Lucite SAR; Plexiglas V-044, V-045, V-052, V-811, V-920)
475,000 psi (Acrylite GP, MS-2 Sheet, 0.25 in.)

Compr. Str.:

1260 kg/cm² (yield) (Lucite L)
6000 psi (Plexiglas DR, DR G, DR M, HFI-10)
10,000 psi (Plexiglas MI-7, MI-7G)
10,500 psi (Plexiglas HFI-7)
14,000 psi (CP-51; Plexiglas VS)
14,000–18,000 psi (Polypenco Cast Acrylic Rod)
14,500 psi (Plexiglas VMj)
15,000 psi (CP-61)
15,800 psi (Plexiglas V-920)
16,000 psi (CP-80, CP-81, CP-82)
17,000 psi (Plexiglas V-044, V-045, V-052, V-811)
18,000 psi (Acrylite GP, MS-2 Sheet, 0.25 in.)

Compr. Mod.:

30,000 kg/cm² (Lucite L)
430,000 psi (Acrylite GP, MS-2 Sheet, 0.25 in.)

Tear Str.:

1.0 lb/mil (Korad A)

180° Peel Adhesion to Steel:

42 oz/in. (Presto Adhesive 2125)
50 oz/in. (Presto Adhesive 2125M88)
60 oz/in. (Presto Adhesive 2127+)

65 oz/in. (Presto Adhesive 2127M6)
Shear Adhesion:
168 h (Presto Adhesive 2125, 2125M88, 2127+, 2127M6)
Shear Str.:
630 kg/cm² (Lucite L)
9000 psi (Acrylite GP, MS-2 Sheet, 0.25 in.)
Impact Str. (Izod):
0.3 ft lb/in. notched (Lucite SAR)
0.4 ft lb/in. notched (Acrylite GP, MS-2 Sheet, 0.25 in.; Plexiglas V-044, V-045, V-052, V-811, V-920, VM, VS)
0.5 ft lb/in. notched (CP-51, CP-61, CP-80, CP-81, CP-82)
0.6 ft lb/in. notched (Plexiglas HIF-7, MI-7, MI-7G)
1.0 ft lb/in. notched (Plexiglas HFI-10)
1.2 ft lb/in. notched (Plexiglas DR, DR G, DR M)
16 ft lb/in. unnotched (XT Polymer 250)
24 ft lb/in. unnotched (XT Polymer 375)
Impact Str. (Charpy):
0.48 kg/cm² unnotched (Lucite L)
Hardness:
Rockwell M38 (Plexiglas HFI-10)
Rockwell M45 (Plexiglas DR, DR G, DR M; XT Polymer 375)
Rockwell M65 (Plexiglas HFI-7)
Rockwell M68 (Plexiglas MI-7, MI-7G; XT Polymer 250)
Rockwell M80–100 (Polypenco Cast Acrylic Rod)
Rockwell M84 (Plexiglas VS)
Rockwell M89 (CP-51; Plexiglas VM)
Rockwell M92 (CP-61)
Rockwell M94 (Acrylite GP, MS-2 Sheet, 0.25 in.)
Rockwell M95 (CP-80, CP-81, CP-82; Plexiglas V-920)
Rockwell M96 (Plexiglas V-044, V-045, V-052)
Rockwell M97 (Plexiglas V-811)
Rockwell M100 (Lucite L)
Rockwell M102 (Lucite SAR)
Shore A 68 ± 5 (Cyanacryl R)
Shore A 73 ± 5 (Cyanacryl L)
Sward 21 (Carboset XL-19)
Sward 24 (Carboset 514A, 514H, 525)
Sward 28 (24 h) (NeoCryl A-602)
Sward 32 (Carboset 531); (7 days) (NeoCryl A-603A)
Sward 34 (7 days) (NeoCryl A-601, A-610)
Sward 35 (Carboset XL-11)
Sward 36 (7 days) (NeoCryl P-920A Glaze)
Sward 38 (7 days) (NeoCryl A-604, S-1004)

Acrylic resin (cont'd.)

Sward 40 (7 days) (NeoCryl B-734, EX-487)
Sward 42 (7 days) (NeoCryl B-723)
Sward 44 (NeoCryl A-620); (7 days) (NeoCryl B-725, B-726, B-750, B-1019, EX-450)
Sward 46 (7 days) (NeoCryl A-624, EX-482, EX-485)
Sward 50 (7 days) (NeoCryl A-630, B-1042, EX-484)
Sward 56 (7 days) (NeoCryl A-621)
Sward 58 (7 days) (NeoCryl B-700, B-705)
Sward 60 (7 days) (NeoCryl B-728, B-1041)
Tukon < 1 (Rhoplex AC-22, AC-34, AC-35, AC-388, AC-707, MV-1)
Tukon 1 (Acrysol WS-32; Rhoplex AC-507)
Tukon 1.0–1.4 (Rhoplex AC-490)
Tukon 1.2 (Rhoplex AC-61)
Tukon 4.5 (Rhoplex AC-73)
Tukon 6 (Acrysol WS-24)
Tukon 8 (Acrysol WS-12), WS-50
Tukon 12 (Acrysol I-94)

Mold Shrinkage:
2–6 mils/in. (Plexiglas HFI-7, MI-7, MI-7G, V-044, V-045, V-052, V-811, V-920, VM, VS)
2–8 mils/in. (Plexiglas DR, DR G, DR M, HFI-10)

Water Absorp.:
0.3% (Plexiglas HFI-7, HFI-10, MI-7, MI-7G, V-044, V-045, V-052, V-811, V-920, VM, VS)
0.4% (Plexiglas DR, DR G, DR M)

THERMAL PROPERTIES:

Conduct.:
0.0005 Cal/s/cm^2/C/cm (Lucite L)
1.2 Btu in./ft^2/h/F (Lucite SAR)
1.3 Btu/h/ft^2/F/in. (Acrylite GP, MS-2 Sheet, 0.25 in.)

Distort. Temp.:
95 C (264 psi) (Lucite L)
167 F (264 psi) (Plexiglas VS)
178 F (264 psi) (Plexiglas HFI-10)
180 F (264 psi) (Plexiglas DR, DR G, DR M)
182 F (264 psi) (Plexiglas VM)
190 F (CP-51); (264 psi) (Plexiglas HFI-7, MI-7, MI-7G)
194 F (264 psi) (XT Polymer 375)
195 F (264 psi) (XT Polymer 250)
196 F (CP-61)
198 F (CP-80, CP-81, CP-82); (264 psi) (Plexiglas V-920)
199 F (264 psi) (Plexiglas V-044, V-045, V-052)
210 F (264 psi) (Acrylite GP, MS-2 Sheet, 0.25 in.)

212 F (264 psi) (Plexiglas V-811)

225 F (264 psi) (Lucite SAR)

Coeff. of Linear Exp.:

7.0 × 10⁻⁵ cm/cm/C (Lucite L)

7.0×10^{-5} cm/cm/C (Lucite L)

3.4×10^{-5} in./in./F (Acrylite GP, MS-2 Sheet, 0.25 in.)

4.0×10^{-5} in./in./F (Lucite SAR)

4.4×10^{-5} in./in./F (XT Polymer 250)

4.5×10^{-5} in./in./F (Polypenco Cast Acrylic Rod)

5.0×10^{-5} in./in./F (XT Polymer 375)

Sp. Heat:

0.35 Cal/g/C (Lucite L)

0.35 Btu (Acrylite GP, MS-2 Sheet, 0.25 in.)

Flamm.:

HB (Plexiglas DR, DR G, DR M, HFI-7, HFI-10, MI-7, MI-7G, V-044, V-045, V-052, V-811, V-920, VM, VS)

0.8 in./min (CP-51, CP-61)

0.9 in./min (CP-80, CP-81)

1.0 in./min (CP-82)

1.2 in./min (Acrylite GP, MS-2 Sheet, 1/8 in.)

Self-Ignit. Temp.:

860 F (Acrylite GP, MS-2 Sheet, 0.25 in.)

ELECTRICAL PROPERTIES:

Dissip. Factor:

0.04 (60 Hz) (Plexiglas DR, DR G, DR M, HFI-10, VM, VS)

0.05 (60 Hz) (CP-51; Plexiglas HFI-7, MI-7, MI-7G, V-044, V-045, V-052, V-811, V-920)

0.06 (60 Hz) (Acrylite GP, MS-2 Sheet, 0.25 in.; CP-61, CP-80, CP-81, CP-82; Lucite L)

Dielec. Str.:

20 kV/mm (Lucite L)

383 V/mil (Plexiglas DR, DR G, DR M, HFI-10)

400 V/mil (Polypenco Cast Acrylic Rod)

430 V/mil (Acrylite GP, MS-2 Sheet, 1/8 in.)

450 V/mil (Plexiglas HFI-7, MI-7, MI-7G)

500 V/mil (CP-51, CP-61, CP-80, CP-81, CP-82; Plexiglas V-044, V-045, V-052, V-811, V-920, VM, VS)

2200 V/mil (Korad A)

Dielec. Const.:

3.3 (60 Hz) (CP-51, CP-61, CP-80, CP-81, CP-82)

3.5 (60 cycles) (Acrylite GP, MS-2 Sheet, 0.25 in.)

3.5–4.5 (60 Hz) (Polypenco Cast AcrylicRod)

3.7 (60 Hz) (Plexiglas V-044, V-045, V-052, V-811, V-920, VM, VS)

3.8 (60 Hz) (Plexiglas HFI-7, MI-7, MI-7G)

Acrylic resin *(cont'd.)*

3.9 (60 Hz) (Plexiglas DR, DR G, DR M, HFI-10)
4.0 (60 Hz) (Lucite L)
4.8 (60 Hz) (Korad A)

Vol. Resist.:

10^{15} ohm-cm (Lucite L; Polypenco Cast Acrylic Rod)
10^{16} ohm-cm (Korad A)
1.6×10^{16} ohm-cm (Acrylite GP, MS-2 Sheet, 0.25 in.)

Surf. Resist.:

1.9×10^{15} ohm (Acrylite GP, MS-2 Sheet, 0.25 in.)\
$> 10^{14}$ ohm (Lucite L)
2.0×10^{14} ohm (Korad A)

Arc Resist.:

No tracking (CP-51, CP-61, CP-80, CP-81, CP-82; Lucite L; Plexiglas DR, DR G, DR M, HFI-7, HFI-10, MI-7, MI-7G, V-044, V-045, V-052, V-811, V-920, VM, VS)

CURING CHARACTERISTICS:

Catalyst:

May be catalyzed with a wide range of acid or latent acid catalysts, e.g., oxalic acid, ammonium chloride, and magensium chloride (Texicryl 13-206)

Typical Cure:

1.5–3 min @ 135–150 C after drying; cure is accelerated by addition of catalyst (Texicryl 13-210)
2–3 min @ 130–140 C with 2% addition of a 10% catalyst sol'n. (Texicryl 13-206)
2–3 min @ 140 C; allow extra time for drying (Texicryl 13-203)
2–3 min @ 140 C after drying (Texicryl 13-205)
20 min @ 350 F—crosslinker required (Acryloid AT-70)
20 min @ 350 F—crosslinker required (Acryloid AT-71, AT-75)
30 min @ 250 F—crosslinker required (Acryloid AT-63)
30 min @ 300 F (Acryloid AT-50, AT-51)
30 min @ 300 F—crosslinker required (Acryloid AT-56, AT-64, WR-97)
45 min @ 75 C or 24 h @ R.T. (Hysol PC20)

TOXICITY/HANDLING:

Combustible (Acrylite GP, MS-2 Sheet)
Combustible; observe fire precautions (CP-51, CP-61, CP-80, CP-81, CP-82)
Flammable—keep away from naked flames (Crystic LS7)

STORAGE/HANDLING:

Store ≤ 20 C in tightly closed containers (Crystic LS7)
Protect from freezing (NeoCryl A-601, A-602, A-604, A-610, A-621, A-630)
pH of thickened mixes and compounds should not fall below pH 7 during storage (Texicryl 13-300, 13-302)

STD. PKGS.:

Avail. in more than 50 standard sizes from 36×48 in. to 120×144 in. (Acrylite GP, MS-2 Sheet)

300 lb net drums, 1200 lb net carbons, rail cars, bulk trucks; also avail. in 55 lb net specialized bags for overseas export (CP-51, CP-61, CP-80, CP-81, CP-82)

20-kg containers (Crystic LS7)

200-kg lacquer-lined metal drums; bulk road tanker (Texicryl 13-003, 13-010, 13-100, 13-101, 13-104, 13-203, 13-205, 13-206, 13-210, 13-439, 13-442)

200-kg polyethylene-lined drums; bulk road tanker (Texicryl 13-011)

200-kg polyethylene or polyethylene-lined fiberboard drums (Texicryl 13-300, 13-302)

Cellulose acetate

SYNONYMS:
Acetate fiber
Acetate film
CA
Cellulose acetate ester
Secondary acetate

TRADENAME EQUIVALENTS:
CA-394-60, -398-3, -398-6, -398-10, -398-30 [Eastman]
Cast Cellulose Acetate Film 904 [Georgia-Pacific]
Cast Cellulose Acetate Sheet 867 [Xcel]
Cellidor A, S 100-25, S 100-30, S 100-33, S200-17, S 200-22, S 200-27, S 200-32
  [Bayer AG]
Crystic Release Agent No. 1 [Scott Bader]
Tenite Cellulosic Acetate [Eastman]

CATEGORY:
Thermoplastic resin

PROCESSING:
Injection molding:
Tenite Cellulosic Acetate

APPLICATIONS:
Architectural applications: building industry (Cellidor A, S 100-25, S 100-30, S 100-33, S200-17, S 200-22, S 200-27, S 200-32)

Automotive applications: (Cellidor A, S 100-25, S 100-30, S 100-33, S200-17, S 200-22, S 200-27, S 200-32)

Consumer products: appliances (Cellidor A, S 100-25, S 100-30, S 100-33, S200-17, S 200-22, S 200-27, S 200-32); furniture (Cellidor A, S 100-25, S 100-30, S 100-33, S200-17, S 200-22, S 200-27, S 200-32); photographic industry (Cellidor A, S 100-25, S 100-30, S 100-33, S200-17, S 200-22, S 200-27, S 200-32); stationery (Cellidor A, S 100-25, S 100-30, S 100-33, S200-17, S 200-22, S 200-27, S 200-32); toiletries (Cellidor A, S 100-25, S 100-30, S 100-33, S200-17, S 200-22, S 200-27, S 200-32); tools (Cellidor A, S 100-25, S 100-30, S 100-33, S200-17, S 200-22, S 200-27, S 200-32); toys (Cellidor A, S 100-25, S 100-30, S 100-33, S200-17, S 200-22, S 200-27, S 200-32)

Electrical/electronic industry: (Cellidor A, S 100-25, S 100-30, S 100-33, S200-17, S 200-22, S 200-27, S 200-32); lighting (Cellidor A, S 100-25, S 100-30, S 100-33, S200-17, S 200-22, S 200-27, S 200-32); radio/TV (Cellidor A, S 100-25, S 100-

30, S 100-33, S200-17, S 200-22, S 200-27, S 200-32)

Functional additives: release agent (Crystic Release Agent No. 1); sealer (Crystic Release Agent No. 1)

Industrial applications: coatings (Crystic Release Agent No. 1); optical parts/instruments (Cellidor A, S 100-25, S 100-30, S 100-33, S200-17, S 200-22, S 200-27, S 200-32); packaging/displays (Cellidor A, S 100-25, S 100-30, S 100-33, S200-17, S 200-22, S 200-27, S 200-32)

PROPERTIES:
Form:

Liquid (Crystic Release Agent No. 1)

Fine, dry powder (CA-394-60, -398-3, -398-6, -398-10, -398-30)

Cylindrical granules, diced granules (Cellidor A, S 100-25, S 100-30, S 100-33, S200-17, S 200-22, S 200-27, S 200-32)

1.2-mil gauge film (Cast Cellulose Acetate Film 904)

5-mil gauge sheet (Cast Cellulose Acetate Sheet 867)

Color:

White (CA-394-60, -398-3, -398-6, -398-10, -398-30)

Colorless or colored (Crystic Release Agent No. 1)

Transparent, translucent, and opaque colors (Cellidor A, S 100-25, S 100-30, S 100-33, S200-17, S 200-22, S 200-27, S 200-32)

Odor:

Low (CA-394-60, -398-3, 398-6, -398-10, -398-30)

Composition:

39.4% acetyl content (CA-394-60)

39.8% acetyl content (CA-398-3, -398-6, -398-10, -398-30)

GENERAL PROPERTIES:
Solubility:

Sol. in acetone (Cellidor A, S 100-25, S 100-30, S 100-33, S200-17, S 200-22, S 200-27, S 200-32)

Sol. in ethyl acetate (Cellidor A, S 100-25, S 100-30, S 100-33, S200-17, S 200-22, S 200-27, S 200-32)

Sol. in methylene chloride (Cellidor A, S 100-25, S 100-30, S 100-33, S200-17, S 200-22, S 200-27, S 200-32)

Sol. in orthochlorphenol (Cellidor A, S 100-25, S 100-30, S 100-33, S200-17, S 200-22, S 200-27, S 200-32)

Sp. Gr.:

1.22–1.31 (Tenite Cellulosic Acetate)

Sp. Vol.:

762–823 cm³/kg (Tenite Cellulosic Acetate)

Density:

1.31 kg/l (CA-398-3, -398-6, -398-10, -398-30)

1.32 kg/l (CA-394-60)

1.16–1.32 mg/m³ (Cellidor A)

Cellulose acetate *(cont'd.)*

    1.26 g/cm³ (Cellidor S 100-33, S 200-32)
    1.27 g/cm³ (Cellidor S 100-30, S 200-27)
    1.28 g/cm³ (Cellidor S 100-25, S 200-22)
    1.29 g/cm³ (Cellidor S 200-17)

Visc.:
    11.4 poise (CA-398-3)
    22.8 poise (CA-398-6)
    38.0 poise (CA-398-10)
    114 poise (CA-398-30)
    228 poise (CA-394-60)

M.P.:
    230–250 C (CA-398-3, -398-6, -398-10, -398-30)
    240–260 C (CA-394-60)

Flash Pt.:
    < 32 C (Crystic Release Agent No. 1)

Stability:
    Resistant to water, carbon tetrachloride, trichloroethylene, perchloroethylene, benzene, xylene, petrol, motor fuel mixture, paraffin, linseed oil, turpentine, ether (Cellidor A, S 100-25, S 100-30, S 100-33, S200-17, S 200-22, S 200-27, S 200-32)

Ref. Index:
    1.49 (Cellidor S 200-17, S 200-22, S 200-27, S 200-32)
    1.50 (Cellidor S 100-25, S 100-30)
    1.51 (Cellidor S 100-33)

MECHANICAL PROPERTIES:

Tens. Str.:
    15.2–51.0 MPa (yield) (Tenite Cellulosic Acetate)
    9000–12,000 psi (Cast Cellulose Acetate Film 904)
    11,500 psi (Cast Cellulose Acetate Sheet 867)

Tens. Elong.:
    15–25% (Cast Cellulose Acetate Film 904)
    40% (Cast Cellulose Acetate Sheet 867)

Tens. Mod.:
    1.4 GPa (Cellidor S 100-33)
    1.5 GPa (Cellidor S 200-32)
    1.8 GPa (Cellidor S 100-30)
    2.0 GPa (Cellidor S 200-27)
    2.2 GPa (Cellidor S 100-25)
    2.7 GPa (Cellidor S 200-22)
    3.0 GPa (Cellidor S 200-17)

Flex. Mod.:
    35 MPa (Cellidor S 100-33)
    37 MPa (Cellidor S 200-32)
    42 MPa (Cellidor S 100-30)

50 MPa (Cellidor S 200-27)
56 MPa (Cellidor S 100-25)
65 MPa (Cellidor S 200-22)
75 MPa (Cellidor S 200-17)

Impact Str. (Izod):

53–390 J/m notched (Tenite Cellulosic Acetate)
50 J/m (Cellidor S 200-17)
90 J/m (Cellidor S 200-22)
110 J/m (Cellidor S 100-25)
200 J/m (Cellidor S 200-27)
250 J/m (Cellidor S 100-30)
260 J/m (Cellidor S 200-32)
300 J/m (Cellidor S 100-33)

Hardness:

Rockwell R73 (Cellidor S 100-33)
Rockwell R77 (Cellidor S 200-32)
Rockwell R90 (Cellidor S 100-30)
Rockwell R100 (Cellidor S 200-27)
Rockwell R105 (Cellidor S 100-25)
Rockwell R110 (Cellidor S 200-22)
Rockwell R120 (Cellidor S 200-17)
Rockwell ≤ R122 (Tenite Cellulosic Acetate)
Tukon 19 (CA-398-3, -398-6, -398-10, -398-30)
Tukon 20 (CA-394-60)

THERMAL PROPERTIES:

Soften. Pt. (Vicat):

60 C (Cellidor S 100-33)
63 C (Cellidor S 200-32)
67 C (Cellidor S 100-30)
75 C (Cellidor S 200-27)
80 C (Cellidor S 100-25)
88 C (Cellidor S 200-22)
100 C (Cellidor S 200-17)

Conduct.:

0.21 W/Km (Cellidor S 100-25, S 200-17, S 200-22)
0.22 W/Km (Cellidor S 100-30, S 100-33, S 200-27, S 200-32)

Distort. Temp.:

43 C (1.81 MPa) (Cellidor S 100-33)
44–91 C (1.82 MPa) (Tenite Cellulosic Acetate)
47 C (1.81 MPa) (Cellidor S 200-32)
52 C (1.81 MPa) (Cellidor S 100-30)
59 C (1.81 MPa) (Cellidor S 200-27)
64 C (1.81 MPa) (Cellidor S 100-25)

Cellulose acetate *(cont'd.)*

73 C (1.81 MPa) (Cellidor S 200-22)
82 C (1.81 MPa) (Cellidor S 200-17)

Coeff. of Linear Exp.:
96×10^{-6} K^{-1} (Cellidor S 100-25)
100×10^{-6} K^{-1} (Cellidor S 200-17)
102×10^{-6} K^{-1} (Cellidor S 100-30)
105×10^{-6} K^{-1} (Cellidor S 200-22)
108×10^{-6} K^{-1} (Cellidor S 100-33)
115×10^{-6} K^{-1} (Cellidor S 200-27)
120×10^{-6} K^{-1} (Cellidor S 200-32)

Sp. Heat:
1.3–1.7 kJ/kgK (Cellidor S 100-25, S 100-30, S 100-33, S 200-17, S 200-22, S 200-27, S 200-32)

Flamm.:
30 mm/min (Cellidor S 100-25, S 100-30, S 100-33, S 200-17, S 200-22, S 200-27, S 200-32)
0.45 in.^{2}/s (Cast Cellulose Acetate Sheet 867)

ELECTRICAL PROPERTIES:

Dissip. Factor:
0.007 (50 Hz) (Cellidor S 100-33, S 200-32)
0.008 (50 Hz) (Cellidor S 100-30, S 200-27)
0.009 (50 Hz) (Cellidor S 100-25, S 200-22)
0.013 (50 Hz) (Cellidor S 200-17)

Dielec. Str.:
11–24 kV/mm (Tenite Cellulosic Acetate)
28.5 kV/mm (50 Hz) (Cellidor S 100-33, S 200-32)
30.5 kV/mm (50 Hz) (Cellidor S 100-30, S 200-27)
31.5 kV/mm (50 Hz) (Cellidor S 100-25, S 200-22)
33 kV/mm (50 Hz) (Cellidor S 200-17)
2300–3000 V/mil (Cast Cellulose Acetate Film 904)

Dielec. Const.:
3.4–7.4 (1 kHz) (Tenite Cellulosic Acetate)
4.8 (Cast Cellulose Acetate Film 904)
5.1 (50 Hz) (Cellidor S 100-25, S 200-17, S 200-22)
5.5 (50 Hz) (Cellidor S 100-30, S 200-27)
5.6 (50 Hz) (Cellidor S 100-33, S 200-32)

Vol. Resist.:
10^{14} ohm-cm (Cellidor S 100-33, S 200-32)
10^{15} ohm-cm (Cellidor S 100-25, S 100-30, S 200-17, S 200-22, S 200-27)

Surf. Resist.:
10^{13} ohm (Cellidor S 100-30, S 100-33, S 200-27, S 200-32)
10^{14} ohm (Cellidor S 100-25, S 200-17, S 200-22)

STORAGE/HANDLING:
Highly flammable; keep away from naked flames; recommended storage ≤ 20 C
(Crystic Release Agent No. 1)
STD. PKGS.:
5-kg, 25-kg, and 200-kg containers (Crystic Release Agent No. 1)

Coumarone-indene resin

SYNONYMS:
Coal tar resin
STRUCTURE:

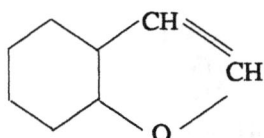 coumarone

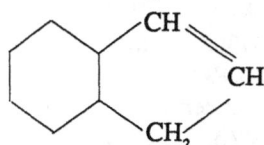 indene

TRADENAME EQUIVALENTS:
Cumar LX-509, P-10, P-25, R-1, R-3 (W-1), R-5 (W-1$^1/_2$), R-6 (W-2, 2$^1/_2$), R-7, R-9,
R-10 (V-1), R-11 (V-1$^1/_2$), R-12 (V-2, V-2$^1/_2$), R-12A (V-3), R-13, R-14 (MH-1),
R-15 (MH-1$^1/_2$), R-16 (MH-2, 2$^1/_2$), R-16A (MH-3), R-17 (RH-17), R-19, R-21, R-
27, R-28, R-29 (25°) [Neville]
Coumin [St. Lawrence Resin Products]
Natro-Rez 10, 15, 25, 50-D [Harwick]
Paradene No. 1, No. 2, No. 33, No. 35 [Neville]
CATEGORY:
Thermosetting resin
APPLICATIONS:
Electrical/electronic industry: wire and cable coatings (Cumar LX-509, P-10, P-25, R-
1, R-3 (W-1), R-5 (W-1$^1/_2$), R-6 (W-2, 2$^1/_2$), R-7, R-9, R-10 (V-1), R-11 (V-1$^1/_2$), R-
12 (V-2, V-2$^1/_2$), R-12A (V-3), R-13, R-14 (MH-1), R-15 (MH-1$^1/_2$), R-16 (MH-2,
2$^1/_2$), R-16A (MH-3), R-17 (RH-17), R-19, R-21, R-27, R-28, R-29 (25°); Paradene
No. 1, No. 2, No 33, No. 35)

Coumarone-indene resin *(cont'd.)*

Food-contact applications: citrus fruit coatings (Cumar LX-509, P-10, P-25, R-1, R-3 (W-1), R-5 (W-$1\frac{1}{2}$), R-6 (W-2, $2\frac{1}{2}$), R-7, R-9, R-10 (V-1), R-11 (V-$1\frac{1}{2}$), R-12 (V-2, V-$2\frac{1}{2}$), R-12A (V-3), R-13, R-14 (MH-1), R-15 (MH-$1\frac{1}{2}$), R-16 (MH-2, $2\frac{1}{2}$), R-16A (MH-3), R-17 (RH-17), R-19, R-21, R-27, R-28, R-29 (25°); Paradene No. 1, No. 2, No 33, No. 35)

Functional additives: plasticizer (Natro-Rez 10, 15, 25); processing aid (Coumin; Natro-Rez 50-D); softener (Natro-Rez 10, 15, 25); tackifier (Coumin; Natro-Rez 50-D)

Industrial applications: adhesives (Cumar LX-509, P-10, P-25, R-1, R-3 (W-1), R-5 (W-$1\frac{1}{2}$), R-6 (W-2, $2\frac{1}{2}$), R-7, R-9, R-10 (V-1), R-11 (V-$1\frac{1}{2}$), R-12 (V-2, V-$2\frac{1}{2}$), R-12A (V-3), R-13, R-14 (MH-1), R-15 (MH-$1\frac{1}{2}$), R-16 (MH-2, $2\frac{1}{2}$), R-16A (MH-3), R-17 (RH-17), R-19, R-21, R-27, R-28, R-29 (25°); Natro-Rez 10, 15, 25; Paradene No. 1, No. 2, No 33, No. 35); caulking compounds (Cumar LX-509, P-10, P-25, R-1, R-3 (W-1), R-5 (W-$1\frac{1}{2}$), R-6 (W-2, $2\frac{1}{2}$), R-7, R-9, R-10 (V-1), R-11 (V-$1\frac{1}{2}$), R-12 (V-2, V-$2\frac{1}{2}$), R-12A (V-3), R-13, R-14 (MH-1), R-15 (MH-$1\frac{1}{2}$), R-16 (MH-2, $2\frac{1}{2}$), R-16A (MH-3), R-17 (RH-17), R-19, R-21, R-27, R-28, R-29 (25°); Paradene No. 1, No. 2, No 33, No. 35); coatings (Cumar LX-509, P-10, P-25, R-1, R-3 (W-1), R-5 (W-$1\frac{1}{2}$), R-6 (W-2, $2\frac{1}{2}$), R-7, R-9, R-10 (V-1), R-11 (V-$1\frac{1}{2}$), R-12 (V-2, V-$2\frac{1}{2}$), R-12A (V-3), R-13, R-14 (MH-1), R-15 (MH-$1\frac{1}{2}$), R-16 (MH-2, $2\frac{1}{2}$), R-16A (MH-3), R-17 (RH-17), R-19, R-21, R-27, R-28, R-29 (25°); Natro-Rez 10, 15, 25; Paradene No. 1, No. 2, No 33, No. 35); paper coatings (Cumar LX-509, P-10, P-25, R-1, R-3 (W-1), R-5 (W-$1\frac{1}{2}$), R-6 (W-2, $2\frac{1}{2}$), R-7, R-9, R-10 (V-1), R-11 (V-$1\frac{1}{2}$), R-12 (V-2, V-$2\frac{1}{2}$), R-12A (V-3), R-13, R-14 (MH-1), R-15 (MH-$1\frac{1}{2}$), R-16 (MH-2, $2\frac{1}{2}$), R-16A (MH-3), R-17 (RH-17), R-19, R-21, R-27, R-28, R-29 (25°); Paradene No. 1, No. 2, No 33, No. 35); rubber cements (Cumar LX-509, P-10, P-25, R-1, R-3 (W-1), R-5 (W-$1\frac{1}{2}$), R-6 (W-2, $2\frac{1}{2}$), R-7, R-9, R-10 (V-1), R-11 (V-$1\frac{1}{2}$), R-12 (V-2, V-$2\frac{1}{2}$), R-12A (V-3), R-13, R-14 (MH-1), R-15 (MH-$1\frac{1}{2}$), R-16 (MH-2, $2\frac{1}{2}$), R-16A (MH-3), R-17 (RH-17), R-19, R-21, R-27, R-28, R-29 (25°); Paradene No. 1, No. 2, No 33, No. 35); rubber mechanical goods (Coumin; Cumar LX-509, P-10, P-25, R-1, R-3 (W-1), R-5 (W-$1\frac{1}{2}$), R-6 (W-2, $2\frac{1}{2}$), R-7, R-9, R-10 (V-1), R-11 (V-$1\frac{1}{2}$), R-12 (V-2, V-$2\frac{1}{2}$), R-12A (V-3), R-13, R-14 (MH-1), R-15 (MH-$1\frac{1}{2}$), R-16 (MH-2, $2\frac{1}{2}$), R-16A (MH-3), R-17 (RH-17), R-19, R-21, R-27, R-28, R-29 (25°); Natro-Rez 10, 15, 25, 50-D; Paradene No. 1, No. 2, No 33, No. 35)

Marine applications: coatings (Cumar LX-509, P-10, P-25, R-1, R-3 (W-1), R-5 (W-$1\frac{1}{2}$), R-6 (W-2, $2\frac{1}{2}$), R-7, R-9, R-10 (V-1), R-11 (V-$1\frac{1}{2}$), R-12 (V-2, V-$2\frac{1}{2}$), R-12A (V-3), R-13, R-14 (MH-1), R-15 (MH-$1\frac{1}{2}$), R-16 (MH-2, $2\frac{1}{2}$), R-16A (MH-3), R-17 (RH-17), R-19, R-21, R-27, R-28, R-29 (25°); Paradene No. 1, No. 2, No 33, No. 35)

Transportation industry: tires (Cumar LX-509, P-10, P-25, R-1, R-3 (W-1), R-5 (W-$1\frac{1}{2}$), R-6 (W-2, $2\frac{1}{2}$), R-7, R-9, R-10 (V-1), R-11 (V-$1\frac{1}{2}$), R-12 (V-2, V-$2\frac{1}{2}$), R-12A (V-3), R-13, R-14 (MH-1), R-15 (MH-$1\frac{1}{2}$), R-16 (MH-2, $2\frac{1}{2}$), R-16A (MH-3), R-17 (RH-17), R-19, R-21, R-27, R-28, R-29 (25°); Paradene No. 1, No. 2, No

33, No. 35); traffic coatings (Cumar LX-509, P-10, P-25, R-1, R-3 (W-1), R-5 (W-$1^1/_2$), R-6 (W-2, $2^1/_2$), R-7, R-9, R-10 (V-1), R-11 (V-$1^1/_2$), R-12 (V-2, V-$2^1/_2$), R-12A (V-3), R-13, R-14 (MH-1), R-15 (MH-$1^1/_2$), R-16 (MH-2, $2^1/_2$), R-16A (MH-3), R-17 (RH-17), R-19, R-21, R-27, R-28, R-29 (25°); Paradene No. 1, No. 2, No. 33, No. 35)

PROPERTIES:

Form:

Liquid (Coumin; Cumar P-10, P-25, R-27, R-28, R-29 (25°))

Pastilles (Coumin)

Semisolid (Natro-Rez 50-D)

Solid (Cumar LX-509, R-1, R-3 (W-1), R-5 (W-1¹/2), R-6 (W-2, W¹/2), R-7, R-9, R-10 (V-1), R-11 (V-$1^1/_2$), R-12 (V-2, V-$2^1/_2$), R-12A (V-3), R-13, R-14 (MH-1), R-15 (MH-$1^1/_2$), R-16 (MH-2, $2^1/_2$), R-16A (MH-3), R-17 (RH-17), R-19, R-21; Paradene No. 1, No. 2, No. 33, No. 35)

Flakes (Cumar LX-509, R-1, R-3 (W-1), R-5 (W-1¹/2), R-6 (W-2, W¹/2), R-7, R-9, R-10 (V-1), R-11 (V-$1^1/_2$), R-12 (V-2, V-$2^1/_2$), R-12A (V-3), R-13, R-14 (MH-1), R-15 (MH-$1^1/_2$), R-16 (MH-2, $2^1/_2$), R-16A (MH-3); Paradene No. 2)

Color:

Dark (Natro-Rez 50-D)

Dark brown (Paradene No. 1, No. 2, No. 33, No. 35)

Gardner 9 (50% in toluene) (Cumar R-1, R-9, R-13)

Gardner 11 (50% in toluene) (Cumar LX-509, R-3 (W-1), R-7, R-10 (V-1), R-14 (MH-1), R-27)

Gardner 12 (50% in toluene) (Cumar R-5 (W-$1^1/_2$), R-11 (V-$1^1/_2$), R-15 (MH-$1^1/_2$))

Gardner 14 (50% in toluene) (Cumar R-6 (W-2, $2^1/_2$), R-12 (V-2, V-$2^1/_2$), R-16 (MH-2, $2^1/_2$), R-28)

Gardner 16 (50% in toluene) (Cumar P-10, P-25, R-12A (V-3), R-16A (MH-3), R-17 (RH-17), R-19, R-21, R-29 (25°))

GENERAL PROPERTIES:

Solubility:

Sol. in aromatic hydrocarbons (Cumar LX-509, P-10, P-25, R-1, R-3 (W-1), R-5 (W-$1^1/_2$), R-6 (W-2, $2^1/_2$), R-7, R-9, R-10 (V-1), R-11 (V-$1^1/_2$), R-12 (V-2, V-$2^1/_2$), R-12A (V-3), R-13, R-14 (MH-1), R-15 (MH-$1^1/_2$), R-16 (MH-2, $2^1/_2$), R-16A (MH-3), R-17 (RH-17), R-19, R-21, R-27, R-28, R-29 (25°); Paradene No. 1, No. 2, No. 33, No. 35)

Sol. in chlorinated hydrocarbons (Cumar LX-509, P-10, P-25, R-1, R-3 (W-1), R-5 (W-$1^1/_2$), R-6 (W-2, $2^1/_2$), R-7, R-9, R-10 (V-1), R-11 (V-$1^1/_2$), R-12 (V-2, V-$2^1/_2$), R-12A (V-3), R-13, R-14 (MH-1), R-15 (MH-$1^1/_2$), R-16 (MH-2, $2^1/_2$), R-16A (MH-3), R-17 (RH-17), R-19, R-21, R-27, R-28, R-29 (25°)); Paradene No. 1, No. 2, No. 33, No. 35

Sol. in esters (Cumar LX-509, P-10, P-25, R-1, R-3 (W-1), R-5 (W-$1^1/_2$), R-6 (W-2, $2^1/_2$), R-7, R-9, R-10 (V-1), R-11 (V-$1^1/_2$), R-12 (V-2, V-$2^1/_2$), R-12A (V-3), R-13, R-14 (MH-1), R-15 (MH-$1^1/_2$), R-16 (MH-2, $2^1/_2$), R-16A (MH-3), R-17 (RH-17),

Coumarone-indene resin *(cont'd.)*

19, R-21, R-27, R-28, R-29 (25°)); Paradene No. 1, No. 2, No. 33, No. 35

Sol. in ethers (Cumar LX-509, P-10, P-25, R-1, R-3 (W-1), R-5 (W-$1^1/_2$), R-6 (W-2, $2^1/_2$), R-7, R-9, R-10 (V-1), R-11 (V-$1^1/_2$), R-12 (V-2, V-$2^1/_2$), R-12A (V-3), R-13, R-14 (MH-1), R-15 (MH-$1^1/_2$), R-16 (MH-2, $2^1/_2$), R-16A (MH-3), R-17 (RH-17), R-19, R-21, R-27, R-28, R-29 (25°)); Paradene No. 1, No. 2, No. 33, No. 35

Sol. in ketones (except acetone) (Cumar LX-509, P-10, P-25, R-1, R-3 (W-1), R-5 (W-$1^1/_2$), R-6 (W-2, $2^1/_2$), R-7, R-9, R-10 (V-1), R-11 (V-$1^1/_2$), R-12 (V-2, V-$2^1/_2$), R-12A (V-3), R-13, R-14 (MH-1), R-15 (MH-$1^1/_2$), R-16 (MH-2, $2^1/_2$), R-16A (MH-3), R-17 (RH-17), R-19, R-21, R-27, R-28, R-29 (25°)); Paradene No. 1, No. 2, No. 33, No. 35

Sol. in naphthenic hydrocarbons (Cumar LX-509, P-10, P-25, R-1, R-3 (W-1), R-5 (W-$1^1/_2$), R-6 (W-2, $2^1/_2$), R-7, R-9, R-10 (V-1), R-11 (V-$1^1/_2$), R-12 (V-2, V-$2^1/_2$), R-12A (V-3), R-13, R-14 (MH-1), R-15 (MH-$1^1/_2$), R-16 (MH-2, $2^1/_2$), R-16A (MH-3), R-17 (RH-17), R-19, R-21, R-27, R-28, R-29 (25°)); Paradene No. 1, No. 2, No. 33, No. 35

Partly sol. in paraffinic hydrocarbons (Cumar LX-509, P-10, P-25, R-1, R-3 (W-1), R-5 (W-$1^1/_2$), R-6 (W-2, $2^1/_2$), R-7, R-9, R-10 (V-1), R-12 (V-2, V-$2^1/_2$), R-13, R-14 (MH-1), R-15 (MH-$1^1/_2$), R-16 (MH-2, $2^1/_2$), R-16A (MH-3), R-17 (RH-17), R-19, R-21, R-27, R-28, R-29 (25°)); Paradene No. 1, No. 2, No. 33, No. 35

Sol. in terpene hydrocarbons (Cumar LX-509, P-10, P-25, R-1, R-3 (W-1), R-5 (W-$1^1/_2$), R-6 (W-2, $2^1/_2$), R-7, R-9, R-10 (V-1), R-11 (V-$1^1/_2$), R-12 (V-2, V-$2^1/_2$), R-12A (V-3), R-13, R-14 (MH-1), R-15 (MH-$1^1/_2$), R-16 (MH-2, $2^1/_2$), R-16A (MH-3), R-17 (RH-17), R-19, R-21, R-27, R-28, R-29 (25°)); Paradene No. 1, No. 2, No. 33, No. 35

M.W.:

290 (Cumar R-27)
330 (Cumar P-10)
345 (Cumar R-28)
350 (Cumar P-25)
400 (Cumar R-17 (RH-17))
440 (Paradene No. 33)
455 (Cumar R-29 (25°)
520 (Paradene No. 35)
530 (Cumar R-21)
580 (Cumar R-7, R-12 (V-2, V-$2^1/_2$); Paradene No. 1)
590 (Cumar R-19)
600 (Cumar R-12A (V-3))
610 (Cumar R-9)
640 (Cumar R-10 (V-1))
675 (Cumar R-14 (MH-1))
680 (Cumar R-15 (MH-$1^1/_2$))
700 (Cumar R-11 (V-$1^1/_2$))
710 (Paradene No. 2)

730 (Cumar R-1)
750 (Cumar R-5 (W-1$^1/_2$))
755 (Cumar R-6 (W-2, 2$^1/_2$))
760 (Cumar R-16A (MH-3))
770 (Cumar R-16 (MH-2, 2$^1/_2$))
800 (Cumar R-3 (W-1))
830 (Cumar R-13)
1090 (Cumar LX-509)

Sp. Gr.:

1.000 (Natro-Rez 15)
1.010 (Natro-Rez 10)
1.020 (Cumar R-19)
1.030 (Cumar R-29 (25°); Natro-Rez 25)
1.035 (Natro-Rez 50-D)
1.040 (Paradene No. 33)
1.050 (Cumar R-21)
1.056 (Paradene No. 35)
1.060 (Cumar P-25, Cumar R-12A (V-3), R-16A (MH-3))
1.075 (Cumar R-16 (MH-2, 2$^1/_2$))
1.080 (Cumar R-27, R-28)
1.095 (Paradene No. 2)
1.100 (Cumar R-14 (MH-1); Paradene No. 1)
1.120 (Cumar R-15 (MH-1$^1/_2$), R-17 (RH-17))
1.130 (Cumar R-5 (W-1$^1/_2$))
1.140 (Cumar LX-509, P-10, R-1, R-3 (W-1), R-6 (W-2, 2$^1/_2$), R-7, R-9, R-10 (V-1), R-11 (V-1$^1/_2$), R-12 (V-2, V-2$^1/_2$))
1.150 (Cumar R-13)

Visc.:

200–250 SUS (210 F) (Natro-Rez 10)
250–300 SUS (210 F) (Natro-Rez 15)

Soften. Pt. (Ring & Ball):

10 C (Cumar P-10, R-27)
25 C (Cumar P-25)
27 C (Paradene No. 33)
28 C (Cumar R-29 (25°))
33 C (Cumar R-28)
35–41 C (Natro-Rez 25)
45 C (Cumar R-21)
52–58 C (Natro-Rez 50-D)
53 C (Paradene No. 35)
58 C (Cumar R-19)
75 C (Paradene No. 1)
76 C (Cumar R-17 (RH-17))

Coumarone-indene resin *(cont'd.)*

93 C (Paradene No. 2)
100 C (Cumar R-13, R-14 (MH-1), R-15 (MH-$1^{1}/_{2}$))
103 C (Cumar R-16 (MH-2, $2^{1}/_{2}$), R-16A (MH-3))
106 C (Cumar R-7)
112 C (Cumar R-9, R-10 (V-1), R-11 (V-$1^{1}/_{2}$), R-12 (V-2, V-$2^{1}/_{2}$), R-12A (V-3))
126 C (Cumar R-1, R-3 (W-1), R-5 (W-$1^{1}/_{2}$), R-6 (W-2, $2^{1}/_{2}$))
155 C (Cumar LX-509)

Flash Pt.:
225 F (COC) (Cumar R-17 (RH-17)
250 F (COC) (Natro-Rez 10)
330 F (COC) (Cumar P-25)
350 F (COC) (Cumar P-10, R-27; Natro-Rez 15, 25)
365 F (COC) (Natro-Rez 50-D)
380 F (COC) (Cumar R-28, R-29 (25°))
390 F (COC) (Paradene No. 33)
400 F (COC) (Paradene No. 1, No. 35)
415 F (COC) (Cumar R-16 (MH-2, $2^{1}/_{2}$))
420 F (COC) (Cumar R-16A (MH-3), R-19)
510 F (COC) (Cumar R-12A (V-3))
515 F (COC) (Cumar R-9, R-11 (V-$1^{1}/_{2}$), R-12 (V-2, V-$2^{1}/_{2}$), R-13, R-15 (MH-$1^{1}/_{2}$))
520 F (COC) (Cumar R-10 (V-1), R-14 (MH-1))
525 F (COC) (Cumar R-21; Paradene No. 2)
540 F (COC) (Cumar R-1, R-3 (W-1), R-5 (W-$1^{1}/_{2}$))
550 F (COC) (Cumar R-6 (W-2, $2^{1}/_{2}$))
560 F (COC) (Cumar LX-509)
580 F (COC) (Cumar R-7)

Fire Pt.:
380 F (COC) (Natro-Rez 10, 15, 25)
405 F (COC) (Natro-Rez 50-D)

Acid No.:
1 max. (Natro-Rez 10, 15, 25, 50-D)

Iodine No.:
31 (Cumar R-16 (MH-2, $2^{1}/_{2}$))
35 (Cumar LX-509)
36 (Cumar P-25)
40 (Cumar R-12 (V-2, V-$2^{1}/_{2}$))
43 (Cumar R-3 (W-1))
44 (Cumar P-10, R-7, R-10 (V-1))
45 (Cumar R-6 (W-2, $2^{1}/_{2}$), R-29 (25°))
50 (Cumar R-5 (W-$1^{1}/_{2}$), R-11 (V-$1^{1}/_{2}$), R-14 (MH-1), R-21)
55 (Cumar R-9, R-15 (MH-$1^{1}/_{2}$), R-16A (MH-3), R-28; Paradene No. 1)
57 (Cumar R-13)
58 (Cumar R-1, R-27)

60 (Cumar R-12A (V-3))

65 (Cumar R-17 (RH-17), R-19; Paradene No. 2, No. 33)

70 (Paradene No. 35)

Stability:

Excellent resistance to alkalis, dilute acids, and moisture (Cumar LX-509, P-10, P-25, R-1, R-3 (W-1), R-6 (W-2, $2^1/_2$), R-7, R-9, R-10 (V-1), R-11 (V-$1^1/_2$), R-12 (V-2, V-$2^1/_2$), R-12A (V-3), R-13, R-14 (MH-1), R-15 (MH-$1^1/_2$), R-16 (MH-2, $2^1/_2$), R-16A (MH-3), R-17 (RH-17), R-19, R-21, R-27, R-28, R-29 (25°); Paradene No. 1, No. 2, No. 33, No. 35)

Epoxy resin

STRUCTURE:
Epoxy or oxirane group structure:

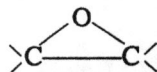

Epichlorohydrin/bisphenol A epoxy resin:

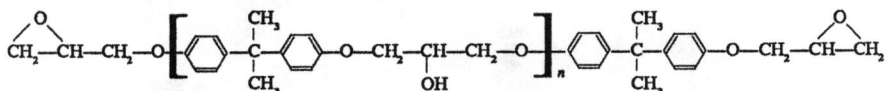

Epoxy novolak resin

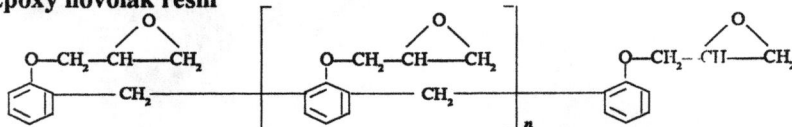

TRADENAME EQUIVALENTS:
Allabond Twenty/twenty Adhesive, Twenty/twenty NM [Bacon Industries]

Barco Bond MB-165, MB-175 [Astro]

D.E.N. 431, 438, 439, 485 [Dow]

D.E.R. 317, 324, 325, 332, 337, 330, 331, 383, 642U, 672U, 652-A75, 661, 661-A80, 662, 663U, 664, 664U, 667, 671-EE75, 671-EEA75, 671-EEK75, 671-MK75, 671-T75, 671-X75, 671-XM75, 673MF, 684EK40, 732, 736 [Dow]

E-231, -3810, -3938, -8405 [Fiberite]

Epotuf Resin 37-127, 37-128, 37-130, 37-134, 37-135, 37-139, 37-140, 37-141, 37-151, 37-170, 37-200, 37-250 [Reichhold]

Heloxy WC-8002, WC-8004

Hysol MG15F, MG15F-01, MG15F-02, MG18, MH19F, MH19F-01, MH19F-02, PC12-007 [Hysol Div./Dexter]

Lekutherm X18, X20, X23, X24, X30S, X201, X227, X256, X257 [Bayer AG]

Norcast 157 [R.H. Carlson]

Polyset EPC50, EPC 68 FR [Morton Int'l.]

Stycast 1210, 1263, 1264, 1266, 1269-A, 1467, 1492, 2057, 2651, 2762, 2762-FT, 2850-FT, 2651-40, 2651-MM, 2741, 2850-GT, 2850-KT, 3050 [Emerson & Cuming]

Tactix 742 Performance Polymer [Dow]
Varcum 6404, 6407, 9413 [Reichhold]
Glass-reinforced:
E-260, -260H, -261H, -264H, -2748, -3800, -3824, -8353, -8354, -8354J, -9405 [Fiberite]
Stycast 1090, 1090-SI [Emerson & Cuming]
Graphite-reinforced:
Magnamite AS/3501-5A, AS/3501-6, AS4/1908, AS4/1919, AS4/2220-3, AS4/3501-5A, AS4/3501-6, AS4/3502, HMS/1908, HMS/3501-5A, HMS/3501-6
Mineral-filled:
E-2068, -9451 [Fiberite]
E-Form 7-72 [Acme Chem. & Insulation]
Polyset 405, 410B, 415-SG [Morton Int'l.]
Mineral/glass-filled:
E-Form 7-51, 7-61, 7-61M, 7-61MM, 7-67 [Acme Chem. & Insulation]
Polyset 317, 707-2 [Morton Int'l.]
Silver-filled:
Tra-Duct 2902, 2924 [Tra-Con]
1-part epoxy:
EP 1760 LV, 1765 TS [Hardman]
Hysol EO1016, EO1017 [Hysol Div./Dexter]
Norcast 971 [R.H. Carlson]
2-part epoxy:
Barco Bond MB-14X, MB-127 [Astro]
C-84 [Bacon Industries]
CI-2, CI-3, CI-6 [Bacon Industries]
DC 2900 HW [Hardman]
EP 2220 TC, 2225 MV, 2230 TS, 2305 LK, 2306 LK, 2400 TC, 2404 GE, 2405 FR, 2408 TS, 2415 TC, 2420 TC, 2740 HT, 2770 TC, 2800 GU, 2805 TC, 2810 TS, 2820 TS, 2825 TS, 2840 TS [Hardman]
Epiceram [Delta Plastics]
Epolite 1301, 1302, 1350, 1353, 1354, 2300, 2302, 2353, 2354, 2360, 3300, 3301, 3302, 3306, 3353, 3354, 3357 [Hexcel/Rezolin]
FA-1, -8, -13, -14, -21 [Bacon Industries]
FFA-2, -5, -9 [Bacon Industries]
Hysol 1C, 3X, 0151, 309, 608, 615, 907, 1105, 9340 Epoxi-Patch Kit [Hysol Div./Dexter]
Hysol C18F, C60, C61, ES4212, ES4312, ES4412, R8-2038/HD3404, R8-2038/HD3475, R9-2039/HD3404, R9-2039/HD3469, R9-2039/HD3561, R9-2039/HD3615, R9-2039/HD3719 [Hysol Div./Dexter]
Interset Concrete Bonder, Concrete Topping Resin #2, Deep Penetrating Sealer, Epoxy Coal Tar, Epoxy Gel, Injection Resin (F) [Int'l. Thermoset]
LCA-1, -4, -4LV, -9, -14, -20, -21, -27 [Bacon Industries]

Epoxy resin *(cont'd.)*

Norbond 406, 2162, 2162 Clear Amber [R.H. Carlson]

Norcast 23, 176X, 424, 3215, 3216, 3220-D, 3222, 3222 LV, Norcast 3231/Norcure 135 or 135-M, Norcast 3231/Norcure 112, Norcast 3253/Norcure 112, Norcast 3253/Norcure 135M, Norcast 3255, Norcast 3343, Norcast 3370, Norcast 34301 M, Norcast 3420 LC, Norcast 3425 HF, Norcast 3425-M, Norcast 3511, Norcast 3705, Norcast 4200-1, 4200-2, 4200-05, 4915, 5500, 5503, 5509-M, Norcast 5615-10/Norcure 178-M, 7086, 7123, 7124, 7146, 7146-F, 7633, 9100, 9101, 9102, 9311, 9383, 9390, T-30, X-9201, Norcast 141/Catalyst 0.31, Norcast 141/Norcure 122, Norcast 154/Norcure 134, Norcast 154/Norcure 137, Norcast 0364/Norcure 142, Norcast 1750/Norcure 112, Norcast 1750/Norcure 112, Norcast 1750/Norcure 135-M, Norcast 1750/Norcure 138, Norcast 1750/Norcure 170, Norcast 2070-PC, Norcast 2795/Norcure 3416, Norcast 3217M/Norcure 111 & 112, Norcast 3217M/Norcure 122, Norcast 3217M/Norcure 135, Norcast 3230 SP/Norcure 112 & 170, Norcast 3230 SP/Norcure 3230B & 135M, Norcast 3256/Norcure 112, Norcast 3256/Norcure 135, Norcast 3257/Norcure 112, Norcast 3257/Norcure 135, Norcast 3258/Curing Agent D, Norcast 3258/Norcure 112, Norcast 3259 Resin-D-Type, 3424-M, 3510, 3515, 3705-F, 3706-F, 4917 HSS, Norcast 5005/Norcure 150 [R.H. Carlson]

P-11, -14, -19, -20, -24, -38, -51, -56, -56A, -70, -76, -78, -80C, -81, -82C, -85, -86 [Bacon Industries]

Stycast 3180-M [Emerson & Cuming]

Tra-Bond 2101, 2112, 2113, 2116, 2126, 2129, 2135D, 2151, 2208, 2211, 2215, 2248 [Tra-Con]

Aluminum-filled:

Hysol 6C Epoxi-Patch Kit [Hysol Div./Dexter]

Norcast T-27 [R.H. Carlson]

Tra-Bond 2122 [Tra-Con]

Aluminum oxide-filled:

Norcast 3230-A/Norcure 112, Norcast 3230-A/Norcure 135, Norcast 3230 LV/Norcure 112 [R.H. Carlson]

Carbon-filled:

Norcast 4915 [R.H. Carlson]

Glass-filled:

P-74, -75 [Bacon Industries]

Tra-Bond 2125 [Tra-Con]

Silica-filled:

Hysol C8-4143/HD3404, C8-4143/HD3475, C9-4183/HD3404, C9-4183/HD3469, C9-4183/HD3485, C9-4183/HD3537, C9-4183/HD3561, C9-4183/HD3615, C9-4183/HD3719 [Hysol Div./Dexter]

Silver-filled:

Norcast 4912, 4920 [R.H. Carlson]

Steel-filled:

Tra-Bond 2123 [Tra-Con]

Epoxy/glass laminate:
 Gillfab 1045, 1053 [M.C. Gill]
 Spauldite FR-4, G-10, G-10-900, G-10CR, G-10-773, G-11, G-11-963, G-11 CR, PEG FR, T-525 [Spaulding Fibre]
Epoxy/nylon laminate:
 Gillfab 1109 [M.C. Gill]

TYPES/MODIFICATIONS/SPECIALTY GRADES:
Epoxy-novolacs:
 D.E.N. 431, 438, 439, 485; Epotuf Resin 37-170; Hysol MH19F, MH19F-01, MH19F-02 [Hysol Div./Dexter]
Bisphenol A epoxies:
 D.E.R. 317, 324, 325, 332, 337, 330, 331, 383, 642U, 672U; Lekutherm X18, X20, X23, X24, X30S
Aliphatic epoxies:
 D.E.R. 732, 736
Dimer/trimer acid-modified:
 Heloxy WC-8002 (27%), WC-8004 (37%)
Conductive grade:
 Norcast 4912, 4915, 4917, 4920; Tra-Duct 2902, 2924
Dielectric grade:
 Polyset EPC50, EPC68FR
Impact grade:
 E-231, -261H, -2748, -3938; Epolite 1302
High-impact:
 Tra-Bond 2135D
Flame-retardant:
 C-84; E-Form 7-61, 7-61M, 7-61MM; Epotuf Resin 37-200 (brominated); Hysol C18F; Polyset EPC68FR; Spauldite FR-4, G-10-900, PEG FR; Stycast 1467
Heat-resistant:
 E-231
Nonmagnetic grade:
 Allabond Twenty/twenty NM

CATEGORY:
 Thermosetting resin

PROCESSING:
Autoclave:
 Magnamite AS4/1908, AS4/1919, HMS/1908
Casting:
 C-84; D.E.R. 317, 324, 325, 332, 337, 330, 331; Epolite 3300, 3301, 3302, 3306, 3353, 3354, 3357; Epotuf Resin 37-127, 37-128, 37-130, 37-134, 37-135, 37-139, 37-140, 37-151, 37-170, 37-200; FA-14; Hysol C8-4143/HD3404, C9-4183/HD3404, C9-4183/HD3469, C9-4183/HD3485, C9-4183/HD3561, C9-4183/HD3615, C9-

Epoxy resin (cont'd.)

4183/HD3719, C18F, C60, C61, EO1016, ES4212, ES4312, ES4412, R8-2038/HD3404, R9-2039/HD3404, R9-2039/HD3469, R9-2039/HD3561, R9-2039/HD3615, R9-2039/HD3719; Lekutherm X20, X23, X30S; Norcast 23, 154/Norcure 134, 154/Norcure 137, 971, 1750/Norcure 112, 1750/Norcure 135-M, 1750/Norcure 138, 1750/Norcure 170, 3215, 3216, 3217M/Norcure 111 & 112, 3217M/Norcure 122, 3217M/Norcure 135, 3230SP/Norcure 12 & 170, 3230SP/Norcure 3230B & 135M, 3231/Norcure 135 or 135-M, 3231/Norcure 112, 3253/Norcure 112, 3253/Norcure 135M, 3255, 3257/Norcure 112, 3257/Norcure 135, 3343, 3370, 3511, 3705, 3705-F, 3706-F, 4200-1, 4200-2, 4200-5, 7086, 7123, 7124, 7146, 7146-F, T-30, X-9201; P-85; Stycast 1210, 1263, 1264, 1266, 1269-A, 1467, 1492, 2057, 2651, 2651-40, 2651-MM, 2741, 2762, 2762-FT, 2850-FT, 2850-GT, 2850-KT, 3050

Cellowrap:
Magnamite AS4/1908, AS4/1919, HMS/1908

Compression molding:
E-261H, -2068; Lekutherm X18

Filament winding:
D.E.N. 431, 438, 439; D.E.R. 383; Epotuf Resin 37-139, 37-140, 37-170, 37-200

Hand layup:
Epotuf Resin 37-127, 37-128, 37-130, 37-135

Injection molding:
E-264H, -8354, -8354J

Lamination:
Epolite 2300, 2302, 2353, 2354, 2360; Epotuf Resin 37-127, 37-128, 37-130, , 37-135, 37-140, 37-170, 37-200; Lekutherm X20

Molding:
D.E.N. 431, 438, 439; E-231, -260, -260H, -2748, -3824, -3938, -8353, -8405; E-Form 7-51, 7-61MM, 7-67; Hysol MG15F, MG15F-01, MG15F-02, MG18, MH19F, MH19F-01, MH19F-02; Polyset 410B, 415-SG

Potting (encapsulation):
C-84; D.E.N. 431, 438, 439; D.E.R. 317, 324, 325, 332, 337, 330, 331, 383; E-2068, -3800, -3810, -3824, -8353, -3854, -8405, -9405, -9451; E-Form 7-61, 7-61M, 7-61MM, 7-67, 7-72; Epotuf Resin 37-127, 37-128, 37-130, 37-134, 37-135, 37-139, 37-140, 37-151, 37-170, 37-200; FA-14; Heloxy WC-8004; Hysol C9-4183/HD3485, C18F, C60, C61, EO1017; Lekutherm X20; Norcast 141/Catalyst 0.31, 971, 1750/Norcure 112, 1750/Norcure 135-M, 1705/Norcure 138, 1750/Norcure 170, 3230SP/Norcure 112 & 170, 3230SP/Norcure 3230B & 135M, 3253/Norcure 112, 3253/Norcure 135M, 3255, 3256/Norcure 112, 3256/Norcure 135, 3257/Norcure 112, 3257/Norcure 135, 3258/Curing Agent D, 3258/Norcure 112, 3420LC, 3434-M, 3425-M, 3510, 3511, 3515, 7124, 7146, 7146-F, 7633, 9100, 9383, 9390; P-11, -14, -19, -20, -24, -38, -51, -56, -56A, -70, -76, -78, -80C, -81, –82C, -85, -86; Polyset 317, 405, 410B, 707-2; Stycast 1210, 1269-A, 3050

Epoxy resin (cont'd.)

Transfer molding:
   E-261H, -2068, -8354; E-Form 7-61, 7-61M; Polyset 317, 405, 410B, 415-SG, 707-2

Vacuum bag molding:
   Epolite 2353

Wet lay-up:
   D.E.R. 317, 324, 325, 332, 337, 330, 331; Epolite 2302

APPLICATIONS:

Architectural applications: construction (D.E.R. 732, 736)

Automotive applications: parts (Epolite 1302); primers (D.E.R. 642U, 661, 662, 663U, 664, 664U, 667, 673MF)

Aviation industry: aerospace adhesive (Hysol 907, 1105, 9340 Epoxi-Patch Kit); airplane construction (Lekutherm X20, X227, X256, X257); structural parts (Gillfab 1045, 1053)

Consumer products: appliance primers (D.E.R. 642U, 661, 662, 663U, 664, 664U, 667, 673MF); repair material (Barco Bond MB-165); sporting goods (Lekutherm X20, X257)

Electrical/electronic industry: (Barco Bond MB-165, MB-175; DC 2900 HW; D.E.N. 431; EP 1760LV, 1765TS, 2220TC, 2225MV, 2230TS, 2305LK, 2306LK, 2400TC, 2404GE, 2405FR, 2408TS, 2415TC, 2420TC, 2740HT, 2770TC, 2800GU, 2805TC, 2810TS, 2820TS, 2825TS, 2840TS; P-85); circuitry (Hysol MG15F, MG15F-01, MG15F-02; Lekutherm X30S; Norcast 23, 2070-PC; P-86); coils (E-Form 7-51); connectors (E-2748, -3938); electrical engineering (Lekutherm X18); electrical laminates (D.E.N. 431, 438, 439); electrical packaging (Norcast 3370); insulation (Barco Bond MB-165, MB-175; D.E.R. 661; Gillfab 1045, 1053; Hysol C8-4143/HD3404, C8-4143/HD3475, C9-4183/HD3485, C9-4183/HD3537, C9-4183/HD3404, C9-4183/HD3469, C9-4183/HD3561, C9-4183/HD3615, C18F, EO1016, ES4212, ES4312, ES4412, R8-2038/HD3404, R8-2038/HD3475, R9-2039/HD3404, R9-2039/HD3469, R9-2039/HD3561, R9-2039/HD3615; LCA-27; Lekutherm X20; Norcast 3215, 3216; Tra-Bond 2101, 2151); potting/encapsulation (CI-2, -3, -6; D.E.R. 383; E-2068, -3800, -3810, -8353, -8354, -8405, -9405, -9451; E-Form 7-61, 7-61M, 7-61MM, 7-67, 7-72; Epotuf Resin 37-127, 37-128, 37-130, 37-134, 37-135, 37-139, 37-140, 37-151, 37-170, 37-200; FA-14; Heloxy WC-8004; Hysol C90-4183/HD3485, C18F, C60, C61, EO1017; Norcast 141/Catalyst 0.31, 971, 3253/Norcure 112, 3253/Norcure 135M, 3255, 3420LC, 3425-M, 3511, 5005/Norcure 150, 5500, 5503, 5509-M, 7124, 7146, 7146-F, 7633, 9100, 9383, 9390; P-51, -80C, -86; Polyset 317, 405, 707-2, EPC68FR); prepregs (D.E.R. 642U, 661, 662, 663U, 664, 664U, 667, 673MF)

Food-contact applications: (Hysol 1C Epoxi-Patch Kit)

Functional additives: modifier (D.E.N. 431, 438, 439; D.E.R. 337, 732, 736; Varcum 6404)

Epoxy resin *(cont'd.)*

Industrial applications: adhesives/sealants/bonding agents (Allabond Twenty/twenty Adhesive, Twenty/twenty NM; Barco Bond MB-14X, MB-127, MB-165, MB-175; D.E.N. 431, 438, 439; D.E.R. 317, 324, 325, 332, 337, 330, 331, 732, 736; Epotuf Resin 37-127, 37-128, 37-130, 37-134, 37-135, 37-139, 37-140, 37-141, 37-151, 37-170, 37-250; FA-1, -8, -13, -14, -21; FFA-2, -5, -9; Heloxy WC-8004; Hysol 1C, 3X, 6C, 0151, 615, 907, 1105, 9340 Epoxi-Patch Kits; Interset Concrete Bonder, Deep Penetrating Sealer, Epoxy Coal Tar; LCA-1, -4, -4LV, -9, -14, -20, -21, -27; Norbond 406, 2162, 2162 Clear Amber; Norcast 154/Norcure 134, 154/Norcure 137, 176X, 424, 3222, 3222LV, 3401M, 3705, 3705-F, 3706-F, 4912, 4915, 4917 HSS, 4920, 5500, 5503, 5509-M, 7633; Tra-Bond 2101, 2112, 2113, 2122, 2123, 2125, 2129, 2135D, 2208, 2211, 2215, 2248, 2902, 2924); castings (C-84; CI-2, –3, -6; D.E.R. 317, 324, 325, 332, 337, 330, 331; Lekutherm X20, X30S, 4200-1, 4200-2, 4200-5); chemical process equipment (Lekutherm X20); civil engineering applications (D.E.R. 325, 732, 736); coatings (CI-2, -3, -6; D.E.N. 431, 438, 439; D.E.R. 337, 642U, 652-A75, 661, 661-A80, 662, 663U, 664, 664U, 667, 671-EE75, 671-EEA75, 671-EEK75, 671-MK75, 671-T75, 671-X75, 671-XM75, 673MF, 684-EK40; Epolite 1301, 1302, 1350, 1353, 1354; Epotuf Resin 37-127, 37-128, 37-130, 37-134, 37-135, 37-140, 37-141, 37-151, 37-170, 37-200, 37-250; FA-14; Heloxy WC-8004; Hysol PC12-007; Interset Epoxy Coal Tar; Norcast 176X, 3256/Norcure 112, 3256/Norcure 135, 4917 HSS); composite systems (Tactix 742 Performance Polymer); concrete (Interset Concrete Bonder, Concrete Topping Resin #2, Deep Penetrating Sealer, Epoxy Coal Tar, Epoxy Gel, Injection Resin (F)); corrosion-resistant coatings (Heloxy WC-8002); dies (Epolite 1301, 1302, 2300, 2302, 3300, 3301; Norcast T-30); dip coatings (Norcast 0364/Norcure 142, 2070-PC, 3220-D, 3220-D); dipping/spraying applications (Norcast 2070-PC, 9101, 9102); displays (Hysol MG18); ducts (Gillfab 1053); floor coatings (D.E.R. 642U, 661, 662, 663U, 664, 664U, 667, 673MF; Interset Epoxy Coal Tar); flooring (D.E.R. 732, 736; Epotuf Resin 37-127, 37-128, 37-130, 37-134, 37-135, 37-151, 37-250; Norcast 141/Norcure 122); foundry patterns (Epolite 2300, 3301, 3302, 3306); gel coats (Lekutherm X201); general purpose moldings (E-8353, -8354); grouting (Interset Epoxy Coal Tar); insulation (Norcast X-9201; Polyset EPC50); laminates (Epolite 1301, 1302, 1350, 1353, 1354, 2353, 2354, 2360; Lekutherm X20, X227, X256; Norcast 141/Norcure 122; Tra-Bond 2101); machinery (Lekutherm X18); maintenance coatings (D.E.R. 642U, 652-A75, 661, 661-A80, 662, 663U, 664, 664U, 667, 671-EE75, 671-EEA75, 671-EEK75, 671-MK75, 671-T75, 671-X75, 671-XM75, 673MF, 684-EK40, 732, 736; Heloxy WC-8002); molds (Epolite 1350, 1353, 1354, 2353, 2354, 2360, 3301, 3353, 3354, 3357; Norcast T-27); petroleum industry applications (Hysol 907 Epoxi-Patch Kit); photography (Tra-Bond 2126); pigmented applications (D.E.R. 671-EE75, 671-EEA75, 671-EEK75, 671-MK75, 671-T75, 671-X75, 671-XM75; Heloxy WC-8002); pipe fittings (E-2748); pipes (D.E.N. 431, 438, 439; Lekutherm X20); powder coatings (D.E.R. 672U; Polyset EPC50, EPC68FR); production parts (Epolite 3306); protective coatings (D.E.R. 661); putties (D.E.R. 732, 736); repair material (Barco Bond

MB-165, MB-175; Epiceram; Hysol 3X, 6C, 309, 907 Epoxi-Patch Kit; Interset Epoxy Gel, Injection Resin (F); Tra-Bond 2101); structural adhesive (Hysol 608 Epoxi-Patch Kit); structural applications (D.E.R. 642U, 661, 662, 663U, 664, 664U, 667, 673MF, 732, 736; E-260, -260H, -2748; Magnamite AS/3501-5A, AS/3501-6, As4/1908, AS4/1919, AS4/2220-3, AS4/3501-5A, AS4/3501-6, AS4/3502, HMS/1908, HMS/3501-5A, HMS/3501-6); structural foam (Epiceram); syntactic foam (Stycast 1090, 1090-SI; Tra-Bond 2125); tank linings (D.E.R. 642U, 661, 662, 663U, 664, 664U, 667, 673MF; Interset Epoxy Coal Tar); tanks (Lekutherm X20); textile processing equipment (E-2748); thick sections (Epolite 3357); thin sections (E-Form 7-51; Polyset 707-2); tooling (Epolite 1354, 2302, 2360, 3353, 3354, 3357; Norcast 141/Norcure 122, T-27, T-30); varnishes (D.E.R. 642U, 661, 662, 663U, 664, 664U, 667, 673MF)

Marine applications: boat construction (Lekutherm X227, X256, X257); finishes (Heloxy WC-8002); repairs (Barco Bond MB-165, MB-175)

Military applications: (E-260, -260H); missile components (E-2748)

PROPERTIES:

Form:

Liquid (D.E.R. 317, 324, 325, 332, 330, 331, 383, 652-A75, 661-A80, 671-EE75, 671-EEA75, 671-EEK75, 671-MK75, 671-T75, 671-X75, 671-XM75, 684-EK40, 732, 736; Epolite 2360, 3302, 3306; Epotuf Resin 37-127, 37-128, 37-130, 37-134, 37-135, 37-139, 37-140, 37-141, 37-151, 37-170, 37-200, 37-250; FA-14, -21; Hysol C8-4143/HD3404, C84143/HD3475, C9-4183/HD3404, C9-4183/HD3469, C9-4183/HD3485, C9-4183/HD3537, C9-4183/HD3561, C9-4183/HD3615, C9-4183/HD3719, C18F, C60, C61, EO1016, EO1017, ES4212, ES4312, ES4412, R8-2038/HD3404, R8-2038/HD3475, R9-2039/HD3404, R9-2039/HD3469, R9-2039/HD3561, R9-2039/HD3615, R9-2039/HD3719; Lekutherm X20, X23, X24; Norcast 157; Stycast 1090, 1090-SI, 1210, 1263, 1264, 1266, 1269-A, 1467, 1492, 2057, 2651, 2651-40, 2651-MM, 2741, 2762, 2762-FT, 2850-FT, 2850-GT, 2850-KT, 3050, 3180-M)

Low viscosity (D.E.N. 431)

Thin liquid (Norcast 7633 Part A)

Watery liquid (Norcure 3416)

Pourable liquid (Norcast 3230-A/Norcure 112; Norcast 3230-A/Norcure 135; Norcast 3230LV/Norcure 112; Norcast 3258/Curing Agent D; Norcast 3258/Norcure 112)

High viscosity (D.E.N. 438)

Viscous liquid (Interset Epoxy Coal Tar)

Syrupy liquid (Norcast 2795)

Semisolid (D.E.N. 439; D.E.R. 337)

Gel (Interset Epoxy Gel; Norcast 3420LC, 3425-M)

Light paste (mixed) (Hysol 0151, 608, 615, 907 Epoxi-Patch Kit)

Smooth paste (Norcast 4912)

Paste (Allabond Twenty/twenty Adhesive, Twenty/twenty NM; Barco Bond MB-14X, MB-175; Epolite 1301, 1302, 1350, 1353, 1354, 2300, 2302, 2353, 2354, 3300,

Epoxy resin (cont'd.)

3301, 3353, 3354, 3357; Norbond 406 (both parts); Norcast 4920 (Part A)) (mixed) (Hysol 1C, 3X, 6C, 309, 1105, 9340 Epoxi-Patch Kit)

Granular (E-231, -2068, -2748, -3824, -3938, -8353, -8354, -8405, -9405, -9451; E-Form 7-51, 7-61, 7-61M, 7-61MM, 7-67, 7-72; Polyset 317, 405, 410B, 415-SG, 707-2)

Coarse particles (Varcum 6404, 9413)

Noncrystallizing (Norcast 424, both parts)

Crystalline (Lekutherm X18)

Powder (Hysol MG15F, MG15F-01, MG15F-02, MH19F, MH19F-01, MH19F-02; Polyset EPC50, EPC68FR)

Coarse powder (Varcum 6407)

Solid (D.E.N. 485; D.E.R. 642U, 661, 662, 663U, 664, 664U, 667, 672U, 673MF; Lekutherm X30S)

Rigid solid (cured) (Norcast 7123, 7124)

Sheet (Gillfab 1045, 1053, 1109)

Sheet, tube, rod (Spauldite FR-4, G-10, G-10-900, G-10CR, G-10-773, G-11, G-11-963, G-11CR, PEG FR, T-525

Tape 30.5 cm width (Magnamite AS/3501-5A, AS/3501-6, AS4/1908, AS4/1919, AS4/2220-3, AS4/3501-5A, AS4/3501-6, AS4/3502, HMS/1908, HMS/3501-5A, HMS/3501-6

Color:

Colorless (Norcast 2795, Norcure 3416)

Water-white (Norcast 7086)

Clear (Hysol 0151, 608 Epoxi-Patch Kit; Norcast 3425-M; Norcast 7633 Part B); (mixed) (Barco Bond MB-14X); (cured) (Barco Bond MB-165, MB-175)

Clear, transparent (Tra-Bond 2113, 2129, 2208)

Clear, slight haze (Tra-Bond 2101)

Clear colorless (mixed) (Norcast 3420LC)

Clear or off-white (Interset Epoxy Gel, resin)

Clear to amber (Interset Deep Penetrating Sealer)

Clear; also avail. in colors (Norcast 9100)

Natural (Gillfab 1053)

Natural, light green (Gillfab 1045)

Natural, black, other colors avail. (Norcast 971; Norcast 1750/Norcure 112; Norcast 1750/Norcure 135-M; Norcast 1750/Norcure 138; Norcast 1750/Norcure 170; Norcast 3253/Norcure 112; Norcast 3253/Norcure 135M)

Translucent (cured) (Barco Bond MB-127)

Milky translucent (Tra-Bond 2112, 2116)

White (Epolite 1301, 1350, 2300, 2302; Hysol 1C; Norcast T-30)

White, black, other colors avail. (Norcast 3215. 3216)

Off-white (Norcast 5005/Norcure 150)

Cream (LCA-1)

Pale/light yellow (Norcast 424; Tra-Bond 2215)

Yellow (Norcast 5500, 5503, 5509-M)
Transparent, amber (Norcast 9383)
Clear amber (Norbond 2162 Clear Amber, Parts A and B)
Clear amber or colored (Norcast 7146)
Light amber (Tra-Bond 2135D)
Amber (FA-1; FFA-2, -5, -9)
Dark amber (Interset Epoxy Gel, hardener; Tra-Bond 2211)
Straw (Norcast 157)
Straw, translucent (Tra-Bond 2106T)
Purple (Norcast 7123)
Maroon (cured) (P-24)
Orange (Polyset EPC68FR)
Light blue (Hysol 615 Epoxi-Patch Kit; Tra-Bond 2151)
Royal blue (Gillfab 1109)
Blue (LCA-20; Norcast 7633 Part A; Polyset EPC50, EPC68FR); (cured) (P-19)
Blue-green (FA-8)
Pale/light green (Hysol 907, 1105 Epoxi-Patch Kit; LCA-4, -4LV)
Green (Barco Bond MB-14X/resin; Polyset EPC50, EPC68FR)
Green-tan (LCA-9)
Red (Barco Bond MB-14X/curing agent; Polyset EPC68FR); (cured) (P-20, -70, -80C, -82C)
Green, black (E-2748)
Light brown (FA-13)
Brown (cured) (P-11, P-14, -38, -56, -56A, -81)
Brown or unpigmented (Norcast 9311)
Dark brown (Epolite 2360; Tra-Bond 2248)
Silver (Norcast 4912; Tra-Duct 2902, 2924)
Off-silver (Norcast 4920)
Light gray (Tra-Bond 2125); (mixed) (Norcast 3225)
Gray (Hysol 309, 9340 Epoxi-Patch Kit); (mixed) (Norbond 406)
Dark gray (Hysol 3X Epoxi-Patch Kit)
Gray-black (Norcast 23)
Aluminum (Tra-Bond 2122)
Aluminum gray (Hysol 6C Epoxi-Patch Kit)
Steel gray (Tra-Bond 2123)
Metallic gray (Epolite 3306, 3353, 3354)
Black (C-84; Epolite 1302, 1353, 1354, 2353, 2354, 3300, 3301, 3357; Interset Epoxy Coal Tar; LCA-21; Norcast 7146-F; Polyset 317, 405, 410B, 415-SG, 707-2, EPC68FR; Tra-Bond 2126); (cured) (Allabond Twenty/twenty Adhesive, Twenty/twenty NM; LCA-27; Norcast 3230-A/Norcure 112; Norcast 3230-A/Norcure 135; Norcast 3230LV/Norcure 112; Norcast 3705-F, 3706-F; P-74, -75, -76, -78, -85, –86)
Metallic black (Epolite 3302)

Epoxy resin (cont'd.)

Black, white (Norbond 2162, Parts A and B)

Black, standard colors (E-2068, -8353, -9405, -9451)

Black; other colors avail. on request (Norcast 3256/Norcure 112; Norcast 3256/Norcure 135; Norcast 3257/Norcure 112; Norcast 3257/Norcure 135; Norcast 3258/Curing Agent D; Norcast 3258/Norcure 112; Norcast 9101, 9102)

Opaque black; colors avail. on request (Norcast 7124)

Unpigmented resin avail. in black and other colors (Hysol C8-4143/HD3404, C8-4143/HD3475, C9-4183/HD3404, C9-4183/HD3469, C9-4183/HD3537, C9-4183/HD3561, C9-4183/HD3615, C9-4183/HD3719, R8-2038/HD3404, R8-2038/HD3475, R9-2039/HD3404, R9-2039/HD3469, R9-2039/HD3561, R9-2039/HD3615, R9-2039/HD3719)

Gardner 1 (cured) (P-51)

Gardner 1 max. (D.E.R. 661, 662, 732, 736)

Gardner 2 max. (D.E.R. 383, 642U, 661-A80, 663U, 664U, 671-EE75, 671-EEA75, 671-EEK75, 671-MK75, 671-T75, 671-X75, 671-XM75)

Gardner 3 (Norcast 176X)

Gardner 3 max. (D.E.R. 652-A75, 664, 667, 684-EK40)

Gardner 10 max. (Heloxy WC-8004)

Gardner 13 (Tactix 742 Performance Polymer)

Gardner 14 max. (Heloxy WC-8002)

Odor:

None (Norcast 157; Norcast 3256/Norcure 112; Norcast 3256/Norcure 135; Norcast 3257/Norcure 112; Norcast 3257/Norcure 135; Norcast 3258/Curing Agent D; Norcast 3258/Norcure 112)

Low (Epolite 2302)

Composition:

15–18% solids (Interset Deep Penetrating Sealer)

32–42% resin content (Magnamite AS4/1919)

34 ± 3% resin content (Gillfab 1045)

35 ± 5% resin content (Gillfab 1053)

35–41% resin content (Magnamite AS4/1908, HMS/1908)

40 ± 1% solids in MEK (D.E.R. 684-EK40)

42 ± 3% resin content (Magnamite HMS/3501-5A, HMS/3501-6)

75 ± 1% solids in acetone (D.E.R. 652-A75)

75 ± 1% solids in Dowanol EE (D.E.R. 671-EE75)

75 ± 1% solids in Dowanol EEA (D.E.R. 671-EEA75)

75 ± 1% solids in Dowanol EE/MEK (70:30) (D.E.R. 671-EEK75)

75 ± 1% solids in MIBK (D.E.R. 671-MK75)

75 ± 1% solids in toluene (D.E.R. 671-T75)

75 ± 1% solids in xylene (D.E.R. 671-X75)

75 ± 1% solids in xylene/MIBK (35:65) (D.E.R. 671-XM75)

80 ± 1% solids in acetone (D.E.R. 661-A80)

100% reactive (Lekutherm X201, X227)

100% solids (Hysol 1C, 0151, 309, 608, 615 Epoxi-Patch Kit; Hysol C9-4183/
HD3404, PC12-007, R9-2039/HD3404; Interset Concrete Topping Resin #2;
Norcast 1750/Norcure 112; Norcast 1750/Norcure 135-M; Norcast 1750/Norcure
138; Norcast 1750/Norcure 170; Norcast 3256/Norcure 112; Norcast 3256/Norcure
135; Norcast 3257/Norcure 112; Norcast 3257/Norcure 135; Norcast 3258/Curing
Agent D; Norcast 3258/Norcure 112; Norcast 4912; Norcast 5005/Norcure 150)

GENERAL PROPERTIES:

Solubility:
Very slight sol. in water (D.E.R. 736); insol. in water (D.E.R. 732)

Sp. Gr.:
0.68 (Tra-Bond 2125)
0.70 (cured) (P-74)
0.75–0.80 (Norcast 7123)
0.78 (cured) (P-75; Stycast 1090-SI)
0.80 (Norcast 7124)
0.88 (cured) (Stycast 1090)
0.95 (Norcure 3416)
0.96 (D.E.R. 684-EK40)
0.98–0.99 (Barco Bond MB-127/curing agent)
1.03–1.06 (Heloxy WC-8004)
1.06 (D.E.R. 732; Norcast 3420LC)
1.07 (D.E.R. 652-A75, 671-MK75, 671-T75, 671-X75, 671-XM75)
1.08 (Norcast 424)
1.09 (D.E.R. 661-A80)
1.10 (D.E.R. 671-EEK75; FA-1, -8; FFA-2, -5, -9; Hysol C60, R8-2038/HD3475, R9-
2039/HD3404, R9-2039/HD3469, R9-2039/HD3561, R9-2039/HD3615, R9-
2039/HD3719; Norcast 3424-M; Norcast 5509-M Part A/B; Norcast 9100); (cured)
(Stycast 1264)
1.11 (Tra-Bond 2135D)
1.11-1.13 (Heloxy WC-8002)
1.12 (D.E.R. 671-EEA75)
1.13 (Norcast 2795); (cured) (FA-14)
1.14 (D.E.R. 671-EE75, 736; Norcast 157)
1.15 (Lekutherm X256; Norcast 3510)
1.15–1.25 (Barco Bond MB-127/resin)
1.16 (Lekutherm X227, X257; Tra-Bond 2113)
1.17 (Lekutherm X18, X20, X23, X24, X201)
1.18 (D.E.R. 673MF; Norcast 9390; Tra-Bond 2208); (cured) (CI-2, -3, -6; Stycast
1266)
1.19 (D.E.R. 642U, 661, 662, 663U, 664, 664U, 667; Lekutherm X30S; Tra-Bond
2129)
1.20 (Hysol R8-2038/HD3404; Norcast 5503/Norcure 112; Tra-Bond 2101, 2112);
(cured) (Stycast 1263, 1269-A)

Epoxy resin (cont'd.)

1.21 (Tra-Bond 2211, 2248)
1.22 (FA-13; Tra-Bond 2106T)
1.26 (Tra-Bond 2116)
1.33 (Tra-Bond 2215)
1.35 (Norcast 3259 Resin-D-Type; Norcast 3705-F semirigid formulation)
1.4 (Hysol C9-4183/HD3719; Norcast 3511; Norcast 3706-F semirigid formulation; Norcast 5005/Norcure 150; Norcast 5500/Norcure 112; Polyset EPC50); (cured) (Stycast 2741, 3180-M)
1.41 (Norcast 9101, 9102); (cured) (FA-21)
1.43 (Norcast 3343)
1.45 (Tra-Bond 2122); (cured) (Allabond Twenty/twenty NM; Stycast 2651-40); (mixed) (Norcast 9311)
1.50 (Hysol C8-4143/HD3404, C8-4143/HD3475, C9-4183/HD3404, C9-4183/HD3469, C9-4183/HD3485, C9-4183/HD3537, C9-4183/HD3561, C9-4183/HD3615, C18F, C61, EO1016, ES4212, ES4312, ES4412); (cured) (Allabond Twenty/twenty Adhesive)
1.55 (cured) (Stycast 2057, 2651, 2651-MM, 3050)
1.58 (Tra-Bond 2126)
1.60 (Norcast 3255; Polyset EPC68FR)
1.62–1.65 (cured) (E-9405)
1.63 (cured) (E-3824)
1.65 (cured) (Stycast 1467, 1492)
1.68 (cured) (C-84)
1.70 (cured) (Stycast 1210)
1.70–1.74 (E-2748)
1.71 (LCA-9)
1.72 (Norcast 3370); (cured) (P-86)
1.75 (cured) (Norcast T-27)
1.77 (cured) (P-14)
1.79 (cured) (P-20)
1.80 (cured) (E-8405; LCA-14)
1.80–1.83 (cured) (E-8354)
1.81 (cured) (P-85)
1.83 (LCA-4, -4LV, -21)
1.85 (Norcast 23); (cured) (E-3938; P-11); (@ 3% resin) (Gillfab 1045)
1.85 ± 0.1 (Gillfab 1053)
1.86 (cured) (P-80C, -82C)
1.86–1.92 (cured) (E-261H, -264H)
1.87 (LCA-1); (cured) (E-231; P-19, -70)
1.87–1.91 (cured) (E-2068)
1.89 (cured) (E-260, -260H; P-24, -81)
1.91–1.95 (cured) (E-9451)
1.92 (E-Form 7-61, 7-72); (cured) (P-38)

1.92–1.95 (cured) (E-8353)

1.93 (E-Form 7-61MM)

1.95 (E-Form 7-51, 7-67)

2.0 (E-Form 7-61M; LCA-20; Norcast 3230-A/Norcure 112; Norcast 3230LV/Norcure 112)

2.10 (Norcast 3515)

2.15 (Norcast 3230-A/Norcure 135)

2.17 (cured) (LCA-27)

2.3 (Tra-Bond 2151); (cured) (P-56A; Stycast 2762, 2762-FT, 2850-FT, 2850-FT)

2.35 (Tra-Bond 2123)

2.45 (Tra-Duct 2902)

2.5 (cured) (P-56)

2.65 (Tra-Duct 2924)

2.8 (cured) (Stycast 2850-KT)

2.9–3.1 (Norcast 4912)

Density:

0.97 kg/m² (Gillfab 1109)

2.93 kg/m² (Gillfab 1045)

0.75 g/cm³ (Polyset EPC68FR)

0.85 g/cm³ (Polyset EPC50)

0.9 g/cm³ (Polyset 410B, 415-SG)

1.0 g/cm³ (Polyset 707-2)

1.217 g/cm³ (Epolite 2360)

1.218 g/cm³ (Epolite 1350)

1.273 g/cm³ (Epolite 2300)

1.384 g/cm³ (Epolite 1301, 2353)

1.430 g/cm³ (Epolite 2354)

1.433 g/cm³ (Epolite 2302)

1.495 g/cm³ (Epolite 3306)

1.5 g/cm³ (Norcast 3253/Norcure 112; Norcast 3253/Norcure 135M)

1.522 g/cm³ (Epolite 1353, 3353)

1.550 g/cm³ (Epolite 3302, 3354)

1.584 g/cm³ (Epolite 1354)

1.605 g/cm³ (Epolite 1302)

1.688–1.938 g/cm³ (Epolite 3300)

1.8 g/cm³ (Spauldite G-10, G-10CR, G-10-773, G-11, G-11-963, G-11 CR, T-525); (molded) (Polyset 317, 410B, 415-SG)

1.85 g/cm³ (Spauldite FR-4, G-10-900)

1.96 g/cm³ (molded) (Polyset 707-2)

2.04 g/cm³ (molded) (Polyset 405)

2.24 g/cm³ (Epolite 3301)

2.463 g/cm³ (Epolite 3357)

44 lb/ft³ (P-74)

Epoxy resin *(cont'd.)*

49 lb/ft³ (P-75)

5 lb/gal (Norcast 7124)

7.2–7.6 lb/gal (mixed) (Interset Deep Penetrating Sealer)

8.0 lb/gal (D.E.R. 684-EK40; Norcast 7633 Part A; Norcure 133, 159)

8.4 lb/gal (Norcast 141/Norcure 122)

8.5 lb/gal (Interset Injection Resin (F), curing agent); (mixed) (Interset Epoxy Gel)

8.8–8.9 lb/gal (Interset Concrete Topping Resin #2, hardener)

8.9 lb/gal (D.E.R. 652-A75, 671-MK75, 671-T75, 671-XM75, 732)

9.0–9.2 lb/gal (Interset Epoxy Coal Tar, resin)

9.1 lb/gal (D.E.R. 661-A80, 671-X75)

9.2 lb/gal (D.E.R. 671-EEK75)

9.25 lb/gal (Interset Concrete Bonder, curing agent)

9.3 lb/gal (Interset Concrete Bonder, resin)

9.4 lb/gal (D.E.R. 671-EEA75)

9.5 lb/gal (D.E.R. 671-EE75, 736; Interset Injection Resin (F), resin)

9.7 lb/gal (D.E.R. 383)

9.9 lb/gal (cured) (D.E.R. 642U, 661, 662, 663U, 664, 664U, 667)

9.9–10.0 lb/gal (Interset Epoxy Coal Tar, hardener)

10.1–10.7 lb/gal (Interset Concrete Topping Resin #2, resin)

Visc.:

4 poise (activated) (P-51)

5 poise (CI-2); (mixed) (FA-13)

6 poise (300 F, activated) (P-81)

8.5 poise (CI-6)

9 poise (212 F, activated) (LCA-14); (300 F, activated) (P-38)

12 poise (160 F, activated) (P-75)

13 poise (160 F, activated) (P-56A)

15 poise (160 F, activated) (P-74)

16 poise (160 F, activated) (P-56)

20 poise (CI-3); (160 F, activated) (P-80C, -82C)

22.5 poise (180 F, activated) (P-85)

25 poise (@ 212 F, activated compd.) (P-11, -14, -19, -20)

35 poise (@ 250 F, activated compd.) (P-24)

40 poise (mixed, 160 F) (FA-21; LCA-4LV); (75 F, activated) (P-86); (250 F, activated) (P-70)

55 poise (mixed) (Allabond Twenty/twenty Adhesive)

60 poise (activated compd.) (C-84); (mixed) (FA-8; FFA-2, -5); (mixed, 160 F) (LCA-9, -20, -21)

80 poise (mixed, 160 F) (LCA-4)

90 poise (mixed) (Allabond Twenty/twenty NM)

100 poise (mixed) (FA-1; LCA-1)

190 poise (mixed) (Tra-Bond 2101)

600 poise (C-84 compd.)

> 4000 poise (LCA-27)
5000–10,000 poise (Epotuf Resin 37-200)
3–9 cps (Interset Deep Penetrating Sealer)
8–10,000 cps (mixed) (Norcast 424)
30–60 cps (D.E.R. 736)
30–60,000 cps (Norcast 23, base)
40 cps (150 C) (Tactix 742 Performance Polymer)
50–100 cps (Norcure 3416)
55–100 cps (D.E.R. 732)
100–200 cps (Interset Injection Resin (F), curing agent)
200–500 cps (Norcast 7633)
250 cps (Norcast 3425-M)
275 cps (FA-14)
300 cps (Tra-Bond 2113); (mixed) (Norcast 3424-M)
300–400 cps (mixed) (Interset Injection Resin (F))
350 cps (Hysol C18F)
400 cps (Hysol C9-4183/HD3537); (mixed) (Norcast 176X)
450–500 cps (Norcast 141/Norcure 122)
460–520 cps (120 C) (Lekutherm X30S)
500 cps (Hysol C9-4183/HD3485, R8-2038/HD3404); (mixed) (Stycast 3050)
500–700 cps (Epotuf Resin 37-130; Interset Injection Resin (F), resin)
500–800 cps (Lekutherm X227)
500–1000 cps (Epotuf Resin 37-127, 37-128; Lekutherm X256; Norcast 2795)
600 cps (Hysol R9-2039/HD3561)
900 cps (mixed) (Stycast 1264)
1000 cps (Norcast 5503); (mixed) (Epolite 2360; Stycast 1266)
1000–2000 cps (Epotuf Resin 37-250)
1030 cps (Tra-Bond 2208)
1200 cps (Tra-Bond 2135D); (mixed) (Norcast 7146)
1200–2200 cps (mixed) (Norcast 5615-10/Norcure 178-M)
1300 cps (mixed) (EP 2408TS)
1500 cps (Hysol R8-2038/HD3475); (mixed) (Norcast 3420LC, 4200-5)
1500–2500 cps (Lekutherm X201, X257)
1500–4500 cps (D.E.R. 652-A75)
1600 cps (mixed) (Norcast 4200-1, 4200-2)
1700 cps (mixed) (EP 2800GU)
1800 cps (mixed) (Stycast 1090-SI)
1900 cps (Tra-Bond 2129); (mixed) (EP 2825TS); (initial mixed) (Norcast 9390)
2000 cps (EP 1760LV; Hysol R9-2039/HD3719); (mixed) (EP 2415TC)
2000–10,000 cps (D.E.R. 671-EEK75)
2100–3100 cps (Epotuf Resin 37-134)
2200–10,000 cps (D.E.R. 671-MK75, 671-T75)
2500 cps (initial) (Norcast 3510); (mixed) (Norcast 7124/Norcure 159)

Epoxy resin *(cont'd.)*

2500–9000 cps (D.E.R. 671-XM75)

2600 cps (mixed) (Epolite 2302)

2800 cps (mixed) (EP 2820TS)

3000 cps (Hysol C8-4143/HD3404, R9-2039/HD3404); (mixed) (Epolite 2300; Stycast 1263)

3300 cps (mixed) (EP 2220TC)

3500 cps (mixed) (Stycast 2057)

3500–8500 cps (D.E.R. 661-A80)

4000 cps (Hysol R9-2039/HD3469); (mixed) (Norcast 3215, 3216; Stycast 1269-A, 2651-40)

4000–6000 cps (Lekutherm X18)

4000–20,000 cps (D.E.R. 671-EE75)

4500 cps (mixed) (EP 2810TS)

4500–10,000 cps (D.E.R. 671-X75)

4800 cps (mixed) (EP 2840TS)

5000 cps (Norcast 2070-PC, Part B); (mixed) (Norcast 3259 Resin-D-Type)

5000–6000 cps (Norcast 3253/Norcure 112 mixed)

5000–8500 cps (Epotuf Resin 37-135)

5200 cps (mixed) (EP 2404GE)

5500 cps (Hysol ES4412); (mixed) (Epolite 2354)

6000 cps (mixed) (EP 2306LK; Epolite 3306; Stycast 2651-MM)

6000–7000 cps (mixed) (Norcast 3253/Norcure 135M)

6000–11,000 cps (100 F) (Heloxy WC-8002)

6000–30,000 cps (D.E.R. 671-EEA75)

6700–7100 cps (cured) (Norcast T-27)

7000 cps (Hysol C8-4143/HD3475)

7000–10,000 cps (Epotuf Resin 37-139)

7100 cps (mixed) (EP 2225MV)

7500 cps (mixed) (Norcast 7146-F)

8000 cps (Hysol C60, R9-2039/HD3615; Norcast 3253; Norcast 3253/Norcure 135M; Norcast 5500); (mixed) (EP 2305LK)

9000 cps (mixed) (Norcast 3255)

9000–11,500 cps (D.E.R. 383)

9000–13,000 cps (Lekutherm X20)

9200 cps (mixed) (EP 2230TS)

10,000 cps (Hysol C9-4183/HD3469, C9-4183/HD3561); (mixed) (EP 2400TC, 2405FR, 2770TC; Epolite 3302; Norcast 3705; Stycast 1210)

10,500 cps (Hysol C9-4183/HD3404)

10,600 cps (mixed) (EP 2805TC)

11,000 cps (Norcast 141/Catalyst 0.31)

11,000–14,000 cps (Epotuf Resin 37-140)

11,800 cps (mixed) (Epolite 2353)

12,000 cps (mixed) (EP 2420TC)

13,600 cps (mixed) (EP 2740HT)
14,000 cps (mixed) (Norcast 3343; Stycast 1467)
14,000–16,000 cps (Lekutherm X23); (initial) (Norcast 3511)
14,000–18,000 cps (Lekutherm X24)
15,000 cps (mixed) (Epolite 1354; Norcast 3425HF; Stycast 3180-M)
15,000–20,000 cps (Norcast 3217M; Norcast 3217M)
15,000–30,000 cps (Norcast 5005/Norcure 150)
16,300 cps (mixed) (DC 2900 HW)
17,500 cps (Tra-Bond 2125)
18,000–32,000 cps (Norcast 2070-PC, Part A)
19,000 cps (Tra-Bond 2211)
20,000 cps (Hysol C9-4183/HD3719); (mixed) (Epolite 1302, 1353, 3301)
20,000–35,000 cps (Epotuf Resin 37-141)
21,000 cps (mixed) (Stycast 1492)
25,000 cps (Hysol C9-4183/HD3615, EO1017); (mixed) (Epolite 3300; Stycast 2651)
25,000–30,000 cps (initial) (Norcast 3515)
25,000–40,000 cps (Norcast 971)
> 27,000 cps (Tra-Bond 2112)
30,000 cps (Hysol C61, ES4312); (mixed) (Stycast 1090)
30,000–50,000 cps (Norcast 1750/Norcure 112; Norcast 1750/Norcure 135-M; Norcast 1750/Norcure 138; Norcast 1750/Norcure 170; Norcast 3256/Norcure 112; Norcast 3256/Norcure 135)
30,000–70,000 cps (Epotuf Resin 37-151; Heloxy WC-8004)
30,000–90,000 cps (Epotuf Resin 37-170)
32,000 cps (Tra-Bond 2123)
33,000 cps (Tra-Bond 2151)
35,000 cps (Hysol ES4212); (mixed) (Norcast 9311)
40,000 cps (mixed) (Epolite 3357; Stycast 2741)
40,000–50,000 cps (Norcast 3257/Norcure 112; Norcast 3257/Norcure 135)
45,000 cps (mixed) (Norcast 9101, 9102)
50,000 cps (Tra-Bond 2126); (mixed) (Stycast 2762)
52,000 cps (mixed) (Epolite 3354)
56,000 cps (EP 1765TS)
60,000 cps (Hysol EO1016; Tra-Bond 2122); (mixed) (Stycast 2762-FT)
75,000 cps (Norcast 3231/Norcure 135)
90,000 cps (Norcast 3231/Norcure 112); (mixed) (Stycast 2850-FT)
100,000 cps (mixed) (Stycast 2850-FT)
> 100,000 cps (Tra-Bond 2116)
100,000–150,000 cps (Norcast 3258/Curing Agent D; Norcast 3258/Norcure 112)
137,000 cps (mixed) (Epolite 3353)
170,000 cps (mixed) (Stycast 2850-KT)
> 200,000 cps (Tra-Bond 2215)
> 250,000 cps (Tra-Bond 2106T)

Epoxy resin (cont'd.)

> 265,000 cps (Tra-Bond 2248)

300,000 cps (Barco Bond MB-127/resin)

U–Y (D.E.R. 684-EK40)

Thixotropic (Norcast 5509-M); (mixed) (Epolite 1301, 1350; Norcast 0364/Norcure 142)

Soften. Pt.:

52 C (Tactix 742 Performance Polymer)

55–60 C (Varcum 9413)

60–70 C (Varcum 6404)

65–75 C (Varcum 6407)

70–80 C (D.E.R. 661)

80–90 C (D.E.R. 662)

85–95 C (D.E.R. 673MF)

87–94 C (D.E.R. 642U)

88–98 C (D.E.R. 663U)

95–105 C (D.E.R. 664, 664U)

115–130 C (D.E.R. 667)

Flash Pt.:

111 C (Lekutherm X24)

140 C (Norcast 141/Norcure 122)

153 C (COC) (Norcast 157)

170 C (COC) (D.E.R. 736)

175 C (Lekutherm X256)

180 C (Lekutherm X227)

205 C (COC) (D.E.R. 732)

210 C (Lekutherm X257)

221 C (Lekutherm X201)

260 C (Lekutherm X18, X23)

274 C (Lekutherm X20)

318 C (Lekutherm X30S)

210 F (COC) (Norcure 3416)

475 F (TOC) (Norcast 2795)

Acid No.:

< 0.2 (Heloxy WC-8002, WC-8004)

Amine No.:

975–1075 (Norcast 141/Norcure 122)

Stability:

Excellent resistance to corrosion under most extreme conditions, to solvents and chemicals (even hot caustic and hot acetone), and to shock (D.E.R. 642U, 651-A75, 661, 661-A80, 662, 663U, 664, 667, 671-EE75, 671-EEA75, 671-EEK75, 671-MK75, 671-T75, 671-X75, 671-XM75, 673MF, 684-EK40)

Excellent chemical resistance and uv stability; post cure improves thermal stability (Norcast 154/Norcure 134)

Epoxy resin (cont'd.)

Excellent chemical resistance, good uv stability (Norcast 154/Norcure 137)
Incompatible with bases (D.E.R. 736)
Avoid amine compounds; incompatible with acid, base (D.E.R. 732)
Excellent heat and chemical resistance (E-2748)
Good chemical resistance (E-260, -260H)
Good chemical and thermal resistance (E-261H)
Good thermal shock, chemical, and heat resistance (E-2068)
Good thermal shock resistance (E-3824)
Outstanding dimensional stability in molded parts (E-Form 7-51)
Outstanding resistance to various solutions, e.g., acetone, freon, various alcohols (Norcast 5005/Norcure 150)
High heat resistance (Epolite 2353)
Not softened by low-boiling alcohols and aromatics, and aliphatic solvents; softened by ketones and chlorinated solvents (FA-1; LCA-4, -4LV, -9)
Not softened by aliphatic solvents; softened by ketones, chlorinated solvents, and low-boiling alcohols and aromatics (FA-8, -13; FFA-2, -5, -9; LCA-20, -21)
Not softened by low-boiling aromatics and aliphatic solvents; softened by ketones, chlorinated solvents, and low-boiling alcohols (LCA-1)
Cured systems exhibit outstanding abrasion and water resistance (Heloxy WC-8002)
Cured systems exhibit good hydrolytic stability (Heloxy WC-8004)
Resistant to acids, bases, oils, aliphatics, and aromatic hydrocarbons (Norcast 7086)
Storage Stability:
4 wks shelf stability @ 75 F; > 30 mo shelf stability if refrigerated below 45 F (Norcast 971)
2 mo. shelf life @ 23 C; 6 mo. @ 0 C (Hysol MG15F, MG15F-01, MG15F-02)
4 mo. shelf life @ 23 C, 12 mo. @ 0 C (Hysol MH19F, MH19F-01, MH19F-02)
6 mos. shelf life @ 40 F (Polyset EPC68FR)
12 mos. storage life @ 4 C (Polyset 317, 405, 410B, 415-SG, 707-2, EPC50)
1 yr min. shelf life in unopened containers (Epolite 1301, 1302, 1350, 1353, 1354, 2300, 2302, 2353, 2354, 2360, 3300, 3301, 3302, 3306, 3353, 3354, 3357)
1 yr shelf life @ –18 C (Magnamite AS/3501-5A, AS/3501-6, AS4/1908, AS4/1919, AS4/2220-3, AS4/3501-5A, AS4/3501-6, AS4/3502, HMS/1908, HMS/3501-5A, HMS/3501-6)
> 1 yr shelf life when stored in unopened containers below 85 F (C-84; FA-21; P-51, -85, -86)
2 yr shelf life when stored in unopened cans below 85 F (CI-2, -3, -6)
> 2 yr shelf life when stored in unopened containers below 85 F (LCA-14, -27; P-80C)
Excellent storage stability (E-Form 7-51, 7-67, 7-72)
Excellent storage stability; 12 mos. @ 77 F (E-Form 7-61, 7-61M)
Ref. Index:
1.51 (P-51)
1.52 (Hysol MG18)

Epoxy resin *(cont'd.)*

Surface Tension:
    24.9 dyne/cm (FA-14)
    42.2 dyne/cm (CI-3)
    48.8 dyne/cm (CI-2)

MECHANICAL PROPERTIES:

Tens. Str.:
    182 kg/cm^2 (Epolite 3300)
    189 kg/cm^2 (Epolite 3357)
    290.5 kg/cm^2 (Epolite 1302)
    329 kg/cm^2 (Epolite 1301)
    385 kg/cm^2 (Epolite 3302)
    441 kg/cm^2 (Epolite 3306)
    525 kg/cm^2 (Epolite 3301)
    700 kg/cm^2 (Polyset 405)
    844 kg/cm^2 (Polyset 415-SG)
    850 kg/cm^2 (Polyset 707-2)
    860 kg/cm^2 (Polyset 410B)
    984 kg/cm^2 (Polyset 317)
    1519 kg/cm^2 (Epolite 2302)
    1750 kg/cm^2 (Epolite 2300)
    5180 kg/cm^2 (Epolite 2360)
    590 kp/cm^2 (Lekutherm X256)
    625 kp/cm^2 (Lekutherm X201)
    630 kp/cm^2 (Lekutherm X257)
    660 kp/cm^2 (Lekutherm X227)
    750 kp/cm^2 (Lekutherm X30S)
    920 kp/cm^2 (Lekutherm X20)
    186 MPa (Gillfab 1109)
    324 MPa (Gillfab 1045)
    650 psi (Norcast 7146)
    700 psi (Norcast 3524-M, 3510)
    750 psi (Norcast 3425-M, 7146-F)
    1400 psi (Hysol C18F)
    1500 psi (Hysol ES4212, ES4312, ES4412)
    1600 psi (Norcast 9100)
    1750 psi (Norcast 3425HF, 3511)
    1800 psi (Hysol R9-2039/HD3719)
    2000 psi (Hysol C9-4183/HD3719, C60; Norcast 3515, 9101, 9102)
    2100 psi (P-51)
    2500 psi (Hysol C61)
    3100 psi (Norbond 406)
    3200 psi (Norcast 3705)
    5000 psi (Norcast 7086)

5200 psi (Norcast 3343; Norcast X-9201)

5500 psi (Interset Concrete Bonder)

5760 psi (Norcast 154/Norcure 137)

6000 psi (Hysol C8-4143/HD3404, C8-4143/HD3475, C9-4183/HD3404; Norcast 1750/Norcure 138; Norcast 3706-F semirigid formulation)

6000 psi min. (Interset Concrete Topping Resin #2); (break) (Interset Epoxy Gel)

6300 psi (Hysol C9-4183/HD3469)

6400 psi (Hysol C9-4813/HD3485)

6800 psi (P-11)

6900 psi (P-19)

7000 psi (Norcast 3253/Norcure 112; Norcast 3705-F semirigid formulation)

7200 psi (Norcast 1750/Norcure 170; Norcast 3215, 3216)

7400 psi (break) (Norcast 3220-D)

7500 psi (E-Form 7-72; Norcast 3256/Norcure 112; Norcast 4200-1, 4200-2, 4200-5; Norcast 5615-10/Norcure 178-M)

7800 psi (E-Form 7-51, 7-67; Hysol R9-2039/HD3561; Norcast 3370)

8000 psi (Hysol EO1016, R8-2038/HD3475; Norcast 1750/Norcure 112; Norcast 9390)

8200 psi (Norcast T-27)

8400 psi (E-8405; Norcast 3259 Resin-D-Type)

8500 psi (Hysol C9-4183/HD3561; Norcast 971)

8600 psi (Hysol R9-2039/HD3469)

8800 psi (Norcast 0364/Norcure 142)

9000 psi (E-9451; Hysol R8-2038/HD3404; Norcast 3217M/Norcure 111 & 112; Norcast T-30)

9000 psi min. (break) (Interset Injection Resin (F))

9500 psi (Norcast 3230SP/Norcure 3230B & 135M)

9550 psi (Norcast 154/Norcure 134)

9600 psi (Hysol C9-4183/HD3615)

9700 psi (Norcast 3230-A/Norcure 112; Norcast 3230LV/Norcure 112)

9800 psi (Norcast 3230-A/Norcure 135)

9950 psi (Hysol C9-4183/HD3537)

10,000 psi (E-2068, -8353, -9405; Hysol R9-2039/HD3404; Norcast 3217M/Norcure 122)

10,100 psi (P-20)

10,250 psi (Norcast 3258/Curing Agent D; Norcast 3258/Norcure 112)

10,300 psi (D.E.R. 383)

10,350 psi (Norcast 3401M; Norcast 5005/Norcure 150)

10,400 psi (Tactix 742 Performance Polymer)

10,500 psi (E-3824, -3938, -8354; Norcast 3230SP/Norcure 112 & 170; Norcast 3255)

11,000 psi (E-231, -2748; Hysol R9-2039/HD3615; Norcast 3217M/Norcure 135; Norcast 3253/Norcure 135M; Norcast 3256/Norcure 135); (break) (E-Form 7-61M, 7-61MM)

Epoxy resin *(cont'd.)*

11,400 psi (break) (E-Form 7-61)
11,500 psi (Norcast 3257/Norcure 112; Norcast 3257/Norcure 135)
11,600 psi (D.E.R. 331)
12,000 psi (Norcast 1750/Norcure 135-M)
12,200 psi (P-14)
12,900 psi (P-24)
13,000 psi (P-70)
13,600 psi (P-38)
14,600 psi (P-81)
17,000 psi (E-264H)
22,000 psi (E-260)
27,000 psi (E-260H)
40,000 psi (Spauldite FR-4, G-10, G-10-900, G-10 CR, G-10-773, G-11, G-11-963, G-11 CR, T-525)
45,100 psi (E-261H)

Tens. Elong.:

0.7% (E-Form 7-61M, 7-61MM)
0.8% (E-Form 7-61)
2% (break) (Lekutherm X227)
2.1% (Norcast 154/Norcure 134)
2.5% (break) (Lekutherm X256, X257)
2.7% (Tactix 742 Performance Polymer)
3.5% (Norcast 5005/Norcure 150)
3.9% (Norcast 154/Norcure 137)
4% (Norcast 4200-1, 4200-2)
4–6% (Interset Injection Resin (F))
4.5% (break) (Lekutherm X201)
4.70% (break) (D.E.R. 331)
7–10% (Norcast 5615-10/Norcure 178-M)
8% (break) (Norcast 3401M)
10% (Interset Concrete Bonder)
10–15% (Interset Epoxy Gel)
26% (Norcast 7086)
50% (break) (Norcast 3425HF, 9101, 9102)
65% (Norcast X-9201)
110% (break) (Norcast 3425-M)
120% (break) (Norcast 3424-M)

Tens. Mod.:

3200 MPa (Gillfab 1109)
27,500 kp/cm² (Lekutherm X201)
31,000 psi (Norcast 3220-D)
35,700 kp/cm² (Lekutherm X227)
36,500 kp/cm² (Lekutherm X257)

40,000 kp/cm² (Lekutherm X256)
360,000 psi (D.E.R. 383)
453,000 psi (Tactix 742 Performance Polymer)
461,000 psi (D.E.R. 331)
10^6 psi (Norcast 3255)
1.84×10^6 psi (Norcast 3370)
2.15×10^6 psi (E-Form 7-61)
2.25×10^6 psi (E-Form 7-61M, 7-61MM)

Flex. Str.:

383 kg/cm² (Epolite 2360)
392 kg/cm² (Epolite 1302)
490 kg/cm² (Epolite 1350)
501 kg/cm² (Epolite 3357)
518 kg/cm² (Epolite 3301)
546 kg/cm² (Epolite 1353)
574 kg/cm² (Epolite 3306)
630 kg/cm² (Epolite 3300, 3302)
651 kg/cm² (Epolite 1354)
770 kg/cm² (Epolite 1301)
1265 kg/cm² (Polyset 707-2)
1300 kg/cm² (Polyset 405)
1400 kg/cm² (Polyset 410B, 415-SG)
1590 kg/cm² (Polyset 317)
2170 kg/cm² (Epolite 2302)
2450 kg/cm² (Epolite 2300)
3605 kg/cm² (Epolite 2354)
5565 kg/cm² (Epolite 2353)
880 kp/cm² (Lekutherm X201)
920 kp/cm² (Lekutherm X20)
1285 kp.cm² (Lekutherm X257)
1350 kp/cm² (Lekutherm X256)
1450 kp/cm² (Lekutherm X30S)
1520 kp/cm² (Lekutherm X18)
1570 kp/cm² (Lekutherm X227)
138 MPa (Gillfab 1109)
483 MPa (Gillfab 1053)
552 MPa (Gillfab 1045)
400 psi (Norcast 9100)
4500 psi (Norcast 7124)
5000 psi (Norcast 7123)
5400 psi (Norcast 3705)
5700 psi (Norcast 3222, 3222LV
6000 psi (Allabond Twenty/twenty Adhesive)

Epoxy resin *(cont'd.)*

7500 psi (Norcast 3705-F semirigid formulation)

7600 psi (P-86)

8000 psi (Norcast 3706-F semirigid formulation; Norcast X-9201)

8500 psi (Norcast 1750/Norcure 138)

8600 psi (yield) (FA-14)

9000 psi (Norcast 1750/Norcure 170)

9135 psi (Norcast 154/Norcure 137)

9500 psi (Norcast 1750/Norcure 112)

10,000 psi (C-84)

10,500 psi (LCA-27)

11,000 psi (E-Form 7-72; Norcast 3230SP/Norcure 112 & 170)

11,200 psi (Norcast T-30)

11,300 psi (Norcast T-27)

11,900 psi (Norcast 3215, 3216)

12,000 psi (E-8405; Norcast 3230-A/Norcure 112; Norcast 3230LV/Norcure 112; Norcast 3230SP/Norcure 3230B & 135M)

12,200 psi (Norcast 3217M/Norcure 122)

12,500 psi (yield) (FA-21)

12,700 psi (Norcast 3370)

13,000 psi (Norcast 3217M/Norcure 111 & 112; Norcast 3256/Norcure 112)

13,800 psi (E-Form 7-51, 7-67)

14,000 psi (E-2068; Norcast 3230-A/Norcure 135; P-85)

14,500 psi (Norcast 3253/Norcure 135M)

14,900 psi (P-80C)

15,000 psi (Norcast 3253/Norcure 112; P-82C)

15,270 psi (Norcast 154/Norcure 134)

15,500 psi (E-3824; Norcast 971; Norcast 1750/Norcure 135-M)

15,680 psi (D.E.R. 331)

15,750 psi (Norcast 3257/Norcure 112)

16,000 psi (E-231, -8353, -9451; Norcast 3217M/Norcure 135; Norcast 3256/Norcure 135; Norcast 3257/Norcure 135)

16,250 psi (Norcast 5005/Norcure 150)

16,450 psi (Norcast 3258/Curing Agent D; Norcast 3258/Norcure 112)

17,000 psi (E-2748); (break) (Norcast 3255)

18,000 psi (E-3938, -8354; Norcast 9390; Tactix 742 Performance Polymer)

18,300 psi (D.E.R. 383)

19,000 psi (E-9405; E-Form 7-61, 7-61M, 7-61MM)

39,300 psi (E-264H)

40,000 psi (Spauldite PEG FR)

50,000 psi (E-260)

55,000 psi (Spauldite FR-4, G-10, G-10-773, G-10-900, G-10CR, G-11, G-11-963, G-11CR)

68,000 psi (E-260H)

    110,600 psi (E-261H)
Flex. Mod.:
  2500 MPa (Gillfab 1109)
  15,900 MPa (Gillfab 1053)
  16,000 psi (Norcast 3220-D)
  23,000 MPa (Gillfab 1045)
  390,000 psi (D.E.R. 383)
  400,000 psi (CI-3, -6; FA-14)
  425,000 psi (D.E.R. 331)
  430,000 psi (Tactic 742 Performance Polymer)
  450,000 psi (FA-21)
  500,000 psi (CI-2)
  600,000 psi (Allabond Twenty/twenty Adhesive)
  1.1×10^6 psi (C-84)
  1.3×10^6 psi (P-86)
  1.4×10^6 psi (E-2748; LCA-27)
  1.5×10^6 psi (E-8405, -9405; P-85)
  1.58×10^6 psi (Norcast 3370)
  1.7×10^6 psi (E-2068)
  2.0×10^6 psi (E-8353, -9451)
  2.19×10^6 psi (P-80C)
  2.46×10^6 psi (P-82C)
Compr. Str.:
  176 kg/cm² (Stycast 1210)
  630 kg/cm² (Stycast 1269-A)
  662 kg/cm² (Stycast 2741)
  770 kg/cm² (Epolite 1350; Stycast 1264)
  777 kg/cm² (Epolite 3300)
  875 kg/cm² (Epolite 3302)
  980 kg/cm² (Epolite 3301, 3306; Stycast 1467)
  1050 kg/cm² (Epolite 1301)
  1085 kg/cm² (Stycast 1263)
  1160 kg/cm² (Stycast 2850-FT)
  1260 kg/cm² (Stycast 2762, 2850-GT)
  1290 kg/cm² (Stycast 2762-FT)
  1295 kg/cm² (Epolite 3357)
  1330 kg/cm² (Epolite 1354)
  1340 kg/cm² (Stycast 2850-KT)
  1484 kg/cm² (Epolite 1353)
  2100 kg/cm² (Polyset 317, 405, 410B)
  2250 kg/cm² (Polyset 707-2)
  7030 kg/cm² (Stycast 1266)
  810 kp/cm² (Lekutherm X201)

Epoxy resin *(cont'd.)*

1285 kp/cm^2 (Lekutherm X30S)

1290 kp/cm^2 (Lekutherm X257)

1430 kp/cm^2 (Lekutherm X20)

1530 kp/cm^2 (Lekutherm X227)

1570 kp/cm^2 (Lekutherm X256)

311 MPa (Gillfab 1045)

2200 psi (Norcast 9100)

8000 psi (Stycast 3180-M)

8500 psi (Norcast 1750/Norcure 138)

9000 psi (Norcast 1750/Norcure 170); (break) (Norcast 3220-D)

10,000 psi (Norcast 1750/Norcure 112; Stycast 1090-SI)

12,000 psi (Norcast 3253/Norcure 112; Stycast 3050)

13,000 psi (Norcast 4200-5)

13,500 psi (Norcast 3215, 3216)

14,400 psi (Norcast 3217M/Norcure 122)

14,500 psi (Norcast 971)

15,000 psi (Norcast 1750/Norcure 135-M; Norcast 3217M/Norcure 111 & 112; Norcast 3256/Norcure 112; Norcast 7124; Stycast 1090)

15,000 psi min. (Interset Concrete Topping Resin #2); (break) (Interset Epoxy Gel)

15,500 psi (Norcast 3259 Resin-D-Type; Stycast 2057, 2651-40, 2651-MM)

15,750 psi (Norcast 3258/Curing Agent D; Norcast 3258/Norcure 112)

16,000 psi (Norcast 3217M/Norcure 135; Norcast 3230SP/Norcure 112 & 170; Norcast 3230SP/Norcure 3230B & 135M; Norcast 3256/Norcure 135; Norcast 3257/Norcure 112; Stycast 1492, 2651)

16,460 psi (D.E.R. 331)

17,000 psi (Norcast 3230-A/Norcure 112; Norcast 3230LV/Norcure 112; Norcast 3257/Norcure 135)

17,500 psi (Norcast 3230-A/Norcure 135)

18,000 psi (Norcast 9390)

18,000 psi min. (break) (Interset Injection Resin (F))

18,500 psi (Norcast T-30)

19,000 psi (Norcast 3253/Norcure 135M; Norcast 3255)

24,000 psi (E-3824; E-Form 7-72)

25,000 psi (E-9405; E-Form 7-51, 7-67)

26,000 psi (E-231)

26,100 psi (yield) (Tactix 742 Performance Polymer)

27,000 psi (E-264H, -8405)

28,000 psi (E-2068, -2748)

28,500 psi (Norcast 3370)

30,000 psi (E-3938, -8354, -9451; E-Form 7-61)

31,000 psi (E-8353)

32,000 psi (E-Form 7-61M, 7-61MM)

37,100 psi (break) (Tactix 742 Performance Polymer)

60,000 psi (Spauldite FR-4, G-10, G-10-773, G-10-900, G-10CR, G-11, G-11-963, G-11CR, T-525)
Compr. Mod.:
7000 kg/cm² (Stycast 1264)
14,000 kg/cm² (Stycast 1090-SI, 1210, 1266)
21,000 kg/cm² (Stycast 3180-M)
28,000 kg/cm² (Stycast 1090, 1467)
35,000 kg/cm² (Stycast 1263, 1269-A)
42,000 kg/cm² (Stycast 3050)
49,000 kg/cm² (Stycast 1492, 2057, 2651-MM)
56,000 kg/cm² (Stycast 2651, 2651-40)
77,000 kg/cm² (Stycast 2850-FT)
84,000 kg/cm² (Stycast 2762, 2762-FT)
91,000 kg/cm² (Stycast 2850-GT)
98,000 kg/cm² (Stycast 2850-KT)
26,000 psi (E-231)
30,000 psi (E-260)
31,000 psi (E-261H)
42,000 psi (E-260H)
260,000 psi (Tactix 742 Performance Polymer)
270,000 psi (D.E.R. 331)
1.85×10^6 psi (Norcast 3370)
Creep:
1 mil/in. (200 F) (P-19, -20, -24)
2 mil/in. (200 F) (P-80C)
5 mil/in. (200 F) (P-38)
8 mil/in. (P-82C)
10 mil/in. (200 F) (P-81)
40 mil/in. (200 F) (P-70)
50 mil/in. (200 F) (P-14)
60 mil/in. (200 F) (P-11)
Bond Str.:
825 psi (FFA-2)
2000 psi (LCA-14)
2200 psi (Al) (LCA-27)
2250 psi (FA-1, -8, -13; FFA-5, -9; LCA-1, -4, -4LV, -9, -20, -21)
Shear Str.:
350 psi (Al) (Barco Bond MB-165)
750 psi (Al) (Barco Bond MB-175)
800 psi (Norcast 4920)
850 psi (Norcast 4917HSS)
1020 psi (concrete) (Interset Concrete Bonder)
1050 psi (concrete) (Interset Injection Resin (F))

Epoxy resin *(cont'd.)*

1630 psi (cured 2 h @ 60 C, Al–Al) (Hysol 309 Epoxi-Patch Kit)

1700 psi (Al–Al) (Norcast 4912)

1900 psi (Norcast 4915)

2100 psi (cured 2 h @ 60 C, Al–Al) (Hysol 6C Epoxi-Patch Kit)

2300 psi (cured 2 h @ 60 C) (Hysol 1C Epoxi-Patch Kit); (cured 2 h @ 60 C, Al–Al) (Hysol 9340 Epoxi-Patch Kit)

2500 psi (cured 2 h @ 60 C, Al–Al) (Hysol 608 Epoxi-Patch Kit)

2600 psi (Allabond Twenty/twenty NM); (cured 2 h @ 60 C, aluminum) (Hysol 3X, 0151 Epoxi-Patch Kit)

2700 psi (Al) (Norcast 7633)

2800 psi (Al) (Allabond Twenty/twenty Adhesive)

3000 psi (Al–Al) Barco Bond MB-127); (cured 2 h @ 60 C, Al–Al) (Hysol 1105 Epoxi-Patch Kit)

3100 psi (2 h cure @ 60 C, Al–Al) (Hysol 615 Epoxi-Patch Kit)

3200 psi (2 h cure @ 60 C, Al–Al) (Hysol 907 Epoxi-Patch Kit)

Peel Str.:

4.5/in. of width (Norbond 406)

Impact Str. (Izod):

0.022 kg m/cm notched (Polyset 405, 410B, 415-SG)

0.028 kg m/cm notched (Polyset 707-2)

0.033 kg m/cm notched (Polyset 317)

1.4 kgm/2.54 cm notched (Gillfab 1053)

8 kp cm/cm² (Lekutherm X227)

9 kp cm/cm² (Lekutherm X256)

13 kp cm/cm² (Lekutherm X257)

20 kp cm/cm² (Lekutherm X20)

24 kp cm/cm² (Lekutherm X18, X30S)

25 kp cm/cm² (Lekutherm X201)

0.22 ft lb/in. (Norcast 3230-A/Norcure 112)

0.25 ft lb/in. (Norcast 3230-A/Norcure 135; Norcast 3230LV/Norcure 112; Norcast 3230SP/Norcure 112 & 170; Stycast 2057, 2651-MM, 2762, 2762-FT, 2850-KT)

0.27 ft lb/in. (Stycast 1090-SI)

0.30 ft lb/in. (Norcast 3215; Norcast 3230SP/Norcure 3230B & 135M; Stycast 1090, 1467, 1492, 2651, 2651-40, 2850-FT, 3050, 3180-M); notched (E-8405, -9451)

0.32 ft lb/in. (Stycast 1263)

0.35 ft lb/in. notched (E-2068, -8353)

0.40 ft lb/in. (Norcast 3256/Norcure 112; Norcast 3258/Curing Agent D; Norcast 3258/Norcure 112; Stycast 2850-GT); notched (E-3824, -9405; E-Form 7-61, 7-61M, 7-67, 7-72)

0.4–0.9 ft lb/in. (Stycast 2741)

0.43 ft lb/in. notched (E-8354)

0.44 ft lb/in. (Norcast 3256/Norcure 135)

0.45 ft lb/in. (Norcast 1750/Norcure 135-M; Norcast 3217M/Norcure 135; Norcast

3257/Norcure 112; Norcast 3257/Norcure 135); notched (E-Form 7-61MM)
0.50 ft lb/in. (Norcast 1750/Norcure 112; Norcast 3216; Norcast 3217M/Norcure 111 & 112; Norcast 3217M/Norcure 122; Stycast 1269-A); notched (D.E.R. 331; E-231, -3938)
0.55 ft lb/in. (Norcast 0364/Norcure 142)
0.6 ft lb/in. (Stycast 1210); notched (Norcast 4200-1, 4200-2, 4200-5)
0.70 ft lb/in. notched (E-2748)
0.75 ft lb/in. (Norcast 1750/Norcure 170)
0.80 ft lb/in. notched (Tra-Bond 2101)
1.0 ft lb/in. (Norcast 1750/Norcure 138)
2.0 ft lb/in. (Stycast 1266)
2.1 ft lb/in. (Stycast 1264)
3.9 ft lb/in. notched (Norcast 3705)
4 ft lb/in. (Norcast 3222, 3222LV)
7 ft lb/in. notched (Spauldite G-10, G-10-773, G-10CR, G-11, G-11-963, G-11CR, T-525)
8 ft lb/in. (Norbond 406)
14 ft lb/in. notched (E-264H)
15 ft lb/in. notched (Gillfab 1045)
30 ft lb/in. notched (E-260, -260H)
53 ft lb/in. notched (E-261H)
Hardness:
Ball Indentation 1245 kp/cm² (Lekutherm X201)
Ball Indentation 1590 kp/cm² (Lekutherm X257)
Ball Indentation 1610 kp/cm² (Lekutherm X18)
Ball Indentation 1630 kp/cm² (Lekutherm X30S)
Ball Indentation 1810 kp/cm² (Lekutherm X20)
Ball Indentation 2120 kp/cm² (Lekutherm X256)
Ball Indentation 2360 kp/cm² (Lekutherm X227)
Barcol 25 min. (Gillfab 1109)
Barcol 60 (Norcast 3253/Norcure 135M; Norcast 3255)
Barcol 60 min. (Gillfab 1045, 1053)
Rockwell M105 (E-260H)
Rockwell M106 (E-Form 7-72)
Rockwell M110 (D.E.R. 331; E-2068, -8405; E-Form 7-67)
Rockwell M111 (Spauldite FR-4, G-10, G-10-773, G-10-900, G-10CR, T-525)
Rockwell M112 (E-Form 7-61; Spauldite G-11, G-11-963, G-11CR)
Rockwell M115 (E-2748, -9405, -9451; E-Form 7-61M, 7-61MM)
Shore A30 (Norcast 3425-M, 7146)
Shore A32–36 (Norcast 3510)
Shore A35 (Norcast 3424-M, 7146-F)
Shore A45–48 (Norcast 3511)
Shore A50 (Norcast 3420LC)

Epoxy resin (cont'd.)

Shore A55–58 (Norcast 3515)
Shore A59 (Norcast 7086)
Shore A85 (EP 2820TS)
Shore D26 (EP 2825TS)
Shore D38 (Norcast 3705)
Shore D38–85 (Epolite 3300)
Shore D40 (Norcast 3222, 3222LV)
Shore D50 (FFA-2; P-51)
Shore D53 (EP 2408TS)
Shore D55 (Norcast 9101, 9102)
Shore D60 (Hysol C18F, C60; Interset Epoxy Gel; Norcast 9100)
Shore D62 (Norcast 9311)
Shore D65 (FFA-5, -9; Hysol ES4212, ES4312, ES4412, R9-2039/HD3719)
Shore D67 (EP 2840TS)
Shore D70 (Hysol C9-4183/HD3719, C61; Norcast 176X, 4200-5; Tra-Bond 2135D)
Shore D75 (Norcast 424, 4200-1, 4200-2; Norcast 5615-10/Norcure 178-M; P-74;
    Stycast 1266; Tra-Bond 2125)
Shore D78 (Stycast 1090-SI, 1264)
Shore D80 (Hysol C8-4143/HD3404, R9-2039/HD3561; Interset Concrete Bonder,
    Injection Resin (F); Norcast 154/Norcure 137; Norcast 0364/Norcure 142; Norcast
    1750/Norcure 138; Norcast 5005/Norcure 150; Norcast 5503/Norcure 112; Norcast
    9383; P-75; Stycast 1210; Tra-Duct 2902)
Shore D82 (EP 2810TS; Stycast 1090, 1263)
Shore D83 (Hysol R9-2039/HD3615)
Shore D84 (FA-13; Hysol R9-2039/HD3469; Tra-Bond 2106T)
Shore D85 (CI-2, -3, -6; FA-1, -8, -14, -21; Hysol C9-4183/HD3485, C9-4183/
    HD3561, R8-2038/HD3404, R8-2038/HD3475, R9-2039/HD3404; LCA-27;
    Norcast 154/Norcure 134; Norcast 1750/Norcure 170; Norcast 3253/Norcure 112;
    Norcast 5500/Norcure 112; Norcast T-27; Stycast 1269-A; Tra-Bond 2101; Tra-
    Duct 2924)
Shore D86 (C-84; EP 2420TC; Epolite 1301, 2353, 2354, 3306; Hysol EO1016; Tra-
    Bond 2122)
Shore D87 (Hysol C9-4183/HD3404, C9-4183/HD3537, C9-4183/HD3615; Stycast
    1467, 2651-40; Tra-Bond 2208)
Shore D88 (Allabond Twenty/twenty Adhesive, Twenty/twenty NM; EP 1760LV,
    2230TS, 2305LK; Epolite 1302, 1350, 1353, 1354, 3301, 3302; Stycast 2057, 2651,
    3050, 3180-M; Tra-Bond 2113, 2123, 2129, 2211, 2215)
Shore D89 (Epolite 2300, 2302, 3353; Hysol C9-4183/HD3469; Stycast 2651-MM)
Shore D90 (DC 2900 HW; EP 1765TS, 2306LK, 2415TC; Epolite 2360, 3357; Hysol
    C8-4143/HD3475; LCA-1, -4, -4LV, -9, -20; Norcast 971; Norcast 1750/Norcure
    112; P-86; Polyset 317, 405, 410B, 415-SG; Tra-Bond 2112, 2116; Tra-Bond 2126,
    2151, 2248)
Shore D91 (EP 2800GU; P-85)

Shore D92 (LCA-14, -21; Stycast 2850-KT)
Shore D93 (EP 2740HT)
Shore D94 (Stycast 2850-FT)
Shore D95 (EP 2220TC, 2405FR; Norcast 1750/Norcure 135-M; Stycast 2850-GT)
Shore D96 (EP 2225MV, 2400TC; Stycast 1492, 2762, 2762-FT)
Shore D97 (EP 2404GE)
Shore D98 (EP 2770TC, 2805TC; Polyset 707-2)

THERMAL PROPERTIES:

Conduct.:

0.126 kcal/mhC (Lekutherm X20)
0.134 kcal/mhC (Lekutherm X30S)
0.01 Cal/s/cm²/C/cm (Stycast 2850-KT)
0.003 Cal/cm/sC (Polyset 707-2)
3.0×10^{-4} Cal/s/cm²/C/cm (Barco Bond MB-127)
3.8×10^{-4} Cal/s/cm²/C/cm (EP 2825TS, 2840TS)
4.0×10^{-4} Cal/s/cm²/C/cm (EP 2820TS)
4.1×10^{-4} Cal/s/cm²/C/cm (Stycast 1090-SI)
4.4×10^{-4} Cal/s/cm²/C/cm (Stycast 1090)
4.6×10^{-4} Cal/s/cm²/C/cm (Hysol 1105 Epoxi-Patch Kit)
4.7×10^{-4} Cal/s/cm²/C/cm (Hysol 0151 Epoxi-Patch Kit)
4.8×10^{-4} Cal/s/cm²/C/cm (Hysol 608 Epoxi-Patch Kit)
5.0×10^{-4} Cal/s/cm²/C/cm (Hysol 309 Epoxi-Patch Kit)
5.1×10^{-4} Cal/s/cm²/C/cm (EP 2230TS)
5.2×10^{-4} Cal/s/cm²/C/cm (EP 2306LK)
6.1×10^{-4} Cal/s/cm²/C/cm (Stycast 1263, 1264, 1266)
6.5×10^{-4} Cal/s/cm²/C/cm (Stycast 1269-A)
6.8×10^{-4} Cal/s/cm²/C/cm (EP 22740HT; Hysol 907 Epoxi-Patch Kit)
6.9×10^{-4} Cal/s/cm²/C/cm (EP 2305LK)
7.2×10^{-4} Cal/s/cm²/C/cm (EP 2404GE)
7.3×10^{-4} Cal/s/cm²/C/cm (EP 2770TC; Hysol 9340 Epoxi-Patch Kit)
7.5×10^{-4} Cal/s/cm²/C/cm (Stycast 1467)
7.8×10^{-4} Cal/s/cm²/C/cm (EP 1765TS)
8×10^{-4} Cal/s/cm²/C/cm (E-2748)
8.2×10^{-4} Cal/s/cm²/C/cm (EP 2220TC; Hysol 3X Epoxi-Patch Kit; Stycast 2741)
8.8×10^{-4} Cal/s/cm²/C/cm (EP 2805TC)
8.9×10^{-4} Cal/s/cm²/C/cm (DC 2900 HW)
9.0×10^{-4} Cal/s/cm²/C/cm (E-3824, -9405; Hysol 615 Epoxi-Patch Kit; Norcast 9101, 9102)
9.5×10^{-4} Cal/s/cm²/C/cm (Stycast 1210, 2057, 3050, 3180-M)
10×10^{-4} Cal/s/cm²C/cm (E-260, -260H, -261H, -264H)
10.3×10^{-4} Cal/s/cm²/C/cm (EP 2400TC, 2415TC)
10.4×10^{-4} Cal/s/cm²/C/cm (Hysol 1C Epoxi-Patch Kit)
10.7×10^{-4} Cal/s/cm²/C/cm (Hysol 6C Epoxi-Patch Kit)

12×10^{-4} Cal/s/cm²C/cm (E-231; EP 2405FR)
13×10^{-4} Cal/s/cm²/C/cm (E-3938; Stycast 2651-40)
13.6×10^{-4} Cal/s/cm²/C/cm (EP 2420TC)
14×10^{-4} Cal/s/cm²/C/cm (Stycast 2651-MM)
15×10^{-4} Cal/s/cm²/C/cm (E-8354, -9451; Polyset 317, 410B, 415-SG)
16×10^{-4} Cal/s/cm²/C/cm (E-2068, -8405; Stycast 2651)
17×10^{-4} Cal/s/cm²C/cm (E-8353; Hysol MH19F)
18×10^{-4} Cal/s/cm²/C/cm (E-Form 7-61MM; Hysol MG15F; Stycast 1492)
25×10^{-4} Cal/s/cm²/C/cm (Hysol MG15F-02, MH19F-02)
32×10^{-4} Cal/s/cm²/C/cm (Hysol MG15F-01; Stycast 2762-FT)
33×10^{-4} Cal/s/cm²/C/cm (Stycast 2762)
34×10^{-4} Cal/s/cm²/C/cm (Hysol MH19F-01; Stycast 2850-FT)
36×10^{-4} Cal/s/cm²/C/cm (Stycast 2850-GT)
0.44 Btu/h/ft²/F/in. (Norcast 3253/Norcure 135M; Norcast 3255)
1.3 Btu/h/ft²/F/in. (Norcast 3424-M, 3425-M, 3510; P-74)
1.4 Btu/h/ft²/F/in. (P-75)
1.5 Btu/h/ft²/F/in. (Norcast 3420LC)
2.0 Btu/h/ft²/F/in. (Norcast 9100; P-11, -19)
2.8 Btu/h/ft²/F/in. (Norcast 3343)
4.0 Btu/h/ft²/F/in. (Norcast 3511; P-14, -20, -24, -38, -70, -81)
5.3 Btu/h/ft²/F/in. (P-80C, -82C)
6.5 Btu/h/ft²/F/in. (Norcast 3230SP/Norcure 112 & 170; Norcast 3230SP/Norcure 3230B & 135M)
7 Btu/h/ft²/F/in. (P-56A, -76)
7.5 Btu/h/ft²/F/in. (Norcast 3230-A/Norcure 112)
7.6 Btu/h/ft²F/in. (104 F) (LCA-27)
8 Btu/h/ft²/F/in. (P-78)
8.5 Btu/h/ft²/F/in. (Norcast 3515)
9.9 Btu/h/ft²/F/in. (Norcast 3230-A/Norcure 135; Norcast 3230LV/Norcure 112)
10 Btu/h/ft²/F/in. (P-56)
25–30 Btu/h/ft²/F/in. (Norcast 3231/Norcure 112; Norcast 3231/Norcure 135 or 135-M)
100 Btu/h/ft²/F/in. (Norcast 4912, 4917HSS, 4920)

Distort. Temp.:
37 C (Lekutherm X201)
44 C (EP 2420TC)
50 C (Epolite 1302; Norcast 154/Norcure 137)
51 C (EP 2230TS)
52.8 C (Epolite 3300)
56.1 C (Epolite 2302)
57.2 C (Epolite 2300)
60 C (EP 2415TC)
64 C (EP 1765TS)

67 C (EP 2305LK)
70 C (Lekutherm X256)
77 C (EP 2405FR)
78 C (EP 2400TC)
79 C (EP 2800GU)
80 C (Lekutherm X227)
84 C (EP 2404GE)
85 C (EP 2306LK)
86 C (EP 2220TC)
87 C (EP 2805TC)
90 C (Lekutherm X257)
100 C (EP 2740HT; Norcast X-9201)
103 C (EP 2770TC)
110 C (18.6 kg/cm²) (Polyset 707-2)
111 C (D.E.R. 331)
112 C (Lekutherm X18)
115 C (Lekutherm X30S)
116 C (EP 2225MV; Lekutherm X20)
140 C (Norcast 3401M); (18.6 kg/cm²) (Polyset 405)
141 C (Epolite 3357); (264 psi) (E-2068; E-Form 7-72)
145 C (18.6 kg/cm²) (Polyset 410B)
146 C (164 psi) (E-9405; E-Form 7-61)
148 C (Norcast 5005/Norcure 150)
150 C (18.6 kg/cm²) (Polyset 317)
154 C (264 psi) (E-2748; E-Form 7-61M)
154.4 C (Epolite 3354)
157 C (Epolite 1354)
158.9 C (Epolite 2354)
163 C (D.E.R. 383)
166 C (264 psi) (E-9451)
168 C (264 psi) (E-Form 7-51)
170 C (264 psi) (E-Form 7-61MM)
173 C (Epolite 1353)
176.7 C (Epolite 3353)
175 C (Norcast 3259 Resin-D-Type)
177 C (264 psi) (E-Form 7-67)
182 C (Epolite 2353)
196.7 C (Epolite 1350)
198 C (Epolite 2360)
132 F (264 psi) (FA-14)
154 F (264 psi) (C-84)
156 F (264 psi) (P-86)
163 F (264 psi) (Allabond Twenty/twenty Adhesive)

Epoxy resin *(cont'd.)*

170 F (264 psi) (LCA-27)

185 F (Barco Bond MB-127)

220 F (Norcast 3230LV/Norcure 112)

223 F (Norcast 3253/Norcure 135M; Norcast 3255)

225 F (Norcast 3230SP/Norcure 112 & 170)

239 F (264 psi) (P-80C)

240 F (Norcast 3230-A/Norcure 112)

242 F (264 psi) (P-82C)

294 F (264 psi) (P-85)

300 F (Norcast 3230SP/Norcure 3230B & 135M)

315 F (264 psi) (E-8405)

320 F (264 psi) (E-3824)

345 F (Norcast 3230-A/Norcure 135)

348 F (264 psi) (FA-21)

425 F (264 psi) (E-8353, -8354)

450 F (264 psi) (E-3938)

482 F (264 psi) (E-261H, -264H)

500 F (264 psi) (E-260, -260H)

500+ F (264 psi) (E-231)

Coeff. of Linear Exp.:

10–14 × 10^{-5} /C (Norcast 9101, 9102)

20 × 10^{-5} /C (Norcast 9100)

1.0 × 10^{-6} /C (Norcast 3215, 3216; Norcast 3230-A/Norcure 112)

1.5 × 10^{-6} /C (Norcast 1750/Norcure 112; Norcast 1750/Norcure 135-M)

2.0 × 10^{-6} /C (Norcast 1750/Norcure 170)

2.5 × 10^{-6} /C (Norcast 1750/Norcure 138)

3.5 × 10^{-6} /C (Norcast 3217M/Norcure 111 & 112; Norcast 3217M/Norcure 122; Norcast 3256/Norcure 112; Norcast 3258/Curing Agent D; Norcast 3258/Norcure 112)

3.6 × 10^{-6} /C (Norcast 3217M/Norcure 135; Norcast 3256/Norcure 135)

4.0 × 10^{-6} /C (Norcast 3257/Norcure 112; Norcast 3257/Norcure 135)

9 × 10^{-6} /C (Norcast 3253/Norcure 112)

25 × 10^{-6} /C (Norcast 3230LV/Norcure 112)

29.2 × 10^{-6} /C (Norcast 3230-A/Norcure 135)

30 × 10^{-6} /C (Norcast 3510)

1.1–3.5 × 10^{-5} in./in./C (Gillfab 1045)

23 × 10^{-6} cm/cm/C (Polyset 410B, 415-SG)

24 × 10^{-6} cm/cm/C (Polyset 317)

25 × 10^{-6} cm/cm/C (Stycast 2850-FT, 2850-GT)

27 × 10^{-6} cm/cm/C (Stycast 2762)

29 × 10^{-6} cm/cm/C (Stycast 2850-KT)

30 × 10^{-6} cm/cm/C (Polyset 405)

38 × 10^{-6} cm/cm/C (Stycast 1492, 2057)

39×10^{-6} cm/cm/C (Stycast 2762-FT)
40×10^{-6} cm/cm/C (Stycast 1090, 2651, 2651-MM, 3050)
45×10^{-6} cm/cm/C (Stycast 2651-40)
48×10^{-6} cm/cm/C (Stycast 1467)
50×10^{-6} cm/cm/C (Stycast 2741)
54×10^{-6} cm/cm/C (Stycast 1090-SI)
55×10^{-6} cm/cm/C (Tra-Bond 2101)
56×10^{-6} cm/cm/C (Stycast 3180-M)
59×10^{-6} cm/cm/C (Stycast 1263)
75×10^{-6} cm/cm/C (Stycast 1269-A)
80×10^{-6} cm/cm/C (Stycast 1266)
120×10^{-6} cm/cm/C (Stycast 1210)
125×10^{-6} cm/cm/C (Stycast 1264)
1.4×10^{-5} in./in./C (E-260, -260H, -261H, -264H)
1.7×10^{-5} in./in./C (E-3938)
1.8×10^{-5} in./in./C (E-231, -2748)
1.88×10^{-5} in./in./C (Norcast 3370)
2.1×10^{-5} in./in./C (E-8353)
2.15×10^{-5} in./in./C (E-Form 7-61MM)
2.2×10^{-5} in./in./C (E-8354)
2.3×10^{-5} in./in./C (E-9451)
2.35×10^{-5} in./in./C (E-Form 7-61M)
2.4×10^{-5} in./in./C (E-9405)
2.5×10^{-5} in./in./C (Hysol MH19F)
2.7×10^{-5} in./in./C (E-8405; E-Form 7-51, 7-67)
2.8×10^{-5} in./in./C (E-2068; Hysol MG15F-02, MH19F-02)
2.9×10^{-5} in./in./C (E-3824; E-Form 7-61, 7-72)
3.0×10^{-5} in./in./C (Hysol EO1017, MG15F)
3.2×10^{-5} in./in./C (Hysol MG15F-01, MH19F-01)
4.3×10^{-5} in./in./C (EP 2220TC)
4.4×10^{-5} in./in./C (EP 2225MV, 2305LK)
5.0×10^{-5} in./in./C (EP 2770TC)
5.1×10^{-5} in./in./C (EP 2805TC)
6.2×10^{-5} in./in./C (EP 2405FR, 2740HT)
6.5×10^{-5} in./in./C (Hysol MG18)
6.8×10^{-5} in./in./C (EP 2404GE)
7.8×10^{-5} in./in./C (EP 2400TC)
9.7×10^{-5} in./in./C (EP 2800GU)
10.5×10^{-5} in./in./C (EP 1765TS)
10.7×10^{-5} in./in./C (EP 2306LK)
11.5×10^{-5} in./in./C (EP 2415TC)
13.5×10^{-5} in./in./C (EP 2420TC)
14×10^{-5} in./in./C (EP 1760LV)

Epoxy resin *(cont'd.)*

14.7 × 10^{-5} in./in./C (DC 2900 HW)
15.2 × 10^{-5} in./in./C (EP 2230TS)
15.6 × 10^{-5} in./in./C (EP 2810TS)
58 × 10^{-6} C^{-1} (Lekutherm X20, X30S)
11 × 10^{-6} in./in./F (LCA-21)
13 × 10^{-6} in./in./F (LCA-20)
14 × 10^{-6} in./in./F (LCA-9)
15–20 × 10^{-6} in./in./F (Norcast 3231/Norcure 112; Norcast 3231/Norcure 135 or 135-M)
17 × 10^{-6} in./in./F (LCA-1, -4, -4LV)
18 × 10^{-6} in./in./F (LCA-14)
30 × 10^{-6} in./in./F (Norcast 3424-M)
33 × 10^{-6} in./in./F (FA-21)
34 × 10^{-6} in./in./F (C-84; FA-14)
35 × 10^{-6} in./in./F (FA-8, -13)
40 × 10^{-6} in./in./F (FA-1)
60 × 10^{-6} in./in./F (FFA-2)
75 × 10^{-6} in./in./F (FFA-5, -9)
12 × 10^{-6} /F (P-38, -81)
13 × 10^{-6} /F (P-80C, -82C)
14 × 10^{-6} /F (P-24, -70)
15 × 10^{-6} /F (Norcast 3515)
16 × 10^{-6} /F (P-20)
17 × 10^{-6} /F (P-76, -85)
18 × 10^{-6} /F (P-19, -56, -56A, -78)
20 × 10^{-6} /F (Norcast 3511; P-14)
22 × 10^{-6} /F (P-74, -86)
23 × 10^{-6} /F (P-11)
24 × 10^{-6} /F (P-75)
40 × 10^{6} F^{-1} (CI-6)
42 × 10^{6} F^{-1} (CI-3)
49 × 10^{6} F^{-1} (CI-2)

Sp. Heat:
0.3 Cal/g (P-56, -56A, -75, -76, -78, -80C, -82C)

Flamm.:
V-0 (E-Form 7-61, 7-61M, 7-61MM; Polyset 317, 405, 410B, 415-SG, EPC68FR)
Nonburning (P-19, -20, -24)
Self-extinguishing (E-8354; E-Form 7-51, 7-67, 7-72; Norcast 9101, 9102; P-11, -14, -38, -56, -56A, -70, -74, -76, -78, -81)
Self-extinguishing in 2 s (Gillfab 1045)
ATB 35 s (C-84)
ATB 220 s (LCA-27)
AEB 5 mm (C-84)

AEB 20 mm (LCA-27)
ELECTRICAL PROPERTIES:
Dissip. Factor:

0.00 (Norcast 3511)

0.001 (60 Hz) (Norcast 3253/Norcure 135M; Norcast 3255); (100 kHz) (EP 2305LK)

0.002 (60 Hz) (Norcast 424; Norcast 5005/Norcure 150); (100 Hz) (Norcast 7086); (1 kHz) (CI-2, -6; P-11, -19; P-76; Polyset 410B)

0.0025 (1 MHz) (Norcast 2070-PC)

0.003 (1 kHz) (CI-3; FA-14; Hysol 309 Epoxi-Patch Kit; Norcast 9100; P-56, -56A; Polyset 415-SG)

0.0033 (1 kHz) (P-86)

0.0035 (P-85)

0.0036 (60 Hz) (Norcast 3220-D)

0.004 (100 kHz) (EP 2810TS)

< 0.005 (1 MHz) (Stycast 1269-A)

0.005 (60 Hz) (Norcast 3257/Norcure 112; Norcast 3258/Curing Agent D; Norcast 3258/Norcure 112); (100 Hz) (Polyset EPC68FR); (1 kHz) (Norcast 9383; Polyset 707-2)

0.006 (60 Hz) (Norcast 3257/Norcure 135); (1 kHz) (P-78; Polyset 317, 405); (100 kHz) (EP 1760LV)

0.0062 (100 Hz) (Norcast 3370)

0.008 (60 Hz) (Stycast 1264); (1 kHz) (C-84; Polyset EPC50; Stycast 2850-FT, 2850-GT); (100 kHz) (EP 2740HT; Hysol R9-2039/HD3404)

0.0084 (1 kHz) (Hysol EO1016)

0.009 (1 kHz) (LCA-27; Norcast 3217M/Norcure 135; Stycast 1210, 2762, 2762-FT); (100 kHz) (Norcast 3259 Resin-D-Type)

0.010 (1 kHz) (E-Form 7-61, 7-61M, 7-61MM; Hysol 0151 Epoxi-Patch Kit; Norcast 3217M/Norcure 111 & 112; Norcast 3217M/Norcure 122; Norcast 3256/Norcure 112; Norcast 3256/Norcure 135; Stycast 1090, 1090-SI; Stycast 1492, 2057, 2651, 2651-40, 2651-MM, 2850-KT); (100 kHz) (DC 2900 HW; EP 2405FR); (1 MC) (E-2068, -2748, -9451; Norcast 3215, 3216)

0.011 (100 kHz) (Hysol R8-2038/HD3475)

0.012 (100 kHz) (EP 2770TC; Hysol C9-41183/HD3485); (1 MHz) (E-231, -3824, –9405; Gillfab 1045)

0.013 (1 MHz) (E-8353)

0.014 (1 MHz) (E-8354)

0.015 (1 kHz) (E-Form 7-72; Norcast 971; P-74); (100 kHz) (Hysol C9-4183/HD3537)

0.016 (60 Hz) (Norcast 141/Catalyst 0.31); (1 kHz) (P-75, -82C); (1 MHz) (E-8405)

0.017 (1 kHz) (P-80C); (100 kHz) (Hysol C8-4143/HD3475, C9-4183/HD3615); (1 MHz) (E-260, -260H, -261H, -264H, -3938)

0.018 (100 kHz) (EP 2306LK); (1 MHz) (Gillfab 1053)

0.019 (1 kHz) (Hysol 608 Epoxi-Patch Kit; P-20)

0.020 (Norcast 3510, 7146); (60 Hz) (Norcast 3424-M, 3425-M); (1 kHz) (Norcast

Epoxy resin *(cont'd.)*

3343; Stycast 1263, 3050; Tra-Bond 2101); (100 kHz) (EP 2220TC, 2225MV, 2420TC, 2800GU, 2805TC; Hysol C9-4183/HD3469, C9-4183/HD3719); (1 MHz) (Stycast 1266)

0.021 (Norcast 0364/Norcure 142); (100 kHz) (EP 2230TS)

0.024 (1 kHz) (Hysol 3X Epoxi-Patch Kit; Stycast 1467)

0.025 (1 kHz) (P-24); (1 MHz) (Spauldite FR-4, G-10, G-10-773, G-10-900, G-10CR, G-11, G-11-963, G-11CR, PEG FR, T-525)

0.026 (1 kHz) (Hysol 907 Epoxi-Patch Kit); (100 kHz) (EP 1765TS)

0.027 (100 kHz) (EP 2404GE; Hysol C9-4183/HD3561, C61, R8-2038/HD3404)

0.029 (1 kHz) (Hysol 1105 Epoxi-Patch Kit; P-51); (100 kHz) (EP 2840TS; Hysol R9-2039/HD3615)

0.030 (1 kHz) (P-14); (100 kHz) (EP 2400TC; Hysol R9-2039/HD3719)

0.031 (1 kHz) (P-70); (100 kHz) (Hysol C9-4183/HD3404)

0.032 (1 kHz) (P-81); (100 kHz) (Hysol C8-4143/HD3404, C60)

0.034 (1 kHz) (P-38)

0.036 (100 kHz) (EP 2415TC)

0.039 (100 kHz) (Hysol R9-2039/HD3469)

0.040 (Norcast 3705); (1 kHz) (Hysol 9340 Epoxi-Patch Kit; Stycast 2741); (100 kHz) (Hysol C18F)

0.048 (100 kHz) (Hysol R9-2039/HD3561)

0.050 (Norcast 71460F); (100 kHz) (EP 2408TS)

0.054 (100 kHz) (EP 2825TS)

0.06 (1 kHz) (Stycast 3180-M)

0.063 (100 kHz) (EP 2820TS; Hysol Es4212, ES4312, ES4412)

0.075 (1 kHz) (Hysol 615 Epoxi-Patch Kit)

0.08 (Norcast 3515); (1 kHz) (Norcast 9311)

0.089 (1 kHz) (Hysol 1C Epoxi-Patch Kit)

0.139 (1 kHz) (Hysol 6C Epoxi-Patch Kit)

0.14 (60 Hz) (D.E.R. 331)

Dielec. Str.:

≈ 300 kV/cm (50 Hz) (Lekutherm X30S)

385 kV/cm (50 Hz) (Lekutherm X20)

300 V/mil (P-74)

> 300 V/mil (P-75, -80C, -82C)

310 V/mil (Norcast 7123)

325 V/mil (Norcast 9383)

330 V/mil (E-8405)

340 V/mil (Norcast 9390)

350 V/mil (E-3824; P-85; Polyset 410B, 415-SG, 707-2)

360 V/mil (Norcast 7146)

370 V/mil (Stycast 1467; Tra-Bond 2101)

375 V/mil (E-260H, -261H, -264H; Norcast 3425-M, 7146-F, 9311; Stycast 1090, 1090-SI)

380 V/mil (E-9451; Norcast 7124; Stycast 2850-GT)

400 V/mil (E-231, -260, -2748, -3938, -9405; Norcast 3343, 3425HF, 3510; Norcast 5005/Norcure 150; Norcast 9100; Stycast 1264, 1266, 1492, 2741, 3050)

410 V/mil (E-8353; E-Form 7-51, 7-61, 7-61M, 7-61MM, 7-67, 7-72; Norcast 23, 3511; Stycast 2762, 2850-KT)

415 V/mil (Norcast 3222, 3222LV, 3515, 3705)

420 V/mil (E-2068; Norcast 3231/Norcure 112; Norcast 3231/Norcure 135 or 135-M; Norcast 3253/Norcure 135M; Norcast 3255)

425 V/mil (E-8354; Norcast 3705-F, 3706-F semirigid formulations; Stycast 2057)

430 V/mil (Norcast 3401M; Stycast 1263, 1269-A, 3180-M)

440 V/mil (Norcast 971; Norcast 3230-A/Norcure 112; Norcast 3230LV/Norcure 112; Norcast 3230SP/Norcure 112 & 170)

450 V/mil (Norcast 1750/Norcure 138; Norcast 1750/Norcure 170; Norcast 3230-A/ Norcure 135; Norcast 3230SP/Norcure 3230B & 135M; Stycast 2651, 2651-40, 2651-MM)

455 V/mil (Norcast 3217M/Norcure 111 & 112; Norcast 3217M/Norcure 122)

460 V/mil (Norcast 141/Catalyst 0.31; Norcast 3215, 3216; Norcast 3217M/Norcure 135; Norcast 3256/Norcure 135; Stycast 1210, 2762-FT)

470 V/mil (Norcast 3253/Norcure 112; Norcast 3258/Curing Agent D; Norcast 3258/ Norcure 112)

475 V/mil (Norcast 1750/Norcure 112; Norcast 1750/Norcure 135-M; Norcast 3256/ Norcure 112; Norcast 3257/Norcure 112; Norcast 3257/Norcure 135; Norcast 9101, 9102)

480 V/mil (Polyset 317)

485 V/mil (Norcast 3259 Resin-D-Type)

500 V/mil (Polyset 405)

550 V/mil (Spauldite FR-4, G-10, G-10-773, G-10-900, G-10CR, G-11, G-11-963, G-11CR, T-525; Stycast 2850-FT)

980 V/mil (Polyset EPC68FR)

1000 V/mil (Polyset EPC50)

1500 V/mil (Barco Bond MB-127)

2000 V/mil (Norcast 2070-PC)

71,000 V/mil (Norcast 176X)

Dielec. Const.:

2.0 (Norcast 7123)

2.51 (1 kHz) (P-74)

2.8 (100 kHz) (EP 1760LV)

2.9 (100 kHz) (EP 2810TS)

3.0 (Norcast 3222, 3222LV)

3.1 (1 kHz) (Stycast 1090-SI)

3.2 (Norcast 3705); (60 Hz) (Norcast 3424-M); (100 kHz) (Hysol R9-2039/HD3719); (1 MHz) (Norcast 2070-PC)

3.3 (Norcast 9383); (60 Hz) (Norcast 176X); (1 kHz) (Stycast 1090); (100 kHz) (EP

Epoxy resin (cont'd.)

2840TS)

3.32 (1 kHz) (Hysol 0151 Epoxi-Patch Kit)

3.35 (1 kHz) (Hysol 309 Epoxi-Patch Kit)

3.4 (100 Hz) (Polyset EPC68FR); (100 kHz) (DC 2900 HW)

3.45 (1 kHz) (Hysol 907 Epoxi-Patch Kit)

3.46 (1 kHz) (Hysol 3X Epoxi-Patch Kit)

3.5 (Norcast 3510); (60 Hz) (Norcast 141/Catalyst 0.31); (1 kHz) (CI-2); (100 kHz) (EP 2800GU; Hysol C9-4183/HD3537, C9-4183/HD3719); (1 MHz) (Stycast 1266)

3.57 (Norcast 0364/Norcure 142)

3.6 (Norcast 4200-5); (60 Hz) (Norcast 3425-M); (1 kHz) (FA-14; Polyset 707-2); (100 kHz) (EP 2230TS); (1 MHz) (E-9405)

3.7 (Norcast 7146); (60 Hz) (Stycast 1264); (1 kHz) (CI-6); (100 kHz) (EP 2770TC)

3.75 (100 Hz) (Norcast 3401M)

3.8 (Norcast 7124); (1 kHz) (CI-3; Hysol C60, C61; Polyset 317); (100 kHz) (EP 2306LK, 2740HT, 2805TC; Hysol C18F, R8-2038/HD3475); (1 MHz) (E-9451; Stycast 1269-A)

3.9 (100 Hz) (Norcast 7086); (1 kHz) (Polyset 405, 410B, 415-SG; Stycast 2762)

3.91 (1 kHz) (P-86)

4.0 (Norcast 3425HF, 3511); (1 kHz) (Norcast 3343; Polyset EPC50); (100 kHz) (EP 2305LK, 2408TS; Hysol R8-2038/HD3404, R9-2039/HD3404, R9-2039/HD3469, R9-2039/HD3615); (1 MHz) (E-2068)

4.03 (60 Hz) (D.E.R. 331)

4.1 (1 kHz) (Hysol 608 Epoxi-Patch Kit; Stycast 2057, 2741); (100 kHz) (Hysol C9-4183/HD3615)

4.13 (1 kHz) (Hysol 9340 Epoxi-Patch Kit)

4.18 (100 Hz) (Norcast 3370)

4.2 (Norcast 3515); (60 Hz) (Norcast 9101, 9102); (1 kHz) (Norcast 3253/Norcure 112; Stycast 1263, 3050, 3180-M); (100 kHz) (EP 2220TC, 2400TC, 2404GE, 2405FR, 2825TS; Hysol C8-4143/HD3404, C8-4143/HD3475, C9-4183/HD3404, C9-4183/HD3469, C9-4183/HD3485, C9-4183/HD3561)

4.25 (1 MHz) (Norcast 3215, 3216)

4.3 (Norcast 3258/Curing Agent D); (60 Hz) (Norcast 3258/Norcure 112; Norcast 5005/Norcure 150); (1 kHz) (C-84; Hysol EO1016; Norcast 971; Norcast 3217M/Norcure 111 & 112; Norcast 3217M/Norcure 122; Tra-Bond 2101); (100 kHz) (EP 1765TS, 2225MV); (10^{10} Hz) (Norcast 3256/Norcure 112)

4.35 (1 kHz) (P-20)

4.4 (Norcast 7146-F); (1 kHz) (Norcast 3217M/Norcure 135; Stycast 1492, 2651-MM); (100 kHz) (EP 2820TS; Hysol ES4212, ES4312, ES4412); (10^{10} Hz) (Norcast 3256/Norcure 135)

4.44 (1 kHz) (P-24)

4.5 (60 Hz) (Norcast 1750/Norcure 138; Norcast 1750/Norcure 170; Norcast 3230-A/Norcure 112; Norcast 3230LV/Norcure 112; Norcast 3230SP/Norcure 112 & 170; Norcast 3230SP/Norcure 3230B & 135M; Norcast 3253/Norcure 135M; Norcast

3255); (1 kHz) (Norcast 9100; Stycast 2651-40); (1 MHz) (E-231, -2748)

4.54 (1 kHz) (P-85)

4.6 (60 Hz) (Norcast 3230-A/Norcure 135); (1 kHz) (Hysol 1105 Epoxi-Patch Kit; Stycast 2651); (100 kHz) (Hysol R9-2039/HD3561; Norcast 3259 Resin-D-Type); (1 MHz) (E-8405)

4.7 (1 MHz) (E-3824)

4.74 (1 kHz) (P-51)

4.75 (60 Hz) (Norcast 1750/Norcure 112; Norcast 1750/Norcure 135-M); (1 MHz) (E-8354)

4.78 (1 kHz) (P-19)

4.79 (1 kHz) (P-14)

4.8 (1 kHz) (Stycast 1467)

4.87 (1 kHz) (P-75, -82C)

4.9 (1 MHz) (E-8353)

5.0 (1 kHz) (E-Form 7-51, 7-61, 7-61M, 7-61MM, 7-67, 7-72; Norcast 9311; P-70); (1 MHz) (E-3938)

5.07 (1 kHz) (P-80C)

5.1 (1 MHz) (Gillfab 1045, 1053)

5.2 (1 kHz) (Hysol 615 Epoxi-Patch Kit; Stycast 1210); (100 kHz) (EP 2415TC); (1 MHz) (Spauldite FR-4, G-10, G-10-773, G-10-900, G-10CR, G-11, G-11-963, G-11CR, PEG FR, T-525)

5.3 (1 kHz) (P-81)

5.31 (1 kHz) (P-38)

5.37 (1 kHz) (P-11)

5.4 (100 kHz) (EP 2420TC)

5.68 (1 kHz) (Allabond Twenty/twenty Adhesive)

5.75 (1 kHz) (LCA-27)

5.76 (1 kHz) (P-56A)

5.8 (1 kHz) (P-56); (1 MHz) (E-260, -260H, -261H, -264H)

5.9 (1 kHz) (P-76)

6.11 (1 kHz) (Hysol 1C Epoxi-Patch Kit)

6.22 (1 kHz) (P-78)

6.3 (60 Hz) (Stycast 2762-FT); (1 kHz) (Stycast 2850-FT, 2850-GT)

7.0 (1 kHz) (Stycast 2850-KT)

8.23 (1 kHz) (Hysol 6C Epoxi-Patch Kit)

Vol. Resist.:

2×10^{-4} ohm-cm (Norcast 4917HSS, 4920)

10^8 megohm-cm (Gillfab 1045, 1053)

10^{11} ohm-cm (EP 2825TS)

10^{12} ohm-cm (EP 2415TC; Stycast 1090, 3180-M)

5×10^{12} ohm-cm (Hysol C9-4183/HD3719, R9-2039/HD3719)

6×10^{12} ohm-cm (EP 2820TS)

10^{13} ohm-cm (Norcast 7123, 7124, 7146, 7146-F; Stycast 1090-SI, 1263, 2741)

Epoxy resin *(cont'd.)*

2×10^{13} ohm-cm (Hysol ES4212, ES4312, ES4412)

3×10^{13} ohm-cm (Hysol R9-2039/HD3469)

3.44×10^{13} ohm-cm (Hysol 615 Epoxi-Patch Kit)

4×10^{13} ohm-cm (EP 2220TC; Hysol C9-4183/HD3469)

7×10^{13} ohm-cm (Hysol C9-4183/HD3485)

8×10^{13} ohm-cm (Hysol R9-2039/HD3404)

8.34×10^{13} ohm-cm (Hysol 6C Epoxi-Patch Kit)

1×10^{14} ohm-cm (E-231, -8405; Norcast 23, 3222, 3222LV, 3424-M, 3510, 3705, 3705-F (semirigid formulation), 3706-F (semirigid formulation); Norcast 7086, 9101, 9102, 9311; Stycast 1210, 1266, 1467, 2651, 2651-MM, 3050)

$> 10^{14}$ ohm-cm (FA-14; P-51)

2×10^{14} ohm-cm (EP 2420TC; P-14, -38, -81)

2.48×10^{14} ohm-cm (Hysol 608 Epoxi-Patch Kit)

3×10^{14} ohm-cm (Hysol C9-4183/HD3404; Norcast 5615-10/Norcure 178-M)

3.16×10^{14} ohm-cm (Hysol 907 Epoxi-Patch Kit)

3.29×10^{14} ohm-cm (Hysol 9340 Epoxi-Patch Kit)

3.38×10^{14} ohm-cm (Hysol 3X Epoxi-Patch Kit)

3.5×10^{14} ohm-cm (Hysol C61)

3.52×10^{14} ohm-cm (Hysol 1C Epoxi-Patch Kit)

3.77×10^{14} ohm-cm (Hysol 0151 Epoxi-Patch Kit)

4×10^{14} ohm-cm (EP 2400TC; Hysol C8-4143/HD3404, C9-4183/HD3561, C9-4183/ HD3615, R9-2039/HD3561, R9-2039/HD3615; P-24)

5×10^{14} ohm-cm (Norcast 3511; P-74, -80C)

6×10^{14} ohm-cm (EP 2404GE, 2805TC; Norcast 3515)

7×10^{14} ohm-cm (Hysol C60)

8×10^{14} ohm-cm (E-2068; EP 2840TS)

8.5×10^{14} ohm-cm (Hysol 1105 Epoxi-Patch Kit)

10^{15} ohm-cm (DC 2900 HW; E-Form 7-67; EP 1765TS, 2770TC; Hysol C18F; Norcast 971; Norcast 3231/Norcure 112; Norcast 3231/Norcure 135 or 135-M; Norcast 3259 Resin-D-Type; Norcast 3425HF; P-75, -82C; Stycast 1264, 2057, 2850-KT)

1.5×10^{15} ohm-cm (Polyset 317, 405, 410B, 415-SG)

2×10^{15} ohm-cm (EP 2405FR; P-20)

2.3×10^{15} ohm-cm (E-3938)

3×10^{15} ohm-cm (Allabond Twenty/twenty Adhesive; EP 2225MV, 2305LK, 2800GU; Hysol C9-4183/HD3537; Norcast 0364/Norcure 142)

4×10^{15} ohm-cm (E-3824, -9405; EP 2230TS, 2408TS; Norcast 9383)

5×10^{15} ohm-cm (EP 2306LK; LCA-27; Norcast 3230SP/Norcure 112 & 170; Norcast 3230SP/Norcure 3230B & 135M; Norcast 9100)

5.3×10^{15} ohm-cm (Norcast 3370)

6×10^{15} ohm-cm (E-9451; EP 1760LV; Hysol EO1016; Norcast 3217M/Norcure 111 & 112; Norcast 3217M/Norcure 122; Polyset 707-2; Tra-Bond 2101)

7×10^{15} ohm-cm (EP 2740HT)

9×10^{15} ohm-cm (E-2748; E-Form 7-51, 7-61, 7-61M, 7-61MM, 7-72; EP 2810TS)

10^{16} ohm-cm (C-84; CI-6; E-260, -260H; Hysol R8-2038/HD3475; Norcast 141/ Catalyst 0.31; Stycast 1492, 2762, 2762-FT, 2850-FT, 2850-GT)

> 1.2×10^{16} ohm-cm (D.E.R. 331)

1.22×10^{16} ohm-cm (Norcast 3220-D)

2×10^{16} ohm-cm (Hysol R8-2038/HD3404; P-86)

2.9×10^{16} ohm-cm (Hysol 309 Epoxi-Patch Kit)

3×10^{16} ohm-cm (Hysol C8-4143/HD3475; Norcast 1750/Norcure 138; P-19)

4×10^{16} ohm-cm (Norcast 1750/Norcure 170; P-85)

4.3×10^{16} ohm-cm (Norcast 3215, 3216)

5×10^{16} ohm-cm (CI-2; Norcast 1750/Norcure 112)

6×10^{16} ohm-cm (Norcast 1750/Norcure 135-M; Norcast 3253/Norcure 112; Norcast 3256/Norcure 112; Norcast 3257/Norcure 112; Norcast 3258/Curing Agent D; Norcast 3258/Norcure 112; P-11)

7×10^{16} ohm-cm (Norcast 3217M/Norcure 135; Norcast 3230-A/Norcure 135; Norcast 3230LV/Norcure 112; Norcast 3256/Norcure 135; Norcast 3257/Norcure 135)

10^{17} ohm-cm (CI-3)

Surf. Resist.:

10^{15} ohm (Polyset 317, 405, 415-SG, 707-2)

15×10^{15} ohm (Polyset 410B)

5.5×10^{15} ohm (Norcast 3220-D)

> 7.85×10^{15} ohm (D.E.R. 331)

1.25×10^{5} megohms (Norcast 2070-PC)

Arc Resist.:

120 s (E-260H, -261H, -264H)

140 s (E-3824, -9405)

180 s (E-231, -260, -2068, -2748, -3938, -8353, -8354, -8405, -9451; Polyset 317, 405)

190 s (Polyset 410B, 415-SG, 707-2)

No failure (Norcast X-9201)

CURING CHARACTERISTICS:

Suggested Curing Agent/Activator/Hardener:

Activator BA-1 (CI-2; P-11, -14, -38, -70, -81)

Activator BA-2 (P-19, -24)

Activator BA-2XM (FA-21)

Activator BA-3 (P-20)

Activator BA-4 (CI-3; FA-1; LCA-1)

Activator BA-5 (FA-8; LCA-4, -4LV, -9)

Activator BA-6A (FFA-2)

Activator BA-11 (FFA-9)

Activator BA-15 (FFA-5)

Activator BA-16 (LCA-14)

Activator BA-21 (P-51)

Activator BA-22 (P-56)

Activator BA-22A (P-56A)

Epoxy resin *(cont'd.)*

Activator BA-39 (FA-13)
Activator BA-40 (LCA-20)
Activator BA-41 (LCA-21)
Activator BA-42 (P-80C)
Activator BA-45 (CI-6; FA-14; P-82C)
Activator BA-47 (P-78)
Activator BA-49 (LCA-27)
Activator BA-55 (P-76)
Activator BA-57 (P-74, -75)
Activator BA-60 (P-85)
Activator BA-62 (P-86)
Activator BA-63 (C-84)
Activator BA-66B (Allabond Twenty/twenty Adhesive, Twenty/twenty NM)
Anhydride-cured (Hysol C9-4183/HD3537, MG15F, MG15F-01, MG15F-02)
D.E.H. 24 (D.E.R. 331, 732, 736)
Lekutherm Hardener H, Desmorapid DB accelerator (Lekutherm X18, X20)
MDA (D.E.R. 383)
Norcure 135 (Norcast 3370)
PSA hardener (Lekutherm X30S)

Mix Ratio:

1A/1B (Norcast 3220-D, 3343, 3420LC, 3705)
1A/1B by wt. or vol. (Norbond 2162 Clear Amber; Norcast 9390)
1A/1B by vol. (Interset Deep Penetrating Sealer resin/curing agent; Interset Epoxy
   Coal Tar resin/hardener; Interset Epoxy Gel resin/hardener; Norcast 176X, 424)
1A/2B (Norcast 9311)
1B/8A (Norcast 3255)
2A/1B by vol. (Interset Concrete Bonder resin/curing agent)
2A/3B (Hysol C60, C61; Norcast 9383)
3A/1B (Norbond 406)
4A/1B by vol. (Interset Concrete Topping Resin #2 resin/hardener; Interset Injection
   Resin (F) resin/curing agent)
7A/5B (Norbond 2162)
100/1.07 (LCA-1/Activator BA-4)
100/2.12 (P-38/Activator BA-1; P-81/Activator BA-1)
100/2.5 (P-70/Activator BA-1)
100/2.75 (Norcast 3231/Norcure 112)
100A/3B (Norcast 4912, 4917 HSS, 4920)
100/3 (Norcast 3230-A/Norcure 112; Norcast 3230LV/Norcure 112; Norcast 3231/
   Norcure 135 or 135-M)
100/3.12 (P-14/Activator BA-1)
100/3.2 (FA-1/Activator BA-4; P-11/Activator BA-1)
100/4 (LCA-27/Activator BA-49)
100/4–5 (Norcast 3230SP/Norcure 112)

100/4.50 (LCA-4/Activator BA-5; LCA-4LV/Activator BA-5; LCA-9/Activator BA-5)

100/5 (CI-3/Activator BA-4; Norcast 3230-A/Norcure 135; Norcast 3370/Norcure 135)

100A/5B (EP 2805TC)

100/5–6 (Norcast 1750/Norcure 112)

100/5–7 (Norcast 3258/Curing Agent D)

100/5.5 (P-76/Activator BA-55)

100A/5.5B (EP 2220TC; Hysol C8-4143/HD3404, C9-4183/HD3404)

100/6 (P-80C/Activator BA-42; P-82C/Activator BA-45)

100A/6B (EP 2225MV, 2404GE; Norcast 3515)

100/6–7 (Norcast 3217M/Norcure 111 & 112; Norcast 3230SP/Norcure 3230B & 135M; Norcast 3253/Norcure 112; Norcast 3256/Norcure 112; Norcast 3258/Norcure 112)

100/6.3 (P-78/Activator BA-47)

100/7 (Epolite 1302 resin/hardener; Norcast 5500/Norcure 112)

100A/7B (EP 2420TC; Hysol C9-4183/HD3485)

100A/7–8B (Norcast 3215 resin/curing agent)

100/7.3 (C-84/Activator BA-63)

100/7.5 (Norcast 141/Catalyst 0.31)

100A/7.5B (Hysol C9-4183/HD3469; Norcast 3511)

100/8 (Norcast 3256/Norcure 135; Norcast T-27/Norcure 122; Norcast T-27/Norcure 135; P-85/Activator BA-60)

100/8–9 (Norcast 1750/Norcure 135-M; Norcast 3217M/Norcure 135; Norcast 3253/Norcure 135M; Norcast 3257/Norcure 112)

100A/8–9B (Norcast 7123)

100A/8.5B (EP 2405FR)

100A/9B (EP 2305LK, 2400TC; Norcast 4915)

100/10 (CI-2/Activator BA-1; Epolite 1301 resin/hardener)

100A/10B (EP 2800GU; Hysol C9-4183/HD3615; Norcast 2795/Norcure 3416; Norcast T-30)

100/11 (Norcast 3257/Norcure 135)

100A/11B (EP 2306LK; Hysol R8-2038/HD3404, R9-2039/HD3404)

100/11.2 (LCA-14/Activator BA-16)

100/12–13 (Norcast 3217M/Norcure 122)

100/12–14 (Norcast 3230SP/Norcure 170; Norcast 7124/Norcure 135)

100/12–15 (Norcast 1750/Norcure 170)

100A/12.5B (Hysol C8-4143/HD3475)

100/13 (Norcast 5503/Norcure 112)

100/13.5 (FA-8/Activator BA-5)

100A/14.5B (Norcast 3510)

100/15 (Norcast 5005/Norcure 150)

100/15–20 (Norcast 1750/Norcure 138)

Epoxy resin *(cont'd.)*

100/15.5 (P-75/Activator BA-57)
100A/15B (EP 2415TC; Hysol C9-4183/HD3561, R9-2039/HD3469)
100/16 (Epolite 1350 resin/hardener)
100/16.6 (P-74/Activator BA-57)
100/17 (Epolite 1354 resin/hardener)
100/18 (Epolite 1353 resin/hardener)
100/19.5 (P-51/Activator BA-21)
100/20 (Norcast 154/Norcure 134)
100A/20B (Hysol R9-2039/HD3615; Norcast 4200-2, 7146-F; Norcast X-9201)
100/22–28 (Norcast 141/Norcure 122)
100/23 (CI-6/Activator BA-45)
100A/23B (Hysol 1105 Epoxi-Patch Kit)
100/24.5 (FA-14/Activator BA-45)
100/25 (Norcast 7124/Norcure 159)
100A/25B (Hysol R8-2038/HD3475; Norcast 3259 Resin-D-Type, 7086, 9101; Tra-Bond 2101 resin/hardener)
100A/25–100B (Norcast 3222, 3222LV)
100/26.3 (P-24/Activator BA-2)
100/26.5 (FA-13/Activator BA-39)
100/27 (D.E.R. 383/MDA; P-86/Activator BA-62)
100/27.4 (LCA-20/Activator BA-40)
100A/30B (Hysol R9-2039/HD3561)
100/30–50 (Norcast 5615-10/Norcure 178-M)
100/32 (P-20/Activator BA-3)
100A/32–34B (Norcast 3425HF)
100A/33B (Hysol 0151 Epoxi-Patch Kit)
100/35 (LCA-21/Activator BA-41)
100/35.2 (P-19/Activator BA-2)
100/38 (Norcast 7633/Norcure 133)
100A/40B (Hysol 309 Epoxi-Patch Kit; Norcast 4200-5; Norcast 5509-M)
100/40–45 (Norcast 154/Norcure 137)
100A/40–150B (Norcast 3222, 3705-F, 3706-F)
100A/43B (Hysol 1C, 6C Epoxi-Patch Kit; Hysol C9-4183/HD3537)
100A/45B (DC 2900 HW; Norcast 4200-1, 9102)
100/50 (Norcast 0364/Norcure 142)
100/50 by vol. (Norcast 3401M/Norcure 300)
100A/50B (EP 2230TS; Hysol C9-4183/HD3719; Norcast 3216, 7146)
100A/60B (EP 2740HT; Norcast 3424-M)
100A/65B (Norcast 9100)
100/70 (FFA-2/Activator BA-6A)
100/71 (P-56/Activator BA-22; P-56A/Activator BA-22A)
100A/75B (Norcast 3425-M)
100A/80B (Hysol 907 Epoxi-Patch Kit; Norcast 2070-PC)

100A/83B (EP 2408TS)

100A/86B (EP 2820TS)

100A/100B (EP 2770TC, 2825TS; Hysol 3X, 608, 615, 9340 Epoxi-Patch Kit; Hysol C9-4183/HD3719, ES4212, ES4312, ES4412)

100/100 (FA-21/Activator BA-2XM; FFA-9/Activator BA-11)

100A/125B (Hysol C18F)

100A/135B (EP 2810TS)

100/150 (FFA-5/Activator BA-15)

100A/200B (EP 2840TS)

Typical/Suggested Cure:

20 s cure @ 250 F (Allabond Twenty/twenty Adhesive)

75–90 s @ 180 C; post cure 4–16 h @ 175 C (Polyset 415-SG)

0.9–1.25 min cure time with post cure of 16 h @ 177 C (Hysol MG15F, MG15F-01, MG15F-02)

1.0–1.5 min cure time @ 177 C with a post cure of 6 h @ 177 C (Hysol MH19F, MH19F-01, MH19F-02)

1–2 min @ 177 C; post cure 4–16 h @ 175 C (Polyset 317, 405, 410B)

1–3 min @ 150 C; hot plate cure 40 s @ 150 C (Polyset 707-2)

2 min @ 300 F (E-Form 7-51, 7-61, 7-61M, 7-61MM, 7-67, 7-72)

4 min @ 150 C plus a post cure of 4 h @ 150 C (Hysol MG18)

10 min @ 100 C (Norcast 3401M)

10 min set, 24 h cure @ 25 C (Hysol 615 Epoxi-Patch Kit)

30 min @ 200 F (LCA-27)

1 h @ 121 C (EP 2825TS)

1 h @ 75 F (FFA-9)

1 h @ 160 F (FFA-5)

1–2 h @ 100 C or 8 h @ 85 C (Norcast 3259 Resin-D-Type)

1–2 h @ 120 C plus 4 h @ 160 C for optimum properties (Norcast X-9201)

2 h @ 60 C or 24–36 h @ 25 C (Hysol C9-4183/HD3719, R9-2039/HD3719)

2 h @ 75 C (Norcast 2070-PC)

2 h @ 75 C or 4 h @ 65 C (Hysol PC12-007)

2 h @ 80 C plus 2 h @ 150 C (Hysol C9-4183/HD3615, R9-2039/HD3615; Norcast 154/Norcure 134)

2 h @ 95 C plus 4 h @ 150 C (Norcast 3370)

2 h @ 100 C (Tra-Bond 2211, 2248; Tra-Duct 2924)

2 h @ 100 C, or 4 h @ 60 C (Norcast 5005/Norcure 150)

2 h @ 100 C, plus 16 h @ 150–160 C (Hysol EO1017)

2 h @ 121 C (EP 2770TC)

2 h @ 150 C, or 16 h @ 93 C (Hysol EO1016)

2 h @ 150 F plus 2–3 h @ 200 F (Norcast 1750/Norcure 135-M)

2 h @ 160 F plus 2 h @ 300 F for small masses; 12 h @ R.T. plus 12 h @ 140 F for larger masses (Norcast 3230-A/Norcure 135)

2 h @ 160 F plus 4 h @ 200 F (FA-13; LCA-20, -21)

Epoxy resin *(cont'd.)*

2 h @ 175 F–200 F plus 2–4 h @ 250–275 F, or overnight @ 190–200 F (Norcast 971)

2 h @ 200 F (CI-3; FA-1, -8; LCA-1, -4, -4LV, -9)

2 h @ 200 F or 12–16 h @ 160 F (Norcast 3257/Norcure 135)

2 h @ 212 F in mold plus 4 h @ 212 F in oven (P-76)

2 h @ 275 F plus 2 h @ 350 F (Norcast 141/Catalyst 0.31)

2–3 h @ 120 C or 6–7 h @ 100 C or 15–20 h @ 75 C (Norcast 9100)

2–3 h @ 120 C or 6–8 h @ 95 C or $1/_2$ h @ 75 C (Norcast 9311)

2–3 h @ 120 C or 5–6 h @ 100 C or 20–24 h @ 70 C (Norcast 9101, 9102)

2–3 h @ 130 F for masses > 1 lb; 2–3 h @ 150–175 F for masses < 1 lb (Norcast 3425HF)

2–4 h @ R.T.; post cure @ 150 F to hasten attainment of full properties (Norcast 3230-A/Norcure 112; Norcast 3230LV/Norcure 112)

2–4 h gel @ 150 F, 2 h @ 150 F, 2 h @ 225 F, 2 h @ 325 F (Epolite 3357)

2–4 h gel @ 150 F; 2 h @ 150 F; 2 h @ 225 F; 2 h @ 325 F; 2 h @ 400 F (Epolite 2360)

2.25 h @ 90 C (EP 2820TS)

3 h @ 77 F plus 2 h @ 250 F (C-84)

3 h @ 60 C or 36–48 h @ R.T. (Hysol ES4212)

3 h @ 120 C, plus 16 h @ 160 C (Hysol C9-4183/HD3537)

3 h @ 121 C (EP 2840 TS)

3 h @ 160 F in mold plus 5 h @ 160 F in oven (P-56A)

3 h @ 200 F for masses < $1/_2$ lb; 2 h @ 175 F plus 2 h @ 225 F for masses > 1 lb (Norcast 3256/Norcure 135)

3–4 h @ 150 F or 2 h @ 200 F (Norcast 3231/Norcure 135 or 135-M)

3–5 h @ R.T.; post cure 1–2 h @ 150–200 F to hasten attainment of full properties (Norcast 3230SP/Norcure 112 & 170)

3.5 h @ 200 F (FFA-2)

4 h @ R.T. with post cure of 2–3 h @ 125 F for masses < 1 lb; overnight @ R.T. with post cure of 3–4 h @ 125 F for masses > 1 lb (Norcast 3424-M)

4 h @ 80 C plus 15 h @ 120 C (Lekutherm X18, X20)

4 h @ 100 C (EP 2800GU, 2805TC)

4 h @ 125 C (Hysol C18F)

4 h @ 180 F (P-85, -86)

4 h @ 212 F (CI-6)

4 h @ 212 F in mold plus 16 h @ 212 F in oven (P-11, -14, -38)

4 h @ 212 F in mold plus 16 h @ 300 F in oven (P-81)

4 h @ 212 F in mold plus 20 h @ 212 F in oven (P-70)

4–6 h @ R.T.; post cure 1–2 h @ 150 F to hasten attainment of optimum properties (Norcast 3258/Norcure 112)

4–6 h @ R.T.; post cure 1–2 h @ 150–200 F to hasten attainment of optium properties (Norcast 3258/Curing Agent D)

4–6 h @ 150–200 F; post cure 2–4 h @ 200–250 F to assure optimum development of properties (Norcast 3230SP/Norcure 3230B & 135M)

5 h @ 150 C (EP 1760LV, 1765TS)

6 h @ 25 C (Tra-Bond 2106T)

6 h @ 100 C, or 16 h @ 75 C (Hysol C9-4183/HD3485)

8 h @ R.T. or cure @ 150 F for faster molding time (Norcast 3256/Norcure 112)

8 h @ R.T. with full cure in 24 h, or 1–2 h @ 150 F (Norcast 3217M/Norcure 111 & 112; Norcast 3217M/Norcure 122; Norcast 3217M/Norcure 135)

8 h @ R.T. with full strength reached in another 24 h, or 30–40 min @ 150 F, or 15–20 min @ 200 F, or 10–15 min @ 225 F (Norcast 3222, 3222LV)

8 h @ 115 C (Tra-Bond 2208)

8 h @ 200 C (Tra-Bond 2215)

8 h @ 160 F (FA-14)

8 h @ 160 F in mold plus 8 h @ 160 F in oven (P-78)

8 h @ 160 F in mold plus 16 h @ 212 F in oven (P-80C)

8 h @ 160 F in mold plus 16 h @ 250 F in oven (P-74)

8 h @ 212 F (CI-2)

8 h @ 212 F in mold, plus 16 h @ 300 F in oven (P-20, -24)

8 h @ 212 F in mold, plus 40 h @ 300 F in oven (P-19)

8 h @ 212 F plus 2 h @ max. use temp. (LCA-14)

8 h @ 212 F plus 16 h @ 300 F (FA-21)

8–12 h @ R.T. (Norcast 4200-1)

12 h @ R.T. or 15 min @ 225 F (Norcast 3705)

12 h @ R.T. or 1 h @ 200 F (Norcast 3343)

12–16 h @ 150–160 F (Norcast 7124/Norcure 135)

12–18 h @ R.T. or 2–3 h @ 70 C (large masses), or 1–2 h @ 100 C (masses < $^1/_2$ lb) (Norcast 3705-F, 3706-F)

15 h @ 150 C (Lekutherm X30S)

16 h @ 110 C (EP 2740HT)

16 h @ 121 C (DC 2900 HW; EP 2810TS)

16 h @ 160 F in mold (P-56)

16–24 h @ 120 C (Hysol C60, C61)

24 h @ R.T. (EP 2220TC, 2225MV, 2230TS, 2305LK, 2306LK, 2400TC, 2404GE, 2405FR, 2415TC, 2420TC; Hysol 1C, 6C, 309, 608, 907, 9340 Epoxi-Patch Kit; Norcast 3220-D, 3510; Tra-Bond 2101, 2112, 2113, 2116, 2122, 2123, 2125, 2126, 2129, 2135D, 2151; Tra-Duct 2902)

24 h @ R.T.; 3 h @ 150 F (Norcast 3511, 3515)

24 h @ R.T. or 1 h @ 100 C (Norcast 4915)

24 h @ R.T. or 1–2 h @ 50 C to accelerate cure (Norcast 154/Norcure 137)

24 h @ R.T. or 2 h @ 60 C (Hysol C8-4143/HD3404, C8-4143/HD3475, C9-4183/ HD3404, C9-4183/HD3561, ES4312, ES4412, R8-2038/HD3404, R8-2038/ HD3475, R9-2039/HD3404, R9-2039/HD3561)

24 h @ R.T. or 2 h @ 120 F (Norcast 4912)

24 h @ 23 C plus 3 h @ 120 C (Lekutherm X227, X256, X257)

24 h @ 23 C plus 5 h @ 80 C (Lekutherm X201)

24 h @ 135 F in mold plus 24 h @ 160 F plus 24 h @ 200 F post cure (P-75, -82C)

24–36 h @ R.T. (Hysol 3X, 0151 Epoxi-Patch Kit)

Epoxy resin (cont'd.)

24–36 h @ R.T. or 1–2 h @ 100 F (Norcast 7123)

24–38 h @ R.T. or 4–6 h @ 150 F (Norcast 1750/Norcure 112; Norcast 1750/Norcure 138; Norcast 1750/Norcure 170)

36–48 h @ R.T. (Hysol 11015 Epoxi-Patch Kit)

48 h @ R.T. (EP 2408TS)

48 h @ 75 F or 4 h @ 140 F (P-51)

48 h @ R.T. or 3–4 h @ 60–70 C or 1–2 h @ 80–100 C (Norcast 7146, 7146-F)

3 days @ R.T. or 10 min @ 300 F for full cure (Barco Bond MB-127)

3–4 deays @ R.T. or 2–3 h @ 70 C (Norcast 3425-M)

R.T. cure (Epolite 1301, 1302, 1350, 2300, 2302, 3300, 3301, 3302, 3306, 3353, 3354)

R.T. and post cure (Epolite 1354)

Gel @ R.T. plus 2 h @ 150 C (Hysol C9-4183/HD3469, R9-2039/HD3469)

Gel @ R.T. then cure 2–6 h @ 125–250 F for accelerated curing (Norcast 4200-2)

Cure several hours or overnight @ R.T. or 2–3 h @ 125 F (Norcast 4920)

Cure @ 70–300 F; full strength obtained in 15 min @ 300 F or 24 h @ 70 F (Norbond 406)

Cures to safe handling strength in 8–12 h @ R.T. (Norbond 2162, 2162 Clear Amber)

Cure @ 350 F, post cure @ 450 F (Magnamite AS4/1908)

Cure overnight @ R.T. (Norcast 3231/Norcure 112; Norcast T-30)

Cure overnight @ R.T.; mild heat will speed cure (Norcast 3253/Norcure 112)

Cure overnight @ R.T. or allow to gel 1 h @ 100 C (Norcast 3253/Norcure 135M; Norcast 3255)

Cure overnight @ R.T. or 1 h @ 160 F (Norcast 3257/Norcure 112)

Cure overnight @ R.T. or 1–2 h @ 100 C (Norcast 23)

Cure overnight @ R.T. or 2 h @ 125 F (Norcast 4917HSS)

Cure overnight @ R.T., or 2 h @ R.T. plus 2 h @ 100 C, or 2 h @ R.T. plus 4 h @ 60 C (Norcast 5500, 5503, 5509-M)

Cure overnight @ R.T. or 4 h @ 90 C or 1 h @ 120 C (Norcast 9383)

Cure overnight @ R.T. (may be accelerated by heat—1 h @ 250 F); or 2 h @ 150 F, plus 4 h @ 350 F—allow to cool slowly to R.T. (Norcast T-27)

Cure overnight or 1 h @ 70 C (Norcast 3216)

Cure at any temp. from 70–150 F; good strength is developed in 2 h @ R.T., 1 h @ 150 F (Norcast 7633)

Dry to the touch in 2 h and usable in 8 h (Norcast 176X)

Handling strength is good after 2 h @ R.T. in 2–6 oz masses; thin films should stand overnight (Norcast 7086)

Post cure 2 h @ 200 F, 3 h @ 325 F (Epolite 2354)

Post cure 3 h @ 100 C (D.E.R. 331, 732, 736)

Initial Gel:

16 h @ 25 C (D.E.R. 331, 732, 736)

Gel Time:

35–55 s (Hysol MG18)

2.5 min @ 177 C (Magnamite AS4/2220-3)

3 min (Hysol 309 Epoxi-Patch Kit)
3–7 min @ 177 C (Magnamite AS/3501-5A, AS4/3501-5A, HMS/3501-5A)
5 min (Hysol 608 Epoxi-Patch Kit)
5–12 min @ 121 C (Magnamite AS4/1919)
6–12 min @ 177 C (Magnamite AS/3501-6, AS4/3501-6, HMS/3501-6)
7–9 min @ 75 F (10 g) (Norcast 5615-10/Norcure 178-M)
7–14 min @ 121 C (Magnamite AS4/1908, HMS/1908)
12–32 min @ 177 C (Magnamite AS4/3502)
15 min (Epolite 1350); @ R.T. (100 g) (Norcast 424)
15–20 min @ 250 F (Norcast 3420LC)
20 min (Epolite 1301, 2302)
22 min (Epolite 2300)
23 min (Epolite 1302)
25 min @ 120 C (Norcast 9100)
30 min (Epolite 3302)
35 min @ 100 C or 20 min @ 120 C (Norcast 9101, 9102)
40 min @ 25 C (Epolite 2354); (100 g) (Norcast 4200-1)
40–45 min @ R.T. (Interset Concrete Bonder)
45 min (Epolite 1354, 3306, 3354)
50–60 min (100 g) (Norcast 4200-5)
70 min (Epolite 3300, 3301)
78 min @ 100 C; 24 min @ 120 C (Hysol EO1017)
1.5 h (1 gal) (Norcast 0364/Norcure 142)
2–3 h @ 25 C (Epolite 3353)
3 h @ 77 F (100 g) (Norcast 4200-2)
4 h (Epolite 1353)
5 h @ 25 C (Epolite 2353)
6–8 h @ 75 F (Norcast 3510)
10 h @ 75 F (Norcast 3511)
12 h @ 75 F (Norcast 3515)

Working Life:
18 min @ R.T. (Interset Injection Resin (F))
> 20 min @ R.T. (Allabond Twenty/twenty Adhesive)
25 min @ R.T. (Interset Concrete Bonder)
35 min (filled) (Interset Concrete Topping Resin #2)
40 min @ 160 F (P-56A, -76)
45 min @ 180 F (P-85)
50 min @ 160 F (P-56)
75 min (Norbond 2162, 2162 Clear Amber)
90 min @ 160 F (P-78)
100 min @ 77 F (LCA-27)
2 h @ 75 F (P-51)
> 16 h @ 75 F (P-86)

Epoxy resin *(cont'd.)*

Tack-Free Time:
 6 h @ 77 F (LCA-27)
Pot Life:
 1.5 min (100 g) (Hysol 309 Epoxi-Patch Kit)
 3 min @ 25 C (100 g) (Hysol 608, 615 Epoxi-Patch Kit)
 5 min (25 g) (Tra-Bond 2106T)
 11 min @ 100 C (200 g) (EP 2800GU)
 12 min @ 100 C (200 g) (EP 2805TC)
 12–14 min ($^1/_2$ lb) (Norcast 1750/Norcure 135-M)
 15 min @ R.T. (200 g) (EP 2405FR); @ 75 F (FFA-9)
 15–20 min (Norcast 7086)
 17 in @ R.T. (200 g) (EP 2400TC)
 18 min @ 150 C (200 g) (EP 1760LV)
 20 min (25 g) (Tra-Bond 2113); (200 g) (Hysol C8-4143/HD3404, C9-4183/HD3404,
   R8-2038/HD3404, R9-2039/HD3404); (1 lb) (Norcast 141/Norcure 122); 20 min
   @ 25 C (100 g) (Hysol 1C Epoxi-Patch Kit); 20 min @ 250 F (P-24)
 25 min (Norcast 3217M/Norcure 122); (25 g) (Tra-Bond 2122, 2123); (200 g) (Hysol
   R8-2038/HD3475); 25 min @ 121 C (200 g) (EP 2810TS); 25 min @ 212 F (P-20)
 30 min (Norbond 406; Norcast 3217M/Norcure 111 & 112; Norcast 7124/Norcure
   159; Norcast 7633); (25 g) (Tra-Bond 2101, 2112, 2116, 2126, 2129); (200 g)
   (Hysol C8-4143/HD3475, C9-4183/HD3469, R9-2039/HD3469); 30 min @ R.T.
   (200 g) (EP 2230TS); @ R.T. (100 g) (Hysol 6C, 907 Epoxi-Patch Kit); @ 121 C
   (200 g) (EP 2825 TS); @ 160 F (LCA-20, -21); 30 min @ 23 C (100 g) (Norcast
   3253/Norcure 135M; Norcast 3255)
 30–50 min (100 g) (Hysol 0151 Epoxi-Patch Kit)
 32 min @ 90 C (200 g) (EP 2820TS)
 35 min @ 160 F (P-74); 35 min (200 g) (Hysol R9-2039/HD3561)
 37 min @ R.T. (200 g) (EP 2306LK)
 40 min @ 150 C (200 g) (EP 1765TS); @ 160 F (LCA-1)
 40–50 min @ R.T. (Norcast 3257/Norcure 112; Norcast 3258/Curing Agent D); ($^1/_2$ lb)
   (Norcast 3256/Norcure 112)
 40–60 min (Norcast 3258/Norcure 112)
 45 min (Norcast 5509-M); (25 g) (Tra-Bond 2125, 2151); 45 min (1 lb) (Norcast 154/
   Norcure 137); 45 min @ R.T. (200 g) (EP 2404GE, 2415TC); 45 min @ 160 F (P-
   75, -82C)
 45–60 min @ R.T. (100 g) (Norcast 3215)
 48 min @ R.T. (200 g) (EP 2225MV)
 50 min (Norcast 5503); 50 min @ R.T. (Interset Epoxy Gel); 50 min @ R.T. (200 g)
   (EP 2305LK); 50 min @ 121 C (200 g) (EP 2840TS); 50 min @ 212 F (P-19)
 < 60 min @ R.T. (Interset Epoxy Coal Tar)
 ≈ 60 min (Norcast 3343, 4912); @ R.T. (Norcast 3401M)
 60 min (Norcast 5500); (25 g) (Tra-Bond 2135D; Tra-Duct 2902); 60 min @ 25 C (100
   g) (Hysol 1105 Epoxi-Patch Kit); 60 min @ 212 F (P-14, -70); 60 min @ 300 F (P-

38, -81)

60–70 min ($^1/_2$ lb) (Norcast 3253/Norcure 112)

60–90 min @ 25 C (Norcast 3230SP/Norcure 112 & 170); (100 g) (Hysol 3X Epoxi-Patch Kit)

61 min @ R.T. (200 g) (EP 2220TC)

63 min @ R.T. (100 g) (EP 2740HT)

73 min @ 121 C (200 g) (DC 2900 HW)

80 min (200 g) (Hysol C9-4183/HD3561); 80 min @ 180 F (P-86)

90 min ($^1/_2$ lb) (Norcast 1750/Norcure 112); 90 min @ 25 C (100 g) (Hysol 9340 Epoxi-Patch Kit); 90 min @ 60 C (Lekutherm X20); 90 min @ 75 F (FA-8; LCA-4, -9); 90 min @ 180 F (P-85); 90 min @ 212 F (P-11)

90–110 min (200 g) (Hysol ES4312, ES4412)

100 min @ 120 C (Lekutherm X30S)

105 min @ R.T. (200 g) (EP 2408TS)

120 min (1 lb) (Norcast 154/Norcure 134); 120 min @ 121 C (200 g) (EP 2770TC)

130 min @ R.T. (200 g) (EP 2420TC)

170 min @ 60 C (Lekutherm X18)

190 min (Hysol PC12-007)

240 min @ 75 F (LCA-4LV)

400 min (Norcast 3217M/Norcure 135)

1 h @ R.T. (400 g) (Norcast T-27)

1 h @ 90 C or 3 h @ 75 C (Norcast 9390)

1 h @ 70 F (Barco Bond MB-127); 1 h @ R.T. (Norcast 9383)

1 h @ 160 F (P-80C)

2 h ($^1/_2$ lb) (Norcast 1750/Norcure 170); (1 qt) (Norcast 3220-D)

≈ 2 h @ R.T. (1 lb) (Norcast 5005/Norcure 150)

2 h @ 75 F (FFA-5)

2 h @ 77 F (100 g) (C-84)

2 h @ 160 F (FA-21)

> 2h @ R.T. (200 g) (Norcast 3222, 3222LV)

2–3 h (Norcast T-30); (100 g) (Norcast 3705)

2.5–3 h ($^1/_2$ lb) (Norcast 1750/Norcure 138)

3 h (25 g) (Tra-Bond 2248); (200 g) (Hysol C9-4183/HD3719, ES4212, R9-2039/HD3719); @ 75 F (FA-1; FFA-2)

3–4 h @ R.T. (Norcast 3424-M)

3.5 h @ 25 C (200 g) (Norcast 2070-PC)

4 h @ R.T. (CI-3); (25 g) (Tra-Duct 2924)

4–6 h (Norcast 3256/Norcure 135; Norcast 3257/Norcure 135)

5 h (25 g) (Tra-Bond 2211)

> 6 h @ R.T. (Norcast 7146)

6–8 h (Norcast 3230SP/Norcure 3230B & 135M)

8 h (Norcast 176X); (200 g) (Hysol C9-4183/HD3615, R9-2039/HD3615); @ R.T. (CI-2; FA-14); @ 160 F (LCA-14)

Epoxy resin *(cont'd.)*

> 8 h @ R.T. (Norcast X-9201)
> 8 h @ R.T. or 2–3 h @ 40 C (Norcast 3370)
14 h @ R.T. (CI-6)
24 h (200 g) (Hysol C9-4183/HD3485); @ 75 F (FA-13)
24–30 h @ R.T. (Norcast 3425-M)
≈ 48 h @ R.T. (Norcast 3425HF)
1–2 days (200 g) (Hysol C9-4183/HD3537)
≤ 2 days (Norcast 3259 Resin-D-Type)
2–3 days (200 g) (Hysol C18F, C60, C61); 2–3 days @ 77 F (Norcast 3420LC, 9311)
4 days (25 g) (Tra-Bond 2215)
5 days (25 g) (Tra-Bond 2208)
10 days @ 25 C, 4 days @ 40 C (200 g) (Hysol EO1017)
2–3 wks @ 20 C (Norcast 141/Catalyst 0.31)
6 mos (200 g) (Hysol EO1016)

Initial Set:
60 s @ R.T. (Barco Bond MB-165)
3 min @ R.T. (Barco Bond MB-175)

80% Bond Set:
5 min @ R.T. (slight heat will hasten cure) (Barco Bond MB-165)
30 min @ R.T. (slight heat will hasten cure) (Barco Bond MB-175)

Hardening Time:
16 h @ 25 C (Epolite 2354)

Peak Exotherm:
36 C (Hysol ES4212)
47 C (Hysol C9-4183/HD3615)
68 C (Hysol C9-4183/HD3719)
100 C (Hysol C8-4143/HD3404, C9-4183/HD3404, R9-2039/HD3615)
115 C (Hysol C9-4183/HD3561)
132 C (Hysol R9-2039/HD3719)
147 C (Norcast 154/Norcure 137)
150 C (Hysol C8-4143/HD3475, C9-4183/HD3469, R8-2038/HD3404, R9-2039/HD3404; Norcast 154/Norcure 134)
190 C (Hysol R9-2039/HD3561)
200 C (Hysol R8-2038/HD3475, R9-2039/HD3469; Norcast 0364/Norcure 142)
240 C (Hysol EO1016)
330 F (Norcast 4200-1)

TOXICITY/HANDLING:

Harmful or fatal if swallowed; may cause eye and skin burns, or delayed dermatitis; avoid breathing of vapor; use good ventilation (Interset Epoxy Coal Tar)

May cause skin sensitization; organic material which will burn when sufficient heat and oxygen are present; evolves toxic by-products when burning (Tactix 742 Performance Polymer)

May cause injury to skin on prolonged/repeated contact; use with adequate ventilation

Epoxy resin (cont'd.)

(CI-3)

May cause skin injury on prolonged/repeated contact; use with adequate ventilation and avoid breathing vapor (P-86)

May cause injury to skin on prolonged/repeated contact; use with adequate ventilation; the activator can cause eye injury and skin burns (Allabond Twenty/twenty Adhesive, Twenty/twenty NM; C-84; CI-6; FA-14; P-51, -75, -80C, -82C, -85)

May cause injury to skin on prolonged/repeated contact; the activator, a powder, causes irritation to the skin, nose, and throat; avoid breathing dust and use with adequate ventilation (FA-21; LCA-14)

May cause skin injury on prolonged/repeated contact; use with adequate ventilation; the activator, an aliphatic amine, may cause skin burns and eye injury (LCA-27)

May cause injury to skin on prolonged/repeated contact; Activator BA-1 is highly flammable, and may cause skin burns and eye injury; use with adequate ventilation (CI-2)

May be irritating to the skin and may cause hypersensitivity; avoid skin contact (Norcast 2795)

May cause serious irritations or burns, or rash or asthmatic-type response on some people; avoid contact with skin and inhalation of vapors (Norcure 3416)

Avoid skin contact (Norcast 4200-2; Norcast 5615-10/Norcure 178-M; Norcast 8=7086)

Avoid skin contact of the uncured resins (Norcast 3220-D)

Avoid skin contact and use with good ventilation (Barco Bond MB-14X, MB-127, MB-165, MB-175; Norbond 2162, 2162 Clear Amber; Norcast 1750/Norcure 112; Norcast 1750/Norcure 135-M; Norcast 1750/Norcure 138; Norcast 1750/Norcure 170; Norcast 3253/Norcure 112; Norcast 4200-5)

Avoid contact with skin or eyes (Norcast 3401M)

Avoid eye and skin contact; use with good ventilation (Norcast 154/Norcure 134; Norcast 154/Norcure 137; Norcast 3217M/Norcure 111 & 112; Norcast 3217M/Norcure 122; Norcast 3217M/Norcure 135)

Avoid contact with eyes and skin; avoid breathing vapors; use with good ventilation (Interset Concrete Bonder, Concrete Topping Resin #2, Epoxy Gel, Injection Resi (F))

Avoid skin contact and prolonged breathing of vapors; use with adequate ventilation (Interset Deep Penetrating Sealer)

Avoid inhalation of powder; use with dust masks and adequate ventilation (Polyset EPC50, EPC68FR)

Use with good ventilation (Norcast 3343)

Observe precautions as for other liquid bisphenol-A epoxy resins; may cause severe burns, sensitization; harmful if swallowed; burning produces carbon monoxide (D.E.R. 383)

Observe safety precautions for handling epoxy resins; use in ventilated areas; avoid prolonged/repeated skin contact (Magnamite AS/3501-5A, AS/3501-6, AS4/1908, AS4/1919, AS4/2220-3, AS4/3501-5A, AS4/3501-6, AS4/3502, HMS/1908,

Epoxy resin (cont'd.)

HMS/3501-5A, HMS/3501-6

Very slightly irritating to eyes; low in acute oral toxicity; may cause skin rashes due to sensitization (D.E.R. 642U, 661, 662, 663U, 664, 664U, 667, 673MF)

May cause irritation to skin and eyes, and sensitization responses; may present a hazard from inhalation (D.E.R. 652-A75, 661-A80, 671-EE75, 671-EEA75, 671-EEK75, 671-MK75, 671-T75, 671-X75, 671-XM75, 684-EK40)

Low acute oral toxicity; may cause moderate eye irritation, minor to moderate skin irritation; vapors may be irritating (D.E.R. 732)

Low single dose oral toxicity; may cause moderate eye irritation; prolonged/repeated exposure may cause moderate skin irritation; may be a sensitizer; vapors from heated material may be irritating (D.E.R. 736)

Resin is harmful if swallowed; may cause skin irritation, sensitization, or dermatitis; hardener is harmful if swallowed; prolonged/repeated skin contact may cause burns, sensitization, or dermatitis; use with adequate ventilation; for industrial use only (Epolite 1301, 1302, 1350, 1353, 1354, 2300, 2302, 2353, 2354, 2360, 3300, 3301, 3302, 3306, 3353, 3354, 3357)

Uncured epoxy resins and hardeners may cause dermatitis, skin sensitization, or other allergenic responses; prevent all contact with skin or eyes (Tra-Bond 2101, 2106T, 2112, 2113, 2116, 2122, 2123, 2125, 2126, 2129, 2135D, 2151, 2208, 2211, 2215, 2248; Tra-Duct 2902, 2924)

Part A may cause skin irritation; Part B may cause skin sensitization and eye irritation; avoid inhalation of vapor, skin and eye contact (Hysol 1C, 6C, 1105, 9340 Epoxi-Patch Kit)

Parts A and B may cause skin sensitization and eye irritation; avoid inhalation of vapors, and skin and eye contact (Hysol 0151 Epoxi-Patch Kit)

Parts A and B may cause skin irritation (Hysol 309 Epoxi-Patch Kit)

Part A may cause skin irritation; Part B may cause skin injury on repeated/prolonged contact (Hysol 608 Epoxi-Patch Kit)

Part B may cause skin injury on prolonged/repeated contact (Hysol 615 Epoxi-Patch Kit)

Part A may cause skin irritation (Hysol 907 Epoxi-Patch Kit)

Inappropriate handling may cause irritation of the skin and mucous membranes (Lekutherm X18, X20, X23, X24, X30S, X201, X227, X256, X257)

No chemical burn hazard to skin; avoid prolonged contact with skin and contact with eyes, nose, etc. (Norcast 157)

STORAGE/HANDLING:

Keep away from heat and open flame (Tra-Bond 2101, 2106T, 2112, 2113, 2116, 2122, 2123, 2125, 2126, 2129, 2135D, 2151, 2208, 2211, 2215, 2248; Tra-Duct 2902, 2924)

Contains solvents—keep away from heat, sparks, and open flames; keep containers closed and upright when not in use (Interset Deep Penetrating Sealer)

Organic material which may burn in appropriate conditions; dusts can be explosive (D.E.R. 642U)

Organic resin burns under appropriate conditions; flammable solvent (D.E.R. 652-A75, 661-A80, 671-EE75, 671-EEA75, 671-EEK75, 671-MK75, 671-T75, 671-X75, 671-XM75, 684-EK40)

Store in closed containers ≤ 65 F (E-231, -260, -264H, -2748, -3824, -3938, -8353, –8354, -8405, -9405)

Store in closed shipment pails @ 10 C or below (Polyset EPC50)

Store in closed containers @ 25 C or below to insure max. storage stability (Polyset EPC68FR)

Store in a cool, dry area (65–80 F) (Epolite 1301)

Avoid contamination by grease, dust, and dirt; place in polyethylene bags for low-temp. storage (Magnamite AS/3501-5A, AS/3501-6, AS4/2220-3, AS4/3501-5A, AS4/3501-6, AS4/3502, HMS/3501-5A, HMS/3501-6)

Allow material to reach ambient temps. when removing from 0 F storage (Magnamite AS4/1908)

Allow material to reach ambient temps. when removing from 0 F storage; protect from contamination by grease, dust, and dirt (Magnamite AS4/1919, HMS/1908)

Part B may crystallize—warm to 200 F to reliquefy (Norcast 9390)

Keep container tightly closed when not in use (P-86)

STD. PKGS.:

Pt containers (FA-14)

Qt containers (CI-3; FA-21)

Qt and gal containers (CI-2)

Qt (3 lb) and gal (12 lb) paint cans, 5 gal (60 lb) open-head pails; Activator BA-63 and 66B avail. in pt (1 lb), qt (2 lb), and gal (8 lb) screw-top cans (Allabond Twenty/twenty Adhesive, Twenty/twenty NM; C-84)

Qt (3 lb) and gal (14 lb net) paint cans; 5 gal (70 lb net) open-head pails; Activator avail. in 4 oz, pt, and gal cans, and 5-gal pails (P-85, -86)

12 qt. Pack containing 12 qt. cans (resin) and 12 $\frac{1}{2}$-pt. cans (hardener); Pail Pack containing 5-gal pail (resin) and 1-gal can (hardener) (Epolite 1301, 1350, 1353, 1354, 2353, 2354, 3302, 3353, 3354)

12 qt. Pack containing 12 qt. cans (resin) and 12 $\frac{1}{2}$-pt. cans (hardener); Pail Pack containing 5-gal pail (resin) and $\frac{1}{4}$-gal can (hardener) (Epolite 1302)

12 qt. Pack containing 12 qt. cans (resin) and 12 $\frac{1}{2}$-pt. cans (hardener); Pail Pack containing 5 gal pail (resin), 1 gal can (hardener), and Drum Pack with 55 gal drum (resin), 15 gal drum (hardener) (Epolite 2300, 2302, 3300, 3301, 3306)

Pail Pack containing 5 gal pail (resin) and 2 gal can (hardener) (Epolite 3357)

Pail Pack containing 5 gal pail (resin) and 5 gal pail (hardener) (Epolite 2360)

Tubes (Barco Bond MB-14X, MB-127)

Self-metered tubes (Barco Bond MB-165, MB-175)

1-gal cans, 5-gal pails, 55-gal drums (Interset Concrete Topping Resin #2, Deep Penetrating Sealer, Epoxy Coal Tar, Epoxy Gel, Injection Resin (F))

2-gal containers (resin) and 1-gal cans (curing agent), and in 5-gal pails (Interset Concrete Bonder)

Epoxy resin (cont'd.)

2.3-oz tube kits, 5.2-lb qt units, 20.5-lb gal units (Hysol 907 Epoxi-Patch Kit)

2.7-oz tube kits, 4.6-lb qt units, 18.0-lb gal units (Hysol 3X Epoxi-Patch Kit)

2.7-oz tube kits, 5.0-lb qt units, 20.0-lb gal units (Hysol 9340 Epoxi-Patch Kit)

2.8 oz tube kits, 5.0-lb qt units, 20.0-lb gal units; also avail. in foil packs (Hysol 608 Epoxi-Patch Kit)

3.2-oz tube kits (Hysol 1105 Epoxi-Patch Kit)

3.2-oz tube kits, 6.0-lb qt. units, 24.0-lb gal units (Hysol 615 Epoxi-Patch Kit)

3.3-oz tube kits, 2.67-lb qt units, 10.7-lb gal units (Hysol 0151 Epoxi-Patch Kit)

3.6-oz tube kits, 3.0-lb qt units, 12.6-lb gal units (Hysol 309 Epoxi-Patch Kit)

4-oz tube kits, 4.3-lb qt units, 17.3-lb gal units (Hysol 6C Epoxi-Patch Kit)

4-oz tube kits, 4.3-lb qt units, 17.3-lb gal units; also avail. in foil packs (Hysol 1C Epoxi-Patch Kit)

50-lb drums (E-8353, -8354, -9405)

Avail. in bulk, kits, and premixed and frozen (LCA-14)

Avail. in bulk, one-shot kits, and premixed and frozen (LCA-27)

22.7-kg rolls in sealed polyethylene bags and cardboard containers; ship with dry-ice refrigeration when temps. over 38 C are anticipated (Magnamite AS4/1908, AS4/1919, HMS/1908)

22.7-kg rolls in sealed polyethylene bags in insulated containers with dry-ice refrigeration (Magnamite AS4/2220-3)

Shipped in sealed polyethylene bags in insulated containers with dry-ice refrigeration (Magnamite HMS/3501-5A, HMS/3501-6)

2-, 6-, and 24-g Bipax kits and qt, gal, and 5-gal kits (Tra-Bond 2125)

2-, 6-, and 25-g Bipax kits and qt, gal, and 5-gal kits (Tra-Bond 2106T, 2135D, 2208, 2211, 2215)

2-, 7-, and 28-g Bipax kits and qt, gal and 5-gal kits (Tra-Bond 2101, 2112, 2113, 2116, 2129, 2248)

2-, 9-, and 36-g Bipax kits and qt, gal, and 5-gal kits (Tra-Bond 2122, 2126)

2- and 10-g Bipax kits and qt, gal, and 5-gal kits (Tra-Duct 2924)

2-, 10-, and 40-g Bipax kits and qt, gal, and 5-gal kits (Tra-Bond 2151)

2-, 11-, and 44-g Bipax kits and qt, gal, and 5-gal kits (Tra-Bond 2123)

2.5- and 10-g Bipax kits and qt, gal, and 5-gal kits (Tra-Duct 2902)

Ethylene/vinyl acetate copolymer (CTFA)

SYNONYMS:
Acetic acid, ethenyl ester, polymer with ethene

EVA copolymer

EMPIRICAL FORMULA:
$(C_4H_6O_2 \cdot C_2H_4)_x$

STRUCTURE:

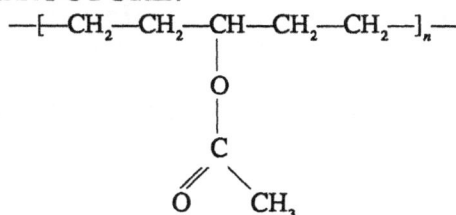

CAS No.:

24937-78-8

TRADENAME EQUIVALENTS:

A-C Copolymer 400A [Allied-Signal]

A-C Polyethylene 400 [Allied-Signal]

BASF Wax EVA 1 [BASF AG]

Elvax 40-P, 150, 210, 220, 230, 240, 250, 260, 265, 310, 350, 360, 410, 420, 450, 460, 470, 550, 560, 565, 650, 660, 670, 750, 760, 770, 3120, 3128, 3135SB, 3135X [DuPont]

EVA 3121, 3134, 3152, 3159, 3167, 3174, 3200, PE-4928 [DuPont]

Levapren 336, 400, 408, 450, 452, 456, KA 8114 [Bayer AG]

Modic E-100H, E-200H, E-300K, E-300S, E-310K [Mitsubishi]

Ultrathene UE 630, UE 631, UE 632, UE 635, UE 637, UE 643, UE 655, UE657 [USI Chem.]

Carbon-reinforced:

PDX-84440 (carbon powder grade) [LNP]

MODIFICATIONS/SPECIALTY GRADES:

Slip additive:

Elvax 3120, 3135SB

Antiblock additive:

Elvax 3120, 3135SB

Paraffin wax additive:

EVA 3200

Statically dissipative:

PDX-84440

CATEGORY:

Thermoplastic resin

PROCESSING:

Blown film:

Elvax 3120, 3128, 3135SB, 3135X; EVA 3121, 3159, 3167, PE-4928; Ultrathene UE 630, UE 631, UE 632, UE 635, UE 637, UE 643, UE 655, UE 657

Cast film:

EVA 3134, 3152; Modic E-100H, E-200H, E-300K, E-300S, E-310K; Ultrathene UE 630, UE 631, UE 632, UE 635, UE 637, UE 643, UE 655, UE 657

Ethylene/vinyl acetate copolymer *(cont'd.)*

Coextrusion:
   Elvax 3120, 3128, 3135SB, 3135X; EVA 3121, 3134, 3152, 3159, 3167, 3200, PE-4928; Modic E-100H, E-200H, E-300K, E-300S, E-310K
Extrusion:
   EVA 3121, 3159, 3167, 3200, PE-4928; Levapren 336, 400, 408, 450, 452, 456, KA 8114
Molding:
   Levapren 336, 400, 408, 450, 452, 456, KA 8114

APPLICATIONS:

Consumer products: cosmetic gels (A-C Polyethylene 400); footwear applications (Elvax 250, 260, 350, 360, 410, 420, 450, 470, 460, 565; Levapren 336, 400, 408, 450, 452, 456, KA 8114)

Electrical/electronic industry: cable and wire jacketing (Levapren 336, 400, 408, 450, 452, 456, KA 8114); electronic components/packaging (PDX-84440); insulation (Levapren 336, 400, 408, 450, 452, 456, KA 8114)

FDA-approved applications: (Elvax 40-P, 150, 310, 350, 360, 3120, 3128, 3135SB, 3135X; EVA 3121, 3134, 3152, 3159, 3167, 3200, PE-4928; Ultrathene UE 630, UE 631, UE 632, UE 635, UE 637, UE 643, UE 655, UE 657)

Food-contact applications: (Elvax 3120, 3128, 3135SB, 3135X; EVA 3121, 3134, 3152, 3159, 3167, 3200, PE-4928); food packaging (Elvax 40-P, 150, 310, 350, 360; Ultrathene UE 630, UE 631, UE 632, UE 635, UE 637, UE 643, UE 655, UE 657)

Functional additives: impact modifier for PVC (Levapren 336, 400, 408, 450, 452, 456, KA 8114); modifier (Elvax 40-P, 150, 310, 350, 360); pigment dispersant (A-C Copolymer 400A)

Industrial applications: adhesives (Elvax 40-P, 150, 260, 310, 350, 360, 410, 420, 450, 470, 460, 650, 660, 670, 750, 760, 770; Levapren 336, 400, 408, 450, 452, 456, KA 8114; Modic E-100H, E-200H, E-300K, E-300S, E-310K; Ultrathene UE 643); asphalt/bituminous compositions (Elvax 40-P, 150; Levapren 336, 400, 408, 450, 452, 456, KA 8114); barrier coating (Elvax 260, 310, 350, 360); caulks and sealants (Elvax 40-P, 150); coatings (BASF Wax EVA 1; Elvax 40-P, 150, 210, 220, 230, 240, 250, 260, 310, 350, 360); color concentrates (A-C Copolymer 400A); elastomers (Elvax 40-P, 150); films (Modic E-100H, E-200H, E-300K, E-300S, E-310K; Ultrathene UE 630, UE 631, UE 632, UE 635, UE 637, UE 643, UE 655, UE 657); heat-sealing applications (Elvax 40-P, 150, 310, 350, 360, 410, 420, 450, 470, 460, 3120, 3128, 3135SB, 3135X; EVA 3121, 3134, 3152, 3159, 3167, 3200, PE-4928); hot-melt applications (Elvax 410, 420, 450, 470, 460; Levapren 336, 400, 408, 450, 452, 456, KA 8114); inks (Elvax 40-P, 150, 410, 420, 450, 470, 460); lacquers (Elvax 410, 420, 450, 470, 460); laminates (Elvax 40-P, 150, 410, 420, 450, 470, 460; Ultrathene UE 632, UE 635, UE 637, UE 643, UE 655); metal coatings (BASF Wax EVA 1; Elvax 40-P, 150); packaging film (Elvax 3120, 3128, 3135SB, 3135X; EVA 3121, 3134, 3152, 3159, 3167, 3200, PE-4928; Ultrathene UE 637, UE 655, UE 657); pallet stretch film (Ultrathene UE 657); paper coatings (Elvax 40-P, 150, 310, 350, 360); paraffin wax systems (Elvax 220, 230, 240); rubber goods (Le-

Ethylene/vinyl acetate copolymer (cont'd.)

vapren 336, 400, 408, 450, 452, 456, KA 8114); textile applications (Elvax 40-P, 150; Levapren 336, 400, 408, 450, 452, 456, KA 8114); waterproof sheeting (Levapren 336, 400, 408, 450, 452, 456, KA 8114); wax blends (Elvax 265, 310, 350, 360); wood primers and sealers (Elvax 40-P, 150)

Transportation industry: traffic paints (Elvax 40-P, 150)

PROPERTIES:

Form:

Powder (A-C Copolymer 400A)

Pellets (Elvax 40-P, 150, 250, 260, 310, 350, 360, 410, 420, 460, 3120, 3128, 3135SB; EVA 3121, 3134, 3152, 3159, 3167, 3174, 3200, PE-4928)

Granular (BASF Wax EVA 1; Levapren 336, 400, 408, 450, 452, 456, KA 8114)

Color:

Nearly transparent (Elvax 250, 260)

Translucent (Elvax 40-P, 150, 310, 350, 360, 410, 420)

White (BASF Wax EVA 1)

Odor:

Slight (Elvax 150, 310, 350, 360, 410, 420)

Composition:

7.5% vinyl acetate (Elvax 3120; EVA 3121)

8–10% vinyl acetate (Elvax 750)

8.5–10.5% vinyl acetate (Elvax 770)

8.8–9.8% vinyl acetate (Elvax 760)

8.9% vinyl acetate (Elvax 3128)

9% vinyl acetate (Ultrathene UE 635, UE 637, UE 655)

9.3% vinyl acetate (EVA PE-4928)

11–13% vinyl acetate (Elvax 650, 660, 670)

12% vinyl acetate (Elvax 3135SB, 3135X; EVA 3134; Ultrathene UE 657)

15% vinyl acetate (Elvax 550, 560, 565; EVA 3152, 3159)

15.3% vinyl acetate (Ultrathene UE 632)

17% vinyl acetate (Ultrathene UE 630)

18% vinyl acetate (Elvax 410, 420, 450. 460, 470; EVA 3167, 3174)

18.5% vinyl acetate (Ultrathene UE 643)

19% vinyl acetate (Ultrathene UE 631)

25% vinyl acetate (Elvax 310, 350, 360)

27.2–28.8% vinyl acetate (Elvax 210, 220, 230, 240, 265)

28% vinyl acetate (Elvax 250, 260)

33% vinyl acetate (Elvax 150)

33 ± 1.5% vinyl acetate (Levapren 336)

40% vinyl acetate (Elvax 40-P)

40 ± 1.5% vinyl acetate (Levapren 400, 408)

45 ± 1.5% vinyl acetate (Levapren 450, 452, 456)

50 ± 1.5% vinyl acetate (Levapren KA 8114)

Ethylene/vinyl acetate copolymer (cont'd.)

GENERAL PROPERTIES:

Solubility:
- Sol. in many organic solvents (Elvax 40-P, 150)
- Sol. in hot ternary systems such as fatty esters/castor oil or fatty esters/silicone oils (A-C Polyethylene 400)

Melt Flow:
- 2–5 (Levapren 400, 450)
- 5–10 (Levapren KA 8114)
- 5–15 (Levapren 452)
- 15–35 (Levapren 336, 456)
- 35-75 (Levapren 408)
- 0.5 g/10 min (Ultrathene UE 657)
- 1.8 g/10 min (Ultrathene UE 630)
- 2.0 g/10 min (Elvax 360)
- 2.2 g/10 min (Ultrathene UE 655)
- 2.3 g/10 min (Modic E-100H; Ultrathene UE 631)
- 2.5 g/10 min (Elvax 460; Modic E-200H)
- 3.2 g/10 min (Ultrathene UE 637)
- 3.3 g/10 min (Modic E-300K)
- 3.5 g/10 min (Modic E-310K)
- 8.2 g/10 min (Ultrathene UE 632)
- 9.8 g/10 min (Ultrathene UE 635, UE 643)
- 11 g/10 min (Modic E-300S)
- 19 g/10 min (Elvax 350)
- 43 g/10 min (Elvax 150)
- 57 g/10 min (Elvax 40-P)
- 150 g/10 min (Elvax 420)
- 400 g/10 min (Elvax 310)
- 500 g/10 min (Elvax 410)
- 0.2–0.4 dg/min (Elvax 670)
- 0.35 dg/min (Elvax 3135SB, 3135X)
- 0.50 dg/min (EVA 3121, 3159)
- 0.6–1.0 dg/min (Elvax 770)
- 0.7 dg/min (Elvax 470; EVA 3167)
- 1.2 dg/min (Elvax 3120)
- 1.5 dg/min (Elvax 565)
- 1.8–2.2 dg/min (Elvax 760)
- 2.0 dg/min (Elvax 3128; EVA PE-4928)
- 2.1–2.9 dg/min (Elvax 660)
- 2.5 dg/min (Elvax 560)
- 2.6–3.4 dg/min (Elvax 265)
- 6 dg/min (Elvax 260)
- 6.3–7.7 dg/min (Elvax 750)

Ethylene/vinyl acetate copolymer (cont'd.)

6.7–9.3 dg/min (Elvax 650)
8 dg/min (Elvax 450, 550; EVA 3134, 3152, 3174)
25 dg/min (Elvax 250)
32 dg/min (EVA 3200)
38–48 dg/min (Elvax 240)
100–120 dg/min (Elvax 230)
134–168 dg/min (Elvax 220)
365–440 dg/min (Elvax 210)
Sp. Gr.:
1.09 (PDX-84440)
Density:
0.926 g/cm³ (Ultrathene UE 635, UE 637)
0.930 g/cm³ (Elvax 3120, 3128; EVA 3121, PE-4928; Modic E-310K; Ultrathene UE 655)
0.932 g/cm³ (Ultrathene UE 632, UE 657)
0.934 g/cm³ (Elvax 410)
0.937 g/cm³ (Elvax 420; Ultrathene UE 630)
0.939 g/cm³ (Ultrathene UE 631, UE 643)
0.940 g/cm³ (Elvax 3135SB, 3135X; EVA 3134, 3152, 3159, 3167, 3174; Modic E-100H, E-200H, E-300K)
0.941 g/cm³ (Elvax 460)
0.948 g/cm³ (Elvax 310, 350)
0.950 g/cm³ (Elvax 360; EVA 3200; Modic E-300S)
0.960 g/cm³ (Levapren 336)
0.980 g/cm³ (Levapren 400, 408, 450, 452, 456)
0.990 g/cm³ (Levapren KA 8114)
935 kg/m³ (Elvax 550, 560, 565)
941 kg/m³ (Elvax 450, 470)
951 kg/m³ (Elvax 250)
955 kg/m³ (Elvax 260)
957 kg/m³ (Elvax 150)
965 kg/m³ (Elvax 40-P)
Visc.:
0.52 (inherent) (Elvax 410)
0.54 (inherent) (Elvax 310)
0.61 (inherent) (Elvax 420)
0.70 (inherent) (Elvax 40-P)
0.78 (inherent) (Elvax 150)
0.84 (inherent) (Elvax 350)
1.02 (inherent) (Elvax 460)
1.05 (inherent) (Elvax 360)
M.P.:
88 C (Modic E-300S)

Ethylene/vinyl acetate copolymer *(cont'd.)*

    90 C (Modic E-310K)
    92 C (Modic E-300K)
    93 C (Modic E-100H, E-200H)
Soften. Pt. (Ring & Ball):
    88 C (Elvax 310, 410)
    99 C (Elvax 420)
    104 C (Elvax 40-P)
    110 C (Elvax 150)
    132 C (Elvax 350)
    188 C (Elvax 360)
    199 C (Elvax 460)
Stability:
    Very good resistance to weathering, ozone, and light; good hot air resistance (Levapren
        336, 400, 408, 450, 452, 456, KA 8114)
Ref. Index:
    1.476 (Elvax 40-P)
    1.482 (Elvax 150)
    1.484 (Elvax 410)
    1.486 (Elvax 310)
    1.489 (Elvax 350)
    1.491 (Elvax 360)
    1.492 (Elvax 420)
    1.493 (Elvax 460)

MECHANICAL PROPERTIES:

Tens. Str.:
    40 kg/cm^2 (yield) (Modic E-300S)
    60 kg/cm^2 (yield) (Modic E-100H, E-200H, E-300K, E-310K)
    110 kg/cm^2 (break) (Modic E-300S)
    150 kg/cm^2 (break) (Modic E-300K, E-310K)
    170 kg/cm^2 (break) (Modic E-200H)
    180 kg/cm^2 (break) (Modic E-100H)
    3.3 MPa (Elvax 310)
    3.8 MPa (Elvax 410)
    4.5 MPa (Elvax 40-P)
    5.9 MPa (Elvax 150)
    6.6 MPa (Elvax 420)
    9.3 MPa (Elvax 250)
    14 MPa (Elvax 350, 550)
    15 MPa (Elvax 260, 450)
    19 MPa (Elvax 460)
    20 MPa (Elvax 560)
    22 MPa (Elvax 565)
    26 MPa (Elvax 360)

Ethylene/vinyl acetate copolymer (cont'd.)

28 MPa (Elvax 470)
1400 psi (PDX-84440)
1680 psi (Ultrathene UE 643)
1700 psi (Ultrathene UE 635)
1850 psi (Ultrathene UE 632)
2000 psi (Ultrathene UE 637)
2020 psi (Ultrathene UE 655)
2100 psi (Ultrathene UE 630)
2150 psi (Ultrathene UE 631)
2480 psi (Ultrathene UE 657)

Tens. Elong.:
500% (PDX-84440)
650% (break) (Elvax 410)
675% (break) (Ultrathene UE 635)
700% (Elvax 550); (break) (Elvax 420; Ultrathene UE 637, UE 657)
710% (break) (Ultrathene UE 631, UE 632)
725% (break) (Ultrathene UE 643)
730% (break) (Ultrathene UE 630)
740% (break) (Ultrathene UE 655)
750% (Elvax 260, 470, 560, 565)
790% (Modic E-200H, E-300K)
800% (Elvax 250, 450; Modic E-310K)
800–1000% (break) (Elvax 310, 350, 360)
810% (Modic E-100H)
850% (Modic E-300S); (break) (Elvax 460)
1050% (break) (Elvax 150)
1450% (break) (Elvax 40-P)

Tens. Mod.:
2.1 MPa (Elvax 40-P)
4.8 MPa (Elvax 150)
14 MPa (Elvax 410)
16 MPa (Elvax 310)
19 MPa (Elvax 420)
24 MPa (Elvax 460)
25 MPa (Elvax 350)
35 MPa (Elvax 360)

Flex. Str.:
290 kg/cm² (Modic E-300S)
380 kg/cm² (Modic E-300K)
420 kg/cm² (Modic E-100H)
430 kg/cm² (Modic E-200H)
450 kg/cm² (Modic E-310K)

Ethylene/vinyl acetate copolymer *(cont'd.)*

Flex. Mod.:
10 MPa (Elvax 250)
14 MPa (Elvax 260)
18 MPa (Elvax 350)
26 MPa (Elvax 360)
40 MPa (Elvax 450)
51 MPa (Elvax 550)
52 MPa (Elvax 470)
56 MPa (Elvax 565)
63 MPa (Elvax 560)

Impact Str. (Charpy):
> 30 kg cm/cm^2 (Modic E-100H, E-200H, E-300K, E-300S, E-310K)

Impact Str. (Dart Drop):
300 g/F$_{50}$ (Ultrathene UE 637, UE 655)
305 g/F$_{50}$ (Ultrathene UE 635)
310 g/F$_{50}$ (Ultrathene UE 632, UE 643)
> 325 g/F$_{50}$ (Ultrathene UE 630, UE 631)
400 g/F$_{50}$ (Ultrathene UE 657)

Impact Str. (Izod):
No break notched (PDX-84440)

Tens. Impact:
505 kJ/m^2 (Elvax 550)
525 kJ/m^2 (Elvax 450)
545 kJ/m^2 (Elvax 560)
610 kJ/m^2 (Elvax 470)
670 kJ/m^2 (Elvax 350)
685 kJ/m^2 (Elvax 250)
695 kJ/m^2 (Elvax 260)
770 kJ/m^2 (Elvax 565)
880 kJ/m^2 (Elvax 360)

Hardness:
Shore A40 (Elvax 40-P)
Shore A65 (Elvax 150)
Shore A70 (Elvax 310)
Shore A80 (Elvax 350, 410)
Shore A84 (Elvax 420)
Shore A85 (Elvax 360)
Shore A90 (Elvax 460)
Shore D30 (Elvax 250)
Shore D34 (Elvax 260)
Shore D38 (Elvax 450)
Shore D40 (Elvax 565)
Shore D41 (Elvax 470, 550)

Ethylene/vinyl acetate copolymer (cont' d.)

Shore D43 (Elvax 560)
THERMAL PROPERTIES:
Soften. Pt. (Vicat):
43 C (Modic E-300S)
59 C (Ultrathene UE 643)
62 C (Modic E-300K)
63 C (Modic E-310K; Ultrathene UE 631)
66 C (Modic E-200H; Ultrathene UE 632)
67 C (Ultrathene UE 630)
69 C (Modic E-100H)
75 C (Ultrathene UE 635)
80 C (Ultrathene UE 637, UE 657)
83 C (Ultrathene UE 655)

Brittle Temp.:
> –70 C (Modic E-100H, E-200H, E-300K, E-300S, E-310K)

Flamm.:
HB (PDX-84440)

ELECTRICAL PROPERTIES:
Vol. Resist.:
10,000 ohm-cm (PDX-84440)

TOXICITY/HANDLING:
Not considered hazardous with respect to inhalation, eye contact, or skin absorption; may generate acetic acid and carbon monoxide if ignited (Elvax 40-P, 150, 310, 350, 360, 410, 420, 460)

Fumes may evolve at temps. above 220 C; use with proper ventilation (Elvax 250, 260, 550, 560, 565)

Fumes may evolve at temps. above 230 C; use with adequate ventilation (Elvax 3120, 3128, 3135SB, 3135X; EVA 3121, 3134, 3152, 3159, 3167, 3174, 3200, PE-4928)

STD. PKGS.:
22.7-kg net multiwall paper bags on 1134-kg shrink-wrapped pallets (Elvax 40-P, 150, 310, 350, 360, 410, 420, 460)

25-kg boxes, polyethylene-lined on disposable 1000-kg pallets (Levapren 336, 400, 408, 450, 452, 456, KA 8114)

Fluorinated ethylene-propylene

SYNONYMS:
FEP resin

Tetrafluoroethylene/hexafluoropropylene copolymer

STRUCTURE:

$$—CF_2—CF_2—CF_2—\underset{\underset{CF_3}{|}}{CF}—$$

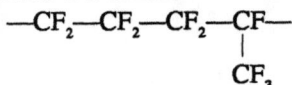

TRADENAME EQUIVALENTS:
Liquinite FEP [LNP]

Teflon 100 FEP, 110 FEP, 120 FEP, 140 FEP, 160 FEP [DuPont]

TL-120 [LNP]

Glass-fortified:

Thermocomp LF-1004 (20% long glass fibers), LF-1004M (20% milled glass) [LNP]

MODIFICATIONS/SPECIALTY GRADES:
Chemically coupled:

Thermocomp LF-1004, LF-1004M

CATEGORY:
Thermoplastic resin

PROCESSING:
Extrusion:

Teflon 100 FEP, 140 FEP, 160 FEP

Injection molding:

Teflon 110 FEP

Molding:

Teflon 100 FEP

Transfer molding:

Teflon 160 FEP

APPLICATIONS:
Automotive applications: (Teflon 110 FEP)

Electrical/electronic industry: (Teflon 110 FEP); cable and wire insulation (Teflon 120 FEP, 140 FEP); cable and wire jacketing (Teflon 100 FEP, 140 FEP); coatings (Liquinite FEP); computer wiring (Teflon 100 FEP); electrical components (Thermocomp LF-1004, LF-1004M)

FDA-approved applications: TL-120

Food-contact applications: (TL-120); equipment lubricants (TL-120); food processing (Teflon 110 FEP)

Functional additives: extreme pressure agent (TL-120); lubricant (TL-120); release agent (TL-120); thickening agent (TL-120)

Industrial applications: bearings and bushings (Thermocomp LF-1004, LF-1004M); bellows (Teflon 160 FEP); chemical process industry (Teflon 110 FEP, 160 FEP); coatings (Liquinite FEP; Teflon 120 FEP); frictional/sliding applications (Liquinite FEP; Teflon 120 FEP); greases and oils (TL-120); impregnation (Teflon 120 FEP); printing inks (TL-120); rubber industry (Teflon 110 FEP); textile applications (Teflon 110 FEP); valve components (Teflon 160 FEP); waxes (TL-120)

Pharmaceutical applications: equipment lubricants (TL-120)

PROPERTIES:

Form:

Dispersion; 0.1–0.25 μ particle size FEP solids (Teflon 120 FEP)

Pellets (Teflon 100 FEP, 140 FEP, 160 FEP)

Powder (1–2 μ) (TL-120)

Color:

Translucent (Teflon 100 FEP, 140 FEP, 160 FEP)

White (TL-120)

Composition:

53–57% solids in water; 5–7% mixture of volatile, nonionic, and anionic wetting agents (Teflon 120 FEP)

GENERAL PROPERTIES:

Sp. Gr.:

2.12–2.17 (Teflon 100 FEP, 110 FEP, 120 FEP, 140 FEP, 160 FEP)

2.12–2.18 (TL-120)

2.15–2.16 (Liquinite FEP)

2.21 (Thermocomp LF-1004, LF-1004M)

Density:

400 g/l (TL-120)

575–600 g/l (Liquinite FEP)

Visc.:

25 cps (Teflon 120 FEP)

M.P.:

260–277 C (Teflon 140 FEP, 160 FEP)

262–282 C (Teflon 100 FEP, 120 FEP)

487–520 F (Teflon 110 FEP)

500–504 F (Liquinite FEP)

Stability:

Excellent weather and chemical resistance (Teflon 140 FEP, 160 FEP)

pH:

10 (Teflon 120 FEP)

Fluorinated ethylene-propylene *(cont'd.)*

Surface Tension:
  21.9 dynes/cm (TL-120)
Coeff. of Friction:
  0.07–0.10 (TL-120)

MECHANICAL PROPERTIES:

Tens. Str.:
  20.7–27.6 MPa (Teflon 100 FEP, 140 FEP, 160 FEP)
  2200 psi (Liquinite FEP)
  2400 psi (Thermocomp LF-1004M)
  3000 psi (Teflon 110 FEP, 120 FEP)
  6000 psi (Thermocomp LF-1004)

Tens. Elong.:
  2–3% (Thermocomp LF-1004)
  5% (Thermocomp LF-1004M)
  10–20% (Liquinite FEP)
  300% (Teflon 100 FEP, 110 FEP, 120 FEP, 160 FEP)
  325% (Teflon 140 FEP)

Flex. Str.:
  4000 psi (Thermocomp LF-1004M)
  10,500 psi (Thermocomp LF-1004)

Flex. Mod.:
  655 MPa (Teflon 100 FEP, 140 FEP, 160 FEP)
  95,000 psi (Teflon 110 FEP, 120 FEP)
  250,000 psi (Thermocomp LF-1004M)
  800,000 psi (Thermocomp LF-1004)

Shear Str.:
  2400 psi (Thermocomp LF-1004M)
  5500 psi (Thermocomp LF-1004)

Impact Str. (Izod):
  No break (Teflon 100 FEP, 140 FEP, 160 FEP)
  3.2 ft lb/in. notched (Thermocomp LF-1004M)
  8.0 ft lb/in. notched (Thermocomp LF-1004)

Hardness:
  Rockwell R45 (Liquinite FEP)
  Shore D55 (Teflon 100 FEP, 110 FEP, 120 FEP, 140 FEP, 160 FEP)

THERMAL PROPERTIES:

Conduct.:
  0.25 W/m • K (Teflon 100 FEP, 140 FEP, 160 FEP)
Distort. Temp.:
  150 F (264 psi) (Thermocomp LF-1004M)
  350 F (264 psi) (Thermocomp LF-1004)

Fluorinated ethylene-propylene (cont'd.)

Coeff. of Linear Exp.:
8.3–10.4 × 10^{-5} C^{-1} (Teflon 100 FEP, 140 FEP, 160 FEP)
2.4 × 10^{-5} in./in./F (Thermocomp LF-1004)
4.0 × 10^{-5} in./in./F (Thermocomp LF-1004M)

Flamm.:
Nonflammable (Teflon 110 FEP, 120 FEP)
V-0 (Liquinite FEP)
AEB 5 mm (Teflon 100 FEP, 140 FEP, 160 FEP)
ATB 5 s (Teflon 100 FEP, 140 FEP, 160 FEP)

ELECTRICAL PROPERTIES:

Dissip. Factor:
0.0001–0.001 (60 to 10^9 Hz) (Teflon 100 FEP, 110 FEP, 140 FEP)
0.0002–0.0007 (60 to 10^9 cps) (Teflon 120 FEP, 160 FEP)

Dielec. Str.:
83 kV/mm (Teflon 100 FEP, 140 FEP, 160 FEP)
2100 V/mil (Teflon 110 FEP, 120 FEP)
7000 V/mil (Liquinite FEP)

Dielec. Const.:
2.1 (60 to 10^9 Hz) (Teflon 100 FEP, 110 FEP, 120 FEP, 140 FEP, 160 FEP)

Vol. Resist.:
> 10^{16} ohm-cm (Teflon 100 FEP, 140 FEP, 160 FEP)
10^{18} ohm-cm (Liquinite FEP)
> 10^{18} ohm-cm (Teflon 110 FEP, 120 FEP)

TOXICITY/HANDLING:

Use with adequate ventilation at processing temps. (Teflon 100 FEP, 110 FEP, 120 FEP, 140 FEP, 160 FEP)

STORAGE/HANDLING:

May settle on standing or prolonged heating—redisperse by mild agitation; store at 4.4–18.3 C; redisperse twice a month by tumbling containers; protect against freezing (Teflon 120 FEP)

STD. PKGS.:

45.4-kg Leverpak containing two 22.7-kg heat-sealed removable bags (Teflon 100 FEP, 140 FEP, 160 FEP)
1-gal Cubitainer in cardboard carton, 5- and 30-gal, epoxy-lined drums (Teflon 120 FEP)

Melamine-formaldehyde resin

SYNONYMS:
Melamine resin
1,3,5-Triazine-2,4,6-triamine, polymer with formaldehyde

EMPIRICAL FORMULA:
$(C_3H_6N_6 \cdot CH_2O)_x$

CAS No.:
9003-08-1

TRADENAME EQUIVALENTS:
Cymel 303, 370, 373, 380, 1141 [Amer. Cyanamid]
Knightset EM-3 [GAF]
Uformite MM-46, MM-47, MM-55, MM-57 [Rohm & Haas]
Cellulose-filled:
M-4536, -7510 [Fiberite]
Plenco 801 White [Plastics Engineering]
Fabric-filled:
M-2015 [Fiberite]
Glass-filled:
M-2037 (chopped glass fiber), -2840, -2880, -3882 [Fiberite]
Mineral-filled:
M-6204 [Fiberite]

TYPES/MODIFICATIONS/SPECIALTY GRADES:
Methylated:
Cymel 303, 370, 373, 380
Alkylated:
Knightset EM-3
Alkylated carboxyl-modified:
Cymel 1141
High-impact:
M-2037, -2880
Plasticized:
M-4536

CATEGORY:
Thermosetting resin

Melamine-formaldehyde resin *(cont'd.)*

PROCESSING:
Compression molding:
   M-2015, -2840, -4536, -6204
Molding:
   M-2037, -2880, -3882; Plenco 801 White
Preform molding:
   M-2840
Transfer molding:
   M-2015, -2840, -4536, -6204

APPLICATIONS:
   Automotive applications: enamels (Uformite MM-47)
   Consumer products: (M-4536); metal furniture coatings (Uformite MM-47)
   Electrical/electronic industry: circuit breakers (M-2037, -2880, -3882, -7510); housings (M-2037, -2880, -3882, -7510); switchgear (M-2037, -2880, -3882, -7510)
   Functional additives: crosslinking agent (Cymel 303, 370, 373, 380, 1141; Knightset EM-3; Uformite MM-47); modifier (Uformite MM-55)
   Industrial applications: appliance finishes/parts (Cymel 303; M-4536; Uformite MM-47, MM-55); coatings (Cymel 303, 370, 373, 380; Uformite MM-46, MM-47, MM-55, MM-57); coil coatings (Cymel 303, 370, 380); decorative color components (M-4536); electrodeposition finishes (Cymel 303, 1141); metal decorating enamels (Cymel 370, 380; Uformite MM-47); molded parts (M-2015; Plenco 801 White); paper coatings (Cymel 303); pigments (Knightset EM-3); varnishes (Cymel 370, 380)

PROPERTIES:
Form:
   Liquid (Knightset EM-3)
   Clear liquid (Cymel 370, 373, 380; Uformite MM-46, MM-47, MM-55, MM-57)
   Clear, viscous liquid (Cymel 303, 1141)
   Granular (M-2840)
   Nodular (M-4536, -6204)
   Fibrous (M-2880, -3882)
   Macerated (M-7510)
   Macerated; also avail. in preforms (M-2015)
Color:
   Colorless (Uformite MM-46, MM-47, MM-55, MM-57)
   Natural (M-2840, -2880)
   White (Plenco 801 White)
   Pastels (M-4536)
   Brown, black (M-6204)
   Black and pastels (M-2015)
   Gardner 1 max. (Cymel 370, 373, 380)
   Gardner 2 max. (Cymel 303)

Melamine-formaldehyde resin *(cont'd.)*

Gardner 3 max. (Cymel 1141)

Composition:
50 ± 2% solids in xylene/butanol (1:4) (Uformite MM-55)
50 ± 2% solids in xylene/butanol (1:9) (Uformite MM-57)
60 ± 2% solids in xylene/butanol (1:1) (Uformite MM-46, MM-47)
80 ± 2% solids in IPA/butanol (Cymel 380)
85 ± 2% solids in water (Cymel 373)
85 ± 2% solids in isobutanol (Cymel 1141)
88 ± 2% solids in IPA/butanol (Cymel 370)
98% min. nonvolatiles (Cymel 303)

GENERAL PROPERTIES:
Solubility:
Sol. in 2-butoxyethanol (Cymel 370, 373, 380)
Sol. in ethanol (Cymel 370, 373, 380)
Sol. in 2-ethoxyethanol (Cymel 370, 373, 380)
Sol. in ethylene glycol (Cymel 370, 373, 380)
Sol. in isobutanol (Cymel 370, 380)
Sol. in isoporone (Cymel 370, 380)
Sol. in isopropanol (Cymel 373)
Sol. in MEK (Cymel 370, 373, 380)
Sol. in methanol (Cymel 370, 373, 380)
Sol. in MIBK (Cymel 370, 380)
Sol. in all common organic solvents (Cymel 1141)
Sol. in propylene glycol (Cymel 370, 373, 380)
Sol. in toluene (Cymel 370, 380)
Partly sol. in water (Cymel 370, 380); limited sol. in water (Cymel 303); disp. in water
   (Cymel 1141); water-emulsifiable (Knightset EM-3)
Sol. in xylene (Cymel 370, 380)

Sp. Gr.:
0.97 (Uformite MM-55)
0.98 (Uformite MM-57)
1.01 (Uformite MM-46)
1.02 (Uformite MM-47)
1.07 (Cymel 1141)
1.15 (Cymel 380)
1.26 (Cymel 373)
1.49–1.53 (molded) (M-2015)
1.60 (Plenco 801 White)
1.63–1.67 (molded) (M-4536)
1.64–1.68 (molded) (M-7510)
1.72–1.73 (molded) (M-6204)
1.79–1.84 (molded) (M-2840)
1.90–1.95 (molded) (M-2880)

1.93–1.97 (molded) (M-2037)
2.10–2.20 (molded) (M-3882)
Density:
0.63 g/cm³ (apparent) (Plenco 801 White)
8.1 lb/gal (Uformite MM-55)
8.2 lb/gal (Uformite MM-57)
8.4 lb/gal (Uformite MM-46)
8.5 lb/gal (Uformite MM-47)
8.95 lb/gal (Cymel 1141)
9.6 lb/gal (Cymel 380)
9.8 lb/gal (Cyme4l 370)
10.0 lb/gal (Cymel 303)
10.5 lb/gal (Cymel 373)
Visc.:
F–K (Uformite MM-55)
H–K (Uformite MM-57)
I–L (Uformite MM-46)
P–T (Uformite MM-47)
V–Z (Cymel 380)
W–Y (Cymel 1141)
$X–Z_2$ (Cymel 303)
$Z–Z_4$ (Cymel 373)
$Z_2–Z_4$ (Cymel 370)
Flash Pt.:
62 F (CC) (Cymel 380)
72 F (CC) (Cymel 370)
93 F (Seta CC) (Cymel 1141)
> 180 F (Cymel 303)
Acid No.:
0–1 (Uformite MM-46, MM-47)
0–1.5 (Uformite MM-55)
8–12 (Uformite MM-57)
22 ± 3 (resin solids) (Cymel 1141)
Stability:
Stable under acidic conditions (Cymel 1141)
Storage Stability:
3 mos. storage life (Plenco 801 White)
MECHANICAL PROPERTIES:
Tens. Str.:
4500 psi (M-7510)
5000 psi (M-6204)
6000 psi (M-4536)
6500 psi (M-3882)

Melamine-formaldehyde resin *(cont'd.)*

6900–8200 psi (Plenco 801 White)
7000 psi (M-2015)
7500 psi (M-2037)
8000 psi (M-2840, -2880)

Flex. Str.:
8000 psi (M-6204)
9000–12,000 psi (Plenco 801 White)
9700 psi (M-7510)
10,000 psi (M-4536)
12,000 psi (M-3882)
14,000 psi (M-2015, -2840)
18,000 psi (M-2880)
20,000 psi (M-2037)

Flex. Mod.:
10^6 psi (Plenco 801 White)
1.4×10^6 psi (M-7510)
1.6×10^6 psi (M-4536)
1.7×10^6 psi (M-2015)
1.9×10^6 psi (M-6204)
2.0×10^6 psi (M-2880)
2.2×10^6 psi (M-2840)
2.3×10^6 psi (M-3882)
2.7×10^6 psi (M-2037)

Compr. Str.:
18,000 psi (M-7510)
26,000 psi (M-6204)
26,800–28,000 psi (Plenco 801 White)
27,000 psi (M-2037)
30,000 psi (M-2880, -3882)
32,000 psi (M-2015, -4536)
42,000 psi (M-2840)

Impact Str. (Izod):
0.04 ft lb/in. notched (M-4536)
0.33–0.34 ft lb/in. (Plenco 801 White)
0.45 ft lb/in. notched (M-6204)
0.6 ft lb/in. notched (M-2840)
0.7 ft lb/in. notched (M-7510)
0.9 ft lb/in. notched (M-3882)
1.0 ft lb/in. notched (M-2015)
5.5 ft lb/in. notched (M-2880)
20.0 ft lb/in. notched (M-2037)

Hardness:
Rockwell M115 (M-4536, -6204, -7510)

Melamine-formaldehyde resin *(cont'd.)*

Rockwell M117 (M-2037)
Rockwell M120 (M-3882)
Rockwell M125 (M-2015, -2840, -2880)

THERMAL PROPERTIES:
Conduct.:
9 × 10⁻⁴ Cal/s/cm²/C/cm (M-2015)
10 × 10⁻⁴ Cal/s/cm²/C/cm (M-2880)
11 × 10⁻⁴ Cal/s/cm²/C/cm (M-2840)
12 × 10⁻⁴ Cal/s/cm²/C/cm (M-4536)
15 × 10⁻⁴ Cal/s/cm²/C/cm (M-6204)

Distort. Temp.:
177 C (264 psi) (M-6204)
193 C (264 psi) (M-4536)
196 C (264 psi) (M-2840)
202 C (264 psi) (M-2015)
227 C (264 psi) (M-2880)
290–300 F (264 psi) (Plenco 801 White)
330 F (264 psi) (M-3882)
400 F (264 psi) (M-7510)
450 F (264 psi) (M-2037)

Coeff. of Linear Exp.:
1.6 × 10⁻⁵ in./in./C (M-2037, -2880)
1.7 × 10⁻⁵ in./in./C (M-2015)
2.0 × 10⁻⁵ in./in./C (M-2840)
2.1 × 10⁻⁵ in./in./C (M-7510)
2.2 × 10⁻⁵ in./in./C (M-3882, -6204)
2.6 × 10⁻⁵ in./in./C (M-4536)

ELECTRICAL PROPERTIES:
Dissip. Factor:
0.02 (M-3882); (1 MC) (M-2840, -2880)
0.03 (M-2037); (1 MC) (M-6204)
0.04 (M-7510)
0.05 (1 MC) (M-4536)
0.07 (1 MC) (M-2015)
17.4–29.8 (1 kHz) (Plenco 801 White)

Dielec. Str.:
200–250 V/mil (Plenco 801 White)
225 V/mil (M-2015)
230 V/mil (M-2037)
250 V/mil (M-4536)
270 V/mil (M-7510)
300 V/mil (M-2840, -2880)
340 V/m8l (M-6204)

Melamine-formaldehyde resin *(cont'd.)*

    360 V/mil (M-3882)
Dielec. Const.:
    5.6 (1 MC) (M-6204)
    6.0 (1 MC) (M-2840)
    6.2 (M-3882); (1 MC) (M-2880)
    6.6 (1 MC) (M-4536)
    8.0 (M-2037); (1 MC) (M-2015)
    8.7 (M-7510)
    10.20–13.42 (1 Khz) (Plenco 801 White)
Arc Resist.:
    130 s (M-7510)
    140 s (M-2015; Plenco 801 White)
    180+ s (M-2037, -2840, -2880, -3882, -4536, -6204)

STORAGE/HANDLING:
    Will polymerize slowly in storage; store in cool, dry area, normally no longer than 3 mos. (Plenco 801 White)

Perfluoroalkoxy resin

SYNONYMS:
PFA resin

STRUCTURE:

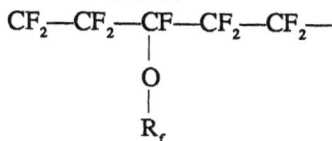

$$CF_2—CF_2—CF—CF_2—CF_2—$$
$$|$$
$$O$$
$$|$$
$$R_f$$

where $R_f = —C_nF_{2n+1}$

TRADENAME EQUIVALENTS:
Stat-Kon FP-P [LNP]
Teflon 340 PFA, 350 PFA [DuPont]
Carbon-reinforced:
Stat-Kon FP-PC-1004 (20% carbon fiber) [LNP]

MODIFICATIONS/SPECIALTY GRADES:
Statically dissipative:
Stat-Kon FP-P

CATEGORY:
Thermoplastic resin

PROCESSING:
Blow molding:
Teflon 340 PFA, 350 PFA
Compression molding:
Teflon 340 PFA, 350 PFA
Extrusion:
Teflon 340 PFA, 350 PFA
Injection molding:
Teflon 340 PFA, 350 PFA

APPLICATIONS:
Electrical/electronic industry: cable and wire jacketing (Teflon 340 PFA, 350 PFA); electrical components (Stat-Kon FP-P); electrical enclosures/packaging (Stat-Kon FP-P); insulation (Teflon 340 PFA, 350 PFA)

Industrial applications: chemical process equipment liner (Teflon 350 PFA); extruded shapes (Teflon 340 PFA); molded parts (Teflon 340 PFA, 350 PFA); tubing/hoses (Teflon 340 PFA, 350 PFA)

Perfluoroalkoxy resin *(cont'd.)*

PROPERTIES:
Form:
   Pellets; 2.5 mm diam. (Teflon 340 PFA, 350 PFA)
Color:
   Translucent (Teflon 340 PFA, 350 PFA)

GENERAL PROPERTIES:
Sp. Gr.:
   2.08 (Stat-Kon FP-P, FP-PC-1004)
   2.12–2.17 (Teflon 340 PFA, 350 PFA)
M.P.:
   302–310 C (Teflon 340 PFA, 350 PFA)
Stability:
   Excellent weather and chemical resistance (Teflon 340 PFA, 350 PFA)

MECHANICAL PROPERTIES:
Tens. Str.:
   13.8 MPa (yield) (Teflon 340 PFA)
   15.2 MPa (yield) (Teflon 350 PFA)
   27.6 MPa (Teflon 340 PFA)
   29.6 MPa (Teflon 350 PFA)
   3200 psi (Stat-Kon FP-P)
   4500 psi (Stat-Kon FP-PC-1004)
Tens. Elong.:
   1.5% (Stat-Kon FP-PC-1004)
   20% (Stat-Kon FP-P)
   300% (break) (Teflon 340 PFA, 350 PFA)
Flex. Str.:
   4100 psi (Stat-Kon FP-P)
   6200 psi (Stat-Kon FP-PC-1004)
Flex. Mod.:
   655 MPa (Teflon 340 PFA)
   689.5 MPa (Teflon 350 PFA)
   150,000 psi (Stat-Kon FP-P)
   1.0×10^6 psi (Stat-Kon FP-PC-1004)
Impact Str. (Izod):
   No break notched (Stat-Kon FP-P)
   2.5 ft lb/in. notched (Stat-Kon FP-PC-1004)
Hardness:
   Shore D60 (Teflon 340 PFA, 350 PFA)
Mold Shrinkage:
   0.009 in./in. (Stat-Kon FP-PC-1004)
   0.030 in./in. (Stat-Kon FP-P)

Water Absorp.:
0.01% (Stat-Kon FP-P, FP-PC-1004)

THERMAL PROPERTIES:

Conduct.:
2.4 Btu/h/ft^2/F/in. (Stat-Kon FP-P)
2.8 Btu/h/ft^2/F/in. (Stat-Kon FP-PC-1004)

Distort. Temp.:
270 F (264 psi) (Stat-Kon FP-P)
> 500 F (264 psi) (Stat-Kon FP-PC-1004)

Coeff. of Linear Exp.:
12.1 × 10^{-5} mm/mm/C (Teflon 340 PFA, 350 PFA)
3.0 × 10^{-5} in./in./F (Stat-Kon FP-PC-1004)
3.1 × 10^{-5} in./in./F (Stat-Kon FP-P)

Flamm.:
V-0 (Stat-Kon FP-P, FP-PC-1004)

ELECTRICAL PROPERTIES:

Dissip. Factor:
< 0.0001–0.0003 (60 to 10^6 Hz) (Teflon 340 PFA)
0.0002–0.0003 (60 to 10^6 Hz) (Teflon 350 PFA)

Dielec. Str.:
78.8 kV/m (Teflon 340 PFA, 350 PFA)

Dielec. Const.:
2.1 (60 to 10^6 Hz) (Teflon 340 PFA, 350 PFA)

Vol. Resist.:
10 ohm-cm (Stat-Kon FP-PC-1004)
100 ohm-cm (Stat-Kon FP-P)
10^{18} ohm-cm (Teflon 350 PFA)
> 10^{18} ohm-cm (Teflon 340 PFA)

Surf. Resist.:
10 ohm/sq. (Stat-Kon FP-PC-1004)
100 ohm/sq. (Stat-Kon FP-P)

TOXICITY/HANDLING:

Use with adequate ventilation at processing temps. (Teflon 340 PFA, 350 PFA)

STD. PKGS.:

11.4-kg Fiberpak and 45.4-kg Leverpak containers with heat-sealed removable poly-ethylene liners (Teflon 340 PFA, 350 PFA)

Polycarbonate resin

SYNONYMS:
PC resin

EMPIRICAL FORMULA:
$(COOC_6H_5C(CH_3)_2C_6H_5O)_n$

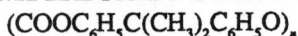

STRUCTURE:
Generalized structure (based on bisphenol A reacted with carbonyl chloride)

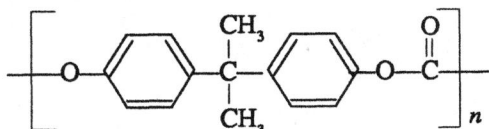

TRADENAME EQUIVALENTS:
Calibre 200-3, 200-6, 200-10, 200-15, 300-3, 300-6, 300-10, 300-15, 700-3, 700-6, 700-10, 700-15, 800-3, 800-6, 800-10, 800-15, 7070 [Dow]

Lexan 8A13, 8A23, 8A33, 8A43, 8B13, 8B23, 8B33, 8B35, 8B36, 8B43, 8B327, 8B328, 8C20, 101, 103, 104, 123, 121, 124, 141, 143, 144, 150, 303, 500, 920, 920A, 940, 940A, 950, 950A, 1500, 2014, 8010-112, 8020, 8030, 8040, 8060, 8800, 8800-112, CF-20 MB-112, CF-23 MB-112, CF-25 MB-112, CF-26 MB-112, CF-60 MB-112, FL-900, FL-910, FL-930, FL-1000, FL-1800, FR60-112, FR60SE-112, FR63-112, FR65-112, FR66-112, FR88-112, S100, S200, SG400, Sheet 9034, Thermoclear, XL Sheet [General Electric]

Makrofol E, G, KG, N, SG, SKG, SN [Bayer AG]

Makrolon 2400, 2403, 2405, 2600, 2603, 2605, 2800, 2803, 2805, 2807, 2808, 2809, 3100, 3103, 3105, 3108, 3109, 3119, 3200, 3203, 3208, 6030, 6550, 6560, 6553, 6555, 6557, 6603, 6655, 6870 [Bayer AG]

Marc Polycarbonate [MRC Enterprises] (reprocessed)

Margard Sheet [General Electric]

Merlon 5300, 6450, 6560, 6870, AL-400, CD-2000, FAR, HMS-3119, M39, M40, M50, MPG-700, MPG-750, SF-600, T70, T85 [Mobay]

Migralube DL-4030, DL-4410, DL-4530 [LNP]

PDX-84368 [LNP]

Polypenco Polycarbonate [Polymer Corp.]

Protect-A-Glaze Sheet [General Electric]

Thermocomp DL-4030 [LNP]

Zelux W (various grades) [Westlake Plastics]

Glass-reinforced:

Calibre 510, 550 [Dow]

Lexan 3412 (20% glass fiber), 3413 (30% glass fiber), 3414 (40% glass fiber) [General Electric]

Makrolon 8020 (20% glass staple fiber), 8025 (20% glass staple fiber), 8030 (30% glass staple fiber), 8035 (30% glass staple fiber), 8315 (10% glass staple fiber), 8320 (20% glass staple fiber), 8324 (20% glass staple fiber), 8344 (35% glass staple fiber), 9310 (10% glass staple fiber), 9410 (10% glass staple fiber), 9415 (10% glass staple fiber) [Bayer AG]

Merlon 8310 (10% glass fiber), 9310 (10% glass fiber), 9510 (10% glass fiber) [Mobay]

RTP 301 (10% glass fiber), 303 (20% glass fiber), 303TFE10 (20% glass fiber), 305 (30% glass fiber), 305TFE15 (30% glass fiber), 307 (40% glass fiber) [Fiberite]

Thermocomp DF-1004 (20% glass fiber), DF-1006 (30% glass fiber), DF-1008 (40% glass fiber), DFA-113 (10% glass fiber), DFL-4036 (30% glass fiber) [LNP]

Glass/carbon-reinforced:

Stat-Kon CDF-1006 (15% carbon fiber, 15% glass fiber), DFD-15 (15% glass fiber) [LNP]

Aluminum flake-filled:

EMI-X DA-30 (30% aluminum flake), DA-35 (35% aluminum flake), DA-40 (40% aluminum flake [LNP]

Carbon-reinforced:

EMI-X DC-1008 (40% carbon fiber) [LNP]

PC-20CF (20% carbon fiber), -30CF (30% carbon fiber), -40CF (40% carbon fiber) [Compounding Technology]

Stat-Kon D (carbon powder), DC-1003FR (15% carbon fiber), DC-1006 (30% carbon fiber), DCL-4033 (15% carbon fiber) [LNP]

Thermocomp DC-1006 (30% PAN carbon fiber), DC-1006PC (30% pitch carbon fiber) [LNP]

Nickel-modified:

PDX-83393 (25% nickel) [LNP]

Stainless steel-modified

PDX-84356 [LNP]

MODIFICATIONS/SPECIALTY GRADES:

Film grade:

Lexan 8A13, 8A23, 8A33, 8A43, 8B13, 8B23, 8B33, 8B35, 8B36, 8B43, 8B327, 8B328, 8C20, 8010-112, 8020, 8030, 8040, 8060, 8800, 8800-112, CF-20 MB-112, CF-23 MB-112, CF-25 MB-112, CF-26 MB-112, CF-60 MB-112, FR60-112, FR60SE-112, FR63-112, FR65-112, FR66-112, FR88-112; Makrofol E, G, KG, N, SG, SKG, SN

Foam grade:

Lexan FL400, FL410, FL478, FL1000

Glazing sheet grade:

Lexan S100, S200, SG400, Sheet 9034, Thermoclear, XL Sheet; Margard Sheet; Protect-A-Glaze Sheet

Antistatic grade:

PDX-84368; Stat-Kon D, DC-1003FR, DC-1006, DCF-1006, DCL-4033, DF-15

Conductive grade:

EMI-X DA-30, DA-35, DA-40, DC-1008PDX-83393, -84356

Polycarbonate resin *(cont'd.)*

Flame-retardant:
> Calibre 510, 550, 700-3, 700-6, 700-10, 700-15; EMI-X DA-30, DA-35, DA-40, DC-1008; Lexan 500, 920, 920A, 940, 940A, 950, 950A, 2014, 3412, 3413, 3414, CF-60 MB-112, FL400, FL410, FL478, FL900, FL910, FL930, FL1000, FL1800, FR60-112, FR60SE-112, FR63-112, FR65-112, FR66-112, FR88-112; Makrofol SKG, SN; Makrolon 6030, 6550, 6553, 6555, 6557, 6560, 6603, 6655, 6870, 9310, 9410, 9415; Merlon 6450, 6560, 6870, 9310, 9510; PDX-84356; Stat-Kon DC-1003FR; Thermocomp DFA-113, DFL-4036, DL-4020FR; Zelux W (flame-resistant grade)

Ignition-resistant:
> Calibre 800-3, 800-6, 800-10, 800-15

UV-stabilized:
> Lexan 8B33, 8B35, 8B36, 103, 123, 143, SG400, Sheet 9034, Thermoclear, XL Sheet; Makrolon 2403, 2603, 2803, 2807, 3103, 3203, 6603, 6553, 6557; Margard Sheet; Zelux W (various grades)

High flow:
> Merlon CD-2000

High gloss:
> Lexan CF-20 MB-112, FR60-112, FR60SE-112, FR88-112, S100

Lubricated:
> **PTFE/TFE-lubricated:**
> > Migralube DL-4030 (15% TFE); RTP 303TFE10 (10% TFE), RTP 305TFE15 (15% TFE); Stat-Kon DCL-4033 (15% PTFE); Thermocomp DFL-4036 (15% TFE), DL-4030 (15% PTFE)

> **Silicone-lubricated:**
> > Migralube DL-4410

> **TFE/silicone-lubricated:**
> > Migralube DL-4530 (15% TFE/silicone)

Surface treated:
> Lexan 8C20 (coated one side), Thermoclear (uv-resistant coating); Margard Sheet (silicone-coated both sides)

Smooth finish:
> Lexan 8A13 (one side), 8A23 (one side), 8A33 (one side), 8A43 (one side), 8010-112 (both sides), 8020 (both sides), 8030 (both sides), 8040 (both sides), 8060 (both sides), 8800 (both sides)

Matte finish:
> Lexan 8A13 (one side), 8A23 (one side), 8A33 (one side), 8A43 (one side), 8B13 (both sides), 8B23 (both sides), 8B33 (both sides), 8B35 (both sides), 8B36 (both sides), 8B43 (both sides), 8B327 (both sides), 8B328 (both sides), FR63-112 (one side), FR65-112 (one side), FR66-112 (one side), S200 (one side)

CATEGORY:
Thermoplastic resin

Polycarbonate resin *(cont'd.)*

PROCESSING:

Extrusion:

Lexan 150, 950; Makrolon 3103, 3105, 3108, 3109, 3119, 3203, 3208, 6030, 6603, 6655, 6870, 8025, 8030, 8035, 8315, 8320, 8324, 8344, 9310, 9410, 9415; Marc Polycarbonate; Merlon 6450, 6560, 6870, HMS-3119, M39, M40, M50, MPG-700, MPG-750

Injection molding:

Calibre 200-3, 200-6, 200-10, 200-15, 300-3, 300-6, 300-10, 300-15, 510, 550, 700-3, 700-6, 700-10, 700-15, 800-3, 800-6, 800-10, 800-15; EMI-X DA-30, DA-35, DA-40, DC-1008; Lexan 101, 103, 104, 123, 124, 143, 144, 121, 141, 303, 500, 920, 920A, 940, 940A, 950, 950A, 1500, 2014, 3412, 3413, 3414, FL900, FL910, FL930, FL1800; Makrolon 2400, 2403, 2405, 2600, 2603, 2605, 2800, 2803, 2805, 2807, 2808, 2809, 3100, 3103, 3105, 3108, 3109, 3119, 3200, 3208, 6030, 6550, 6560, 6553, 6555, 6557, 6603, 6655, 6870, 8020, 8025, 8030, 8035, 8315, 8320, 8324, 8344, 9310, 9410, 9415; Marc Polycarbonate; Merlon 6450, 6560, 6870, 8310, 9310, 9510, CD-2000, M39, M40, M50; RTP 301, 303, 303TFE10, 305, 305TFE15, 307; Thermocomp DF-1004, DF-1006, DF-1008, DFA-113, DFL-4036, DL-4030

Structural foam molding:

Calibre 7070; Lexan FL400, FL410, FL478, FL1000; Merlon SF-600

Blow molding:

Lexan 150; Makrolon 3119; Merlon HMS-3119, MPG-700, MPG-750

Thermoforming:

Lexan FR60-112, FR60SE-112, FR63-112, FR65-112, FR66-112, FR88-112, Sheet 9034, Thermoclear, XL Sheet; Makrofol E, G, KG, N, SG, SKG, SN; Zelux W (thermoforming grade)

Rotational molding:

Merlon 5300

APPLICATIONS:

Architectural applications: glazing for buildings (Margard Sheet); greenhouse glazing (Lexan Thermoclear); overhead and vertical glazing (Protect-A-Glaze Sheet); portholes (Zelux W); protective glazing (Lexan Sheet 9034, Thermoclear; Zelux W); skylights (Lexan Thermoclear, XL Sheet); space enclosures (Lexan Thermoclear, XL Sheet; Margard Sheet); windows (Margard Sheet)

Automotive applications: (Lexan FL410, FL478, FL900, FL910, FL930, FL1000, FL1800); exterior parts (Merlon SF-600); interior door panels (Calibre 7070); interior components (Merlon SF-600); lens (Merlon AL-400)

Aviation industry: (Lexan 920, 920A, 940, 940A, 950, 950A, 2014; Merlon 6450, 6560, 6870, FAR); avionics housings (EMI-X DA-30, DA-35, DA-40, DC-1008; PDX-83393, -84356)

Consumer products: appliances (Calibre 300-3, 300-6, 300-10, 300-15, 700-3, 700-6, 700-10, 700-15, 800-3, 800-6, 800-10, 800-15; Lexan 8A13, 8A23, 8A33, 8A43, 8B13, 8B23, 8B33, 8B35, 8B36, 8B43, 8B327, 8B328, 8C20, 920, 920A, 940,

Polycarbonate resin *(cont'd.)*

940A, 950, 950A, 8010-112, 8020, 8030, 8040, 8060, 8800, CF-20 MB-112, CF-23 MB-112, CF-25 MB-112, CF-26 MB-112, CF-60 MB-112, FL410, FL478, FL900, FL910, FL930, FL1000, FL1800, FR60-112, FR60SE-112, FR63-112, FR65-112, FR66-112, FR88-112; Merlon 6450, 6560, 6870); compact disks (Merlon CD-2000); hardware (Lexan 920, 950); housewares (Calibre 300-3, 300-6, 300-10, 300-15; Makrolon 2400, 2403, 2405, 2600, 2603, 2605, 2800, 2803, 2805, 2807, 2808, 2809, 3100, 3103, 3105, 3119, 3200, 3203, 6030, 6550, 6553, 6555, 6557, 6560, 6603, 6655, 6870, 8020, 8030, 8320, 8344, 9310, 9410); outdoor applications (Lexan 103, 123, 143; Zelux W); recreational equipment (Calibre 300-3, 300-6, 300-10, 300-15; Makrolon 2400, 2403, 2405, 2600, 2603, 2605, 2800, 2803, 2805, 2807, 2808, 2809, 3100, 3103, 3105, 3119, 3200, 3203, 6030, 6550, 6560, 6603, 8344, 9310, 9410); safety applications (Makrolon 2400, 2403, 2405, 2600, 2603, 2605, 2800, 2803, 2805, 2807, 2808, 2809, 3100, 3103, 3105, 3119, 3200, 3203, 6030, 6550, 6553, 6555, 6557, 6560, 6603, 6655, 6870, 8020, 8030, 8320, 8344, 9310, 9410); smoke detectors (Stat-Kon D, DC-1003FR, DC-1006, DCF-1006, DCL-4033, DF-15)

Electrical/electronic industry: (Calibre 510, 550, 800-3, 800-6, 800-10, 800-15; EMI-X DA-30, DA-35, DA-40, DC-1008; Lexan 920, 920A, 940, 940A, 950, 950A, 2014, 8800-112; Makrolon 2400, 2403, 2405, 2600, 2603, 2605, 2800, 2803, 2805, 2807, 2808, 2809, 3100, 3103, 3105, 3119, 3200, 3203, 6030, 6550, 6553, 6555, 6557, 6560, 6603, 6655, 6870, 8020, 8030, 8320, 8344, 9310, 9410; Merlon 6450, 6560, 6870, 8310, 9310, 9510; PDX-83393, -84356, -84368; Thermocomp DF-1004, DF-1006, DF-1008); business machines and office equipment (Calibre 300-3, 300-6, 300-10, 300-15, 800-3, 800-6, 800-10, 800-15, 7070; EMI-X DA-30, DA-35, DA-40, DC-1008; Lexan 8A13, 8A23, 8A33, 8A43, 8B13, 8B23, 8B33, 8B35, 8B36, 8B43, 8B327, 8B328, 8C20, 920, 920A, 940, 940A, 950, 950A, 2014, 8010-112, 8020, 8030, 8040, 8060, CF-20 MB-112, CF-23 MB-112, CF-25 MB-112, CF-26 MB-112, CF-60 MB-112, FL410, FL478, FL900, FL910, FL930, FL1000, FL1800, FR60-112, FR60SE-112, FR63-112, FR65-112, FR66-112, FR88-112; Merlon SF-600; PDX-83393, -84356; RTP 301, 303, 305, 307; Thermocomp DFA-113); cable and wire jacketing (Makrofol E, G, KG, N, SG, SKG, SN); capacitors (Makrofol KG, SG, SKG); computers (Lexan 920, 920A, 940A, 950, 950A; Stat-Kon D, DC-1003FR, DC-1006, DCF-1006, DCL-4033, DF-15; Thermocomp DF-1004, DF-1006, DF-1008); electrical components (Stat-Kon D, DC-1003FR, DC-1006, DCF-1006, DCL-4033, DF-15); electrical enclosures/packaging (Stat-Kon D, DC-1003FR, DC-1006, DCF-1006, DCL-4033, DF-15); housings (Calibre 700-3, 700-6, 700-10, 700-15; EMI-X DA-30, DA-35, DA-40, DC-1008; Merlon SF-600; PDX-83393, -84356; Thermocomp DFA-113); insulation (Lexan FR60-112, FR60SE-112, FR63-112, FR65-112, FR66-112, FR88-112; Makrofol E, G, KG, N, SG, SKG, SN); lighting applications (Lexan 103, 123, 143, 303, 920; Makrolon 2400, 2403, 2405, 2600, 2603, 2605, 2800, 2803, 2805, 2807, 2808, 2809, 3100, 3103, 3105, 3119, 3200, 3203, 6030, 6550, 6553, 6555, 6557, 6560, 6603, 6655, 6870, 8020, 8030, 8320, 8344, 9310, 9410); optical memory substrates (Merlon

Polycarbonate resin *(cont'd.)*

CD-2000); power tools (RTP 301, 303, 305, 307); radio/TV (Makrofol E, G, KG, N, SG, SKG, SN); speaker membranes (Makrofol E, G, KG, N, SG, SKG, SN); telecommunications (Lexan 920, 940, 950, 2014, FL410, FL478, FL900, FL910, FL930, FL1000, FL1800, FR60-112, FR60SE-112, FR63-112, FR65-112, FR66-112, FR88-112; Makrofol E, G, KG, N, SG, SKG, SN; Merlon 6450, 6560, 6870); TV graphic overlays (Lexan FR60-112, FR60SE-112, FR63-112, FR65-112, FR66-112, FR88-112)

FDA-approved applications: (Calibre 200-3, 200-6, 200-10, 200-15; Lexan 8A43, 8B43, 104, 124, 144, 8040; Makrolon 2808, 3108, 3208; Merlon M39, M40, M50; Zelux W)

Food-contact applications: (Calibre 200-3, 200-6, 200-10, 200-15; Lexan 104, 124, 144; Zelux W)

Industrial applications: (Lexan 8800-112); bearing applications (Thermocomp DFL-4036); cams (RTP 301, 303, 303TFE10, 305, 305TFE15, 307; Thermocomp DL-4030); caps, troughs, sleeves, cups (Makrofol E, N, SN); coils (Polypenco Polycarbonate); frictional/sliding applications (Migralube DL-4030, DL-4410, DL-4530; RTP 303TFE10, 305TFE15; Thermocomp DFL-4036, DL-4030); gears (RTP 301, 303, 303TFE10, 305, 305TFE15, 307; Thermocomp DFL-4036, DL-4030); graphic applications (Lexan CF-20 MB-112, CF-23 MB-112, CF-25 MB-112, CF-26 MB-112, CF-60 MB-112); housings (Calibre 7070; Lexan 920A, 940A, 950A; Thermocomp DF-1004, DF-1006, DF-1008); insulators (Polypenco Polycarbonate; Thermocomp DF-1004, DF-1006, DF-1008); intricate/hard-to-fill moldings (Lexan 121, FL400); material handling (Lexan FL410, FL478, FL900, FL910, FL930, FL1000, FL1800); mechanical engineering (Makrolon 2400, 2403, 2405, 2600, 2603, 2605, 2800, 2803, 2805, 2807, 2808, 2809, 3100, 3103, 3105, 3119, 3200, 3203, 6030, 6550, 6553, 6555, 6557, 6560, 6603, 6655, 6870, 8020, 8030, 8320, 8344, 9310, 9410; Merlon 8310, 9310, 9510; Thermocomp DF-1004, DF-1006, DF-1008); office supplies (Makrolon 2400, 2403, 2405, 2600, 2603, 2605, 2800, 2803, 2805, 2807, 2808, 2809, 3100, 3103, 3105, 3119, 3200, 3203, 6030, 6550, 6553, 6555, 6557, 6560, 6603, 6655, 6870, 8020, 8030, 8320, 8344, 9310, 9410); opaque applications (Calibre 800-3, 800-6, 800-10, 800-15); optical parts (Calibre 700-3, 700-6, 700-10, 700-15; Makrolon 2400, 2403, 2405, 2600, 2603, 2605, 2800, 2803, 2805, 2807, 2808, 2809, 3100, 3103, 3105, 3119, 3200, 3203, 6030, 6550, 6553, 6555, 6557, 6560, 6603, 6655, 6870, 8020, 8030, 8320, 8344, 9310, 9410; Polypenco Polycarbonate; Zelux W); packaging/displays (Lexan CF-20 MB-112, CF-23 MB-112, CF-25 MB-112, CF-26 MB-112, CF-60 MB-112; Merlon MPG-700, MPG-750); photographic and film equipment (Makrolon 2400, 2403, 2405, 2600, 2603, 2605, 2800, 2803, 2805, 2807, 2808, 2809, 3100, 3103, 3105, 3119, 3200, 3203, 6030, 6550, 6553, 6555, 6557, 6560, 6603, 6655, 6870, 8020, 8030, 8320, 8344, 9310, 9410); precision engineering (Makrolon 2400, 2403, 2405, 2600, 2603, 2605, 2800, 2803, 2805, 2807, 2808, 2809, 3100, 3103, 3105, 3119, 3200, 3203, 6030, 6550, 6553, 6555, 6557, 6560, 6603, 6655, 6870, 8020, 8030, 8320, 8344, 9310, 9410); process engineering (Makrolon 2400, 2403, 2405,

Polycarbonate resin *(cont'd.)*

2600, 2603, 2605, 2800, 2803, 2805, 2807, 2808, 2809, 3100, 3103, 3105, 3119, 3200, 3203, 6030, 6550, 6553, 6555, 6557, 6560, 6603, 6655, 6870, 8020, 8030, 8320, 8344, 9310, 9410); product trim (Lexan CF-20 MB-112, CF-23 MB-112, CF-25 MB-112, CF-26 MB-112, CF-60 MB-112); profiles (Merlon HMS-3119); safety equipment (Lexan 940, 2014); service parts (Calibre 510, 550); sheet (Merlon HMS-3119); shrink coils for cylindrical components (Makrofol G, SG); signs (Lexan S100, S200, SG400); structural components (Lexan FL900, FL910, FL930, FL1800; Merlon 8310, 9310, 9510; Polypenco Polycarbonate; Thermocomp DF-1004, DF-1006, DF-1008); thick sections (Lexan 101)

Medical applications: (Calibre 200-3, 200-6, 200-10, 200-15; Zelux W)

Transportation industry: (Calibre 300-3, 300-6, 300-10, 300-15, 510, 550, 700-3, 700-6, 700-10, 700-15, 800-3, 800-6, 800-10, 800-15, 7070; Lexan 8A13, 8A23, 8A33, 8A43, 8B13, 8B23, 8B33, 8B35, 8B36, 8B43, 8B327, 8B328, 8C20, 8010-112, 8020, 8030, 8040, 8060, CF-20 MB-112, CF-23 MB-112, CF-25 MB-112, CF-26 MB-112, CF-60 MB-112); windows in combat helicopters (Zelux W)

PROPERTIES:

Form:

Cast film (Makrofol G, KG, N, SG, SKG, SN)

Extruded film (Makrofol E)

Film; 0.0005–0.005 mil gauges (Lexan 8800)

Film; 0.003–0.030 mil gauges (Lexan 8A13, 8A23, 8A33, 8A43, 8B13, 8B23, 8B33, 8B43, 8B328, 8C20, 8010-12, 8030, 8040)

Film; 0.005–0.020 mil gauges (Lexan 8B35)

Film; 0.005–0.030 mil gauges in colored (Lexan 80200

Film; 0.007–0.030 mil gauges (Lexan 8020)

Film; 0.010–0.020 mil gauges (Lexan 8B36, 8B327, 8B328)

Film; 0.010–0.030 mil gauges (Lexan 8C20)

Film; 0.013–0.13 mm gauge (Lexan 8800-112)

Film; 0.015–0.030 mil gauges (Lexan 8060)

Film; 10–20 mil gauges (Lexan CF-25 MB-112, CF-26 MB-112, FR65-112, FR66-112)

Film; 10–25 mil gauges (Lexan CF-23 MB-112, FR63-112, FR88-112)

Film; 10–30 mil gauges (Lexan CF-20 MB-112, FR60-112)

Film; 25–30 mil gauges (Lexan CF-60 MB-112)

Film; 30–80 mil gauges (Lexan FR60SE-112)

Sheet (Marc Polycarbonate; Zelux W)

Sheet in standard thicknesses: 6, 8, 10, 16 mm (Lexan Thermoclear)

Sheet in standard thicknesses: 0.060, 0.080, 0.075, 0.093, 0.125, 0.187, 0.250 in. (Lexan S100)

Sheet in standard thicknesses: 0.080, 0.093, 0.125, 0.187, 0.250, 0.375, 0.5 in. (Lexan Sheet 9034)

Sheet in standard thicknesses: 0.080, 0.125, 0.187, 0.250 in.; industrial thicknesses: 0.125, 0.187, 0.25 in. (Protect-A-Glaze Sheet)

Polycarbonate resin *(cont'd.)*

Sheet in standard thicknesses: 0.093, 0.125, 0.187, 0.250 in. (Lexan S300)
Sheet in standard thicknesses: 0.093, 0.125, 0.187 in. (Lexan SG400)
Sheet in standard thicknesses: 0.125, 0.187, 0.250, 0.375, 0.5 in. (Lexan XL Sheet; Margard Sheet)
Powder (Merlon 5300)
Pellets (Lexan 101, 103, 104, 121, 123, 124, 141, 143, 144, 150, 303, 500, 940, 2014, 3412, 3413, 3414; Marc Polycarbonate)
Cylindrical granules (Makrolon 2400, 3100, 3119, 3200, 6030, 6550, 6560, 6603, 6655, 6870, 8020, 8030, 8320, 8344, 9310, 9410)
Rod (Polypenco Polycarbonate)
Color:
Colorless (Makrofol E)
Transparent (Polypenco Polycarbonate)
Transparent, translucent, and/or opaque colors avail. (Makrolon 2400, 3100, 3119, 3200, 6030, 6550, 6560, 6603, 6655, 6870, 8020, 8030, 8320, 8344, 9310, 9410)
Transparent, opaque, colorable (Merlon 6450, HMS-3119, M39, M40, M50)
Opaque, colorable (Merlon 6560, 8310, 9310, SF-600)
Transparent, colorable (Merlon 6870)
Clear (Lexan 8010-112, 8800-112)
Clear, custom colors (Lexan Thermoclear)
Clear, white, red, yellow, blue, black, green, custom colors (Lexan S100, S200, SG400)
Clear, amber, blue, blue/green, smoke (standard); gray (industrial) (Protect-A-Glaze Sheet)
Clear, bronze, gray, custom colors (Lexan Sheet 9034, XL Sheet)
Clear, bronze, gray, light green, custom colors (Margard Sheet)
Natural and transparent (Lexan 121)
Natural and clear tints (Merlon 5300, CD-2000)
Natural clear and colors (Lexan 150)
Natural, black (RTP 301, 303, 303TFE10, 305, 305TFE15, 307)
Natural, black, and custom colors (Lexan 940)
Golden yellow (Makrofol G, N)
Yellowish green (Makrofol KG)
Turquoise (Makrofol SKG)
Red (Makrofol SG, SN)
Standard black and custom colors (Lexan 950)

GENERAL PROPERTIES:
Solubility:
Swelling effect in acetone (Makrolon 2400, 2600, 2800, 3100, 3119, 3200, 6030, 6550, 6560, 6603, 6655, 6870, 8020, 8030, 8320, 8344, 9310, 9410)
Swelling effect in aromatic hydrocarbons (Makrofol E, G, KG, N, SG, SKG, SN)
Swelling effect in benzene (Makrolon 2400, 2600, 2800, 3100, 3119, 3200 6030, 6550, 6560, 6603, 6655, 6870, 8020, 8030, 8320, 8344, 9310, 9410)

Polycarbonate resin *(cont'd.)*

Swelling effect in carbon tetrachloride (Makrolon 2400, 2600, 2800, 3100, 3119, 3200
    6030, 6550, 6560, 6603, 6655, 6870, 8020, 8030, 8320, 8344, 9310, 9410)
Swelling effect in chloroform (Makrofol E, G, KG, N, SG, SKG, SN)
Sol. in methylene chloride (Makrofol E, G, KG, N, SG, SKG, SN)
Sol. in many commercial solvents (Makrolon 2400, 2600, 2800, 3100, 3119, 3200
    6030, 6550, 6560, 6603, 6655, 6870, 8020, 8030, 8320, 8344, 9310, 9410)

Melt Flow:
3 g/10 min (Calibre 200-3, 300-3, 700-3, 800-3)
6 g/10 min (Calibre 200-6, 300-6, 700-6, 800-6)
10 g/10 min (Calibre 200-10, 300-10, 700-10, 800-10)
15 g/10 min (Calibre 200-15, 300-15, 700-15)
55–60 g/10 min (Merlon CD-2000)
6.3 dg/min (Marc Polycarbonate)

Sp. Gr.:
0.90 (Lexan FL478 (foam molded), FL900, FL1000 (foam molded))
0.95 (Lexan FL400, FL910, FL1800)
1.00 (Lexan FL410 (foam molded))
1.05 (Calibre 7070; Lexan FL930)
1.18 (Migralube DL-4410)
1.20 (Calibre 200-3, 200-6, 200-10, 200-15, 300-3, 300-6, 300-10, 300-15, 700-3, 700-
    6, 700-10, 700-15; Lexan 8A13, 8A23, 8A33, 8A43, 8B13, 8B23, 8B33, 8B35,
    8B36, 8B43, 8B327, 8B328, 8C20, 101, 103, 104, 121, 123, 124, 141, 143, 144, 303,
    8010-112, 8020, 8030, 8040, 8060, 8800-112, CF-20 MB-112, CF-23 MB-112, CF-
    25 MB-112, CF-26 MB-112, CF-60 MB-112, FL1000 (solid); Polypenco Polycar-
    bonate; Zelux W)
1.21 (Calibre 800-3, 800-6, 800-10, 800-15; Lexan 920, 920A, 940, 940A, 950, 950A)
1.23 (PDX-84368)
1.24 (Lexan 2014)
1.25 (Lexan 500, FL410 (solid); Thermocomp DFA-113)
1.26 (Lexan FL478 (solid); Migralube DL-4530; PDX-84356; Thermocomp DL-
    4020FR)
1.27 (Calibre 510, 550; RTP 301)
1.28 (Migralube DL-4030; PC-20CF; Stat-Kon D, DC-1003FR; Thermocomp DL-
    4030)
1.32 (Lexan FR60-112, FR60SE-112, FR63-112, FR65-112, FR66-112, FR88-112)
1.33 (PC-30CF; Stat-Kon DC-1006; Thermocomp DC-1006)
1.34 (RTP 303; Thermocomp DF-1004)
1.35 (Lexan 3412; Thermocomp DC-1006PC)
1.36 (Stat-Kon DCL-4033)
1.37 (PC-40CF)
1.38 (EMI-X DC-1008; Stat-Kon DCF-1006)
1.39 (Stat-Kon DF-15)
1.40 (PDX-83393)

Polycarbonate resin *(cont'd.)*

1.41 (RTP 303TFE10)
1.43 (Lexan 3413; RTP 305; Thermocomp DF-1006)
1.44 (EMI-X DA-30)
1.49 (EMI-X DA-35)
1.52 (Lexan 3414; RTP 307; Thermocomp DF-1008)
1.54 (EMI-X DA-40)
1.55 (RTP 305TFE15; Thermocomp DFL-4036)

Sp. Vol.

0.83 cm³/g (Lexan 920, 920A, 940A, 950A; Merlon CD-2000)
18.2 in.³/lb (Lexan 3414; Thermocomp DF-1008)
19.3 in.³/lb (Lexan 3413)
19.4 in.³/lb (Thermocomp DF-1006)
20.5 in.³/lb (Lexan 3412)
20.7 in.³/lb (Thermocomp DF-1004)
21.6 in.³/lb (Thermocomp DL-4030)
21.8 in.³/lb (Calibre 510, 550)
22.2 in.³/lb (Lexan 500)
22.3 in.³/lb (Lexan 2014)
22.9 in.³/lb (Calibre 800-3, 800-6, 800-10, 800-15; Lexan 940, 950)
23.1 in.³/lb (Calibre 200-3, 200-6, 200-10, 200-15, 300-3, 300-6, 300-10, 300-15, 700-3, 700-6, 700-10, 700-15; Lexan 101, 103, 104, 121, 123, 124, 141, 143, 144, 303)
26.4 in.³/lb (Calibre 7070)

Density:

1.20 mg/m³ (Makrolon 2400, 2600, 2800, 3100, 3119, 3200, 6550, 6560, 6655, 6870; Merlon 6450, 6560, 6870, HMS-3119, M39, M40, M50)
1.24 mg/m³ (Merlon SF-600)
1.25 mg/m³ (Makrolon 6030, 6603)
1.27 mg/m³ (Makrolon 9310, 9410; Merlon 8310)
1.28 mg/m³ (Merlon 9310)
1.35 mg/m³ (Makrolon 8020, 8320)
1.44 mg/m³ (Makrolon 8030)
1.51 mg/m³ (Makrolon 8344)
1.20 g/cm³ (Makrofol E, G, N; Merlon CD-2000)
1.21 g/cm³ (Makrofol KG)
1.28 g/cm³ (Makrofol SG, SN)
1.29 g/cm³ (Makrofol SKG)
0.038 lb/in.³ (Calibre 7070)
0.043 lb/in.³ (Calibre 200-3, 200-6, 200-10, 200-15, 300-3, 300-6, 300-10, 300-15, 700-3, 700-6, 700-10, 700-15)
0.044 lb/in.³ (Calibre 800-3, 800-6, 800-10, 800-15)
0.046 lb/in.³ (Calibre 510, 550)

M.P.:

250 F (PC-20CF, -30CF, -40CF)

Polycarbonate resin (cont'd.)

514 ± 9 F (Polypenco Polycarbonate)

Stability:

Resistant to dilute acids, saturated aliphatic and cycloaliphatic hydrocarbons, and alcohols; destroyed by alkaline sol'ns., ammonia, and amines (Makrofol E, G, KG, N, SG, SKG, SN)

Good dimensional stability; resistant to mineral acids up to high concs., many organic acids, oxidants and reducing agents, neutral and acid salt sol'ns., greases and oils, saturated aliphatic and cycloaliphatic hydrocarbons and alcohols, except for methanol; destroyed by alkaline sol'ns., ammonia gas and its sol'ns., and amines (Makrolon 2400, 2600, 2800, 3100, 3119, 3200, 6030, 6550, 6560, 6603, 6655, 6870, 8020, 8030, 8320, 8344, 9310, 9410)

Resistant to hydrolysis (Makrolon 2809, 3109, 3119, 3208)

Excellent chemical, sunlight, and weather resistance (Polypenco Polycarbonate)

Fair resistance to acids and bases; poor resistance to solvents (Thermocomp DC-1006, DC-1006PC)

The compound is resistant to attack by acids, salt sol'ns., alcohols, and ethers at R.T.; however, the base resin is attacked by alkaline sol'ns. and most aromatic and chlorinated solvnets (Thermocomp DF-1004, DF-1006, DF-1008)

Resistant to water, dilute organic and inorganic acids, oxidizing and reducing agents, neutral and acid salts, minerals, fats and oils, and aliphatic and cyclic hydrocarbons; affected by alkalis, amines, ketones, esters, aromatic hydrocarbons, methylene chloride, ethylene dichloride, cresol, dioxane (Zelux W)

Ref. Index:

1.584 (Makrolon 2400, 2600, 2800, 3100, 3119, 3200, 6030, 6550, 6560, 66036655, 6870)

1.586 (Calibre 200-3, 200-6, 200-10, 200-15, 300-3, 300-6, 300-10, 300-15, 700-3, 700-6, 700-10, 700-15; Lexan 101, 103, 104, 121, 123, 124, 141, 143, 144, 303, 920, 920A, 940, 940A, 950A)

1.64 (Zelux W)

Transmittance:

69.7% (Zelux W)

84–88% (Calibre 700-3, 700-6, 700-10, 700-15)

85% (Lexan 920, 920A, 940A, 950A, CF-25 MB-112, CF-26 MB-112)

86% (Lexan Sheet 9034, XL Sheet; Margard Sheet)

87–91% (Calibre 200-3, 200-6, 200-10, 200-15, 300-3, 300-6, 300-10, 300-15)

90% (Lexan CF-20 MB-112, CF-23 MB-112, CF-60 MB-112, FR60-112, FR60SE-112, FR63-112, FR65-112, FR66-112, FR88-112; Merlon CD-2000)

> 90% (Lexan 8800-112)

Haze: < 0.5% (Lexan 8800-112)

0.7–1.5% (Calibre 200-3, 200-6, 200-10, 200-15, 300-3, 300-6, 300-10, 300-15, 700-3, 700-6, 700-10)

< 1% (Lexan CF-20 MB-112, CF-60 MB-112, FR60-112, FR60SE-112, FR63-112, FR65-112, FR66-112, FR88-112)

1–3% (Lexan 920, 920A, 940A, 950A)

3.2% (Zelux W)

40% (Lexan CF-23 MB-112)

85% (Lexan CF-25 MB-112)

95% (Lexan CF-26 MB-112)

60° Gloss:

11 (Lexan CF-26 MB-112)

14 (Lexan CF-25 MB-112)

32 (Lexan CF-23 MB-112)

95 (Lexan CF-20 MB-112, CF-60 MB-112)

MECHANICAL PROPERTIES:

Tens. Str.:

34 MPa (Lexan FL478)

36 MPa (Lexan FL1000)

50 MPa (Lexan FL410)

55 MPa (break) (Makrolon 8020)

62 MPa (break) (Merlon CD-2000); (yield) (Lexan 920, 920A, 940A, 950A, 8800-112)

> 65 MPa (break) (Makrolon 2400, 2600, 2800, 3100, 3119, 3200, 6030, 6550, 6560, 6603, 6655, 6870)

65.5 MPa (break) (Merlon 8310, 9310)

66 MPa (break) (Merlon 6560)

68 MPa (break) (Merlon 6450, HMS-3119)

70 MPa (break) (Makrolon 8030, 9310, 9410; Merlon M39, M40, M50)

71 MPa (break) (Merlon 6870)

76 MPa (break) (Lexan 8800-112)

> 80 MPa (Makrofol E, N, SN)

100 MPa (break) (Makrolon 8320, 8344)

> 100 MPa (Makrofol G, SG)

> 220 MPa (Makrofol KG, SKG)

6100 psi (Lexan FL400, FL900)

6300 psi (Lexan FL910, FL1800)

6400 psi (EMI-X DA-40)

6500 psi (EMI-X DA-35; Migralube DL-4530); (yield) (Calibre 7070)

6600 psi (EMI-X DA-30)

7000 psi (Migralube DL-4030; Thermocomp DL-4030)

7000–9100 psi (break) (Calibre 510, 550)

7500 psi (Thermocomp DL-4020FR)

8400–8800 psi (yield) (Lexan 8A13, 8A23, 8A33, 8A43, 8B13, 8B23, 8B33, 8B35, 8B36, 8B43, 8B327, 8B328, 8C20, 8010-112, 8020, 8030, 8040, 8060)

8500 psi (Migralube DL-4410); (break) (Calibre 800-3, 800-6, 800-10, 800-15)

8600–9300 psi (break) (Lexan 8A13, 8A23, 8A33, 8A43, 8B13, 8B23, 8B33, 8B35, 8B36, 8B43, 8B327, 8B328, 8C20)

8800 psi (break) (Lexan FR60-112, FR60SE-112, FR63-112, FR65-112, FR66-112,

Polycarbonate resin *(cont'd.)*

FR88-112)

9000 psi (PDX-84356; Zelux W); (yield) (Lexan 101, 103, 104, 121, 123, 124, 141, 143, 144, 303, 940, 950, 2014)

9000–10,500 psi (Polypenco Polycarbonate)

9153 psi (Marc Polycarbonate)

9400 psi (Stat-Kon D); (yield) (Lexan CF-20 MB-112, CF-23 MB-112, CF-25 MB-112, CF-26 MB-112, CF-60 MB-112)

9500 psi (Thermocomp DFA-113); (break) (Calibre 700-3, 700-6, 700-10, 700-15)

9600 psi (RTP 301); (yield) (Lexan 500)

9920 psi (Lexan FL930)

10,000 psi (yield) (Lexan FR60-112, FR60SE-112, FR63-112, FR65-112, FR66-112, FR88-112)

10,200 psi (break) (Lexan CF-20 MB-112, CF-23 MB-112, CF-25 MB-112, CF-26 MB-112, CF-60 MB-112)

10,300 psi (break) (Calibre 200-10, 200-15, 300-10, 300-15)

10,500 psi (Stat-Kon DF-15); (break) (Calibre 200-3, 200-6, 300-3, 300-6)

11,600 psi (Thermocomp DC-1006PC)

12,000 psi (RTP 303TFE10)

15,000 psi (PDX-83393, -84368; RTP 305TFE15)

16,000 psi (Thermocomp DF-1004); (break) (Lexan 3412)

17,000 psi (RTP 303; Stat-Kon DCL-4033)

17,500 psi (Thermocomp DFL-4036)

18,500 psi (Thermocomp DF-1006)

19,000 psi (break) (Lexan 3413)

20,000 psi (RTP 305; Stat-Kon DC-1003FR, DCF-1006)

21,000 psi (RTP 307; Thermocomp DF-1008)

22,800 psi (PC-20CF)

23,000 psi (break) (Lexan 3413)

24,000 psi (Thermocomp DC-1006)

24,600 psi (PC-30CF)

25,000 psi (Stat-Kon DC-1006)

26,000 psi (EMI-X DC-1008; PC-40CF)

Tens. Elong.:

1.5% (RTP 305TFE15)

1.9% (PC-40CF)

2% (EMI-X DA-40; RTP 303TFE10, 305, 307); (break) (Lexan FL930)

2.5% (EMI-X DA-35; RTP 303)

2.7% (PC-30CF)

3% (EMI-X DA-30, DC-1008); (break) (Lexan FL478, FL910, FL1800)

3–5% (break) (Lexan 3413, 3414)

3.2% (PC-20CF)

3.4% (Stat-Kon DCF-1006)

3.5% (PDX-83393; Stat-Kon DCL-4033); (break) (Makrolon 8030)

Polycarbonate resin (cont'd.)

3.6% (break) (Lexan FL410)
3.7% (Stat-Kon DC-1006)
3.8% (break) (Makrolon 8320, 8344)
4% (PDX-84368; Stat-Kon DC-1003FR, DF-15); (break) (Lexan FL1000)
4–6% (Thermocomp DF-1004, DF-1006, DF-1008, DL-4030); (break) (Lexan 3412)
4–8% (break) (Calibre 510, 550)
4.6% (break) (Lexan FL900)
5% (break) (Meerlon 8310)
5.5% (Stat-Kon D)
6% (PDX-84356); (break) (Calibre 7070)
6–8% (yield) (Lexan 101, 103, 104, 121, 123, 124, 141, 143, 144, 303, 500, 940, 950, 2014)
7% (RTP 301); (break) (Makrolon 8020, 9410)
8% (Lexan FL400); (break) (Makrolon 9310)
10% (Migralube DL-4030, DL-4530)
10–20% (break) (Lexan 500)
12% (break) (Merlon 9310); (yield) (Marc Polycarbonate)
25–50% (Lexan FR60-112, FR60SE-112, FR63-112, FR65-112, FR66-112, FR88-112)
≈ 40% (break) (Makrofol KG, SKG)
≈ 50% (break) (Makrofol G, SG)
50–100% (Polypenco Polycarbonate)
85–105% (Lexan 8A13, 8A23, 8A33, 8A43, 8B13, 8B23, 8B33, 8B35, 8B36, 8B43, 8B327, 8B328, 8C20, 8010-112, 8020, 8030, 8040, 8060, CF-20 MB-112, CF-23 MB-112, CF-25 MB-112, CF-26 MB-112, CF-60 MB-112)
90% (Migralube DL-4410); (break) (Lexan 920, 920A, 940, 940A, 950, 950A; Makrolon 3119; Merlon CD-2000)
100% (break) (Calibre 800-3, 800-6, 800-10, 800-15; Makrofol E, N, SN; Merlon 6560, HMS-3119)
100–150% (Lexan 8800-112)
110% (Lexan Sheet 9034, XL Sheet; Margard Sheet); (break) (Lexan 101, 103, 104, 121, 123, 124, 141, 143, 144, 303, 2014; Merlon 6870)
> 110% (break) (Makrolon 2400, 2600, 2800, 3100, 3200, 6030, 6550, 6560, 6603, 6655, 6870)
120% (break) (Calibre 700-3, 700-6, 700-10, 700-15; Merlon 6450, M39, M40, M50)
150% (break) (Calibre 200-3, 200-6, 200-10, 200-15, 300-3, 300-6, 300-10, 300-15)
Tens. Mod.:
1900 MPa (Lexan 8800-112)
2200 MPa (Lexan 920, 920A, 940A, 950A)
2300 GPa (Makrolon 2400, 2600, 2800, 3100, 3119, 3200, 6030, 6550, 6560, 6603, 6655, 6870)
3500 GPa (Makrolon 9310, 9410)
3900 GPa (Makrolon 8020)

Polycarbonate resin *(cont'd.)*

5500 GPa (Makrolon 8030)
6000 GPa (Makrolon 8320)
9500 GPa (Makrolon 8344)
31,900 psi (PC-20CF)
35,900 psi (PC-30CF)
36,400 psi (PC-40CF)
290,000–300,000 psi (Lexan 8A13, 8A23, 8A33, 8A43, 8B13, 8B23, 8B33, 8B35, 8B36, 8B43, 8B327, 8B328, 8C20, 8010-112, 8020, 8030, 8040, 8060)
300,000 psi (Lexan FL900)
310,000 psi (Thermocomp DL-4030)
312,000 psi (Lexan FL910, FL1800)
320,000 psi (Lexan FR60-112, FR60SE-112, FR63-112, FR65-112, FR66-112, FR88-112; Polypenco Polycarbonate)
325,000 psi (Calibre 700-3, 700-6, 700-10, 700-15; Lexan 940, 950)
330,000 psi (Calibre 800-3, 800-6, 800-10, 800-15, 7070)
340,000 psi (Calibre 200-15, 300-15)
345,000 psi (Lexan 101, 103, 104, 121, 123, 124, 141, 143, 144, 303, 2014, Sheet 9034, XL Sheet; Margard Sheet)
350,000 psi (Calibre 200-6, 200-10, 300-6, 300-10)
360,000 psi (Calibre 200-3, 300-3)
364,000 psi (Lexan FL930)
450,000 psi (Lexan 500)
450,000–530,000 psi (Calibre 510, 550)
600,000 psi (RTP 301)
860,000 psi (Lexan 3412)
10^6 psi (Thermocomp DF-1004)
1.2×10^6 psi (Lexan 3413; RTP 303, 303TFE10)
1.375×10^6 psi (Thermocomp DF-1006)
1.4×10^6 psi (RTP 305, 305TFE15)
1.68×10^6 psi (Lexan 3414)
1.7×10^6 psi (RTP 307; Thermocomp DF-1008)

Flex. Str.:
61 MPa (Lexan FL478)
66 MPa (break) (Merlon SF-600)
69 MPa (Lexan FL1000)
76 MPa (Lexan FL410)
86 MPa (break) (Merlon HMS-3119, M39, M40, M50)
91 MPa (Lexan 920, 920A, 940A, 950A); (break) (Merlon 6450, 6560)
92 MPa (break) (Merlon 6870)
103.4 MPa (break) (Merlon 8310, 9310)
120 MPa (Makrolon 8020)
130 MPa (Makrolon 8030, 9310, 9410)
160 MPa (Makrolon 8320, 8344)

Polycarbonate resin (cont'd.)

9000 psi (Migralube DL-4530)
9300 psi (Lexan FL400)
10,000 psi (Migralube DL-4030)
10,500 psi (Thermocomp DL-4020FR, DL-4030)
11,000 psi (Migralube DL-4410)
11,500–12,500 psi (Polypenco Polycarbonate)
11,630 psi (Lexan FL900)
12,300 psi (Lexan FL1800)
12,500 psi (Calibre 7070; EMI-X DA-40)
12,540 psi (Lexan FL910)
13,000 psi (EMI-X DA-30, DA-35)
13,200 psi (Lexan 940, 950)
13,500 psi (Lexan 101, 103, 104, 121, 123, 124, 141, 143, 144, 303, 2014; Zelux W)
13,520 psi (Marc Polycarbonate)
13,700–15,800 psi (Calibre 510, 550)
13,800 psi (Stat-Kon D)
14,000 psi (Calibre 200-3, 200-6, 200-10, 200-15, 300-3, 300-6, 300-10, 300-15, 800-3, 800-6, 800-10, 800-15; PDX-84356)
14,500 psi (Calibre 700-3, 700-6, 700-10, 700-15)
15,000 psi (Lexan 500)
16,000 psi (RT 301; Thermocomp DFA-113)
16,700 psi (Lexan FL930)
17,000 psi (Stat-Kon DF-15; Thermocomp DC-1006PC)
18,000 psi (RTP 303TFE10)
19,000 psi (Lexan 3412; RTP 303)
22,000 psi (PDX-83393; RTP 305TFE15)
23,000 psi (Lexan 3413; PDX-84368; RTP 305)
25,000 psi (Thermocomp DF-1004)
26,000 psi (RTP 307; Thermocomp DFL-4036)
27,000 psi (Lexan 3414; Stat-Kon DCL-4033)
28,000 psi (Stat-Kon DC-1003FR, DCF-1006; Thermocomp DF-1006)
30,000 psi (Thermocomp DF-1008)
36,000 psi (Thermocomp DC-1006)
38,000 psi (EMI-X DC-1008; Stat-Kon DC-1006)
1.5×10^6 psi (PC-20CF)
1.9×10^6 psi (PC-30CF)
2.1×10^6 psi (PC-40CF)
Flex. Mod.:
2.0 GPa (Merlon SF-600)
2.20 GPa (Merlon M39, M40, M50)
2.30 GPa (Merlon 6870, HMS-3119)
2.40 GPa (Merlon 6560)
2.48 GPa (Merlon 6450)

179

Polycarbonate resin *(cont'd.)*

3.44 GPa (Merlon 8310, 9310)

> 95 MPa (Makrolon 3100, 3119, 3200, 6030, 6550, 6560, 6603, 6655, 6870)

2200 MPa (Lexan 920, 920A, 940A, 950A)

2206 MPa (Lexan FL1000)

2506 MPa (Lexan FL478)

2791 MPa (Lexan FL410)

255,000 psi (Lexan FL400)

280,000 psi (Migralube DL-4530)

300,000 psi (Migralube DL-4030; Thermocomp DL-4020FR, DL-4030)

310,000–500,000 psi (Polypenco Polycarbonate)

325,000 psi (Lexan 940, 950; Migralube DL-4410)

330,952 psi (Marc Polycarbonate)

340,000 psi (Lexan 101, 103, 104, 121, 123, 124, 141, 143, 144, 303)

345,000 psi (Lexan 2014; Zelux W)

350,000 psi (Calibre 200-3, 200-6, 200-10, 200-15, 300-3, 300-6, 300-15, 700-3, 700-6, 700-10, 700-15)

357,000 psi (Lexan FL900)

360,000 psi (Calibre 800-3, 800-6, 800-10, 800-15)

375,000 psi (Lexan FL1800)

383,000 psi (Lexan FL910)

400,000 psi (Calibre 7070)

450,000 psi (PDX-84356; Stat-Kon D)

460,000–550,000 psi (Calibre 510, 550)

500,000 psi (Lexan 500; RTP 301; Thermocomp DFA-113)

700,000 psi (EMI-X DA-30)

800,000 psi (EMI-X DA-35; Lexan 3412; RTP 303; Stat-Kon DF-15)

820,000 psi (Lexan FL930)

850,000 psi (Thermocomp DF-1004)

900,000 psi (PDX-84368; RTP 303TFE10)

950,000 psi (EMI-X DA-40)

1.1×10^6 psi (Lexan 3413; PDX-83393; RTP 305; Thermocomp DC-1006PC)

1.2×10^6 psi (RTP 305TFE15; Stat-Kon DC-1003FR; Thermocomp DF-1006, DFL-4036)

1.3×10^6 psi (Stat-Kon DCL-4033)

1.4×10^6 psi (Lexan 3414; RTP 307; Stat-Kon DCF-1006)

1.5×10^6 psi (Thermocomp DF-1008)

1.9×10^6 psi (Thermocomp DC-1006)

2.1×10^6 psi (Stat-Kon DC-1006)

2.3×10^6 psi (EMI-X DC-1008)

Compr. Str.:

71.15 MPa (yield) (Merlon 6450)

72.0 MPa (yield) (Merlon M39, M40, M50)

> 80 MPa (Makrolon 2400, 2600, 2800, 3100, 3119, 3200, 6030, 6550, 6560, 6603,

6655, 6870)
86 MPa (Lexan 920, 920A, 940A, 950A)
93.1 MPa (yield) (Merlon 8310, 9310)
100 MPa (Makrolon 8020)
105 MPa (Makrolon 9310, 9410)
110 MPa (Makrolon 8030)
125 MPa (Makrolon 8320, 8344)
5200 psi (Lexan FL900)
6800 psi (Lexan FL1800)
7030 psi (Lexan FL910)
8600 psi (PC-20CF)
9500 psi (Thermocomp DL-4030)
10,000 psi (Calibre 200-3, 200-6, 200-10, 200-15, 300-3, 300-6, 300-10, 300-15, 700-3, 700-6, 700-10, 700-15, 800-3, 800-6, 800-10, 800-15; PC-30CF)
10,000–14,000 psi (Polypenco Polycarbonate)
11,000 psi (RTP 303TFE10)
11,800 psi (Lexan FL930)
12,000 psi (Calibre 510, 550)
12,500 psi (Lexan 101, 103, 104, 121, 123, 124, 141, 143, 144, 303, 940, 950, 2014; Zelux W)
14,000 psi (Lexan 500; RTP 301, 305TFE15)
16,000 psi (Lexan 3412)
18,000 psi (Lexan 3413; RTP 303)
20,000 psi (RTP 305; Thermocomp DF-1004)
21,000 psi (Lexan 3414)
22,000 psi (RTP 307; Thermocomp DF-1006)
24,000 psi (Thermocomp DF-1008)
Compr. Mod.:
2200 MPa (Lexan 920, 920A, 940A, 950A)
325,000 psi (Lexan 940, 950)
345,000 psi (Lexan 101, 103, 104, 121, 123, 124, 141, 143, 144, 303, 2014)
520,000 psi (Lexan 500)
760,000 psi (Lexan 3412)
1.13×10^6 psi (Lexan 3413)
1.5×10^6 psi (Lexan 3414)
Creep:
0.06% strain (500 psi) (Lexan FL930)
0.156% strain (500 psi) (Lexan FL910, FL1800)
0.187% strain (500 psi) (Lexan FL900)
Tear Str. (Elmendorf, propagation):
7700–9600 N/m (Lexan 8800-112)
44–55 lb/in. (Lexan 8A13, 8A23, 8A33, 8A43, 8B13, 8B23, 8B33, 8B35, 8B36, 8B43, 8B327, 8B328, 8C20, 8010-112, 8020, 8030, 8040, 8060, CF-20 MB-112, CF-23

Polycarbonate resin *(cont'd.)*

MB-112, CF-25 MB-112, CF-26 MB-112, CF-60 MB-112)

73 lb/in. (Lexan FR60-112, FR60SE-112, FR63-112, FR65-112, FR66-112, FR88-112)

Shear Str.:

70 MPa (break) (Lexan 920, 920A, 940A, 950A)

4475 psi (break) (Lexan FL900)

4600 psi (break) (Lexan FL910, FL1800)

6000 psi (yield) (Lexan 101, 103, 104, 121, 123, 124, 141, 143, 144, 303, 940, 950, 2014)

8000 psi (Thermocomp DC-1006PC)

8200 psi (Thermocomp DL-4030)

8500 psi (yield) (Lexan 500)

8600 psi (Thermocomp DF-1004)

9000 psi (break) (Calibre 200-3, 200-6, 200-10, 200-15, 300-3, 300-6, 300-10, 700-3, 700-6, 700-10, 700-15, 800-3, 800-6, 800-10, 800-15)

9400 psi (Thermocomp DF-1006)

9900 psi (Thermocomp DF-1008)

10,000 psi (Thermocomp DC-1006); (yield) (Lexan 3412)

10,500 psi (yield) (Lexan 3413)

11,000 psi (yield) (Lexan 3414)

Shear Mod.:

780 MPa (Lexan 920, 920A, 940A, 950A)

Abrasion Resistance (Taber):

10 mg/1000 cycles (Lexan 920, 920A, 940A, 950A)

25 mg/1000 cycles (Thermocomp DL-4030)

Impact Str. (Izod):

80 J/m notched (Makrolon 8020, 8030)

85.4 J/m notched (Merlon 8310, 9310)

105 J/m notched (Makrolon 9310, 9410)

112 J/m notched (−40 C) (Merlon SF-600)

120 J/m notched (Makrolon 8320, 8344)

600 J/m notched (Merlon CD-2000)

640 J/m notched (Lexan 920, 920A, 940A, 950A)

> 700 J/m (Makrolon 2400); notched (Makrolon 6030, 6603)

740 J/m notched (Merlon 6560)

> 750 J/m notched (Makrolon 6550, 6560, 6655, 6870)

800 J/m notched (Merlon HMS-3119)

> 800 J/m (Makrolon 2600, 2800)

850 J/m notched (Merlon 6450, M39, M40, M50)

> 850 J/m notched (Makrolon 3100, 3119, 3200)

930 J/m notched (Merlon 6870)

0.5 ft lb/in. notched (Stat-Kon D)

1.0 ft lb/in. notched (Stat-Kon DF-15; Thermocomp DC-1006PC)

1.2 ft lb/in. notched (PDX-84356)
1.3 ft lb/in. nothced (EMI-X DA-40)
1.4 ft lb/in. notched (EMI-X DA-35; RTP 303; Stat-Kon DCL-4033)
1.5 ft lb/in. (PC-40CF); notched (EMI-X DA-30; PDX-83393; RTP 303TFE10; Stat-Kon DC-1003FR)
1.7 ft lb/in. (PC-20CF); notched (PDX-84368; RTP 305)
1.8 ft lb/in. (PC-30CF); notched (EMI-X DC-1008; RTP 305TFE15; Thermocomp DC-1006)
1.9 ft lb/in. notched (Stat-Kon DC-1006)
2.0 ft lb/in. notched (Lexan 500, 3412, 3413; Thermocomp DFL-4036)
2.0–3.0 ft lb/in. notched (Thermocomp DFA-113)
2.0–4.0 ft lb/in. notched (Calibre 510, 550)
2.2 ft lb/in. notched (RTP 307; Stat-Kon DCF-1006)
2.5 ft lb/in. notched (Lexan 3414; Migralube DL-4030; Thermocomp DL-4020FR, DL-4030)
2.6 ft lb/in. notched (RTP 301)
3.4 ft lb/in. notched (Thermocomp DF-1004)
3.5 ft lb/in. notched (Migralube DL-4530)
3.7 ft lb/in. notched (Thermocomp DF-1006)
4.0 ft lb/in. notched (Thermocomp DF-1008)
6.5 ft lb/in. unnotched (Lexan FL1000)
7.1 ft lb/in. unnotched (Lexan FL410)
7.3 ft lb/in. unnotched (Lexan FL478)
10 ft lb/in. notched (Calibre 800-15)
12 ft lb/in. notched (Calibre 800-3, 800-6, 800-10; Lexan 940, 950, 2014)
12–16 ft lb/in. notched (Lexan 101, 103, 104, 121, 123, 124, 141, 143, 144, 303, Sheet 9034, XL Sheet; Margard Sheet; Polypenco Polycarbonate; Zelux W)
14.4 ft lb/in. notched (Marc Polycarbonate)
16 ft lb/in. notched (Calibre 200-15, 300-15, 700-15; Migralube DL-4410)
17 ft lb/in. notched (Calibre 200-6, 200-10, 300-6, 300-10, 700-6, 700-10)
18 ft lb/in. notched (Calibre 200-3, 300-3, 700-3)
19 ft lb/in. unnotched (Lexan FL400)
Impact Str. (Gardner):
60 in.-lb (Lexan FR60-112, FR60SE-112, FR63-112, FR65-112, FR66-112, FR88-112)
Tens. Impact:
525 kJ/m^2 (Lexan 920, 920A, 940A, 950A)
30 ft lb/in.^{2} (Lexan 3412)
32 ft lb/in.^{2} (Lexan 3413)
35 ft lb/in.^{2} (Lexan 3414)
60 ft lb/in.^{2} (Thermocomp DF-1004)
70 ft lb/in.^{2} (Calibre 510, 550; Thermocomp DF-1006)
75 ft lb/in.^{2} (Lexan 500)

Polycarbonate resin *(cont'd.)*

80 ft lb/in.^{2} (Thermocomp DF-1008)

220 ft lb/in.^{2} (Calibre 200-15, 300-15, 700-15)

225 ft lb/in.^{2} (Lexan 2014)

225–300 ft lb/in.^{2} (Lexan 101, 103, 104, 121, 123, 124, 141, 143, 144, 303)

240 ft lb/in.^{2} (Calibre 800-15)

250 ft lb/in.^{2} (Calibre 200-10, 300-10, 700-10; Lexan 940, 950)

260 ft lb/in.^{2} (Calibre 800-3, 800-6, 800-10)

280 ft lb/in.^{2} (Calibre 200-6, 300-6, 700-6)

300 ft lb/in.^{2} (Calibre 200-3, 300-3, 700-3)

Hardness:

Ball Indentation 110 N/mm^2 (Makrolon 2400, 2600, 2800, 3100, 3119, 3200, 6030, 6550, 6560, 6603, 6655, 6870)

Ball Indentation 125 N/mm^2 (Makrolon 9310, 9410)

Ball Indentation 140 N/mm^2 (Makrolon 8020, 8320)

Ball Indentation 145 N/mm^2 (Makrolon 8030)

Ball Indentation 155 N/mm^2 (Makrolon 8344)

Rockwell M62 (Merlon M39, M40, M50)

Rockwell M65 (Merlon 6450, CD-2000)

Rockwell M68 (Merlon 6560)

Rockwell M70 (Lexan 920, 920A, 940A, 950A)

Rockwell M72 (Merlon 6870)

Rockwell M75 (Merlon 8310, 9310, HMS-3119; Thermocomp DL-4030)

Rockwell M85 (PC-20CF)

Rockwell M92 (Thermocomp DF-1004)

Rockwell M95 (Thermocomp DF-1006)

Rockwell M97 (Thermocomp DF-1008)

Rockwell R110–120 (Polypenco Polycarbonate)

Rockwell R117 (RTP 303TFE10)

Rockwell R118 (Calibre 200-3, 200-6, 200-10, 200-15, 300-3, 300-6, 300-10, 300-15; Lexan 101, 103, 104, 121, 123, 124, 141, 143, 144, 303, 940, 950; RTP 301, 303, 305TFE15)

Rockwell R119 (Lexan 3414; RTP 305, 307)

Rockwell R120 (Lexan 2014, 3413)

Rockwell R122 (Calibre 510, 550, 800-3, 800-6, 800-10, 800-15; Lexan 3412)

Rockwell R123 (Calibre 700-3, 700-6, 700-10, 700-15)

Rockwell R124 (Lexan 500)

Shore D77 (Marc Polycarbonate)

Mold Shrinkage:

0.005–0.007 mm/mm (Lexan 920, 920A, 940A, 950A)

0.001 in./in. (EMI-X DA-30)

0.0015 in./in. (PDX-83393; Stat-Kon DC-1006, DCF-1006)

0.002 in./in. (Stat-Kon DC-1003FR, DCL-4033, DF-15)

0.002–0.003 in./in. (EMI-X DA-30)

0.002–0.005 in./in. (Calibre 510, 550)
0.003 in./in. (EMI-X DA-30, DA-35)
0.003–0.004 in./in. (PDX-84368)
0.004 in./in. (PDX-84356; Stat-Kon D)
0.004–0.006 in./in. (Lexan FL410, FL478, FL1000)
0.004–0.007 in./in. (Calibre 7070)
0.005–0.007 in./in. (Calibre 200-3, 200-6, 200-10, 200-15, 300-6, 300-10, 300-15, 700-3, 700-6, 700-10, 700-15, 800-3, 800-6, 800-10, 800-15)
0.006 in./in. (Thermocomp DL-4030)
0.007–0.008 in./in. (Lexan FL400)

Water Absorp.:
0.06% (EMI-X DA-40)
0.07% (EMI-X DA-35, DC-1008; Stat-Kon DCF-1006)
0.08% (EMI-X DA-30; Stat-Kon DC-1006)
0.09% (PDX-83393; Stat-Kon DCL-4033)
0.10% (Stat-Kon DC-1003FR)
0.11% (PDX-84368)
0.12% (PDX-84356; Stat-Kon DF-15; Thermocomp DL-4030)
0.15% (Calibre 200-3, 200-6, 200-10, 200-15, 300-6, 300-10, 300-15, 700-3, 700-6, 700-10, 700-15, 800-3, 800-6, 800-10, 800-15; Lexan 920, 920A, 940A, 950A)
0.18% (Stat-Kon D)
0.25% (Calibre 510, 550)
0.28% (Lexan FR60-112, FR60SE-112, FR63-112, FR65-112, FR66-112, FR88-112)
0.35% (Lexan 8A13, 8A23, 8A33, 8A43, 8B13, 8B23, 8B33, 8B35, 8B36, 8B43, 8B327, 8B328, 8C20, 8010-112, 8020, 8030, 8040, 8060, 8800-112, CF-20 MB-112, CF-23 MB-112, CF-25 MB-112, CF-26 MB-112, CF-60 MB-112)

THERMAL PROPERTIES:
Soften. Pt. (Vicat):
115–130 C (Makrolon 2600, 2800)
140 C (Merlon CD-2000)
148 C (Makrolon 3119, 6550, 6560, 6655, 6870)
150 C (Makrolon 3100, 3200, 8020, 8030, 8320, 8344, 9410)
152 C (Lexan 920, 920A, 940A, 950A)
153 C (Makrolon 6030, 6603, 9310)
300 F (Calibre 800-15)
305 F (Calibre 800-6, 800-10; Lexan 940, 950)
305–315 F (Lexan 101, 103, 104, 121, 123, 124, 141, 143, 144, 303)
309 F (Calibre 200-15, 300-15, 700-15)
310 F (Calibre 800-3; Lexan 500)
310–315 F (Lexan 2014)
312 F (Calibre 200-10, 300-10, 700-10)
314 F (Calibre 200-6, 300-6, 700-6)
316 F (Calibre 200-3, 300-3, 700-3)

Polycarbonate resin *(cont'd.)*

    320 F (Calibre 510, 550)

    330 F (Lexan 3412, 3413, 3414)

    540–600 F (PC-20CF, -30CF, -40CF)

Conduct.:

    0.19 W/Km (Lexan 920, 920A, 940A, 950A)

    0.21 W/Km (Makrolon 2400, 2600, 2800, 3100, 3119, 3200, 6030, 6550, 6560, 6603, 6655, 6870)

    0.23 W/Km (Makrolon 8020, 8320, 9310, 9410)

    0.24 W/Km (Makrolon 8030)

    0.25 W/Km (Makrolon 8344)

    0.6 Btu/h/ft^2/F/in. (PC-20CF)

    0.90 Btu/h/ft^2/F/in. (Lexan FL910)

    0.98 Btu/h/ft^2/F/.in. (Lexan FL1800)

    1.05 Btu/h/ft^2/F/in. (Lexan FL900)

    1.10 Btu/h/ft^2/F/in. (Calibre 7070)

    1.12 Btu/h/ft^2/F/in. (Lexan FL930)

    1.35 Btu/h/ft^2/in./F (Calibre 200-3, 200-6, 200-10, 200-15, 300-3, 300-6, 300-10, 300-15, 700-3, 700-6, 700-10, 700-15, 800-3, 800-6, 800-10, 800-15; Lexan 101, 103, 104, 121, 123, 124, 141, 143, 144, 303, 940, 950, 2014; Thermocomp DL-4030)

    1.41 Btu/h/ft^2/F/in. (Lexan 500)

    1.47 Btu/h/ft^2/F/in. (Lexan 3412)

    1.50 Btu/h/ft^2/F/in. (Lexan 3413; RTP 301)

    1.53 Btu/h/ft^2/F/in. (Lexan 3414)

    1.6 Btu/h/ft^2/F/in. (Stat-Kon D)

    2.0 Btu/h/ft^2/F/in. (RTP 303, 303TFE10)

    2.2 Btu/h/ft^2/F/in. (RTP 305)

    2.3 Btu/h/ft^2/F/in. (RTP 305TFE15; Stat-Kon DF-15; Thermocomp DF-1004)

    2.4 Btu/h/ft^2/F/in. (RTP 307)

    2.5 Btu/h/ft^2/F/in. (Thermocomp DF-1006)

    2.7 Btu/h/ft^2/F/in. (Thermocomp DF-1008)

    2.9 Btu/h/ft^2/F/in. (PDX-84368)

    3.6 Btu/h/ft^2/F/in. (Stat-Kon DCL-4033)

    3.8 Btu/h/ft^2/F/in. (Stat-Kon DC-1003FR)

    4.0 Btu/h/ft^2/F/in. (Stat-Kon DCF-1006)

    4.7 Btu/h/ft^2/F/in. (Thermocomp DC-1006PC)

    4.9 Btu/h/ft^2/F/in. (Stat-Kon DC-1006; Thermocomp DC-1006)

    4.95 Btu/h/ft^2/F/in. (PC-30CF)

    5.3 Btu/h/ft^2/F/in. (EMI-X DC-1008)

    6.5 Btu/h/ft^2/F/in. (EMI-X DA-30)

    6.6 Btu/h/ft^2/F/in. (EMI-X DA-35)

    6.8 Btu/h/ft^2/F/in. (EMI-X DA-40)

Distort. Temp.:

    127 C (1.82 MPa) (Lexan FL1000)

128 C (1.82 MPa) (Lexan FL478; Marc Polycarbonate)

131 C (1.82 MPa) (Merlon 6560, HMS-3119)

132 C (1.82 MPa) (Lexan 920, 920A, 940A, 950A, FL410; Makrolon 2400; Merlon 6870, M39, M40, M50)

133 C (1.82 MPa) (Merlon 6450, SF-600)

135 C (264 psi) (Lexan 8A13, 8A23, 8A33, 8A43, 8B13, 8B23, 8B33, 8B35, 8B36, 8B43, 8B327, 8B328, 8C20, 8010-112, 8020, 8030, 8040, 8060, 8800-112, CF-20 MB-112, CF-23 MB-112, CF-25 MB-112, CF-26 MB-112, CF-60 MB-112, FR60-112, FR60SE-112, FR63-112, FR65-112, FR66-112, FR88-112)

138 C (1.81 MPa) (Makrolon 2600, 2800, 3100, 3119, 3200, 6030, 6550, 6560, 6603, 6655, 6870)

142 C (1.82 MPa) (Merlon 8310, 9310)

147 C (Makrolon 8020, 8030, 8320); (1.81 MPa) (Makrolon 8344, 9410)

150 C (Makrofol E, G, N, SG, SN); (1.81 MPa) (Makrolon 9310)

240 C (Makrofol KG, SKG)

255 F (264 psi) (Lexan FL400)

260 F (Calibre 200-15, 300-15, 700-15)

260–280 F (264 psi) (Lexan 101, 103, 104, 121, 123, 124, 141, 143, 144, 303)

263 F (Calibre 200-10, 300-10, 700-10)

265 F (Calibre 200-6, 300-6, 700-6); (264 psi) (Migralube DL-4410; Thermocomp DL-4020FR)

266 F (Calibre 800-3, 800-6, 800-10, 800-15)

270 F (Calibre 200-3, 300-3, 700-3); (264 psi) (Lexan 940, 950, 2014, FL910, FL1800, Sheet 9034, XL Sheet; Margard Sheet; Stat-Kon D)

273 F (Calibre 7070)

275 F (264 psi) (Migralube DL-4030, DL-4530; Thermocomp DL-4030)

280 F (264 psi) (Lexan FL900; RTP 301, 303TFE10)

280–290 F (264 psi) (Polypenco Polycarbonate)

282 F (264 psi) (Lexan FL930)

285 F (Calibre 510, 550); (264 psi) (PDX-84356)

288 F (264 psi) (Lexan 500)

290 F (Thermocomp DC-1006PC); (264 psi) (EMI-X DA-30, DA-35, DA-40; PDX-84368; RTP 303; Stat-Kon DC-1003FR, DCL-4033; Thermocomp DFA-113, DFL-4036)

294 F (264 psi) (PC-20CF)

295 F (264 psi) (Lexan 3412, 3413, 3414; PC-30CF; RTP 305TFE15; Stat-Kon DF-15)

296 F (264 psi) (PC-40CF)

300 F (264 psi) (EMI-X DC-1008; PDX-83393; RTP 305, 307; Stat-Kon DC-1006, DCF-1006; Thermocomp DC-1006, DF-1004, DF-1006, DF-1008)

Brittle Temp.:

–135 C (Lexan 8A13, 8A23, 8A33, 8A43, 8B13, 8B23, 8B33, 8B35, 8B36, 8B43, 8B327, 8B328, 8C20, 8010-112, 8020, 8030, 8040, 8060, 8800-112, CF-20 MB-

Polycarbonate resin (cont'd.)

112, CF-23 MB-112, CF-25 MB-112, CF-26 MB-112, CF-60 MB-112)

Coeff. of Linear Exp.:

20×10^{-6} K^{-1} (Makrolon 8344)

27×10^{-6} K^{-1} (Makrolon 8030, 8320)

32×10^{-6} K^{-1} (Makrolon 9310, 9410)

45×10^{-6} K^{-1} (Makrolon 8020)

65×10^{-6} K^{-1} (Makrolon 2400, 2600, 2800, 3100, 3119, 3200, 6030, 6550, 6560, 6603, 6655, 6870)

6.75×10^{-5} m/m/C (Lexan 920, 920A, 940A, 950A, 8800-112, CF-20 MB-112, CF-23 MB-112, CF-25 MB-112, CF-26 MB-112, CF-60 MB-112)

3.7×10^{-5} mm/mm/C (Merlon 8310, 9310)

7.0×10^{-5} mm/mm/C (Merlon M39, M40, M50)

0.6×10^{-5} in./in./F (PC-30CF)

0.9×10^{-5} in./in./F (Lexan FL930; Stat-Kon DC-1006)

0.93×10^{-5} in./in./F (Lexan 3414)

1.0×10^{-5} in./in./F (Stat-Kon DCF-1006; Thermocomp DF-1008)

1.2×10^{-5} in./in./F (PDX-83393)

1.21×10^{-5} in./in./F (Lexan 3413)

1.3×10^{-5} in./in./F (Stat-Kon DC-1003FR; Thermocomp DF-1006)

1.4×10^{-5} in./in./F (Stat-Kon DCL-4033)

1.49×10^{-5} in./in./F (Lexan 3412)

1.5×10^{-5} in./in./F (PDX-84368; Thermocomp DF-1004, DFL-4036)

1.79×10^{-5} in./in./F (Lexan 500)

1.8×10^{-5} in./in./F (Lexan FL910; Stat-Kon DF-15; Thermocomp DFA-113)

2.0×10^{-5} in./in./F (Lexan FL900, FL1800)

2.1×10^{-5} in./in./F (Calibre 510, 550)

3.5×10^{-5} in./in./F (Stat-Kon D)

3.7×10^{-5} in./in./F (Migralube DL-4410; Zelux W)

3.75×10^{-5} in./in./F (Lexan 101, 103, 104, 121, 123, 124, 141, 143, 144, 303, 940, 950, 2014)

3.8×10^{-5} in./in./F (Calibre 200-3, 200-6, 200-10, 200-15, 300-3, 300-6, 300-10, 300-15, 700-3, 700-6, 700-10, 700-15, 800-3, 800-6, 800-10, 800-15)

3.9×10^{-5} in./in./F (Migralube DL-4030, DL-4530; Polypenco Polycarbonate; Thermocomp DL-4030)

4.0×10^{-5} in./in./F (Thermocomp DL-4020FR)

6.75×10^{-5} in./in./C (Lexan 8A13, 8A23, 8A33, 8A43, 8B13, 8B23, 8B33, 8B35, 8B36, 8B43, 8B327, 8B328, 8C20, 8010-112, 8020, 8030, 8040, 8060)

120×10^{-5} in./in./F (PC-20CF)

Sp. Heat:

1.09 kJ/kgK (Makrolon 8030, 8344)

1.13 kJ/kgK (Makrolon 8020, 8320, 9310, 9410)

1.17 kJ/kgK (Makrolon 2400, 2600, 2800, 3100, 3119, 3200, 6030, 6550, 6560, 6603, 6655, 6870)

Polycarbonate resin *(cont'd.)*

0.277 Btu/lb/F (Lexan FL900)
0.281 Btu/lb/F (Lexan FL1800)
0.287 Btu/lb/F (Lexan FL910)
0.29 Btu/lb/F (Calibre 510, 550)
0.292 Btu/lb/F (Lexan FL930)
0.30 Btu/lb/F (Calibre 200-3, 200-6, 200-10, 200-15, 300-3, 300-6, 300-10, 300-15, 700-3, 700-6, 700-10, 700-15, 800-3, 800-6, 800-10, 800-15; Lexan 8A13, 8A23, 8A33, 8A43, 8B13, 8B23, 8B33, 8B35, 8B36, 8B43, 8B327, 8B328, 8C20; Lexan 101, 103, 104, 121, 123, 124, 141, 143, 144, 303, 8010-112, 8020, 8030, 8040, 8060, 8800-112, CF-20 MB-112, CF-23 MB-112, CF-25 MB-112, CF-26 MB-112, CF-60 MB-112)

Flamm.:
V-0/5V (Calibre 550, 800-3, 800-6, 800-10, 800-15, 7070; Lexan FL400, FL410, FL478, FL900, FL910, FL930, FL1000, FL1800; Merlon 9310, SF-600)
V-0 (Calibre 510, 700-3, 700-6, 700-10, 700-15; EMI-X DA-30, DA-35, DA-40; Lexan 500, 920, 920A, 940, 940A, 950, 950A, 2014, 3412, 3413, 3414, 8060, CF-60 MB-112, FR60-112, FR60SE-112, FR63-112, FR65-112, FR66-112, FR88-112; Makrolon 6030, 6550, 6553, 6555, 6557, 6560, 6603, 6870, 9310, 9410, 9415; Merlon 6450, 6560, 6870, 9510; PDX-84356; Stat-Kon DC-1003FR; Thermocomp DFA-113)
V-0/V-1 (Makrolon 8344)
VTF-0 (Makrofol SG, SKG, SN)
V-1 (EMI-X DC-1008; Makrolon 8030, 8320; PDX-83393, -84368; RTP 301, 303TFE10, 305TFE15; Stat-Kon DC-1006, DCF-1006, DCL-4033, DF-15; Thermocomp DC-1006, DC-1006PC, DF-1004, DF-1006, DF-1008, DFL-4036, DL-4030)
VE1 (RTP 303, 305, 307)
V-2/V-1 (Makrolon 6655)
V-2 (Calibre 200-3, 200-6, 200-10, 200-15, 300-3, 300-6, 300-10, 300-15; Lexan 8A13, 8A23, 8A33, 8A43, 8B13, 8B23, 8B33, 8B35, 8B36, 8B43, 8B327, 8B328, 8C20, 101, 103, 104, 121, 123, 124, 141, 143, 144, 303, 8010-112, 8020, 8030, 8040; Makrolon 2400, 2600, 2800, 3100, 3200, 8020; Marc Polycarbonate; Merlon 8310, HMS-3119, M39, M40, M50; Stat-Kon D; Thermocomp DL-4020FR)

Self-Ignit. Temp.:
> 1000 F (Lexan Sheet 9034, XL Sheet; Margard Sheet)

ELECTRICAL PROPERTIES:
Dissip. Factor:
0.0004 (60 Hz) (Merlon 6450)
0.0006 (60 Hz) (Merlon 6870, M39, M40, M50)
0.0006–0.010 (60 to 10^6 Hz) (Thermocomp DL-4020FR)
0.0008 (50 Hz) (Makrolon 2400); (60 Hz) (Lexan 500)
0.0008–0.0075 (60 to 10^6 Hz) (Thermocomp DFA-113)
0.0009 (50 Hz) (Makrolon 2600, 2800, 3100, 3119, 3200, 6550, 6560, 6655, 6870,

Polycarbonate resin *(cont'd.)*

    8020, 8320, 8344, 9410); (60 Hz) (Lexan 101, 103, 104, 121, 123, 124, 141, 143,
        144, 303, 920, 920A, 940, 940A, 950, 950A, 2014, 3412)
    < 0.001 (60 Hz) (Calibre 200-3, 200-6, 200-10, 200-15, 300-3, 300-6, 300-10, 300-15);
        (100 Hz) (Calibre 700-3, 700-6, 700-10, 700-15, 800-3, 800-6, 800-10, 800-15)
    0.001 (50 Hz) (Makrofol E, G, KG, N, SG, SKG, SN; Makrolon 8030, 9310)
    0.0011 (60 Hz) (Lexan 3413); (1000 cycles) (Thermocomp DF-1004)
    0.0012 (60 Hz) (Merlon 8310, 9310); (1000 cycles) (Thermocomp DF-1006)
    0.0013 (60 Hz) (Lexan 3414)
    0.0014 (1 kHz) (Thermocomp DF-1008)
    0.0015 (Makrolon 6030, 6603)
    0.002 (100 Hz) (Calibre 510, 550)
    0.0029 (100 Hz) (Lexan FL910)
    0.0034 (100 Hz) (Lexan FL930)
    0.0041 (100 Hz) (Lexan FL1800)
    0.007 (1 MC) (RTP 305TFE15, 307)
    0.0075 (1 MC) (RTP 305)
    0.008 (1 MC) (RTP 303, 303TFE10)
    0.01 (1 MC) (RTP 301)
    0.10–0.23 (60 Hz) (Lexan 8A13, 8A23, 8A33, 8A43, 8B13, 8B23, 8B33, 8B35, 8B36,
        8B43, 8B327, 8B328, 8C20, 8010-112, 8020, 8030, 8040, 8060, 8800-112, CF-20
        MB-112, CF-23 MB-112, CF-25 MB-112, CF-26 MB-112, CF-60 MB-112)
    0.26 (60 Hz) (Lexan FR60-112, FR60SE-112, FR63-112, FR65-112, FR66-112,
        FR88-112)
Dielec. Str.:
    > 16 kV/mm (Merlon 6450, 6560, 6870, HMS-3119, M39, M40, M50)
    16.7 kV/mm (Lexan 920, 920A, 940A, 950A)
    20.9 kV/mm (Merlon 8310, 9310)
    > 30 kV/mm (Makrolon 2400, 2600, 2800, 3100, 3119, 3200, 6030, 6550, 6560, 6603,
        6655, 6870, 8020, 8030, 8320, 8344, 9310, 9410)
    276 kV/mm (Lexan 8800-112)
    350 kV/mm (Makrofol E, G, KG, N, SG, SKG, SN)
    245 V/mil (Lexan FL910)
    253 V/mil (Lexa FL1800)
    279 V/mil (Lexan FL930)
    380 V/mil (Lexan 101, 103, 104, 121, 123, 124, 141, 143, 144, 303, 940, 950, 2014)
    399 V/mil (Calibre 200-3, 200-6, 200-10, 200-15, 300-3, 300-6, 300-10, 300-15)
    400 V/mil (Polypenco Polycarbonate; Thermocomp DL-4020FR)
    405 V/mil (Calibre 800-3, 800-6, 800-10, 800-15)
    425 V/mil (Calibre 700-3, 700-6, 700-10, 700-15)
    450 V/mil (Lexan 500, 3414; RTP 305TFE15; Thermocomp DFA-113)
    455 V/mil (RTP 307)
    460 V/mil (RTP 303TFE10)
    465 V/mil (Thermocomp DF-1008)

470 V/mil (Calibre 510, 550; RTP 305)

475 V/mil (Lexan 3413)

480 V/mil (Thermocomp DF-1006)

490 V/mil (Lexan 3412; RTP 301, 303; Thermocomp DF-1004)

1400 V/mil (Lexan 8A13, 8A23, 8A33, 8A43, 8B13, 8B23, 8B33, 8B35, 8B36, 8B43, 8B327, 8B328, 8C20, 8010-112, 8020, 8030, 8040, 8060)

1520 V/mil (Lexan FR60-112, FR60SE-112, FR63-112, FR65-112, FR66-112, FR88-112)

1600 V/mil (Lexan CF-20 MB-112, CF-23 MB-112, CF-25 MB-112, CF-26 MB-112, CF-60 MB-112)

Dielec. Const.:

2.41 (100 Hz) (Lexan FL910)

2.44 (100 Hz) (Lexan FL1800)

2.70 (100 Hz) (Lexan FL930)

2.9 (60 Hz) (Lexan FR60-112, FR60SE-112, FR63-112, FR65-112, FR66-112, FR88-112)

2.93 (60 Hz) (Merlon 6870)

2.95 (60 Hz) (Merlon 6450)

2.96 (60 Hz) (Merlon M39, M40, M50)

2.99 (60 Hz) (Lexan 8A13, 8A23, 8A33, 8A43, 8B13, 8B23, 8B33, 8B35, 8B36, 8B43, 8B327, 8B328, 8C20, 8010-112, 8020, 8030, 8040, 8060, 8800-112, CF-20 MB-112, CF-23 MB-112, CF-25 MB-112, CF-26 MB-112, CF-60 MB-112)

3.00 (60 Hz) (Calibre 200-3, 200-6, 200-10, 200-15, 300-3, 300-6, 300-10, 300-15); (100 Hz) (Calibre 700-3, 700-6, 700-10, 700-15, 800-3, 800-6, 800-10, 800-15); (1 MC) (RTP 301)

3.01 (60 Hz) (Lexan 920, 920A, 940, 940A, 950, 950A)

3.08 (60 Hz) (Merlon 8310, 9310)

3.10 (60 Hz) (Lexan 500)

3.15–2.96 (60 to 10^6 Hz) (Thermocomp DL-4020FR)

3.15–3.05 (60 to 10^6 Hz) (Thermocomp DFA-113)

3.17 (60 Hz) (Lexan 101, 103, 104, 121, 123, 124, 141, 143, 144, 303, 2014, 3412; Polypenco Polycarbonate)

3.2 (100 Hz) (Calibre 510, 550); (1 MC) (RTP 303)

3.3 (1 MC) (RTP 303TFE10)

3.31 (1 kHz) (Thermocomp DF-1004)

3.35 (60 Hz) (Lexan 3413)

3.4 (1 MC) (RTP 305, 305TFE15)

3.51 (1 kHz) (Thermocomp DF-1006)

3.53 (60 Hz) (Lexan 3414)

3.6 (1 MC) (RTP 307)

3.68 (1 kHz) (Thermocomp DF-1008)

Vol. Resist.:

10 ohm-cm (EMI-X DA-40; PDX-83393)

Polycarbonate resin *(cont'd.)*

100 ohm-cm (EMI-X DA-35, DC-1008; PDX-84356)

1000 ohm-cm (EMI-X DA-30; Stat-Kon DC-1006, DCL-4033, DF-15)

10,000 ohm-cm (Stat-Kon D, DC-1003FR, DCF-1006)

6×10^{15} ohm-cm (Thermocomp DF-1004, DF-1006, DF-1008)

10^{16} ohm-cm (Lexan 8A13, 8A23, 8A33, 8A43, 8B13, 8B23, 8B33, 8B35, 8B36, 8B43, 8B327, 8B328, 8C20, 8010-112, 8020, 8030, 8040, 8060, 8800-112, CF-20 MB-112, CF-23 MB-112, CF-25 MB-112, CF-26 MB-112, CF-60 MB-112, FR60-112, FR60SE-112, FR63-112, FR65-112, FR66-112, FR88-112; RTP 301, 303, 303TFE10, 305, 305TFE15, 307; Thermocomp DL-4030)

> 10^{16} ohm-cm (Calibre 200-3, 200-6, 200-10, 200-15, 300-3, 300-6, 300-10, 300-15; Lexan 101, 103, 104, 121, 123, 124, 141, 143, 144, 303, 500, 920, 920A, 940, 940A, 950, 950A, 2014, 3412, 3413, 3414; Makrolon 2400, 2600, 2800, 3100, 3119, 3200, 6030, 6550, 6560, 6603, 6655, 6870, 8020, 8030, 8320, 8344, 9310, 9410)

2.1×10^{16} ohm-cm (Polypenco Polycarbonate)

10^{17} ohm-cm (Makrofol E, G, N, SG, SN)

2×10^{17} ohm-cm (Makrofol KG, SKG)

Surf. Resist.:

10 ohm/sq. (EMI-X DA-40; PDX-83393)

100 ohm/sq. (EMI-X DA-35, DC-1008; PDX-84356; Stat-Kon DC-1006, DCL-4033)

1000 ohm/sq. (EMI-X DA-30; Stat-Kon DC-1003FR, DCF-1006, DF-15)

3500 ohm/sq. (Thermocomp DC-1006)

10,000 ohm/sq. (Stat-Kon D; Thermocomp DC-1006PC)

10^9–10^{12} ohm/sq. (PDX-84368)

10^{15} ohm (Lexan 8800-112)

10^{16} ohm/sq. (Thermocomp DL-4030)

Arc Resist.:

5 s (Lexan 3412, 3413, 3414, FL910)

5–10 s (Lexan 500)

10 s (Lexan 920, 920A, 940, 940A, 950, 950A)

10–11 s (Lexan 101, 103, 104, 121, 123, 124, 141, 143, 144, 303, 2014)

15 s (Lexan FL930)

30–90 s (Merlon 6450, 6560, 6870, 8310, 9310, M39, M40, M50)

53 s (Lexan FL1800)

100 s (Merlon HMS-3119)

120 s (RTP 301, 303, 303TFE10, 305, 305TFE15, 307; Thermocomp DF-1004, DF-1006, DF-1008)

STORAGE/HANDLING:

Avoid exposure to moisture (Thermocomp DF-1004, DF-1006, DF-1008, DL-4030)

STRUCTURE:
PET:

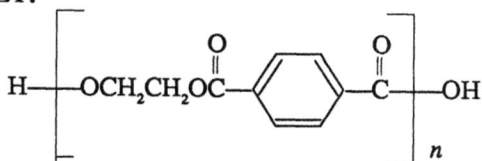

PBT:

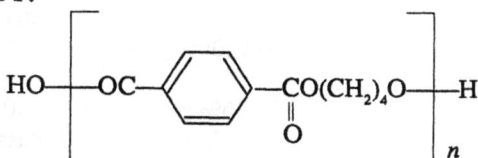

TRADENAME EQUIVALENTS:
Malon 1070, 1080, 1090 [M.A. Industries]

Mamax 2020 [M.A. Industries]

PA-55-004, -55-011, -55-012, 55-013, 55-024 [Polymer Applications]

Carbon-reinforced:

Stat-Kon WC-1002 (10% carbon fiber), WC-1006 (30% carbon fiber) [LNP]

Saturated polyester:

Terlan 6090, 6100, 6300, 6430, 6490, 6530, 6600, 6676 [Terrell]

Polyester elastomer:

Gaflex 355ZS, 372, 372ZS, 540ZS, 547, 547ZS, 555, 555ZS, 572ZS [GAF]

HTG-4275, -4450 [DuPont]

Hytrel Series, 4056, 5526, 5555HS, 6346, 7246, G-4074, G-4075, G-4766, G-4774, G-5544 [DuPont]

Glass-reinforced polyester elastomer:

RTP 1500.5 (5% glass fiber), 1501 (10% glass fiber), 1503 (30% glass fiber), 1505 (30% glass fiber), 1507 (40% glass fiber) [Fiberite]

Thermocomp YF-1002 (10% glass fiber), YF-1004 (20% glass fiber), YF-1006 (30% glass fiber), YF-1008 (40% glass fiber) [LNP]

Liquid crystal polyester (LCP):

Xydar G-330, G-345, G-430, G-445, M-350, M-450, MG-350, MG-450 [Amoco]

Polyarylate:

Arylef U 100 [Solvay]

Durel 400, P400, P410, P430 [Occidental]

Polybutylene terephthalate (PBT):

Celanex 2000, 2002, 2012 [Hoechst-Celanese]

Gafite 1600A, 1602F, 1602Z, 1700A, LW X-4612R [GAF]

Pocan B 1305, B 1505 [Bayer AG]

Ultradur B 2550, B 4500, B 4520, B 4550, KR 4015, KR 4036, KR 4070, KR 4071 [Badische]

Polyester, thermoplastic *(cont'd.)*

Valox 310-SEO, 325, 340, 357, FV-600, FV-608, FV-609, FV-650, FV-699 [General Electric]

Aluminum flake-filled PBT:

EMI-X WA-40 (40% aluminum flake) [LNP]

Carbon fiber-reinforced PBT:

EMI-X WC-1008 (40% PAN carbon fiber) [LNP]

Thermocomp WC-1006 (30% carbon fiber) [LNP]

Glass-reinforced PBT:

Celanex 3200 (15% glass fiber), 3210 (18% glass fiber), 3211 (18% glass fiber), 3300 (30% glass fiber), 3310 (30% glass fiber), 3311 (30% glass fiber), 3400 (40% glass fiber) [Hoechst-Celanese]

Gafite 1432F (15% glass), 1432Z (15% glass), 1452F (26% glass), 1462Z (30% glass), 1465F (30% glass), 1482Z (40% glass), 1614G, 1632F (15% glass), 1632Z (15% glass), 1642Z (20% glass), 1662Z (30% glass), LW 4424R (40% glass spheres/fiberglass), LW X-4424R (40% glass spheres/fiberglass), LW X-4424S (26% glass spheres/fiberglass) [GAF]

Pocan B 3225 (20% glass), B 3235 (30% glass fiber), B 4225 (20% glass), B 4235, B 5335 (30% glass fiber) [Bayer AG]

RTP 1001 (10% glass fiber), 1003 (20% glass fiber), 1005 (30% glass fiber), 1005FR (30% glass fiber), 1007 (40% glass fiber) [Fiberite]

Thermocomp WF-1004 (20% glass fiber), WF-1006 (30% glass fiber), WF-1006FR (30% glass fiber), WF-1008 (40% glass fiber) [LNP]

Ultradur B 4300 G2 (10% glass fiber), B 4300 G4 (20% glass fiber), B 4300 G6 (30% glass fiber), B 4300 G10 (50% glass fiber), B 4300 K4 (20% glass beads), B 4300 K6 (30% glass beads), B 4305 G2 (10% glass fiber), B 4305 G4 (20% glass fiber), B 4305 G6 (30% glass fiber), KR 4025 (20% glass fiber), KR 4035 (30% glass fiber) [Badische]

Valox 420 (30% glass), 420-SEO (30% glass), DR-48 (15% glass), DR-51 (15% glass) [General Electric]

Glass/mica-reinforced PBT:

Gafite LW X-5632Z (fiberglass/mica) [GAF]

Glass/mineral-reinforced PBT:

RTP 1007 GB 10, 1025, 1026, 1027, 1028 [Fiberite]

Ultradur KR 4011 (30% glass fiber/mineral) [Badische]

Valox 701 (35% glass/mineral), 730 (35% glass/mineral), 750 (45% glass/mineral), 752 (45% glass/mineral) [General Electric]

Mica-reinforced PBT:

Gafite LW6362R, LW 6443B, LW 6443F, LW 6443R, LW 7342R, LW X-6443R, LW X-7443R [GAF]

Mineral-reinforced PBT:

Ultradur KR 4001 (25% mineral) [Badische]

Valox 744 (10% mineral), 745 (25% mineral), 760 (25% mineral) [General Electric]

Polyester, thermoplastic *(cont'd.)*

Polyethylene terephthalate (PET):
Kodapak PET Polyester [Eastman]
Mylar Series, Type A, AB, C, D, EL, HS, KB, M, M461, MO, PB, S, T, VB, WC, WC 11, WC 11 G, WC 22 [DuPont]
Tenite PET Polyester 6857 [Eastman]
Glass-reinforced PET:
Petlon 3530 (30% short glass fibers), 3550 (50% short glass fibers), 4530 (30% short glass fibers) [Mobay]
RTP 1105FR (30% glass fibers) [Fiberite]
Rynite 430 (30% glass), 530 (30% glass fiber), 545 (45% glass fiber), 555, FR-530 (30% glass fiber) [DuPont]
Glass/mica-reinforced:
Rynite 935 (35% glass/mica), 940 (40% glass/mica), 940FB (40% glass/mica) [DuPont]

MODIFICATIONS/SPECIALTY GRADES:
Conductive grade (EMI shielding):
EMI-X WA-40, WC-1008; Stat-Kon WC-1002, WC-1006
High-flow:
Celanex 3200, 3210
High-impact:
Valox 340
Impact-modified:
Celanex 2002; Gafite LW 4424R, LW 6362R, LW 6443B, LW 6443F, LW 6443R, LW 76342R, LW X-4424R, LW X-4424S, LW X-4612R, LW X-6443R, LW X-7443R; Ultradur KR 4070, KR 4071; Valox 357, 745
Flame-retardant:
Arylef U 100; Celanex 2012, 3210, 3211, 3310, 3311; Durel 400, P400, P410, P430; Gafite 1432F, 1452F, 1465F, 1602F, 1614G, 1632F, LW 6443F; HTG-4450; Petlon 4530; Pocan B3235, B 4225, B 4235; RTP 1005FR, 1105FR; Rynite FR-530; Thermocomp WF-1006FR; Ultradur B 4305 G2, B 4305 G4, B 4305 G6, KR 4015, KR 4025, KR 4035; Valox 310-SEO, 357, 420-SEO, 750, 752, 760, DR-48, FV-600, FV-699; Xydar G-330, G-345, G-430, G-445, M-350, M-450, MG-350, MG-450
Heat-stabilized:
Pocan B1305, B3225, B4225, B5335
Plasticized:
PA-55-011

CATEGORY:
Thermoplastic resin

PROCESSING:
Extrusion:
Arylef U 100; Gaflex 540ZS, 547, 547ZS, 555, 555ZS, 572ZS; Hytrel 4056, 5526,

Polyester, thermoplastic *(cont'd.)*

5556, 6346, 7246, G-4774, G-5544; Tenite PET Polyester 6857; Ultradur B 2550, B 4550, KR 4036

Injection molding:
Arylef U 100; Celanex 2000, 2002, 2012, 3200, 3210, 3211, 3300, 3310, 3311, 3400; EMI-X WA-40, WC-1008; Gafite 1432F, 1432Z, 1452F, 1462Z, 1465F, 1482Z, 1600A, 1602F, 1602Z, 1614G, 1632F, 1632Z, 1642Z, 1662Z, 1700A, LW 4424R, LW 6362R, LW 6443B, LW 6443F, LW 6443R, LW 7342R, LW X-4424R, LW X-4424S, LW X-4612R, LW X-5632Z, LW X-6443R, LW X-7443R; Gaflex 540ZS, 547, 547ZS, 555, 555ZS, 572ZS; Hytrel 4056, 5526, 5556, 6346, 7246, G-4074, G-4774, G-5544; Petlon 3530, 3550, 4530; Pocan B1305, B1505, B3225, B3235, B4225, B4235, B5335; RTP 1001, 1003, 1005, 1005FR, 1007, 1007 GB 10, 1025, 1026, 10267, 1028, 1105FR, 1500.5, 1501, 1503, 1505, 1507; Rynite 430, 530, 545, 555, 935, 940, FR-530; Thermocomp WC-1006, WF-1004, WF-1006, WF-1006FR, WF-1008; Ultradur B 4300 G2, B 4300 G4, B 4300 G6, B 4300 G10, B 4300 K4, B 4300 K6, B 4305 G2, B 4305 G4, B 4305 G6, B 4500, B 4520, KR 4001, KR 4011, KR 4015, KR 4025, KR 4035, KR 4070, KR 4071; Valox 310-SEO, 325, 340, 357, 420, 420-SEO, 701, 730, 744, 745, 750, 752, 760, DR-48, DR-51; Xydar G-330, G-345, G-430, G-445, M-350, M-450, MG-350, MG-450

Foam injection molding:
Rynite 940FB; Valox FV-600, FV-608, FV-609, FV-650, FV-699

Blow molding:
Arylef U 100; Gaflex 355ZS, 372, 372ZS; HTG-4275

Film processing:
Tenite PET Polyester 6857

APPLICATIONS:

Architectural applications: glazing (Durel 400, P400, P410, P430); solar energy industry (Durel 400, P400, P410, P430)

Automotive applications: (HTG-4275; Hytrel Series, 4056, 5526, 5555HS, 5556, 6346, 7246; Petlon 3530, 3550; RTP 1001, 1003, 1005, 1007; Rynite 530, 545, 555, 935; Thermocomp WC-1004, WF-1006, WF-1008; Ultradur B 4300 K4, KR 4071; Valox 420, DR-48, DR-51); consoles (Ultradur B 4300 G6); exterior parts (Gafite 1432F, 1432Z, 1452F, 1462Z, 1600A, 1602F, 1602Z, 1642Z, 1662Z, LW 4424R, LW 6362R, LW 6443B, LW 6443F, LW 6443R, LW 7342R, LW X-4424R, LW X-4424S, LW X-4612R, LW X-5632Z, LW X-6443R, LW X-7443R; Ultradur B 4300 G4); ignition parts (Valox 701, 730, 744, 760); light housings (Gafite LW 4424R, LW 6362R, LW 6443B, LW 6443F, LW 6443R, LW 7342R); snowmobile components (Thermocomp YF-1002, YF-1004, YF-1006, YF-1008); under-the-hood parts (Gafite 1432F, 1432Z, 1465F, 1452F, 1462Z, 1482Z, 1600A, 1602F, 1602Z, 1614G, 1632F, 1632Z, 1642Z, 1662Z, 1700A; Rynite 530, 545, 555, 935); wiper blade supports (Rynite 430; Ultradur B 4300 G6)

Aviation industry: aerospace (Xydar G-330, G-345, G-430, G-445, M-350, M-450, MG-350, MG-450); avionics housings (EMI-X WA-40, WC-1008)

Consumer products: appliances (Durel 400, P400, P410, P430; Thermocomp WC-

1004, WF-1006, WF-1008; Ultradur B 4520); appliance housings (Gafite 1432F, 1432Z, 1452F, 1462Z, 1600A, 1602F, 1602Z, 1642Z, 1662Z; Ultradur KR 4001); appliance parts (Gafite 1432F, 1432Z, 1452F, 1462Z, 1465F, 1482Z, 1600A, 1602F, 1602Z, 1614G, 1632F, 1642Z, 1662Z, 1700A; Ultradur B 4300 G2); carbon ribbon (Mylar Series, Type A, AB, C, D, EL, HS, KB, M, M461, MO, PB, S, T, VB, WC, WC 11, WC 11 G, WC 22); electrical goods (Arylef U 100); hardware (Gafite 1432F, 1432Z, 1452F, 1462Z, 1465F, 1482Z, 1600A, 1602F, 1602Z, 1614G, 1632F, 1632Z, 1642Z, 1662Z, 1700A); office supplies (Mylar Type S); photographic applications (Mylar Type D); sporting goods (HTG-4275; Hytrel Series, 4056, 5526, 5555HS, 5556, 6346, 7246; Thermocomp YF-1002, YF-1004, YF-1006, YF-1008; Ultradur KR 4070); tools (Thermocomp YF-1002, YF-1004, YF-1006, YF-1008)

Electrical/electronic industry: (Arylef U 100; Celanex 3211, 3311; Durel 400, P400, P410, P430; EMI-X WA-40, WC-1008; Gafite 1432F, 1432Z, 1452F, 1462Z, 1465F, 1482Z, 1600A, 1602F, 1602Z, 1614G, 1632F, 1632Z, 1642Z, 1662Z, 1700A; Petlon 4530; Rynite FR-530; Thermocomp WC-1004, WF-1006, WF-1008; Valox 750, 752); business machines and office equipment (Durel 400, P400, P410, P430; EMI-X WA-40, WC-1008; Gafite 1432F, 1432Z, 1452F, 1462Z, 1600A, 1602F, 1602Z, 1642Z, 1662Z, LW X-4424R, LW X-4424S, LW X-4612R, LW X-5632Z, LW X-6443R, LW X-7443R; Rynite 530, 545, 555, 935; Ultradur B 4520; Valox 701, 730, 744, 760); cable and wire jacketing (Gaflex 355ZS, 372, 372ZS, 540ZS, 547, 547ZS, 555, 555ZS, 572ZS; HTG-4275; Hytrel Series, 4056, 5526, 5555HS, 5556, 6346, 7246, G-4074, G-4774, G-5544; Mylar Series, Type A, AB, C, D, EL, HS, KB, M, M461, MO, PB, S, T, VB, WC, WC 11, WC 11 G, WC 22); capacitors (Mylar Series, Type A, AB, C, D, EL, HS, KB, M, M461, MO, PB, S, T, VB, WC, WC 11, WC 11 G, WC 22; Ultradur KR 4015); circuit breakers (Valox 701, 730, 744, 760); computers (Rynite 530, 545, 555, 935); connectors (Ultradur B 4305 G2, B 4305 G4, B 4305 G6, KR 4015, KR 4025, KR 4035; Valox 420, DR-48, DR-51); electrical components (HTG-4275; Hytrel Series, 4056, 5526, 5555HS, 5556, 6346, 7246; Xydar G-330, G-345, G-430, G-445, M-350, M-450, MG-350, MG-450); electrical enclosures/packaging (Gaflex 355ZS, 372, 372ZS, 540ZS, 547, 547ZS, 555, 555ZS, 572ZS; Stat-Kon WC-1002, WC-1006; Valox FV-608, FV-609, FV-699); housings (Ultradur B 4300 G4, B 4305 G4, KR 4015, KR 4025, KR 4035); insulation (Mylar Type WC); lighting equipment (Arylef U 100; Durel 400, P400, P410, P430; Ultradur B 4300 G4, B 4305 G6); magnetic tape (Mylar Series, Type A, AB, C, D, EL, HS, KB, M, M461, MO, PB, S, T, VB, WC, WC 11, WC 11 G, WC 22); microfilm (Mylar Series, Type A, AB, C, D, EL, HS, KB, M, M461, MO, PB, S, T, VB, WC, WC 11, WC 11 G, WC 22); motor controls (Valox 701, 730, 744, 760); switchgear (Ultradur B 4305 G2, B 4305 G6, KR 4015; Valox 420, DR-48, DR-51); TV components (Valox 701, 730, 744, 760)

Food-contact applications: food packaging (Mylar Type HS; Tenite PET Polyester 6857; Ultradur B 2550)

Polyester, thermoplastic (cont'd.)

Functional additives: plastics modification (HTG-4450); resin modification (PA-55-004, -55-013)

Industrial applications: adhesives (PA-55-024; Terlan 6090, 6100, 6300, 6430, 6480, 6490, 6530, 6600, 6676); barrier coating (Mylar Series, Type A, AB, C, D, EL, HS, KB, M, M461, MO, PB, S, T, VB, WC, WC 11, WC 11 G, WC 22); belting (Gaflex 355ZS, 372, 372ZS, 540ZS, 547, 547ZS, 555, 555ZS, 572ZS; HTG-4275; Hytrel Series, 4056, 5526, 5555HS, 5556, 6346, 7246); bottles and containers (Kodapak PET Polyester); cams (Ultradur B 4300 G2, B 4300 G4, B 4500); ceramic and metal substitutes (Xydar G-330, G-345, G-430, G-445, M-350, M-450, MG-350, MG-450); chemical process equipment (Xydar G-330, G-345, G-430, G-445, M-350, M-450, MG-350, MG-450); connectors (RTP 1001, 1003, 1005, 1007; Rynite FR-530); critical/performance parts (RTP 1007 GB 10, 1025, 1026, 1027, 1028); engineering parts (Ultradur B 4300 G2, B 4300 G4, B 4300 G6, B 4300 G10, B 4300 K4, B 4300 K6, B 4500, KR 4070); films (Hytrel G-4074; Mylar Series, Type A, AB, C, D, EL, HS, KB, M, M461, MO, PB, S, T, VB, WC, WC 11, WC 11 G, WC 22; Tenite PET Polyester 6857; Ultradur B 4550); frames (Rynite 940, 940FB); gaskets/seals (Gaflex 355ZS, 372, 372ZS, 540ZS, 547, 547ZS, 555, 555ZS, 572ZS; HTG-4275; Hytrel Series, 4056, 5526, 5555HS, 5556, 6346, 7246; RTP 1500.5, 1501, 1503, 1505, 1507; Thermocomp YF-1002, YF-1004, YF-1006, YF-1008); gears (HTG-4275; Hytrel Series, 4056, 5526, 5555HS, 5556, 6346, 7246; Rynite 530, 545, 555, 935; Thermocomp YF-1002, YF-1004, YF-1006, YF-1008); graphic arts applications (Mylar Type D); hot stamping (Mylar Series, Type A, AB, C, D, EL, HS, KB, M, M461, MO, PB, S, T, VB, WC, WC 11, WC 11 G, WC 22); housings (Rynite 430, 530, 545, 555, 935; Ultradur B 4300 K4, B 4300 K6, B 4520; Ultradur KR 4011, KR 4070); hydraulic controls (Gafite 1432F, 1432Z, 1452F, 1462Z, 1465F, 1482Z, 1600A, 1602F, 1602Z, 1614G, 1632F, 1632Z, 1642Z, 1662Z, 1700A); inks (PA-55-011, -55-012); intricate/hard-to-fill moldings (Rynite FR-530); lamination (Mylar Series, Type A, AB, C, D, EL, HS, KB, M, M461, MO, PB, S, T, VB, WC, WC 11, WC 11 G, WC 22); large parts (Valox 745); low-warpage parts (Gafite LW X-4424R, LW X-4424S, LW X-4612R, LW X-5632Z, LW X-6443R, LW X-7443R; Valox 745)); mechanical engineering (Petlon 3530, 3550); molded parts (Hytrel G-4074, G-4774, G-5544); monofilament (Rynite 530, 545, 555, 935); motor insulation (Mylar Type MO); noise dampening (Gaflex 355ZS, 372, 372ZS, 540ZS, 547, 547ZS, 555, 555ZS, 572ZS); optical parts/instruments (Arylef U 100; Ultradur B 4300 K4); paper coatings (Ultradur B 2550); precision parts (Rynite 530, 545, 555, 935; Ultradur B 4300 K4); pressure-sensitive tapes (Mylar Type M461); process control instruments (Valox FV-608, FV-609, FV-699); profiles (Hytrel G-4774, G-5544); pumps (Gafite 14465F, 1482Z, 1614G, 1632F, 1632Z, 1700A; Gaflex 355ZS, 372, 372ZS, 540ZS, 547, 547ZS, 555, 555ZS, 572ZS; Rynite 430); rubber tires (PA-55-011, 55-012); safety equipment (Xydar G-330, G-345, G-430, G-445, M-350, M-450, MG-350, MG-450); sanding pads (RTP 1500.5, 1501, 1503, 1505, 1507); sheet (Hytrel G-4074; Ultradur B 4550); structural components (Rynite 940, 940FB); structural foam (Valox FV-600,

Polyester, thermoplastic *(cont'd.)*

FV-608, FV-609, FV-650, FV-699); thin sections (Rynite FR-530; Valox DR-48, DR-51); tubing/hoses (Gaflex 355ZS, 372, 372ZS, 540ZS, 547, 547ZS, 555, 555ZS, 572ZS; HTG-4275; Hytrel Series, 4056, 5526, 5555HS, 5556, 6346, 7246, G-4074, G-4774, G-5544); valves (Ultradur B 4500)

Medical applications: (Arylef U 100); packaging (Ultradur B 4550)

Transportation industry: (Durel 400, P400, P410, P430; Valox FV-608, FV-609, FV-699; Xydar G-330, G-345, G-430, G-445, M-350, M-450, MG-350, MG-450)

PROPERTIES:

Form:

Solid (Arylef U 100; Celanex 2000, 2012, 3200, 3210, 3211, 3300, 3310, 3311; Kodapak PET Polyester)

Pellets (Durel 400, P400, P410, P430; Gafite 1432F, 1432Z, 1452F, 1462Z, 1465F, 1482Z, 1600A, 1602F, 1602Z, 1614G, 1632F, 1632Z, 1642Z, 1662Z, 1700A, LW 4424R, LW 6362R, LW 6443B, LW 6443F, LW 6443R; Hytrel Series, G-4074, G-4774, G-5544; Valox 310-SEO, 325, 420, 420-SEO, 750, DR-48, DR-51)

Cylindrical granules (Pocan B 1505)

Crushed or granulated (Terlan 6090, 6100, 6300, 6430, 6480)

Flakes or crushed (PA-55-004, -55-011, -55-012, -55-013)

Flakes or lumps (PA-55-024)

35-mesh powder (Hytrel Series, G-4074, G-4774, G-5544)

Color:

Clear, green (Kodapak PET Polyester)

Transparent (Durel 400)

Natural and colors (Gafite 1432F, 1432Z, 1452F, 1462Z, 1600A, 1602F, 1602Z, 1642Z, 1662Z)

Natural and custom colors (Mamax 2020)

Natural, black (RTP 1001, 1005FR, 1007 GB 10, 1025, 1026, 1105FR, 1500.5, 1501, 1503, 1505, 1507)

Natural, black, and standard colors (Pocan B 1505)

Supplied in colored form (Ultradur B 4300 G2, B 4300 G4, B 4300 G6, B 4300 G10, B 4300 K4, B 4300 K6, B 4305 G2, B 4305 G4, B 4305 G6, B 4520, KR 4001, KR 4011, KR 4015, KR 4025, KR 4035, KR 4070, KR 4071)

Opaque (Durel P400, P410, P430)

Opaque, colorable (Petlon 3530, 3550, 4530)

Amber (PA-55-004, -55-011, -55-012, -55-013, -55-024)

GENERAL PROPERTIES:

Solubility:

Sol. in ammonia (PA-55-004, -55-011, -55-012, -55-013)

Melt Flow:

3–6 g/10 min (220 C) (Gaflex 355ZS)

5 g/10 min (190 C) (Hytrel 4056)

5.2 g/10 min (190 C) (Hytrel G-4074)

7 g/10 min (220 C) (Hytrel 5556, 6346)

Polyester, thermoplastic *(cont'd.)*

    9 g/10 min (220 C) (Hytrel 5555HS)
    10–15 g/10 min (220 C) (Gaflex 372, 372ZS, 555, 555ZS)
    11–15 g/10 min (220 C) (Gaflex 547, 547ZS)
    12 g/10 min (230 C) (Hytrel G-4774)
    12.5 g/10 min (240 C) (Hytrel 7246); (230 C) (Hytrel G-5544)
    16–22 g/10 min (220 C) (Gaflex 540ZS)
    18 g/10 min (220 C) (Hytrel 5526)
    28–33 g/10 min (240 C) (Gaflex 572ZS)

Sp. Gr.:
    1.07–1.10 (PA-55-024)
    1.13–1.15 (PA-55-011)
    1.14–1.16 (PA-55-004, -55-012, -55-013)
    1.16 (HTG-4275; Hytrel G-4766)
    1.17 (Hytrel 4056)
    1.18 (Hytrel G-4074, G-4075)
    1.2 (Durel 400; Hytrel 5526, 5555HS, 5556, G-4774; Terlan 6090; Valox FV-600, FV-650)
    1.21 (Terlan 6490, 6600)
    1.22 (Durel P400; Hytrel 6346, G-5544; Terlan 6100, 6300, 6430)
    1.23 (RTP 1500.5; Terlan 6480)
    1.24 (Terlan 6530)
    1.25 (Hytrel 7246; Valox 340)
    1.26 (RTP 1501)
    1.27 (Terlan 6676; Thermocomp YF-1002; Valox FV-699 solid)
    1.29 (Durel P410)
    1.30 (Gafite LW X-4612R; Valox FV-609 solid)
    1.31 (Celanex 2000, 2002; Gafite 1600A, 1602Z; Valox 325)
    1.34 (RTP 1503; Thermocomp YF-1004)
    1.36 (Stat-Kon WC-1002; Valox 357, 744)
    1.38 (RTP 1001)
    1.395 (Mylar Type A)
    1.41 (Celanex 3200; Stat-Kon WC-1006; Valox 310-SEO, DR-51)
    1.42 (Gafite 1432Z; Thermocomp YF-1006)
    1.43 (Celanex 2012; Durel P430; RTP 1505; Thermocomp WF-1004)
    1.44 (Gafite 1602F)
    1.45 (Gafite 1642Z, LW X-4424S; RTP 1003)
    1.46 (Valox 745)
    1.47 (Thermocomp WC-1006; Valox 760)
    1.48 (EMI-X WC-1008)
    1.49 (Rynite 430)
    1.50 (Valox FV-608 solid)
    1.52 (Gafite 1462Z, 1662Z; RTP 1507; Thermocomp WF-1006, YF-1008)
    1.53 (RTP 1005; Valox 420, DR-48)

1.54 (Celanex 3300)
1.55 (Gafite LW X-4424R; Valox 730)
1.56 (Gafite LW X-5632Z; Rynite 530)
1.57 (Gafite LW X-6443R)
1.58 (Gafite 1432F; Rynite 935; Valox 701)
1.59 (Gafite LW X-7443R)
1.60 (Celanex 3211)
1.61 (Celanex 3400)
1.62 (Celanex 3210; RTP 1025, 1026, 1027; Rynite 940, 940FB; Thermocomp WF-1008; Valox 420-SEO; Xydar G-330, G-430)
1.63 (RTP 1007)
1.65 (Celanex 3311)
1.66 (EMI-X WA-40; RTP 1005FR)
1.67 (Rynite FR-530; Thermocomp WF-1006FR)
1.68 (Gafite 1452F; RTP 1105FR)
1.69 (Celanex 3310; Rynite 545)
1.71 (RTP 1028)
1.72 (RTP 1007 GB 10)
1.75 (Xydar G-345, G-445)
1.80 (Rynite 555; Valox 750; Xydar MG-350, MG-450)
1.85 (Xydar M-350, M-450)
1.86 (Valox 752)

Sp. Vol.:

14.9 in.^{3}/lb (Valox 752)
15.4 in.^{3}/lb (Valox 750)
17.1 in.^{3}/lb (Valox 420-SEO)
17.5 in.^{3}/lb (Valox 701)
17.9 in.^{3}/lb (Valox 730)
18.2 in.^{3}/lb (Valox 420)
18.7 in.^{3}/lb (Valox DR-48)
18.8 in.^{3}/lb (Valox 760)
19.0 in.^{3}/lb (Valox 745)
19.6 in.^{3}/lb (Valox 310-SEO, DR-51)
20.4 in.^{3}/lb (Valox 744)
20.5 in.^{3}/lb (Valox 357)
21.1 in.^{3}/lb (Valox 325)
22.3 in.^{3}/lb (Valox 340)

Density:

1.21 mg/m^3 (Arylef U 100)
1.57 mg/m^3 (Petlon 3530)
1.67 mg/m^3 (Petlon 4530)
1.77 mg/m^3 (Petlon 3550)
1.30 mg/cm^3 (Pocan B 1505)

Polyester, thermoplastic *(cont'd.)*

1.55 mg/cm³ (Pocan B 3235)
1.65 mg/cm³ (Pocan B 4235)
1.27 g/ml (Mamax 2020)
1.38 g/ml (Malon 1070, 1080, 1090)
1.2 g/cm³ (Ultradur KR 4070, KR 4071)
1.23 g/cm³ (Ultradur KR 4001)
1.25 g/cm³ (Ultradur KR 4011)
1.30 g/cm³ (Pocan B 1305; Ultradur B 2550, B 4500, B 4520, B 4550, KR 4036)
1.33 g/cm³ (Tenite PET Polyester 6857)
1.38 g/cm³ (Ultradur B 4300 G2)
1.45 g/cm³ (Ultradur B 4300 G4, B 4300 K4, KR 4015)
1.46 g/cm³ (Pocan B 3225)
1.49 g/cm³ (Ultradur B 4305 G2)
1.53 g/cm³ (Ultradur B 4300 G6, B 4300 K6)
1.55 g/cm³ (Pocan B 5335; Ultradur KR 4025)
1.57 g/cm³ (Pocan B 4225; Ultradur B 4305 G4)
1.65 g/cm³ (Ultradur B 4305 G6, KR 4035)
1.71 g/cm³ (Ultradur B 4300 G10)

Visc.:
9500 cps (221 C) (Terlan 6090)
10,000 cps (232 C) (Terlan 6480)
18,000 cps (232 C) (Terlan 6676)
40,000 cps (232 C) (Terlan 6430)
40,000 cps (232 C) (Terlan 6600)
45,000 cps (232 C) (Terlan 6100, 6300)
100,000 cps (210 C) (Terlan 6490)
120,000 cps (210 C) (Terlan 6530)

M.P.:
85–105 C (PA-55-024)
90–110 C (PA-55-011)
120–140 C (PA-55-004, -55-013)
135–155 C (PA-55-012)
166 C (Gaflex 540ZS)
168 C (Hytrel 4056)
173 C (Hytrel G-4074)
178 C (Gaflex 547, 547ZS)
184 C (Gaflex 355ZS, 555, 555ZS)
206 C (Hytrel 6346)
208 C (Hytrel G-4774)
211 C (Hytrel 5526, 5555HS, 5556)
215 C (Hytrel G-5544)
218 C (Hytrel 7246)
228 C (Gaflex 372, 372ZS, 572ZS)

Polyester, thermoplastic (cont'd.)

250 C (Mylar Type A; Rynite 430, 940, 940FB)
252 C (Rynite 935)
254 C (Rynite 530, 545, FR-530)
255 C (Rynite 555)
417 F (Celanex 3211, 3311)
428–437 F (Ultradur B 2550, B 4300 G2, B 4300 G4, B 4300 G6, B 4300 G10, B 4300
     K4, B 4300 K6, B 4305 G2, B 4305 G4, B 4305 G6, B 4500, B 4520, B 4550, KR
     4001, KR 4011, KR 4015, KR 4025, KR 4035, KR 4036, KR 4070, KR 4071)
442 F (Celanex 2000, 2002, 2012, 3200, 3210, 3300, 3310, 3400)
480 F (Malon 1070)
485 F (Malon 1080)
490 F (Malon 1090)
Softening Pt. (R&B):
91 C (Terlan 6600)
93 C (Terlan 6490)
113 C (Terlan 6090)
129 C (Terlan 6100)
138 C (Terlan 6480)
141 C (Terlan 6300, 6430)
148 C (Terlan 6530)
182 C (Terlan 6676)
Acid No.:
10–30 (PA-55-024)
100–140 (PA-55-013)
140–170 (PA-55-011)
200–220 (PA-55-012)
Stability:
Good dimensional stability; excellent R.T. resistance to water, weak acids and bases, detergents, aliphatic hydrocarbons, chlorinated aliphatic hydrocarbons, oils, fats, ethers, alcohols, and glycols (Celanex 2000)

Excellent R.T. resistance to water, weak acids and bases, detergents, aliphatic hydrocarbons, chlorinated aliphatic hydrocarbons, oils, fats, ethers, alcohols, and glycols (Celanex 2012, 3200, 3210, 3211, 3300, 3310, 3311)

Excellent resistance to organic solvents, incl. aliphatic hydrocarbons, gasoline, carbon tetrachloride, alcohols, ketones, ethylene glycol, fluorinated hydrocarbons, oils, and fats; resistant to salt sol'ns., soap sol'ns., and dilute acids and bases; attacked by strong acids and bases; excellent dimensional stability under load (Gafite 1432F, 1432Z, 1452F, 1462Z, 1600A, 1602F, 1602Z, 1642Z, 1662Z)

Good dimensional stability and broad chemical resistance (Petlon 3530, 3550, 4530)

Excellent resistance to a wide variety of fluids such as gasoline, motor oil, transmission fluid, hydrocarbons, and organic solvents; some absorption by ketones and esters causes plasticization and small dimensiaonl changes; good resistance to acids and bases at R.T.; attacked by strong and weak acids and bases at elevated temps.

Polyester, thermoplastic *(cont'd.)*

(Rynite 530, 545, 555, 935)
Ref. Index:
  1.61 (Arylef U 100)
  1.64 (Mylar Type A)

MECHANICAL PROPERTIES:

Tens. Str.:
  520 kg/cm² (yield) (Gafite LW X-4612R)
  560 kg/cm² (yield) (Gafite 1600A, 1602F, 1602Z, LW X-4424R, LW X-4424S)
  800 kg/cm² (yield) (Gafite LW X-7443R)
  825 kg/cm² (yield) (Gafite LW X-6443R)
  860 kg/cm² (yield) (Gafite LW X-5632Z)
  880 kg/cm² (yield) (Gafite 1432Z)
  940 kg/cm² (yield) (Gafite 1432F)
  1100 kg/cm² (yield) (Gafite 1642Z)
  1200 kg/cm² (yield) (Gafite 1452F)
  1300 kg/cm² (Gafite 1462Z) (yield) (Gafite 1662Z)
  13.8 MPa (break) (Hytrel G-4074)
  15 MPa (Hytrel G-4075)
  20.6 MPa (break) (Hytrel G-4774)
  21 MPa (Hytrel G-4766)
  25 MPa (Hytrel 4056)
  31 MPa (break) (Hytrel G-5544)
  35 MPa (break) (Pocan B 1505)
  37 MPa (break) (Pocan B 1305)
  38 MPa (Hytrel 5526, 5555HS, 5556, 6436, 7246)
  42 MPa (Valox FV-699)
  52 MPa (Valox FV-609)
  55 MPa (yield) (Pocan B 1505)
  57 MPa (yield) (Pocan B 1305)
  68.9 MPa (Durel 400)
  71 MPa (Arylef U 100)
  90 MPa (break) (Pocan B 5335)
  91 MPa (Valox FV-608)
  96.5 MPa (Rynite 935)
  117 MPa (break) (Petlon 4530)
  120 MPa (break) (Pocan B 4225); (yield and break) (Pocan BH 4235)
  123 MPa (break) (Petlon 3530)
  124 MPa (Rynite 430)
  126.3 MPa (Rynite 940, 940FB)
  130 MPa (break) (Pocan B 3225)
  140 MPa (yield and break) (Pocan B 3235)
  152 MPa (Rynite FR-530); (break) (Petlon 3550)
  158 MPa (Rynite 530)

193 MPa (Rynite 545)
196 MPa (Rynite 555)
100 psi (break) (Terlan 6490)
400 psi (break) (Terlan 6090, 6300)
600 psi (break) (Terlan 6100)
1400 psi (break) (Terlan 6600)
1500 psi (break) (Terlan 6530)
1800 psi (break) (Terlan 6480)
2000 psi (break) (Terlan 6430, 6676)
4100 psi (RTP 1500.5)
4800 psi (RTP 1501)
5000 psi (Thermocomp YF-1002)
5075 psi (yield) (Ultradur KR 4071)
6500 psi (Valox 340)
6525 psi (yield) (Ultradur KR 4070)
6800 psi (RTP 1028); (yield) (Mamax 2020)
7000 psi (Valox 357)
7100 psi (Valox 745)
7200 psi (RTP 1503)
7250 psi (break) (Ultradur B 4300 K4, B 4300 K6)
7500 psi (EMI-X WA-40; Gafite 1700A; Thermocomp YF-1004; Valox 325, 744)
8000 psi (Celanex 2012; Gafite 1614G, LW 4424R; RTP 1505; Valox 310-SEO, 760)
8100 psi (break) (Celanex 2002)
8300 psi (Celanex 2000)
8500 psi (Malon 1070, 1080, 1090 films)
8700 psi (yield) (Ultradur B 2550, B 4500, B 4520, B 4550, KR 4015, KR 4036);
    (break) (Ultradur KR 4001)
8800 psi (RTP 1507)
9000 psi (RTP 1027)
9500 psi (Valox 730)
9910 psi (Valox FV-600)
9920 psi (Valox FV-650)
10,000 psi (RTP 1026; Thermocomp YF-1006)
10,100 psi (Durel P400)
11,000 psi (RTP 1001)
11,096 psi (Xydar M-450)
11,202 psi (Xydar M-350)
11,400 psi (Gafite LSW 7342R)
11,700 psi (Gafite LW 6443B, LW 6443R)
12,000 psi (Gafite LW 6443F; Valox 701)
12,500 psi (Thermocomp YF-1008; Valox 750, 752)
13,000 psi (Celanex 3200; Gafite 1632Z; Valox DR-48, DR-51)
13,050 psi (break) (Ultradur B 4300 G2)

13,100 psi (Durel P410)
13,500 psi (Gafite 1632F)
13,622 psi (Xydar MG-450)
13,775 psi (break) (Ultradur KR 4011)
13,983 psi (Xydar MG-350)
14,000 psi (Gafite LW 6362R)
14,500 psi (break) (Ultradur B 4305 G2)
15,000 psi (Stat-Kon WC-1002)
15,470 psi (Xydar G-345)
15,950 psi (break) (Ultradur B 4300 G4, KR 4025)
16,000 psi (RTP 1025)
16,050 psi (Xydar G-445)
16,500 psi (Celanex 3210; RTP 1003)
17,000 psi (Celanex 3211; Valox 420-SEO)
17,300 psi (Valox 420)
17,355 psi (Xydar G-330)
17,400 psi (break) (Ultradur KR 4035)
17,500 psi (Thermocomp WF-1004, WF-1006FR)
18,000 psi (Gafite 1465F; RTP 1005)
18,027 psi (Xydar G-430)
18,850 psi (break) (Ultradur B 4305 G4)
19,500 psi (Celanex 3300, 3310, 3311; RTP 1105FR)
19,575 psi (break) (Ultradur B 4300 G6)
19,700 psi (Durel P430)
20,000 psi (Gafite 1482Z; Thermocomp WF-1006)
21,000 psi (RTP 1005FR, 1007 GB 10)
21,300 psi (break) (Celanex 3400)
21,500 psi (RTP 1007)
21,750 psi (break) (Ultradur B 4300 G10, B 4305 G6)
22,000 psi (Stat-Kon WC-1006; Thermocomp WC-1006, WF-1008)
25,000 psi (EMI-X WC-1008); (break) (Mylar Type A)
Tens. Elong.:
0.85% (break) (Xydar G-445)
1.0% (EMI-X WC-1008); (break) (Xydar G-430, MG-450)
1–2% (Celanex 3310, 3311)
1.3% (break) (Valox FV-600)
1.37% (break) (Xydar M-450)
1.38% (break) (Xydar G-345)
1.5% (EMI-X WA-40; Stat-Kon WC-1006)
1.57% (break) (Xydar G-330)
1.6% (break) (Rynite 555)
1.67% (break) (Xydar MG-350)
1.9% (break) (Rynite 940, 940FB)

2% (Celanex 3200, 3300; RTP 1005FR, 1007, 1028, 1105FR); (break) (Celanex 3400; Gafite 1452F; Petlon 3550); (yield and break) (Pocan B 4235)

2–3% (Celanex 3210, 3211; Thermocomp WC-1006)

2.1% (break) (Rynite 545)

2.2% (Durel P430; RTP 1007 GB 10); (break) (Pocan B 4225; Rynite 935)

2.24% (break) (Xydar M-350)

2.3% (break) (Rynite FR-530; Valox FV-650)

2.5% (RTP 1005, 1025, 1026, 1027); (break) (Ultradur B 4300 G10)

2.7% (break) (Rynite 530); (yield and break) (Pocan B 3235)

3% (RTP 1003); (break) (Gafite LW X-4424R, LW X-4424S, LW X-5632Z, LW X-6443R, LW X-7443R; Petlon 3530, 4530; Pocan B 5335; Ultradur B 4300 G2, B 4300 G4, B 4300 G6, B 4305 G2, B 4305 G4, B 4305 G6; Valox 420, 420-SEO)

3.1% (break) (Pocan B 3225)

3.2% (break) (Rynite 430)

3.5% (RTP 1001; Stat-Kon WC-1002)

3.5–4.5% (Thermocomp WF-1006, WF-1008)

3.7% (yield) (Pocan B 1305, B 1505)

3.9% (Durel P410)

4% (break) (Gafite 1462Z, 1662Z; Ultradur B 4300 K4, B 4300 K6, KR 4025, KR 4035; Valox 701, 730, 752, FV-608, FV-699)

4–5% (Thermocomp WF-1004)

5% (Celanex 2000); (break) (Gafite 1642Z; Ultradur KR 4015; Valox DR-48, DR-51)

6% (break) (Gafite 1432F, 1432Z, LW X-4612R)

7% (RTP 1507)

7–8% (Thermocomp YF-1008)

7.5% (Durel P400)

8% (yield) (Durel 400); (break) (Valox FV-609)

8–10% (Thermocomp YF-1006)

9% (RTP 1505)

10% (break) (Ultradur KR 4011)

10+% (RTP 1500.5, 1501)

12% (RTP 1503)

14% (break) (Ultradur KR 4001)

15% (Celanex 2012); (break) (Valox 760)

15–16% (Thermocomp YF-1004)

> 20% (break) (Pocan B 1305)

25% (break) (Gafite 1602F)

30% (break) (Valox 745)

50% (Arylef U 100); (break) (Valox 357)

60% (break) (Ultradur B 4520)

80% (break) (Ultradur KR 4070; Valox 310-SEO)

100% (Thermocomp YF-1002); (break) (Terlan 6090, 6676; Ultradur KR 4071)

120% (break) (Mylar Type A)

Polyester, thermoplastic *(cont'd.)*

    150% (break) (Celanex 2002; Ultradur B 2550; Valox 744)
    170% (break) (Hytrel G-4074)
    175% (break) (Valox 340)
    200% (break) (Terlan 6600; Ultradur B 4500, B 4550, KR 4036)
    > 200% (break) (Gafite 1600A, 1602Z; Pocan B 1505)
    240% (break) (Hytrel G-4075)
    275% (break) (Hytrel G-4774)
    300% (break) (Terlan 6430; Valox 325)
    350% (break) (Hytrel 6346, 7246)
    375% (break) (Hytrel G-5544)
    400% (break) (Terlan 6480)
    450% (break) (Hytrel 4056, 5526, 5555HS, 5556)
    500% (break) (Hytrel G-4766; Malon 1070, 1080, 1090; Terlan 6300)
    600% (break) (Terlan 6100)
    700% (break) (Terlan 6530)
    2000% (break) (Terlan 6490)

Tens. Mod.:
    120,000 psi (RTP 1500.5)
    170,000 psi (RTP 1501)
    320,000 psi (RTP 1503)
    344,000 psi (Valox 325)
    350,000 psi (Mamax 2020)
    408,000 psi (Valox 310-SEO)
    510,000 psi (RTP 1505)
    650,000 psi (RTP 1507)
    719,000 psi (Valox DR-51)
    820,000 psi (Valox FV-650)
    860,000 psi (Valox DR-48)
    1.028×10^6 psi (Valox FV-600)
    1.3×10^6 psi (Valox 420)
    1.34×10^6 psi (Xydar M-450)
    1.43×10^6 psi (Valox 420-SEO)
    1.56×10^6 psi (Xydar M-350)
    1.59×10^6 psi (Xydar G-345, G-445)
    1.6×10^6 psi (Valox 750)
    1.7×10^6 psi (Xydar G-330)
    1.74×10^6 psi (Xydar MG-450)
    1.96×10^6 psi (Xydar MG-350)
    2.34×10^6 psi (Xydar G-430)

Flex. Str.:
    850 kg/cm² (Gafite 1600A, 1602Z)
    870 kg/cm² (Gafite LW X-4612R)
    910 kg/cm² (Gafite 1602F)

920 kg/cm² (Gafite LW X-4424R, LW X-4424S)
1200 kg/cm² (Gafite LW X-6443R)
1300 kg/cm² (Gafite LW X-7443R)
1420 kg/cm² (Gafite LW X-5632Z)
1600 kg/cm² (Gafite 1432F, 1432Z)
1770 kg/cm² (Gafite 1642Z)
1810 kg/cm² (Gafite 1452F)
2020 kg/cm² (Gafite 1462Z, 1662Z)
65 MPa (Valox FV-699)
81 MPa (Arylef U 100)
84 MPa (Valox FV-609)
85 MPa (limiting) (Pocan B 1505)
103.4 MPa (Durel 400)
139 MPa (Valox FV-608)
145 MPa (Pocan B 5335)
148 MPa (Rynite 935)
168 MPa (break) (Petlon 4530)
180 MPa (break) (Petlon 3530)
185 MPa (Pocan B 4225)
190 MPa (Pocan B 3225)
193.2 MPa (Rynite 430)
198 MPa (Rynite 940, 940FB)
200 MPa (Pocan B 3235, B 4235)
207 MPa (break) (Petlon 3550)
221 MPa (Rynite FR-530)
231 MPa (Rynite 530)
283 MPa (Rynite 545)
310 MPa (Rynite 555)
5200 psi (RTP 1500.5)
6400 psi (RTP 1501)
8500 psi (Thermocomp YF-1002)
8700 psi (Mamax 2020)
9800 psi (RTP 1503)
10,150 psi (Ultradur KR 4015)
10,500 psi (Thermocomp YF-1004; Valox 340)
10,800 psi (RTP 1028)
11,000 psi (RTP 1505)
12,000 psi (Valox 325)
12,400 psi (Celanex 2000)
12,500 psi (Celanex 2002; EMI-X WA-40; RTP 1507; Valox 744)
13,000 psi (Thermocomp YF-1006; Valox 745)
13,500 psi (Valox 357)
14,000 psi (Celanex 2012)

Polyester, thermoplastic *(cont'd.)*

14,500 psi (Ultradur B 4300 K4)
14,700 psi (Valox 310-SEO)
14,870 psi (Xydar M-450)
15,000 psi (RTP 1027; Thermocomp YF-1008; Valox 760)
15,128 psi (Xydar M-350)
16,000 psi (Valox 730)
16,675 psi (Ultradur B 4300 K6)
17,500 psi (RTP 1026)
17,900 psi (Valox FV-600)
18,500 psi (Valox FV-650)
18,755 psi (Xydar MG-450)
19,000 psi (RTP 1001)
19,330 psi (Xydar MG-350)
19,500 psi (Valox 750, 7652)
20,000 psi (Valox 701, DR-51)
20,300 psi (Ultradur B 4300 G2)
21,000 psi (Valox DR-48); (break) (Celanex 3200)
21,715 psi (Xydar G-445)
21,743 psi (Xydar G-345)
22,475 psi (Ultradur KR 4025)
23,000 psi (RTP 1003)
23,103 psi (Xydar G-430)
23,155 psi (Xydar G-330)
23,200 psi (Ultradur B 4305 G2)
24,000 psi (Thermocomp WF-1004, WF-1006FR); (break) (Celanex 3210)
24,650 psi (Ultradur B 4300 G4)
25,000 psi (RTP 1025)
26,000 psi (Stat-Kon WC-1002)
26,100 psi (Ultradur KR 4035)
26,500 psi (break) (Celanex 3211)
27,000 psi (RTP 1005; Valox 420-SEO)
27,500 psi (RTP 1005FR; Valox 420)
27,550 psi (Ultradur B 4305 G4)
28,000 psi (Thermocomp WF-1006); (break) (Celanex 3300, 3310)
29,000 psi (Ultradur B 4300 G6); (break) (Celanex 3311)
30,000 psi (Celanex 3400; RTP 1007, 1105FR)
31,000 psi (RTP 1007 GB 10; Thermocomp WC-1006)
31,900 psi (Ultradur B 4305 G6)
32,000 psi (Thermocomp WF-1008)
32,625 psi (Ultradur B 4300 G10)
33,000 psi (Stat-Kon WC-1006)
35,000 psi (EMI-X WC-1008)

Flex. Mod.:

23,000 kg/cm² (Gafite 1600A, 1602Z)
26,000 kg/cm² (Gafite 1602F)
47,000 kg/cm² (Gafite 1432Z)
56,000 kg/cm² (Gafite 1432F)
64,000 kg/cm² (Gafite 1642Z)
85,000 kg/cm² (Gafite 1452F, 1462Z, 1662Z)
8960 MPa (Rynite 530)
9645 MPa (Rynite 935)
10,343 MPa (Rynite FR-530)
13,780 MPa (Rynite 545)
17,915 MPa (Rynite 555)
9 GPa (Petlon 3530)
9.7 GPa (Petlon 4530)
14.1 GPa (Petlon 3550)
100,000 psi (RTP 1500.5)
110,000 psi (Thermocomp YF-1002)
130,000 psi (RTP 1501)
185,000 psi (Thermocomp YF-1004)
250,000 psi (RTP 1503)
260,000 psi (Malon 1070, 1080, 1090)
261,000 psi (Ultradur KR 4071)
275,000 psi (Ultradur KR 4070; Valox 340)
320,000 psi (Thermocomp YF-1006)
340,000 psi (Valox 325, 357)
350,000 psi (Mamax 2020; RTP 1505)
360,000 psi (Celanex 2000)
370,000 psi (Celanex 2002)
377,000 psi (Ultradur B 4500, B 4520, B 4550, KR 4036)
380,000 psi (Thermocomp YF-1008; Valox 310-SEO)
390,000 psi (Valox 744)
391,500 psi (Ultradur B 2550)
400,000 psi (Celanex 2012)
435,000 psi (Ultradur KR 4015)
500,000 psi (RTP 1507; Valox 745, 760)
507,500 psi (Ultradur B 4300 K4)
580,000 psi (Ultradur B 4300 K6)
623,000 psi (Ultradur KR 4001)
652,500 psi (Ultradur B 4300 G2)
670,000 psi (Valox DR-51)
700,000 psi (Celanex 3200)
730,000 psi (Valox DR-48)
841,000 psi (Ultradur B 4305 G2)

Polyester, thermoplastic *(cont'd.)*

850,000 psi (Valox 701)
900,000 psi (Valox 730, FV-650)
10^6 psi (EMI-X WA-40; Valox FV-600)
1.015×10^6 psi (Ultradur B 4300 G4)
1.0875×10^6 psi (Ultradur KR 4025)
1.1×10^6 psi (Celanex 3210; Valox 420, 420-SEO)
1.102×10^6 psi (Ultradur KR 4011)
1.16×10^6 psi (Ultradur B 4305 G4)
1.2×10^6 psi (Celanex 3211, 3300; Valox 750, 752)
1.45×10^6 psi (Ultradur B 4300 G6, B 4305 G6, KR 4035)
1.5×10^6 psi (Celanex 3310, 3311, 3400)
1.56×10^6 psi (Xydar M-350)
1.68×10^6 psi (Xydar M-450)
1.78×10^6 psi (Xydar G-330)
1.79×10^6 psi (Xydar MG-350)
2×10^6 psi (Xydar G-345)
2.06×10^6 psi (Xydar MG-450)
2.16×10^6 psi (Xydar G-430)
2.46×10^6 psi (Xydar G-445)
2.465×10^6 psi (Ultradur B 4300 G10)
2.5×10^6 psi (EMI-X WC-1008)

Compr. Str.:
99 MPa (yield) (Petlon 4530)
113 MPa (yield) (Petlon 3530)
127 MPa (yield) (Petlon 3550)
141 MPa (Rynite 935)
172 MPa (Rynite 530, FR-530)
179 MPa (Rynite 545)
196 MPa (Rynite 555)
5000 psi (RTP 1503)
5200 psi (RTP 1500.5)
5300 psi (RTP 1501)
5800 psi (Valox 357)
7000 psi (RTP 1505)
7500 psi (RTP 1507)
8600 psi (Valox 340)
9900 psi (Valox 760)
10,400 psi (Valox 745)
11,000 psi (Valox 730, FV-650)
11,300 psi (Valox FV-600)
13,000 psi (Valox 325)
14,500 psi (Valox 310-SEO)
15,000 psi (Valox 752, DR-48, DR-51)

18,000 psi (Celanex 3300, 3310; Valox 420, 420-SEO)
19,000 psi (Valox 701)
22,000 psi (Celanex 3311)

Creep:
0.064% strain (500 psi) (Valox FV-650)
0.58% strain (500 psi) (Valox FV-600)

Tear Str.:
70 kN/m (Hytrel G-4075)
81 kN/m (Hytrel G-4074)
94 kN/m (Hytrel G-4774)
95 kN/m (Hytrel G-4766)
122 kN/m (Hytrel G-5544)

Shear Str.:
53.7 MPa (Rynite 935)
65.5 MPa (Rynite FR-530)
79.2 MPa (Rynite 530)
82.7 MPa (Rynite 555)
86.1 MPa (Rynite 545)
5000 psi (break) (Valox FV-650)
5300 psi (Valox 340); (break) (Valox FV-600)
6400 psi (Valox 760)
6700 psi (Valox 730, 745)
6800 psi (Valox 752)
7700 psi (Valox 310-SEO, 325)
8000 psi (Valox 701, DR-48, DR-51)
8100 psi (Celanex 3300)
8900 psi (Celanex 3310; Valox 420)
9000 psi (Valox 420-SEO)
9700 psi (Celanex 3311)

Impact Str. (Izod):
4.5 cm kg/cm notched (Gafite 1602F)
4.9 cm kg/cm notched (Gafite LW X-5632Z)
5.4 cm kg/cm notched (Gafite 1432F, 1600A, 1602Z)
6.0 cm kg/cm notched (Gafite 1432Z)
7.0 cm kg/cm notched (Gafite 1452F)
7.6 cm kg/cm notched (Gafite LW X-4424R, LW X-4424S)
9.0 cm kg/cm notched (Gafite 1462Z)
9.2 cm kg/cm notched (Gafite 1642Z)
9.7 cm kg/cm notched (Gafite LW X-7443R)
10.3 cm kg/cm notched (Gafite LW X-6443R)
10.9 cm kg/cm notched (Gafite 1662Z)
18.4 cm kg/cm notched (Gafite LW X-4612R)
2.1 J/m notched (Hytrel 7246)

Polyester, thermoplastic *(cont'd.)*

10.6 J/m (Hytrel 4056); notched (Hytrel 5526, 5555HS, 5556, 6346)

64.1 J/m (Rynite 935)

74.7 J/m (Rynite 940, 940FB)

83 J/m notched (Petlon 4530)

85.4 J/m (Rynite FR-530)

91 J/m notched (Petlon 3530)

101 J/m (Rynite 530)

123 J/m (Rynite 555)

125 J/m notched (Petlon 3550)

128 J/m (Rynite 545)

138.8 J/m (Rynite 430)

220 J/m notched (Arylef U 100)

294 J/m notched (Durel 400)

2 kJ/m² notched (Pocan B 1305)

3–4 kJ/m² notched (Pocan B 1505)

5 kJ/m² notched (Pocan B 5335)

6 kJ/m² notched (Pocan B 4225)

7 kJ/m² notched (Pocan B 3225)

7–8 kJ/m² notched (Pocan B 4235)

8–9 kJ/m² notched (Pocan B 3235)

6.0 kg cm/mil (Mylar Type A)

0.5 ft lb/in. notched (Celanex 2012; RTP 1027; Ultradur B 4300 K6)

0.6 ft lb/in. notched (Ultradur B 4300 K4)

0.7 ft lb/in. notched (Celanex 2000; Gafite 1614G; RTP 1026; Ultradur KR 4015; Valox 310-SEO, 701, 752)

0.75 ft lb/in. notched (RTP 1028)

0.8 ft lb/in. notched (Ultradur B 2550, B 4305 G2; Valox 750)

0.87 ft lb/in. notched (Ultradur B 4300 G2)

0.9 ft lb/in. notched (Celanex 2002, 3200; Stat-Kon WC-1002; Xydar M-450)

1.0 ft lb/in. notched (Celanex 3210; EMI-X WA-40; RTP 1003, 1105FR; Ultradur B 4305 G4, B 4305 G6, B 4520, KR 4025, KR 4035; Valox 325, 730, 744, DR-48)

1.0–1.2 ft lb/in. notched (Ultradur B 4500, B 4550, KR 4036)

1.1 ft lb/in. notched (RTP 1001; Valox DR-51)

1.15 ft lb/in. notched (RTP 1025)

1.2 ft lb/in. notched (Celanex 3211; Gafite 1632F; Stat-Kon WC-1006; Thermocomp WC-1006, WF-1006FR; Ultradur B 4300 G4; Valox 760; Xydar MG-450)

1.3 ft lb/in. notched (Celanex 3310; Gafite LW 6443B; Valox 420-SEO; Xydar M-350)

1.4 ft lb/in. notched (Celanex 3311; Gafite 1465F, 1632Z, 1700A, LW 4424R, LW 6443F; RTP 1005; Ultradur B 4300 G10; Xydar G-445, MG-350)

1.5 ft lb/in. notched (Durel P410; RTP 1005FR)

1.7 ft lb/in. notched (Celanex 3300; Xydar G-430)

1.8 ft lb/in. notched (EMI-X WC-1008; Gafite LW 7342R; RTP 1007; Valox 420, 745; Xydar G-345)

1.8–2.0 ft lb/in. notched (Thermocomp WF-1004)
1.9 ft lb/in. notched (Celanex 3400; Gafite LW 6362R, LW 6443R)
2.0 ft lb/in. notched (Durel P430; RTP 1007 GB 10; Ultradur B 4300 G6)
2.1 ft lb/in. notched (Gafite 1482Z)
2.4–2.8 ft lb/in. notched (Thermocomp WF-1006)
2.8–3.0 ft lb/in. notched (Thermocomp WF-1008)
2.9 ft lb/in. notched (Xydar G-330)
3.0 ft lb/in. (Mamax 2020)
3.3 ft lb/in. notched (Ultradur KR 4070)
3.6 ft lb/in. notched (Durel P400)
4.0 ft lb/in. notched (Thermocomp YF-1008)
5.0 ft lb/in. notched (Thermocomp YF-1006)
5.5 ft lb/in. notched (RTP 1507)
5.9 ft lb/in. notched (RTP 1501)
6.0 ft lb/in. notched (RTP 1505)
6.2 ft lb/in. notched (RTP 1500.5)
6.5 ft lb/in. notched (RTP 1503)
8.0 ft lb/in. notched (Thermocomp YF-1004; Valox 357)
9.0 ft lb/in. notched (Thermocomp YF-1002)
9.7 ft lb/in. unnotched (Valox FV-608)
16 ft lb/in. notched (Valox 340)
No break notched (Ultradur KR 4071)
Impact Str. (Falling Ball):
15 J (Valox FV-609)
23 J (Valox FV-699)
Hardness:
Ball Indentation 110 MPa (Pocan B 1505)
Ball Indentation 125 MPa (Pocan B 1305)
Ball Indentation 170 MPa (Pocan B 5335)
Ball Indentation 175 MPa (Pocan B 3225)
Ball Indentation 190 MPa (Pocan B 3235)
Ball Indentation 200 MPa (Pocan B 4225)
Ball Indentation 210 MPa (Pocan B 4235)
Rockwell M33 (Gafite LW X-4424R, LW X-4424S)
Rockwell M41 (Gafite LW X-4612R)
Rockwell M50 (Gafite LW X-7443R)
Rockwell M57 (Gafite LW X-6443R)
Rockwell M72 (Gafite 1600A, 1602Z)
Rockwell M75 (Celanex 2000; Gafite LW X-5632Z)
Rockwell M78 (Celanex 2002)
Rockwell M80 (Celanex 2012; Petlon 4530)
Rockwell M81 (Gafite 1602F)
Rockwell M82 (Rynite 430; Thermocomp WF-1004)

Polyester, thermoplastic *(cont'd.)*

Rockwell M84 (Thermocomp WF-1006)
Rockwell M86 (Thermocomp WF-1008)
Rockwell M89 (Gafite 1642Z; Petlon M89)
Rockwell M90 (Celanex 3300, 3310; Gafite 1432F, 1432Z, 1452F, 1462Z, 1662Z)
Rockwell M93 (Celanex 3211, 3400)
Rockwell M94 (Celanex 3311)
Rockwell M98 (Petlon 3550)
Rockwell M100 (Rynite 530, 545, FR-530)
Rockwell R70 (RTP 1500.5)
Rockwell R80 (RTP 1501)
Rockwell R83 (RTP 1503)
Rockwell R85 (RTP 1505, 1507)
Rockwell R101 (Valox 730)
Rockwell R112 (Valox 745)
Rockwell R113 (Valox 340, 760)
Rockwell R114 (Valox 752)
Rockwell R117 (RTP 1001, 1003, 1026, 1027; Valox 325, 357)
Rockwell R118 (RTP 1025; Rynite 940, 940FB; Valox 420, 420-SEO, DR-48, DR-51)
Rockwell R119 (Valox 701)
Rockwell R120 (RTP 1005, 1005FR, 1007, 1007 GB 10, 1028, 1105FR; Valox 310-SEO)
Rockwell R125 (Arylef U 100)
Shore D40 (Gaflex 540ZS; Hytrel 4056, G-4074)
Shore D47 (Gaflex 547, 547ZS; Hytrel G-4774)
Shore D55 (Gaflex 355ZS, 555, 555ZS; HTG-4275; Hytrel 5526, 5555HS, 5556, G-5544)
Shore D63 (Hytrel 6346)
Shore D70 (Mamax 2020)
Shore D72 (Gaflex 372, 372ZS, 572ZS; Hytrel 7246)

Mold Shrinkage:
0.002 in./in. (EMI-X WC-1008)
0.002–0.003 in./in. (Ultradur B 4300 G10)
0.003–0.005 in./in. (Ultradur B 4300 G6, B 4305 G6, KR 4035)
0.004–0.005 in./in. (Ultradur KR 4011)
0.006–0.008 in./in. (Ultradur B 4300 G4, B 4305 G4, KR 4025)
0.008 in./in. (EMI-X WA-40)
0.008–0.010 in./in. (Ultradur B 4300 G2, B 4305 G2)
0.009–0.020 in./in. (Ultradur B 4300 K6)
0.009–0.022 in./in. (Ultradur B 4300 K4, B 4500, B 4550, KR 4036)
0.010–0.013 in./in. (Ultradur KR 4001)
0.011–0.012 in./in. (Ultradur KR 4015)
0.011–0.022 in./in. (Ultradur B 2550, B 4520)
0.017–0.021 in./in. (Ultradur KR 4070, KR 4071)

0.07 (Mamax 2020)
Water Absorp.:
  0.03% (EMI-X WC-1008)
  0.04% (EMI-X WA-40)
  0.09% (24 h) (Celanex 2002)
  < 0.1% (24 h) (Xydar G-330, G-345, G-430, G-445, M-350, M-450, MG-350, MG-450)
  0.3% (Ultradur B 4300 G6, B 4300 G10, B 4300 K6, KR 4070, KR 4071)
  0.4% (Ultradur B 4300 G2, B 4300 G4, B 4300 K4, B 4305 G2, B 4305 G4, B 4305 G6, KR 4015, KR 4025, KR 4035)
  0.5% (Ultradur B 2550, B 4500, B 4520, B 4550, KR 4001, KR 4011, KR 4036)

THERMAL PROPERTIES:
Soften. Pt. (Vicat):
  112 C (Hytrel G-4075)
  122 C (Hytrel G-4074)
  159 C (Hytrel G-4766)
  165 C (Pocan B 1505)
  174 C (Hytrel G-4774)
  180 C (Pocan B 1305)
  196 C (Hytrel G-5544)
  205 C (Pocan B 4225, B 5335)
  210 C (Pocan B 3225, B 4235)
  215 C (Pocan B 3235)
Conduct.:
  0.25 W/m•K (Rynite FR-530)
  0.29 W/m•K (Rynite 530)
  0.47 W/m•K (Rynite 545)
  1.0 Btu/h/ft²/F/in. (RTP 1500.5, 1501)
  1.1 Btu/h/ft²/F/in. (RTP 1503; Valox 325, FV-600, FV-650)
  1.2 Btu/h/ft²/F/in. (Valox 310-SEO, DR-48, DR-51)
  1.3 Btu/h/ft²/F/in. (RTP 1505; Valox 420, 420-SEO)
  1.4 Btu/h/ft²/F/in. (RTP 1507)
  1.54 Btu/h/ft²/F/in. (Celanex 3300)
  1.63 Btu/h/ft²/F/in. (Celanex 3210)
  1.71 Btu/h/ft²/F/in. (Celanex 3310)
  6.7 Btu/h/ft²/F/in. (EMI-X WA-40)
  7.1 Btu/h/ft²/F/in. (EMI-X WC-1008)
Distort. Temp.:
  54 C (0.46 MPa) (Hytrel G-4075)
  60 C (264 psi) (Mamax 2020)
  70 C (1.81 MPa) (Pocan B 1305, B 1505)
  77 C (184 kg/cm²) (Gafite LW X-4612R)
  102 C (1.82 MPa) (Valox FV-609)

Polyester, thermoplastic *(cont'd.)*

108 C (1.82 MPa) (Valox FV-699)
146 C (1.82 MPa) (Valox FV-608)
154 C (264 psi) (Durel P400); (66 psi) (Gafite 1700A); (4.6 kg/cm²) (Gafite 1600A, 1602Z)
160 C (1.8 MPa) (Durel 400; Pocan B 5335)
165 C (184 kg/cm²) (Gafite LW X-4424R, LW X-7443R)
166 C (4.6 kg/cm²) (Gafite 1602F); (66 psi) (Gafite 1614G)
168 C (184 kg/cm²) (Gafite LW X-4424S)
171 C (264 psi) (Durel P410)
175 C (1.8 MPa) (Arylef U 100; Durel P430)
180 C (1.81 MPa) (Pocan B 4225)
181 C (184 kg/cm²) (Gafite LW X-5632Z)
195 C (184 kg/cm²) (Gafite LW X-6443R); (1.81 MPa) (Pocan B 4235)
200 C (1.81 MPa) (Pocan B 3225, B 3235)
210 C (66 psi) (Gafite 1632F, LW 4424R, LW 6443F, LW 7342R); (1.8 MPa) (Rynite 430)
211 C (1.8 MPa) (Rynite 940, 940FB)
212 C (4.6 kg/cm²) (Gafite 1432F, 1432Z); (66 psi) (Gafite 1632Z)
213 C (4.6 kg/cm²) (Gafite 1642Z); (66 psi) (Gafite LW 6443B)
215 C (66 psi) (Gafite LW 6443R); (1.8 MPa) (Rynite 935)
217 C (4.6 kg/cm²) (Gafite 1452F, 1462Z, 1662Z); (66 psi) (Gafite 1465F)
218 C (66 psi) (Gafite LW 6362R)
219 C (66 psi) (Gafite 1482Z)
223 C (1.82 MPa) (Petlon 4530)
224 C (1.8 MPa) (Rynite 530, FR-530)
226 C (1.8 MPa) (Rynite 545)
228 C (1.82 MPa) (Petlon 3530)
229 C (1.8 MPa) (Rynite 555)
230 C (1.82 MPa) (Petlon 3550)
122 F (264 psi) (Ultradur KR 4070, KR 4071)
123 F (264 psi) (Celanex 2000)
130 F (264 psi) (Valox 325)
131 F (264 psi) (Celanex 2002)
135 F (264 psi) (Celanex 2012)
150 F (264 psi) (Valox 744)
153 F (264 psi) (Ultradur B 2550, B 4500, B 4520, B 4550, KR 4036)
158 F (264 psi) (Ultradur B 4300 K4)
160 F (264 psi) (Valox 310-SEO, 340)
170 F (264 psi) (Valox 760)
176 F (264 psi) (Ultradur KR 4015)
190 F (264 psi) (Valox 745)
194 F (264 psi) (Ultradur KR 4001)
203 F (264 psi) (Ultradur B 4300 K6)

210 F (264 psi) (Valox 357)
240 F (264 psi) (RTP 1500.5)
270 F (264 psi) (RTP 1501)
290 F (264 psi) (Thermocomp YF-1002)
330 F (264 psi) (RTP 1503, 1505)
335 F (264 psi) (RTP 1507)
340 F (264 psi) (RTP 1001; Thermocomp YF-1004, YF-1006, YF-1008; Valox 730, FV-600)
360 F (264 psi) (Valox DR-48, FV-650)
365 F (264 psi) (Celanex 3211)
374 F (264 psi) (Ultradur B 4305 G2)
375 F (264 psi) (Valox DR-51)
378 F (264 psi) (Celanex 3200, 3311)
380 F (264 psi) (EMI-X WA-40; RTP 1027, 1028; Stat-Kon WC-1002; Valox 750, 752)
392 F (264 psi) (Ultradur B 4300 G2, KR 4025)
395 F (264 psi) (Celanex 3210)
400 F (264 psi) (RTP 1003, 1026; Thermocomp WF-1006FR; Valox 420-SEO)
401 F (264 psi) (Ultradur B 4300 G4, B 4305 G4, KR 4035)
403 F (264 psi) (Celanex 3300)
405 F (264 psi) (RTP 1025; Valox 420)
406 F (264 psi) (Celanex 3310)
408 F (264 psi) (Celanex 3400)
410 F (264 psi) (RTP 1005FR, 1007 GB 10; Thermocomp WF-1004; Ultradur B 4305 G6, KR 4011)
415 F (264 psi) (RTP 1005, 1007)
419 F (264 psi) (Ultradur B 4300 G6, B 4300 G10)
420 F (264 psi) (Valox 701)
430 F (264 psi) (EMI-X WC-1008; Stat-Kon WC-1006; Thermocomp WC-1006, WF-1006)
440 F (264 psi) (RTP 1105FR)
450 F (264 psi) (Thermocomp WF-1008)
475 F (264 psi) (Xydar M-350)
510 F (264 psi) (Xydar G-330)
515 F (264 psi) (Xydar G-345, MG-350)
545 F (264 psi) (Xydar M-450)
560 F (264 psi) (Xydar MG-450)
585 F (264 psi) (Xydar G-445)
595 F (264 psi) (Xydar G-430)
Coeff. of Linear Exp.:
2.4×10^{-5} mm/mm/C (Petlon 3550)
3.0×10^{-5} mm/mm/C (Petlon 3530)
3.6×10^{-5} mm/mm/C (Petlon 4530)

Polyester, thermoplastic *(cont'd.)*

2.3 × 10⁻⁵ m/m/C (Rynite 545)
2.9 × 10⁻⁵ m/m/C (Rynite 530)
0.4 × 10⁻⁵ in./in./F (EMI-X WC-1008)
1.0 × 10⁻⁵ in./in./F (Valox 701)
1.3 × 10⁻⁵ in./in./F (Celanex 3210, 3300, 3310)
1.4 × 10⁻⁵ in./in./F (Valox 420-SEO)
1.5 × 10⁻⁵ in./in./F (Valox 420)
1.6 × 10⁻⁵ in./in./F (Valox 750)
1.7 × 10⁻⁵ in./in./F (Valox 752)
2.0 × 10⁻⁵ in./in./F (Thermocomp YF-1008)
2.2 × 10⁻⁵ in./in./F (Valox DR-48)
2.5 × 10⁻⁵ in./in./F (Valox 730, DR-51)
3.0 × 10⁻⁵ in./in./F (Thermocomp YF-1006)
3.6 × 10⁻⁵ in./in./F (Valox 310-SEO)
4.0 × 10⁻⁵ in./in./F (Valox 325)
4.1 × 10⁻⁵ in./in./F (Valox 760)
4.2 × 10⁻⁵ in./in./F (Valox 745)
4.5 × 10⁻⁵ in./in./F (Valox 744, FV-600, FV-650)
5.0 × 10⁻⁵ in./in./F (Thermocomp YF-1004)
5.7 × 10⁻⁵ in./in./F (Valox 357)
7.0 × 10⁻⁵ in./in./F (Thermocomp YF-1002)
7.3 × 10⁻⁵ in./in./F (Valox 340)

Sp. Heat:
0.22 Btu/lb/F (Celanex 3211)
0.23 Btu/lb/F (Celanex 3311)
0.27 Btu/lb/F (Celanex 3310)
0.282 Btu/lb/F (Valox FV-650)
0.30 Btu/lb/F (Celanex 3300)

Flamm.:
V-0/5V (Petlon 4530; Valox FV-600, FV-699)
V-0 (Arylef U 100; Celanex 2012, 3210, 3211, 3310, 3311; Durel 400, P400, P410,
P430; Gafite 1432F, 1452F, 1465F, 1602F, 1614G, 1632F, LW 6443F; Pocan B
4225, B 4235; RTP 1005FR, 1105FR; Rynite FR-530; Thermocomp WF-1006FR;
Ultradur B 4305 G2, B 4305 G4, B 4305 G6, KR 4015, KR 4025, KR 4035; Valox
310-SEO, 357, 420-SEO, 750, 752, 760, DR-48; Xydar G-330, G-345, G-430, G-
445, M-350, M-450, MG-350, MG-450)
V-2 (Gafite LW 6443B)
HB (Celanex 2000, 2002, 3200, 3300; EMI-X WA-40, WC-1008; Gafite 1432Z,
1462Z, 1482Z, 1600A, 1602Z, 1632Z, 1642Z, 1662Z, 1700A, LW 4424R, LW
6362R, LW 6443R, LW 7342R; Petlon 3530, 3550; Pocan B 1305, B 3225, B 5335;
RTP 1001, 1003, 1005, 1007, 1007 GB 10, 1025, 1026, 1027, 1028; RTP 1500.5,
1501, 1503, 1505, 1507; Rynite 430, 530, 545, 935, 940, 940FB; Stat-Kon WC-
1002, WC-1006; Thermocomp WC-1006, WF-1004, WF-1006, WF-1008, YF-

1002, YF-1004, YF-1006, YF-1008; Ultradur B 2550, B 4300 G2, B 4300 G4, B 4300 G6, B 4300 G10, B 4300 K4, B 4300 K6, B 4500, B 4520, B 4550, KR 4001, KR 4011, KR 4036, KR 4070, KR 4071; Valox 325, 340, 420, 730, 744, 745, DR-51, FV-608, FV-609, FV-650)

ELECTRICAL PROPERTIES:
Dissip. Factor:
0.001 (1 kHz) (Gafite 1600A, 1602Z)
0.0012 (100 Hz) (Celanex 3311)
0.0015 (100 Hz) (Celanex 3300)
0.002 (100 Hz) (Celanex 2012, 3210; Valox 325, 420, 420-SEO, 701, 730, 745, 752, 760, DR-48, DR-51); (1 kHz) (Gafite 1432F, 1432Z, 1452F, 1462Z, 1602F, 1642Z, 1662Z)
0.0021 (100 Hz) (Celanex 3400)
0.0025 (60 Hz) (Petlon 3530)
0.003 (100 Hz) (Valox 310-SEO, 357)
0.004 (100 Hz) (Valox 340, FV-600, FV-650)
0.0041 (100 Hz) (Celanex 3211)
0.005 (1 kHz) (Rynite 530, 545)
0.0064 (100 Hz) (Celanex 3310)
0.007 (1 kHz) (Rynite FR-530)
0.008 (1 kHz) (Rynite 935)
0.0096 (60 Hz) (Petlon 4530)
0.010 (100 Hz) (Xydar G-330, G-345, G-430, G-445, M-350, M-450, MG-450)
0.012 (1 MHz) (Ultradur KR 4015, KR 4025, KR 4035)
0.013 (100 Hz) (Xydar MG-350); (1 MHz) (Ultradur B 4300 G10, B 4300 K6)
0.015 (1 MHz) (Ultradur B 4300 G2, B 4300 G4, B 4300 G6, B 4300 K4, B 4305 G2, B 4305 G4, B 4305 G6)
0.020 (100 Hz) (Celanex 2000); (1 MHz) (Ultradur B 2550, B 4500, B 4520, B 4550, KR 4036, KR 4070, KR 4071)
0.024 (1 MC) (RTP 1507)
0.025 (1 MC) (RTP 1505)
0.027 (1 MC) (RTP 1503)
0.031 (1 MC) (RTP 1501)
0.032 (1 MC) (RTP 1500.5)
Dielec. Str.:
13.8 kV/mm (Gafite 1602F)
15.8 kV/mm (Gafite 1452F, 1600A, 1602Z)
15.9 kV/mm (Petlon 4530)
16.2 kV/mm (Gafite 1432F; Petlon 3530)
16.5 kV/mm (Durel 400; Petlon 3550)
16.9 kV/mm (Rynite FR-530)
17.7 kV/mm (Gafite 1432Z)
19 kV/mm (Gafite 1642Z)

Polyester, thermoplastic *(cont'd.)*

20.5 kV/mm (Gafite 1462Z)
20.9 kV/mm (Gafite 1662Z)
21.3 kV/mm (Rynite 545)
21.5 kV/mm (Rynite 940, 940FB)
21.7 kV/mm (Rynite 430, 530)
22.5 kV/mm (Rynite 935)
23 kV/mm (Pocan B 1305)
24 kV/mm (Pocan B 1505)
27 kV/mm (Pocan B 5335)
28 kV/mm (Pocan B 4225)
29 kV/mm (Pocan B 4235)
30 kV/mm (Arylef U 100)
33 kV/mm (Pocan B 3225)
34 kV/mm (Pocan B 3235)
383 V/mil (Durel P410)
400 V/mil (Gafite LW X-4612R; Valox DR-48)
409 V/mil (Durel P430)
410 V/mil (Ultradur B 2550, B 4500, B 4520, B 4550, KR 4036)
420 V/mil (Celanex 2000, 2002)
430 V/mil (Ultradur KR 4015; Valox DR-51)
450 V/mil (Celanex 2012; Ultradur B 4305 G2)
460 V/mil (Celanex 3200; Gafite LW X-4424S; Ultradur B 4300 G2)
470 V/mil (Ultradur B 4305 G4, KR 4025)
480 V/mil (Celanex 3210, 3400; RTP 1500.5, 1501; Thermocomp WF-1006FR)
490 V/mil (Celanex 3310; Gafite LW X-4424R)
500 V/mil (RTP 1001, 1003, 1005, 10105FR, 1007, 1007 GB 10, 1025, 1026, 1027,
    1028, 1105FR; Ultradur B 4305 G6, KR 4035)
500+ V/mil (RTP 1503, 1505, 1507)
510 V/mil (Thermocomp WF-1006; Ultradur B 4300 G4, B 4300 K4)
530 V/mil (Celanex 3311; Ultradur B 4300 G6)
560 V/mil (Celanex 3300; Valox 744)
570 V/mil (Celanex 3211; Valox 340)
580 V/mil (Valox 760)
590 V/mil (Valox 325)
600 V/mil (Gafite LW X-6443R, LW X-6443R, LW X-7443R; Valox 357)
613 V/mil (Durel P400)
630 V/mil (Valox 745)
650 V/mil (Valox 701)
690 V/mil (Valox 752)
745 V/mil (Valox 750)
750 V/mil (Valox 310-SEO, 420, 420-SEO)
825 V/mil (Valox 730)
900 V/mil (Xydar G-330, G-345, G-430, G-445, M-350, M-450, MG-450)

920 V/mil (Xydar MG-350)
7500 V/mil (Mylar Type A)
Dielec. Const.:
2.96 (100 Hz) (Valox FV-600)
3.09 (100 Hz) (Valox FV-650)
3.1 (50 Hz) (Ultradur B 4305 G2); (100 Hz) (Valox 760)
3.2 (100 Hz) (Celanex 2000, 2012; Valox 357)
3.24 (1 kHz) (Gafite 1600A, 1602F, 1602Z)
3.3 (50 Hz) (Ultradur B 2550, B 4305 G4, B 4500, B 4520, B 4550, KR 4036, KR 4070, KR 4071); (100 Hz) (Valox 310-SEO, 325, 340, 745)
3.38 (1 kHz) (Gafite 1432Z)
3.4 (100 Hz) (Xydar G-430)
3.45 (1 kHz) (Gafite 1642Z)
3.48 (1 kHz) (Gafite 1432F)
3.5 (50 Hz) (Ultradur B 4305 G6, KR 4015, KR 4025, KR 4035)
3.6 (50 Hz) (Ultradur B 4300 G2, B 4300 K4, B 4300 K6); (100 Hz) (Valox DR-48, DR-51); (1 kHz) (Rynite 530, FR-530)
3.66 (1 kHz) (Gafite 1462Z)
3.69 (1 kHz) (Gafite 1662Z)
3.7 (50 Hz) (Ultradur B 4300 G4); (100 Hz) (Celanex 3211, 3300; Valox 701, 730)
3.78 (1 kHz) (Gafite 1452F)
3.8 (50 Hz) (Ultradur B 4300 G6); (100 Hz) (Celanex 3210, 3311; Valox 420, 420-SEO); (1 kHz) (Rynite 935)
3.9 (50 Hz) (Ultradur B 4300 G10); (100 Hz) (Celanex 3310, 3400; Xydar G-330)
4.0 (100 Hz) (Valox 752); (1 kHz) (Rynite 545)
4.10 (100 Hz) (Xydar G-345, G-445, M-450, MG-450)
4.18 (100 Hz) (Xydar MG-350)
4.20 (60 Hz) (Petlon 4530); (100 Hz) (Xydar M-350)
4.21 (60 Hz) (Petlon 3530)
4.39 (1 MC) (RTP 1500.5)
4.45 (1 MC) (RTP 1503)
4.49 (1 MC) (RTP 1501)
4.60 (1 MC) (RTP 1505)
4.75 (1 MC) (RTP 1507)
Vol. Resist.:
100 ohm-cm (EMI-X WA-40, WC-1008; Stat-Kon WC-1006)
100,000 ohm-cm (Stat-Kon WC-1002)
10^{13} ohm-cm (RTP 1503, 1505, 1507)
10^{14} ohm-cm (Pocan B 5335)
10^{15} ohm-cm (Celanex 2000, 2002, 2012; RTP1005FR, 1105FR, 1500.5, 1501; Rynite 430, 530, 545, 935, 940, 940FB, FR-530)
> 10^{15} ohm-cm (Gafite LW X-4424R, LW X-4424S, LW X-4612R, LW X-5632Z, LW X-6443R, LW X-7443R)

Polyester, thermoplastic *(cont'd.)*

1.7×10^{15} ohm-cm (Celanex 3400)

3.8×10^{15} ohm-cm (Valox 340)

4×10^{15} ohm-cm (Celanex 3211, 3311; Valox 730, 744)

4.9×10^{15} ohm-cm (Pocan B 1505)

5×10^{15} ohm-cm (Celanex 3210, 3310)

9×10^{15} ohm-cm (Valox 752)

10^{16} ohm-cm (Celanex 3300; Pocan B 1305, B 3225, B 4225; RTP 12001, 1003, 1005, 1007, 1007 GB 10, 1025, 1026, 1027, 1028; Ultradur B 2550, B 4300 G2, B 4300 G4, B 4300 G6, B 4300 G10, B 4300 K4, B 4300 K6, B 4305 G2, B 4305 G4, B 4305 G6, B 4500, B 4520, B 4550, KR 4001, KR 4011, KR 4015, KR 4025, KR 4035, KR 4036, KR 4070, KR 4071)

$> 10^{16}$ ohm-cm (Gafite 1432F, 1432Z, 1452F, 1462Z, 1600A, 1602F, 1602Z, 1642Z, 1662Z)

1.8×10^{16} ohm-cm (Pocan B 4235)

1.9×10^{16} ohm-cm (Valox 701)

2×10^{16} ohm-cm (Durel 400; Valox 750)

2.2×10^{16} ohm-cm (Valox 357)

2.8×10^{16} ohm-cm (Valox 760)

3×10^{16} ohm-cm (Arylef U 100; Durel P400, P430)

3.2×10^{16} ohm-cm (Valox 420)

3.3×10^{16} ohm-cm (Valox DR-51)

3.4×10^{16} ohm-cm (Valox 420-SEO)

3.6×10^{16} ohm-cm (Valox DR-48)

4×10^{16} ohm-cm (Valox 310-SEO, 325)

5×10^{16} ohm-cm (Durel P410; Pocan B 3235)

6.9×10^{16} ohm-cm (Valox 745)

2.5×10^{17} ohm-cm (Valox FV-650)

2.6×10^{17} ohm-cm (Valox FV-600)

10^{18} ohm-cm (Mylar Type A)

Surf. Resist.:

100 ohm/sq. (EMI-X WA-40, WC-1008)

10^{13} ohm (Ultradur B 4300 G4, B 4300 G6, B 4300 G10, B 4300 K6, B 4305 G2, B 4305 G4, B 4305 G6, KR 4001, KR 4011, KR 4035, KR 4070, KR 4071)

$> 10^{13}$ ohm (Ultradur B 2550, B 4300 G2, B 4300 K4, B 4500, B 4520, B 4550, KR 4015, KR 4025, KR 4036)

Arc Resist.:

28 s (Valox 420-SEO, DR-48)

30 s (Valox FV-600)

40 s (RTP 1503)

50 s (RTP 1505)

60 s (RTP 1507)

63 s (Gafite 1602F; RTP 1500.5; Valox 310-SEO)

70 s (Rynite FR-530)

Polyester, thermoplastic *(cont'd.)*

71 s (Valox 357)
72 s (Rynite 530)
75 s (Celanex 2000; Valox 760)
80 s (RTP 1501)
81 s (Petlon 4530)
85 s (Gafite 1432F)
101 s (Valox 730)
109 s (Petlon 3550)
110 s (Celanex 2012)
123 s (Rynite 935)
124 s (Valox 745)
125 s (Celanex 3210; Gafite 1432Z, 1452F; Valox 750)
126 s (Rynite 545; Valox 744, 752)
128 s (Celanex 3311)
129 s (Valox DR-51)
130 s (Celanex 3200, 3211, 3300, 3310; Gafite 1462Z, 1642Z, 1662Z)
146 s (Valox 340, 420)
184 s (Valox 325)
190 s (Gafite 1600A, 1602Z)

STD. PKGS.:

50-lb to truck-load quantities (Gafite 1432F, 1432Z, 1452F, 1462Z, 1600A, 1602F, 1602Z, 1642Z, 1662Z)

Polyester, thermoset

STRUCTURE:
Unsaturated polyester:

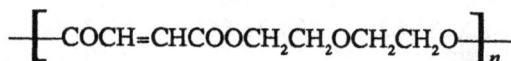

$$-\left[-COCH{=}CHCOOCH_2CH_2OCH_2CH_2O-\right]_n-$$

TRADENAME EQUIVALENTS:
Alkyd resins:
Durez 23410, 24150, 24337, 24668, 27962, 29276 [Occidental]
Chempol 13-1905, 13-2211 [Freeman]
Neolyn 20, 35D, 40 [Hercules]
Plenco 1529 Black [Plastics Engineering]
Glass/mineral-filled:
Plenco 1523 Black Pellets, 1530 [Plastics Engineering]

Polyester, thermoset (cont'd.)

Mineral-filled:
Plenco 1528 [Plastics Engineering]

Unsaturated polyester:
ANE, ANE-D30, ANE-K10, ANE-K20, ANE-P10, ANE-P20 [Ciech Plastofarb]

Aropol 7020, 7021, 7030, 7131, 7141, 7221, 7229, 7240, 7240T-15, 7240WT-12, 7241-12, 7241T-15, 7242T-15, 7280, 7320, 7329, 7362, 7420, 7530T-16, 7532, 8101, 8110, 8310, 8319, 8321, 8329, 8329-25, 8339T-12, 8349-15, 8420, 8420-09, 8343T-11, 8343T-12, 8343T-21, 8343T-22 [Ashland]

Chempol 19-4807 [Freeman]

Crystic 2-406 PA, 2-489PA, 2-491 PA, 57, 113PA, 189LV, 189MV, 191LV, 193 HR, 195, 196, 196E, 196 PALV, 198, 272, 272E, 302, 306, 323 PA, 328PA, 345PA, 328PALV, 345PALV, 346PA, 385E, 385PA, 387, 392, 392E, 405E, 405PA, 406PA, 471PALV, 474PA, 489PA, 491, 491E, 491PA, 625E, 625MV, 17449 [Scott Bader]

Crystic Fireguard 75PA [Scott Bader]

Crystic Fireguard Gelcoat 70PA [Scott Bader]

Crystic Gelcoat 33E, 39PA, 46PA, 47PA, 65E, 65PA, 68E, 68PA [Scott Bader]

Crystic Prefil F [Scott Bader]

Crystic Pregel 17, 27 [Scott Bader]

Crystic Stopper [Scott Bader]

Dion Cor-Res 6693FR, 6693HV, 6694, 6694HV, 7000 [Koppers]

Dion FR 6114, 6124, 6125, 6127, 6308, 6399, 6604T, 6655T, 6657, 6692T, 8300 [Koppers]

Dion-Iso 6246, 6314, 6315HV, 6631T, 8101, 8200 [Koppers]

Durez 31375, 31572 [Occidental]

Hetron 32A, 197AT, 325FS, 470P, 700 Series, 700DMA, 27196 [Ashland]

Impco RC-2, RC-2/140 Resins [Impco]

Leguval E60, E81, E90, F33, F30, F30S, F35, K25R, K26, K41, K70, N30, N50, N50S, W16, W16T, W20, W35, W41 [Bayer AG]

Paraplex AL-111, G-60, RG-10 [Rohm & Haas]

Polylite 31-000, 31-001, 31-006, 31-039, 31-402, 31-439, 31-447, 31-450, 31-451, 31-546, 31-571, 31-585, 31-587, 31-820, 31-822, 31-830, 31-851, 31-866, 32-032, 32-033, 32-035, 32-037, 32-039, 32-050, 32-111, 32-112, 32-120, 32-122, 32-125, 32-127, 32-128, 32-129, 32-180, 32-300, 32-344, 32-345, 32-356, 32-357, 32-358, 32-367, 32-402, 32-516, 32-528, 32-737, 32-738, 32-739, 32-740, 32-773, 33-089, 33-092, 33-084, 33-086, 33-401, 33-011, 33-031, 33-049, 33-059, 33-067, 33-071, 33-072, 33-081, 33-404, 33-411, 33-441, 33-450, 33-770, 33-773, 34-308 [Reichhold]

Stypol 40-4033, 40-4074, 40-5727, 40-5732, 40-7016, 40-8006, 83-2000 [Freeman]

Vestopal 110, 120L, 120T, 121L, 123L, 140L, 141L, 145, 145T, 150, 150B, 150T, 151, 152, 160, 160T, 221L, 250, 251, 252, 310, 320, 330, 400, 510, 511, 530, 710, 730, 810, 811, 830 [Huls AG]

Cellulose-filled:
Polylite 30002-1 [Reichhold]

Glass/mineral-filled:
Polylite 30001-1, 30001-2, 30001-3, 30003-1, 30003-2 [Reichhold]
Glass/mineral/cellulose-filled:
Polylite 30004-3, 30004-4 [Reichhold]
Graphite-reinforced:
Magnamite AS4/1655 [Hercules]

MODIFICATIONS/SPECIALTY GRADES:
Acrylic-modified:
Crystic 193 HR, 195; Polylite 32-032
Vinyl-toluene modified:
Aropol 7362; Chempol 13-1905, 13-2211; Polylite 31-585
Styrenated:
Aropol 7020, 7021, 7030, 7131, 7141, 7221, , 7229, 7241-12, 7320, 7329, 7420, 7530T-16, 8110, 8310, 8319, 8321, 8329, 8329-25, 8339T-12, 8349-15, 8420, 8420-09, 8343T-11, 8343T-12; Chempol 19-4807; Hetron 32A, 197AT, 470P; Leguval E60, E81, E90, F30, F30S, F33, F35, K25R, K26, K41, K70, N30, N50, N50S, W16, W16T, W20, W35, W41; Stypol 40-5727, 40-5732, 40-7106, 83-2000; Vestopal 110, 120L, 120T, 121L, 123L, 140L, 141L, 145, 145T, 150, 150B, 151, 152, 160, 221L, 250, 251, 252, 310, 320, 330, 400, 510, 511, 530, 710, 730, 810, 811, 830
DAP-modified:
Polylite 31-402
Silica-modified:
Crystic Pregel 17, 27
Oil-modified:
Paraplex AL-111, RG-10
Flame/fire-retardant:
Crystic 302 (Het acid), 306 (Het acid), 323PA, 328PA, 328PALV, 345PA, 345PALV, 346PA, 385E (Het acid), 385PA (Het acid), 387; Crystic Fireguard 75PA; Crystic Fireguard Gelcoat 70PA; Crystic Gelcoat 39PA, 46PA, 46PA; Crystic Prefil F (antimony trioxide/chlorinated); Dion Cor-Res 6693FR (antimony trioxide), 6693HV (antimony trioxide); Dion FR 6114, FR 6124 (antimony trioxide), FR 6125, FR 6127, FR 6308 (antimony trioxide), FR 6399, FR 6604T (antimony trioxide), FR 6655T, FR 6657, FR 6692T, FR 8300 (antimony trioxide); Durez 24150, 27962, 31375, 31572; Hetron 32A, 197AT, 325FS, 470P, 27196 (chlorinated); Leguval F30 (Het acid), F30S (Het acid), F33 (chlorinated), F35 (chlorinated); Plenco 1528; Polylite 33-441 (halogenated), 33-450 (halogenated); Vestopal 221L, 250, 251, 252
Corrosion-resistant:
Aropol 7362; Dion Cor-Res 6693FR, 6693HV, 6694, 6694HV, 7000; Dion FR 6308, 6604T; Dion-Iso 6246, 6314, 6315HV, 6631T, 8101, 8200; Hetron 197AT, 700DMA; Vestopal 150, 400

Polyester, thermoset *(cont'd.)*

UV-stabilized:
Crystic 191LV, 193HR, 195, 306, 385E, 385PA, 387; Crystic Gelcoat 39PA, 46PA, 47PA; Dion FR 6114, 6125, 6127; Leguval F30S, N50S; Polylite 32-516, 32-528, 32-739, 32-740, 34-308; Vestopal 120L, 121L, 123L, 140L, 141L, 221L

CATEGORY:
Thermosetting resin

PROCESSING:

Autoclaving:
Magnamite AS4/1655

BMC molding:
Aropol 7020, 7030; Dion Cor-Res 6693HV, 6694HV; Dion FR 6308, 6399; Dion-Iso 6314

Casting:
ANE, ANE-D30; Impco RC-2, RC-2/140 Resins; Polylite 32-032, 32-033, 32-035, 32-037, 32-039, 32-050, 32-111, 32-112, 32-120, 32-122, 32-125, 32-127, 32-128, 32-129, 32-180, 32-344, 32-356, 32-357, 32-358; Stypol 40-5727, 40-5732; Vestopal 310, 320, 330, 710, 730, 810, 811, 830

Centrifugal casting:
Crystic 272, 272, 272E; Vestopal 145, 150, 152, 160

Cold press molding:
Crystic 196 PALV

Compression molding:
Dion Cor-Res 6693HV, 6694HV; Dion FR 6308, 6399; Dion-Iso 6314; Durez 23410, 24337, 24668, 27962, 29276, 31375, 31572; Plenco 1523 Black Pellets, 1528, 1530; Polylite 30003-1, 30003-2

Contact molding:
Aropol 7221, 7240, 7241-12; Crystic 189LV, 189MV, 2-406PA, 272, 272E, 323PA, 345PALV, 385E, 406PA, 471PALV

Extrusion:
Crystic 272, 272, 272E

Filament winding:
Aropol 7221, 7241-12, 7532; Crystic 272, 272, 272E; Dion Cor-Res 6693FR, 6694, 7000; Dion FR 6308, 6604T, 6692T; Dion-Iso 6246, 6631T; Stypol 83-2000; Vestopal 145, 150, 152, 160, 310, 320, 510, 530

General purpose molding:
Crystic 328PA, 17449; Leguval F30, F33, F35, K41, N30, W20, W35, W41; Polylite 31-820, 30002-1, 30004-3, 30004-4

Hand layup:
Aropol 7240T-15, 7240WT-12, 7241-12, 7241T-15, 7242T-15, 7530T-16, 8329-25, 8339T-12, 8349-15; Crystic 196, 196E, 17449; Dion Cor-Res 6693FR, 6694, 7000; Dion FR 6604T, 6655T, 6657, 6692T; Dion-Iso 6246, 6631T; Hetron 470P, 27196, 27196; Leguval F30, F35, K25R, K26, N30, N50, N50S, W16, W16T, W35; Polylite 33-770; Stypol 40-4033, 40-4074; Vestopal 110, 120L, 120T, 145, 150,

145T, 150B, 150T, 152, 160, 160T, 250, 252, 310, 320, 400, 510, 530

Injection molding:
Crystic 196PALV, 471PALV; Durez 23410, 24150, 24337, 24668, 27962, 29276, 31375, 31572; Plenco 1523 Black Pellets, 1528, 1530; Polylite 30001-1, 30001-2, 30001-3; Stypol 40-8006

Lamination:
ANE, ANE-D30; Aropol 7021; Crystic 113PA, 191LV, 196, 196E, 198, 2-406PA, 2-489PA, 2-491PA, 302, 306, 323PA, 345PA, 328PALV, 392. 302E, 405PA, 406PA, 474PA, 489PA, 491, 491E, 491PA, 625E, 625MV; Crystic Fireguard 75PA; Crystic Fireguard Gelcoat 70PA; Crystic Pregel 17; Polylite 32-740, 33-011, 33-031, 33-049, 33-059, 33-067, 33-071, 33-072, 33-081, 33-089, 33-092, 33-084, 33-086, 33-402, 33-770; Vestopal 121L, 123L, 140L, 141L, 160, 221L, 250, 252

Low-pressure molding:
Crystic 189LV

Machine molding:
Crystic 196, 196E

Mat molding:
Aropol 7229, 7320, 7329, 7362, 7420; Vestopal 145, 150, 150B, 152, 250, 252, 310, 320, 510, 530

Matched-die molding:
Aropol 8110, 8310, 8319, 8321, 8420; Polylite 31-006, 31-039, 31-402, 31-447, 31-450, 31-451, 31-546, 31-571, 31-585, 31-587, 31-830; Vestopal 145, 150, 150B, 151, 152, 250, 251, 252, 310, 320, 510, 511, 530

Open molding:
Polylite 31-000, 31-001, 32-739, 32-740, 33-011, 33-031, 33-049, 33-059, 33-067, 33-071, 33-072, 33-081, 33-089, 33-092, 33-084, 33-086, 33-401, 33-402, 33-404, 33-411, 33-441, 33-450, 33-770

Plunger molding:
Polylite 30001-1, 30001-2, 30001-3, 30003-1, 30003-2

Preform molding:
Aropol 7229, 7320, 7329, 7362, 7420, 8110, 8310, 8319, 8321, 8329; Vestopal 145, 150, 150B, 152, 250, 252, 310, 320, 510, 530

Premix molding:
Aropol 7229, 7320, 7329, 7362, 7420; Dion-Iso 6315HV; Polylite 31-006, 31-039, 31-402, 31-447, 31-450, 31-451, 31-546, 31-571, 31-585, 31-587, 31-830

Press molding:
Aropol 7221, 7229, 7240, 7241-12; Leguval K25R, K26; Magnamite AS4/1655; Vestopal 145, 150, 150B, 152, 250, 252, 310, 320, 510, 530

Pultrusion:
Aropol 7320, 7362, 8110, 8310, 8319, 8321, 8329, 8420; Dion FR 8300; Dion-Iso 6314, 8101, 8200; Stypol 40-7016; Vestopal 120L, 121L, 123L, 140L, 141L, 160, 221L, 250, 252

Polyester, thermoset *(cont'd.)*

Rigidized vacuum forming:
Aropol 8349-15

SMC molding:
Aropol 7020, 7030; Crystic 191LV, 193 HR, 195, 306, 387, 474PA; Dion Cor-Res 6693HV, 6694HV; Dion FR 6308, 6399; Dion-Iso 6314; Leguval F30S, F33, F35, N30, W20, W41; Vestopal 251, 511

Sprayup:
Aropol 7240T-15, 7240WT-12, 7241-12, 7241T-15, 7242T-15, 7530T-16, 8329-25, 8339T-12, 8349-15, 8343T-11, 8343T-12, 8343T-21, 8343T-22; Crystic 196E; Crystic Gelcoat 68PA; Dion Cor-Res 6693FR, 6694, 7000; Dion FR 6604T, 6655T, 6657, 6692T; Dion-Iso 6246, 6631T; Hetron 325FS, 470P, 27196; Leguval F30, F35, N30, N50, W16, W16T, W35; Polylite 33-089, 33-092; Stypol 40-4033, 40-4074; Vestopal 120T, 145T, 150B, 150T, 152, 160T, 330, 710, 730, 810, 811, 830

Transfer molding:
Durez 23410, 24150, 24337, 24668, 27962, 29276, 31375, 31572; Plenco 1523 Back Pellets, 1528, 1529 Black, 1530

Vacuum bag molding:
Aropol 8329-25

Vacuum impregnation:
Crystic 196 PALV

APPLICATIONS:

Agriculture industry: silos (ANE, ANE-D30; Aropol 8343T-11, 8343T-12, 8343T-21, 8343T-22)

Architectural applications: (Crystic Fireguard Gelcoat 70PA; Crystic Gelcoat 46PA, 47PA, 65PA); awnings/canopies (Crystic 385E, 385PA; Crystic Gelcoat 39PA); building components (Crystic 302, 328PALV, 345PA, 346PA); building panels (Crystic 196E, 302; Dion FR 6114, FR 6124, FR 6125, FR 6127, FR 6692T); construction applications (Aropol 7530T-16; Dion-Cor Res 6693FR, 6693HV, 6694HV; Dion FR 6114, FR 6124, FR 6125, FR 6127, FR 6604T, FR 6655T, FR 6657, FR 6692T, FR 8300; Dion-Iso 6315HV, 6631T, 8101, 8200); decorative panels (Hetron 470P); factory-built homes (Dion FR 6604T); fiberglass panels (Polylite 32-516, 32-528); light domes (Leguval F30S, N50S; Vestopal 120L, 121L, 123L, 140L, 141L, 160, 221L); modular structures (Hetron 470P); roof structures (Crystic 385E, 385PA; Crystic Gelcoat 39PA; Vestopal 120L, 121L, 123L, 140L, 141L, 160, 221L); sheeting (Aropol 8420); shower enclosures (Aropol 8343T-11, 8343T-12, 8343T-21, 8343T-22); structural applications (Crystic 346PA)

Automotive applications: (ANE-D30; Crystic 323PA, 471PALV); autobody patching (Polylite 32-345, 32-367); engine casings (ANE, -D30, -K10, K-20); ignition parts (Polylite 30004-3, 30004-4); vehicle bodies (ANE-K10, K-20; Crystic 196, 196E, 196PALV, 2-406PA, 406PA; Crystic Gelcoat 33E, 68E, 68PA; Crystic Stopper)

Aviation industry: (Dion FR 6692T; Magnamite AS4/1655); aircraft parts (Impco RC-2, RC-2/140 Resins); avionics housings (Crystic 17449); ducts (Crystic 17449)

Polyester, thermoset *(cont'd.)*

Consumer products: appliances (Aropol 7530T-16, 8339T-12); bowling ball cores (Aropl 8329); furniture (Aropol 8110, 8310, 8319, 8321, 8329; Polylite 32-344, 32-356, 32-357, 32-358, 32-738); handicrafts (Polylite 32-032); hardware (Impco RC-2, RC-2/140 Resins); sanitaryware (Hetron 470P); surfboards (Polylite 32-737, 32-739, 32-740)

Electrical/electronic industry: (Aropol 7021, 7362; Dion Cor-Res 6693HV; Dion FR 6399, FR 8300; Dion-Iso 6315HV; Durez 23410, 24150, 24337, 24668, 27962, 29276; Impco RC-2, RC-2/140 Resins; Plenco 1523 Black Pellets, 1528, 1529 Black, 1530); coils (Stypol 83-20000; compressors (Impco RC-2, RC-2/140 Resins); computers (Dion FR 6308, FR 6604T); housings (Dion FR 6308, FR 6604T); impregnation (Impco RC-2, RC-2/140 Resins; Stypol 83-2000); insulation (Crystic 17449); potting (Aropol 8329); switchgear (Crystic 17449)

FDA-approved applications: (Aropol 7242T-15)

Food-contact applications: (Crystic 392); food packaging (Crystic 196, 196PALV, 2-491PA, 491, 491E, 491PA)

Functional additives: blending resin (Dion FR 6399; Polylite 31-820, 31-830, 31-851); corrosion control (Dion Cor-Res 6693FR, 6693HV, 6694HV; Dion FR 6308, FR 6604T; Dion-Iso 6246, 6314, 6315HV, 6631T, 8101, 8200; Hetron 700DMA; Vestopal 150, 400); dispersant (Crystic 57); filler (Crystic Stopper; Vestopal 310, 320, 710, 730, 810, 811, 830); plasticizer (ANE-P10, P-20; Neolyn 40; Paraplex 5-B, G-60); plastics modifier (Crystic 198; Crystic Prefil F; Neolyn 35D); stabilizer (Paraplex G-60); thixotropic agent (Crystic Pregel 17, 27); vehicle (Polylite 31-822); wetting vehicle (Aropol 8101)

Industrial applications: adhesives (Neolyn 20, 35D, 40; Polylite 32-356); aerosols (Chempol 13-1905, 13-2211); casting and potting compounds (Vestopal 310, 320, 330); castings (ANE; Aropol 8420); chemical process industry (ANE; Crystic 392, 17449; Dion Cor-Res 6693FR, 6694, 7000; Polylite 31-439); chemical plant equipment and parts (ANE, -D30, -K10, K-20; Crystic 2-491PA, 491, 491E, 491PA); clay pipe seals (Polylite 31-866); coatings (Chempol 13-1905, 13-2211, 19-4807; Crystic Fireguard 75PA; Neolyn 35D; Paraplex AL-111, G-60, RG-10; Polylite 32-737, 32-738, 32-773, 33-773; Vestopal 110, 120T, 145, 145T, 150T, 152, 160, 160T, 250, 252, 310, 320, 400, 510, 530); concrete (Leguval K25R, K26, N50, W16, W35; Vestopal 110, 120L, 145, 250, 252, 310, 320, 400, 510, 530); containers (Crystic 196, 196E, 196PALV, 2-491PA, 491, 491E, 491PA; Leguval K41, N30, N50, W16, W35, W41; Vestopal 120T, 145, 145T, 150B, 150T, 152, 160, 160T, 250, 252, 310, 320, 400, 510, 530); decorative castings (Polylite 32-180, 32-344, 32-356, 32-357, 32-358); encapsulation (Polylite 32-033); engines (Impco RC-2, RC-2/140 Resins); filled compounds (Aropol 7221, 7240, 7241-12; Leguval E81, E90, K25R, K26, K70; Vestoal 310, 320); fittings (Impco RC-2, RC-2/140 Resins); flooring (ANE-K10, K-200; Polylite 32-300; Vestopal 110, 120L, 145, 250, 252, 310, 320, 400, 510, 530); floor tile (Neolyn 20); fumestacks (Crystic 198, 474PA; Hetron 197AT, 700 Series); gears (Impco RC-2, RC-2/140 Resins); gel coats (Aropol 7131, 7141, 7362, 8420, 2-489PA, 489PA; Crystic Fireguard Gelcoat

Polyester, thermoset *(cont'd.)*

70PA; Crystic Gelcoat 33E, 39PA, 46PA, 47PA, 65E, 65PA, 68E, 68PA; Crystic Pregel 17; Dion-Iso 6315HV; Polylite 34-308; Stypol 40-4033; Vestopal 110, 120T, 145, 145T, 150T, 152, 160, 160T, 250, 252, 310, 320, 400, 510, 530); general purpose moldings (Crystic 328PA, 406PA; Polylite 31-000, 31-001, 31-006, 31-039, 31-820); housings (ANE, ANE-D30; Aropol 7530T-16; Crystic 302; Impco RC-2, RC-2/140 Resins; Leguval F30, F33, F35, K41, N30, N50, W16, W35, W41); hydraulic systems (Impco RC-2, RC-2/140 Resins); industrial finishes (Chempol 13-2211); inks (Neolyn 20, 40); lacquers (Neolyn 20, 40; Paraplex 5-B); laminates (ANE; Aropol 7021; Crystic 113PA, 191LV, 196, 196E, 196PALV, 198, 2-406PA, 2-489PA, 2-491PA, 302, 306, 323PA, 328PALV, 345PA, 392, 405PA, 406PA, 474PA, 489PA, 491, 491E, 491PA, 625E, 625MV; Crystic Fireguard 75PA, Gelcoat 70PA; Crystic Pregel 17; Polylite 32-739, 32-740, 33-089, 33-092, 33-011, 33-031, 33-049, 33-059, 33-067, 33-071, 33-072, 33-081, 33-084, 33-086, 33-402, 33-770; Vestopal 120L, 121L, 123L, 140L, 141L, 160, 221L, 250, 252); large parts (Crystic 625E, 625MV); machinery (ANE-D30, -K10, K-20; Crystic 196E); marble (Aropol 8420-09; Polylite 32-050; Stypol 40-5727, 40-5732); paper coatings (Chempol 13-1905); particle board (Chempol 19-4807); pigments (Aropol 8101; Crystic 57; Polylite 31-822); pipes (ANE-D30, -K10, K-20; Aropol 7240T-15, 7241T-15, 7242T-15, 7532; Crystic 198, 2-491PA, 272E, 474PA, 491, 491E, 491PA; Dion Cor-Res 6693FR, 6694, 7000; Dion-Iso 6246; Hetron 700 Series; Leguval K41, N30, N50, W16, W35, W41; Polylite 31-439; Vestopal 120T, 145, 145T, 150B, 150T, 152, 160, 160T, 250, 252, 310, 320, 400, 510, 530); plastisols/organosols (Neolyn 20, 40); pollution control (Polylite 31-439); preimpregnation of fiberglass, paper, asbestos (Aropol 7280); profiles (Vestopal 120L, 121L, 123L, 140L, 141L, 150, 160); pumps (ANE-D30; Impco RC-2, RC-2/140 Resins); putties (Leguval E60, E81, E90, K25R, K26, K70, N50S); rods (Dion-Iso 8200); sand grout compounds (Aropol 7221, 7240, 7241-12); sealing applications (ANE-K10, K-20); sheet (Aropol 7141; Crystic 191LV, 193HR, 195, 306, 387; Dion-Iso 8200; Leguval F30S, N50S); sheet metal structures (Crystic Stopper); SMC/BMC moldings (Aropol 7030); tanks (ANE, ANE-D30; Aropol 7240T-15, 7240WT-12, 7241T-15, 7242T-15, 7532; Crystic 196E, 198, 2-491PA, 272E, 474PA, 491, 491E, 491PA; Dion Cor-Res 6693FR, 6694, 7000; Polylite 31-439); technical parts (Crystic 302, 323PA, 406PA, 471PALV; Dion FR 6308; Vestopal 110, 120L, 120T, 145, 145T, 150B, 150T, 152, 160, 160T, 250, 252, 310, 320, 510, 530); textile applications (Paraplex RG-10); thick sections (Dion-Iso 8200); thin sections (Polylite 33-031; Stypol 40-4033, 40-4074, 40-8006); tooling (Polylite 33-401, 33-402, 33-404, 33-411); trays (Aropol 8110, 8310, 8319, 8321, 8329); tropical applications (Crystic 196E, 272E, 2-491PA, 385E, 392E, 491E, 625E; Crystic Gelcoat 68E); tubing (Vestopal 150B); utility poles (Aropol 7532); ventilation ducts/hoods (ANE; Aropol 7240T-15, 7240WT-12, 7241T-15, 7242T-15, 7530T-16, 272E, 17449; Dion Cor-Res 6693FR, 6694, 7000; Dion FR 6604T; Hetron 197AT, 700 Series)
Marine applications: (Aropol 7530T-16, 8339T-12, 8349-15; Crystic 302, 471PALV, 625MV; Dion FR 6604T, FR 6655T, FR 6692T, FR 8300; Dion-Iso 6246, 6631T,

8101, 8200); boat coverings (Polylite 32-737, 32-773); boat hulls (ANE-D30, -K10, K-20; Crystic 189LV, 189MV, 2-406PA, 2-489PA, 2-491PA, 406PA, 489PA, 491, 491E, 491PA; Crystic Gelcoat 33E, 65E, 68E, 68PA; Leguval F30, F33, F35, K41, N30, N50, W16, W35, W41); boats (Aropol 8343T-11, 8343T-12, 8343T-21, 8343T-22; Crystic 196E, 302, 405PA; Dion FR 6604T, FR 6692T; Vestopal 110, 120T, 145, 145T, 150T, 152, 160T, 250, 252, 310, 320); gel coats (Aropol 7131, 7141); outboard motor cowlings (Aropol 8110, 8310, 8319, 8321)

Medical applications: (Plenco 1529 Black); dental applications (Plenco 1529 Black)

Transportation industry: (Aropol 7530T-16, 8339T-12, 8349-15; Crystic 346PA; Dion Cor-Res 6693FR, 6693HV; Dion FR 6124, FR 6399, FR 6604T, FR 6655T, FR 6657, FR 6692T, FR 8300; Dion-Iso 6246, 6314, 6315HV, 6631T, 8101, 8200); DOT-approved seating/walls (Hetron 325FS); railway coaches (Crystic 196, 196E, 196PALV); road surfacing (Polylite 32-300); tank trucks (ANE-D30)

PROPERTIES:

Form:

Liquid (ANE, -D30, -K10, -K20; Aropol 7021, 7030, 7131, 7141, 7221, 7229, 7240, 7240T-15, 7240WT-12, 7280, 7320, 7329, 7362, 7420, 7532, 8101, 8110, 8310, 8319, 8321, 8329, 8329-25, 8420, 8420-09, 8343T-11, 8343T-12, 8343T-21, 8343T-22; Chempol 13-1905, 13-2211, 19-4807; Crystic 57, 189LV, 189MV, 191LV, 193HR, 195, 196, 196E, 196PALV, 198, 2-406PA, 2-489PA, 272, 272E, 302, 306, 323PA, 328PALV, 345PALV, 346PA, 392, 392E, 405E, 489PA, 625E, 625MV, 17449; Crystic Fireguard 75PA; Hetron 32A, 700DMA, 27196; Impco RC-2, RC-2/140 Resins)

Clear liquid (Hetron 325FS; Stypol 40-5727, 40-5732; Stypol 83-2000)

Turbid/cloudy liquid (Aropol 7241-12, 7241T-15, 7242T-15, 7530T-16, 8339T-12, 8349-15; Crystic 385E, 385PA, 387, 405PA, 406PA, 471PALV, 474PA, 491, 491E, 491PA, 2-491PA; Crystic Gelcoat 33E, 65E, 68E, 68PA; Hetron 197AT, 470P)

Low-viscosity liquid (Dion Cor-Res 6693FR, 6694, 7000; Dion FR 6114, FR 6124, FR 6125, FR 6127, FR 6604T, FR 6655T, FR 6657, FR 6692T, FR 8300; Dion-Iso 6246, 6631T, 8101, 8200; Paraplex G-60)

Medium-viscosity liquid (Dion FR 6308)

High-viscosity liquid (Dion Cor-Res 6693HV, 6694HV; Dion FR 6399; Dion-Iso 6314, 6315HV)

Viscous liquid (Vestopal 110, 120L, 121L, 123L, 140L, 141L, 145, 150, 150B, 151, 152, 160, 221L, 250, 251, 252, 310, 320, 330, 400, 510, 511, 530, 710, 730, 810, 811, 830)

Tough, viscous material (Paraplex 5-B)

Thixotropic, viscous liquid (Vestopal 120T, 145T)

Thixotropic (Polylite 32-740, 33-011, 33-031, 33-049, 33-067, 33-071, 33-072, 33-081, , 33-084, 33-086, 33-089, 33-092, 33-401, 33-402, 33-404, 33-411, 33-441, 33-450, 33-770; Vestopal 150T, 160T)

2 part: liquid/powder (Crystic Stopper)

Polyester, thermoset *(cont'd.)*

Paste (Crystic Prefil F; Crystic Pregel 17, 27)
Solid (Aropol 7020; Neolyn 20, 35D)
Powder (ANE, -D30, -K10, -K20)
Flakes (Neolyn 35D)
Granules (ANE, D-30, -K10, -K20; Durez 23410, 24150, 24337, 24668, 27962, 29276, 31375, 31572)
Free-flowing granular (Polylite 30001-1, 30001-2, 30001-3, 30002-1, 30003-1, 30003-2, 30004-3, 30004-4)
Pellets (Plenco 1523 Black Pellets)
30.5 cm width graphite prepreg tape (Magnamite AS4/1655)

Color:
Water clear (Crystic 191LV)
Water-white to pale blue (Crystic 195)
Colorless to yellowish (Leguval E60)
Opaque white (Crystic 323PA)
Opaque white, black, or gray (Crystic Fireguard 75PA)
Opaque pinkish (Crystic 328PALV, 345PALV, 346PA)
Pink (Crystic 385PA)
Pinkish (Crystic 196PALV, 2-489PA, 489PA)
Pinkish-mauve (Crystic 198)
Mauvish (Crystic 405PA, 406PA, 471PALV, 474PA, 491PA, 2-491PA)
Pale yellow (Crystic Pregel 17, 27)
Light amber (Impco RC-2, RC-2/140 Resins)
Amber (Stypol 83-2000)
Light straw (Crystic 193HR, 196, 196E, 198, 272, 272E, 306, 387, 392, 392E, 625E, 625MV)
Straw (Crystic 189LV, 189MV, 302, 17449)
Tan (Polylite 30004-4)
Cobalt-milky (Aropol 8343T-21)
Cobalt (Aropol 7530T-16, 8339T-12, 8343T-11, 8343T-12, 8343T-22; Hetron 325FS, 470P)
Light blue-green (Stypol 40-5727, 40-5732)
Blue (Aropol 7241-12, 7241T-15, 7242T-15, 8329-25, 8349-15; Polylite 3001-3, 30004-3)
Purple-red (Aropol 8420-09)
Gray (Polylite 30001-2, 30003-2)
Black (Durez 23410, 24150, 24337, 24668, 27962, 29276, 31375, 31572; Plenco 1523 Black Pellets, 1529 Black; Polylite 30001-1, 30002-1, 30003-1)
Avail. in a range of colors (Crystic Gelcoat 68E, 68PA)
APHA 4 max. (Aropol 7030)
APHA 75 max. (Aropol 8321)
Gardner 1 max. (Aropol 8110, 8310, 8420)
Gardner 1–2 (Vestopal 110, 120L, 121L, 123L, 140L, 141L, 145, 150, 151, 152, 160,

Polyester, thermoset *(cont'd.)*

221L, 251, 252, 310, 330, 400, 810, 811, 830)
Gardner 2 max. (Aropol 7021, 7131, 7141, 7240, 7280, 8101; Paraplex G-60)
Gardner 2–3 (Vestopal 320)
Gardner 3 max. (Crystic 57)
Gardner 3–4 (Vestopal 510, 511, 530)
Gardner 3–6 (Paraplex 5-B)
Gardner 3–7 (Paraplex RG-10)
Gardner 4 (Aropol 7532; Hetron 32A, 700DMA)
Gardner 4 max. (Aropol 7020, 7221, 7320, 7420)
Gardner 5 max. (Aropol 7229, 7329; Stypol 40-7106)
Gardner < 6 (Vestopal 150B)
Gardner 6 max. (Aropol 7362, 8319, 8329; Hetron 197AT)
Gardner 6 (Vestopal 710, 730)

Odor:

Low (Neolyn 20, 35D, 40)

Composition:

35–42% resin (Magnamite AS4/1655)
50 ± 1% solids in VM&P naphtha (Chempol 13-1905)
50 ± 1% solids in toluol/VM&P naphtha (5/95) (Chempol 13-2211)
55 ± 1% solids in styrene (Chempol 19-4807)
80 ± 2% solids in toluol (Paraplex 5-B)
100% solids (Aropol 8101; Impco RC-2, RC-2/140 Resins; Paraplex AL-111, B-60, RG-10)
60% nonvolatile (Aropol 7532; Hetron 470P)
62.5% nonvolatiles (Hetron 197AT)
20% volatiles (Crystic Fireguard 75PA)
21% volatiles (Crystic 346PA)
24% volatiles (Crystic 328PALV)
25% volatiles (Crystic 345PALV)
27% volatiles (Crystic 302)
31% volatiles (Crystic 306)
32% volatiles (Crystic 387)
33% volatiles (Crystic 189MV, 196, 196E, 323PA, 17449; Crystic Gelcoat 33E)
34% volatiles (Crystic 385E, 385PA, 625E, 625MV)
36% volatiles (Crystic 191LV, 198; Crystic Gelcoat 65E)
37% volatiles (Crystic 405PA)
38% volatiles (Crystic 189LV, 195, 196PALV, 474PA)
40% volatiles (Crystic 193HR, 272E, 406PA, 491, 491E)
41% volatiles (Crystic 2-406PA, 272, 392, 392E, 471PALV)
42% volatiles (Crystic 2-491PA, 491PA; Crystic Gelcoat 68E, 68PA)
44% volatiles (Crystic 2-489PA, 489PA)
28% acetone (Aropol 7280)
25% styrene (Leguval E60, E81, F30, F33; Vestopal 320)

Polyester, thermoset *(cont'd.)*

25–27% styrene (Aropol 8110, 8319)
27% styrene (Vestopal 310)
27–29% styrene (Aropol 7229)
28% styrene (Vestopal 250)
28–30% styrene (Aropol 7329, 8310)
29–31% styrene (Aropol 7020, 7320, 7420, 8321)
30% styrene (Leguval E90, F30S, F35, K25R, K26, N30; Vestopal 110, 140L, 141L, 145, 221L, 251, 252, 330)
30 ± 1% styrene (Aropol 8420)
30–32% styrene (Aropol 8329-25)
33% styrene (Aropol 8420-09; Vestopal 120L, 140L, 141L, 160, 710, 810)
33.5–36.5% styrene (Aropol 7021)
34% styrene (Vestopal 150, 150B, 151, 730, 830)
34–36% styrene (Aropol 8329)
35% styrene (Aropol 7221; Leguval K70, N50, N50S, W16, W16T, W20; Vestopal 400, 811)
36.5% styrene (Vestopal 121L)
37–39% styrene (Aropol 7030)
38% styrene (Vestopal 120T, 510, 511)
39% styrene (Vestopal 145T)
40% styrene (Aropol 7131; Leguval K41, W35, W41; Vestopal 530)
41% styrene (Vestopal 152)
42% styrene (Aropol 7530T-16, 8339T-12)
43% styrene (Vestopal 123L)
44% styrene (Aropol 8343T-11, 8343T-12)
44–46% styrene (Aropol 7141)
45–57% styrene (Aropol 8349-15)
50% styrene (Hetron 700DMA)
33.5–37.5% vinyl toluene (Aropol 7362)

GENERAL PROPERTIES:

Solubility:
Sol. in acetone (Neolyn 35D)
Sol. in low m.w. alcohols (Neolyn 40)
Sol. in aromatic hydrocarbon solvents (Neolyn 20, 40)
Sol. 22–24% in butanol (Chempol 13-1905); sol. 22–25% in butanol (Chempol 13-2211)
Sol. in butyl acetate (Neolyn 35D)
Sol. in carbon tetrachloride (Neolyn 35D)
Sol. in chlorinated hydrocarbon solvents (Neolyn 20, 40)
Sol. in esters (Neolyn 20, 40)
Sol. in ethyl acetate (Neolyn 35D)
Sol. in ethylene dichloride (Neolyn 35D)
Sol. in ketones (Neolyn 20, 40)

Polyester, thermoset *(cont'd.)*

Sol. in MEK (Neolyn 35D)

Sol. 16–18% in min. spirits (Chempol 13-1905, 13-2211)

Miscible and compatible with oleoresinous varnishes and alkyd-type resins (Paraplex 5-B)

Sol. in toluene (Neolyn 35D)

Insol. in water (Neolyn 35D, 40)

Sol. < 5% in xylol (Chempol 13-1905, 13-2211)

Sp. Gr.:

1.012 (Hetron 700DMA)

1.02 (Dion Cor-Res 6694, 7000)

1.06 (Dion Cor-Res 6694HV; Dion-Iso 8200)

1.07 (Crystic 392, 392E; Stypol 40-5727)

1.07–1.09 (Aropol 7362)

1.08 (Dion-Iso 6631T); (55% in styrene sol'n.) (ANE-D30, -K10, -K20)

1.09 (Stypol 40-4033, 40-4074; Stypol 40-8006); (55% in styrene sol'n.) (ANE)

1.10 (Crystic 2-406PA, 272, 272E, 406PA, 474PA; Crystic Gelcoat 68E, 68PA; Dion FR 6657; Dion-Iso 6246, 8101)

1.10–1.15 (Aropol 8319)

1.11 (Crystic 191LV, 193HR, 196PALV, 198, 2-489PA, 2-491PA, 471PALV, 489PA, 491, 491E, 491PA, 17449; Crystic Gelcoat 65E; Stypol 40-5732)

1.12 (Crystic 195, 196, 196E, 405PA)

1.13 (Crystic 625E, 625MV; Dion-Iso 6314, 6315HV)

1.15 (Crystic 189LV, 385E, 385PA; Crystic Gelcoat 33E); (cured) (Aropol 7131, 7141)

1.16 (Dion Cor-Res 6693FR); (cured) (Stypol 40-5732)

1.17 (Crystic 189MV; Dion FR 6127); (cured) (Aropol 7240WT-12, 7280)

1.18 (Crystic 306); (cured) (Stypol 40-4033, 40-4074, 40-8006)

1.19 (cured) (Aropol 7240, 7530T-16)

1.196 (Impco RC-2, RC-2/140 Resins)

1.20 (Crystic 57, 323PA; Dion FR 6604T)

1.23 (Hetron 325FS); (cured) (Aropol 8329-25, 8339T-12, 8349-15, 8343T-11, 8343T-12, 8343T-21, 8343T-22)

1.24 (Dion FR 6308)

1.25 (Dion FR 6692T; Hetron 197AT, 27196); (cured) (Aropol 8329)

1.26 (Dion FR 8300)

1.28 (Dion FR 6655T)

1.30 (Dion Cor-Res 6693HV; Dion FR 6114, FR 6124, FR 6125; Hetron 32A)

1.32 (Crystic 387)

1.33 (Crystic 302)

1.35 (Crystic Fireguard 75PA)

1.37 (Dion FR 6399)1.38 (Crystic 328PALV)

1.45 (Crystic 345PALV)

1.48 (Crystic 346PA)

Polyester, thermoset *(cont'd.)*

1.77 (Plenco 1523 Black Pellets)
1.90 (Durez 29276)
1.92 (Durez 23410)
1.97 (Durez 31375, 31572; Plenco 1530)
1.98 (Durez 24337)
2.0 (Durez 27962; Plenco 1528)
2.08 (Durez 24668)
2.2 (Durez 24150)
2.88 (Plenco 1529 Black)

Density:

0.65 g/cm³ (Plenco 1523 Black Pellets)
0.75 g/cm³ (Durez 31572)
0.76 g/cm³ (Durez 23410)
0.77 g/cm³ (Plenco 1530)
0.80 g/cm³ (Durez 24668, 27962, 29276, 31375)
0.83 g/cm³ (Durez 24337)
0.92 g/cm³ (Plenco 1528)
1.05 g/cm³ (Durez 24150)
1.06 g/cm³ (Leguval E60)
1.07 g/cm³ (Leguval W35; Vestopal 510, 511)
1.09 g/cm³ (Leguval E90, K41, W41; Vestopal 160T)
1.10 g/cm³ (Vestopal 123L, 141L, 145T, 152, 160, 530)
1.11 g/cm³ (Leguval W20; Vestopal 120T, 121L, 140L, 400)
1.12 g/cm³ (Leguval K70, W16T; Vestopal 145, 150T, 811)
1.13 g/cm³ (Leguval E81, K25R, K26, N50, N50S, W16; Vestopal 120L, 150, 150B, 151, 320, 330, 710, 730, 810, 830)
1.15 g/cm³ (Leguval N30)
1.17 g/cm³ (Vestopal 110, 251, 252, 310)
1.18 g/cm³ (Leguval F35)
1.21 g/cm³ (Leguval F33)
1.22 g/cm³ (Vestopal 250)
1.30 g/cm³ (Leguval F30S)
1.31 g/cm³ (Vestopal 221L)
1.35 g/cm³ (Leguval F30)
1.40 g/cm³ (Plenco 1529 Black)
1.06 kg/l (Neolyn 40)
1.18 kg/l (Neolyn 20)
1.208 kg/l (Neolyn 35D)
7.25–7.45 lb/gal (Chempol 13-1905)
7.3–7.5 lb/gal (Chempol 13-2211)
8.0 lb/gal (Paraplex AL-111, RG-10)
8.2 lb/gal (Paraplex 5-B, G-60)
8.9 lb/gal (Aropol 7240WT-12; Stypol 40-5727)

9.0 lb/gal (Aropol 7240)
9.1 lb/gal (Aropol 7530T-16; Stypol 83-2000)
9.2 lb/gal (Aropol 8329)
9.2–9.4 lb/gal (Chempol 19-4807)
9.3 lb/gal (Aropol 7021; Stypol 40-5732)
9.35–9.55 lb/gal (Aropol 8319)
9.4 lb/gal (Aropol 8101)
9.6 lb/gal (Aropol 8110, 8420, 8420-09)
9.7 lb/gal (Aropol 8310, 8321)
10.7 lb/gal (Hetron 470P)
Visc.:
0.15 Pa·s (Vestopal 123L)
0.25 Pa·s (Vestopal 152, 330)
0.3 Pa·s (Vestopal 145T)
0.4 Pa·s (Vestopal 121L, 141L, 710, 810, 811)
0.5 Pa·s (Vestopal 120T, 150T, 160T)
0.55 Pa·s (Vestopal 221L)
0.6 Pa·s (ANE, -D30, -K10, -K20140L)
0.8 Pa·s (Vestopal 120L, 160)
0.95 Pa·s (Vestopal 320)
1.0 Pa·s (Vestopal 110, 150, 150B, 151, 310, 510, 511, 530, 730, 830)
1.1 Pa·s (Vestopal 145)
1.4 Pa·s (Vestopal 250, 251, 252)
4.5 Pa·s (Vestopal 400)
200–350 cSt (Impco RC-2, RC-2/140 Resins)
1.7 poise (Crystic 195)
1.8 poise (Crystic 193HR); (121 C) (Neolyn 40)
2.0 poise (Crystic 387)
2.7 poise (Crystic Gelcoat 68E, 68PA)
2.9 poise (160 C) (Neolyn 20)
3.0 poise (Crystic 471PALV)
3.5 poise (Crystic 191LV, 196PALV, 272, 272E, 306)
3.6 poise (Crystic 189LV)
3.8 poise (Crystic 17449)
4.0 poise (Crystic 2-406PA, 385E)
4.1 poise (Crystic 2-489PA, 489PA)
5.0 poise (Crystic 2-491PA, 385PA, 405PA, 474PA, 491PA)
6–10 poise (Crystic 57)
6.5 poise (Crystic 198)
8 poise (Crystic 323PA, 392, 392E, 491, 491E, 625E, 625MV)
9 poise (Crystic 189MV, 196, 196E)
10.5 poise (Crystic 302)
11 poise (Crystic 406PA)

Polyester, thermoset (cont'd.)

13 poise (Crystic 328PALV)
18 poise (Crystic 345PALV)
20 poise (Crystic 346PA)
45–85 cps (Aropol 8349-15)
50 cps (Dion FR 6657)
50–150 cps (Polylite 32-180)
100 cps (Dion FR 6127)
100–150 cps (Polylite 31-866; Stypol 40-8006)
115 cps (Stypol 83-2000)
120 cps (Hetron 325FS)
150–250 cps (Polylite 32-738)
200–300 cps (Polylite 32-356, 32-357)
200–400 cps (Polylite 32-039)
250–300 cps (Polylite 32-344)
250–400 cps (Polylite 32-122, 32-737, 33-089, 33-092)
250–450 cps (Polylite 32-032)
275–425 cps (Polylite 32-033, 32-035)
300 cps (Aropol 7240)
300–360 cps (Polylite 32-112)
300–400 cps (Polylite 31-450, 32-345, 33-067, 33-071, 33-072, 33-081)
300–450 cps (Aropol 7141; Polylite 31-851)
300–500 cps (Polylite 33-049)
300–650 cps (Polylite 33-401, 33-402, 33-404, 33-411)
325 cps (Aropol 8343T-11, 8343T-12, 8343T-21, 8343T-22)
340 cps (55% in styrene sol'n.) (ANE-K20)
340–400 cps (Polylite 32-111)
350 cps (Aropol 8339T-12)
350–450 cps (Aropol 7241-12, 8329; Polylite 32-037, 33-084; Stypol 40-4033)
350–500 cps (Polylite 33-031, 33-441)
350–550 cps (Polylite 33-450)
380 cps (55% in styrene sol'n.) (ANE-D30)
400 cps (Dion Cor-Res 6694, 7000)
400–475 cps (Polylite 32-367)
400–500 cps (Polylite 32-050, 32-358, 33-059)
400–550 cps (Polylite 33-086)
400–700 cps (Polylite 33-011)
420 cps (55% in styrene sol'n.) (ANE-K10)
430 cps (55% in styrene sol'n.) (ANE)
450 cps (Aropol 7240T-15, 7241T-15, 7242T-15, 7530T-16, 7532; Hetron 470P)
450–500 cps (Polylite 32-739)
500 cps (Aropol 7240WT-12; Dion Cor-Res 6693FR; Dion FR 6604T, FR 6655T, FR 6692T, FR 8300; Dion-Iso 6246, 6631T, 8101, 8200; Hetron 197AT, 700DMA, 27196)

500–600 cps (Stypol 40-4074)
500–700 cps (Polylite 32-740)
500–800 cps (Polylite 33-773)
575–675 cps (Polylite 32-125)
600 cps (Dion FR 6114, FR 6124, FR 6125)
600–800 cps (Polylite 32-128, 32-773)
600–900 cps (Polylite 31-830, 32-300)
650 cps (Leguval N50, N50S)
700 cps (Leguval F30S)
700–900 cps (Polylite 32-127)
750–1050 cps (Polylite 31-000, 31-001)
800–1200 cps (Chempol 19-4807)
900 cps (Aropol 7131)
900–1200 cps (Polylite 34-308; Stypol 40-5727)
1000 cps (Leguval E81)
1000–1100 cps (Aropol 8329-25)
1000–1400 cps (Aropol 7030)
1100 cps (Aropol 8420-09; Leguval K70)
1100–1400 cps (Polylite 31-820)
1200 cps (Leguval E90)
1200–1500 cps (Stypol 40-5732)
1200–1700 cps (Aropol 8101)
1300–1700 cps (Aropol 7021)
1400 cps (Leguval E60, W16)
1500 cps (Leguval W20; Stypol 40-7016)
1500–2000 cps (Polylite 31-439, 31-571, 33-770)
1600 cps (Leguval W35)
1700 cps (Aropol 7221)
1700–2200 cps (Aropol 7329)
1800–2400 cps (Aropol 8319)
1800–2500 cps (Polylite 31-546, 31-587)
1900 cps (Aropol 8420; Leguval N30)
1900–2400 cps (Aropol 7229)
2000 cps (Aropol 7280; Dion FR 6308)
2000–2400 cps (Aropol 8310)
2000–4000 cps (Polylite 31-822)
2100–2400 cps (Polylite 31-451)
2100–2600 cps (Polylite 31-585)
2100–2700 cps (Aropol 7420, 8321)
2200 cps (Hetron 32A)
2500 cps (Leguval F35)
2500–3000 cps (Aropol 7320)
2700–3700 cps (Polylite 31-006, 31-039, 31-447, 32-516, 32-528)

Polyester, thermoset *(cont'd.)*

2800 cps (Leguval F30)
2900 cps (Leguval K25R)
3000 cps (Leguval K26)
3000–3400 cps (Polylite 32-129)
3000–3500 cps (Polylite 32-120)
3250–4150 cps (Aropol 7362)
3300–3900 cps (Aropol 8110)
3500 cps (Dion Cor-Res 6693HV; Dion FR 6399; Leguval F33, K41, W41)
4000–6000 cps (Polylite 31-402)
4500 cps (Dion-Iso 6315HV)
6000–9000 cps (Aropol 7020)
7000 cps (Dion-Iso 6314)
8000 cps (Dion Cor-Res 6694HV)
J–O (Paraplex G-60)
W–Z (Chempol 13-1905)
Y–Z (Chempol 13-2211)
Z1 (Paraplex 5-B)
Z3 (Neolyn 35D)
Z8 (Paraplex AL-111)
Z10 (Paraplex RG-10)

Soften. Pt.:

75 C (Drop) (Neolyn 20)
87 C (Neolyn 35D)
90 C (ANE, solid)

Flash Pt.:

31 C (Crystic 405E; Crystic Gelcoat 33E); (TCC) (Chempol 19-4807)
< 32 C (Crystic 189LV, 189MV, 191LV, 193HR, 195, 196, 196E, 198, 2-406PA, 2-489PA, 2-491PA, 272, 272E, 302, 306, 323PA, 328PALV, 345PALV, 346PA, 385E, 385PA, 387, 392, 392E, 405PA, 406PA, 471PALV, 474PA, 489PA, 491, 491E, 491PA, 625E, 625MV, 17449; Crystic Fireguard 75PA; Crystic Gelcoat 65E, 68E, 68PA; Crystic Prefil F; Crystic Pregel 17, 27)
35 C (Vestopal 110, 120L, 120T, 121L, 123L, 140L, 141L, 145, 145T, 150, 150B, 150T, 151, 152, 160, 160T, 221L, 250, 251, 252, 310, 320, 330, 400, 510, 511, 530, 710, 730, 810, 811, 830)
192 C (COC) (Neolyn 20)
201 C (COC) (Neolyn 40)
217 C (COC) (Neolyn 35D)
35 F (TCC) (Chempol 13-1905, 13-2211)
93 F (Impco RC-2, RC-2/140 Resins)
275 F (PM) (Aropol 8101)

Fire Pt.:

97 F (Impco RC-2, RC-2/140 Resins)

Polyester, thermoset *(cont'd.)*

Acid No.:

0–8 (Paraplex RG-10)

0–15 (Paraplex AL-111)

1 max. (Paraplex G-60)

2–6 (Chempol 13-1905)

4–7 (Chempol 13-2211)

8 (Neolyn 35D)

8–12 (solids) (Aropol 8101)

10 (Hetron 700DMA; Neolyn 20); (solids) (Aropol 7240, 7240T-15, 7240WT-12, 7280)

12 (Crystic 328PALV, 346PA)

13 (Crystic 345PALV, 392E; Neolyn 40)

14 (Crystic 392)

14–22 (Crystic 57)

16 (Crystic 2-406PA)

16–20 (Aropol 7241-12)

17 (Crystic 406PA, 625E, 625MV)

17–23 (solids) (Aropol 7229, 7329, 8319, 8329, 8329-25, 8349-15)

18 (Crystic 2-491PA, 272, 272E, 323PA, 491, 491E, 491PA; Crystic Fireguard 75PA; Crystic Gelcoat 68E, 68PA); (solids) (Aropol 7221, 7241T-15, 7530T-16)

18–24 (solids) (Aropol 8310)

19 (Crystic 196PALV, 2-489PA, 471PALV, 489PA; Crystic Gelcoat 33E, 65E); (solids) (Aropol 7532)

20 (Crystic 387, 17449); (solids) (Aropol 7242T-15, 8339T-12); (plastic) (Aropol 8343T-11, 8343T-12, 8343T-21, 8343T-22)

20–24 (solids) (Aropol 7141)

20–28 (Chempol 19-4807)

21 (Crystic 196, 196E); (plastic) (Aropol 8420); (solids) (Aropol 8420-09)

22 (solids) (Aropol 7131)

22–28 (solids) (Aropol 7362)

23 (Crystic 474PA)

23–27 (solids) (Aropol 7021, 7420)

24 (Crystic 189LV, 198, 302, 306, 405PA)

24–30 (solids) (Aropol 8110)

25 (solids) (Hetron 197AT)

26 (Crystic 189MV)

26–30 (solids) (Aropol 7020, 7030)

27 (Crystic 385E, 385PA)

28–32 (solids) (Aropol 7320)

30 (Crystic 193HR)

31 (Crystic 195)

31–37 (solids) (Aropol 8321)

32 (Crystic 191LV)

Polyester, thermoset *(cont'd.)*

    37 (solid) (ANE)

    47–60 (Paraplex 5-B)

    50 (Hetron 27196)

Iodine No.:

    1.0 (Leguval N50, N50S, W16, W16T)

    1.5 (Leguval N30, W20)

    3.0 (Leguval W35)

    10 (Leguval W41)

    15 (Leguval K70)

    20 (Leguval K41)

Stability:

    Resistant to alcohols, organic and inorganic acids, hydrocarbon solvents, fuels, aq. solutions of inorganic salts, elevated temps., etc. (ANE)

    Stable to chemically aggressive media (esp. acids) and to elevated temps. (ANE-D30)

    Resistant to inorganic and organic acids, dilute alkalis, aq. solutions of inorganic salts (ANE-K10, -K20)

    Resistant to oil, grease, petrol, and similar solvents (Crystic Stopper)

    Good chemical resistance (Aropol 7280)

    Fully cured sheeting has excellent weather resistance (Crystic 191LV)

    Fully cured laminates made with powder bonded "E" glass chopped strand mat have excellent weather resistance (Crystic 195)

    Does not discolor; resistant to fuels, oils, alcohols, glycols, solvents, salts, and mild acids; resistant to constant operating temps. from –65 to 350 F and to intermittent temps. from –80 to 450 F, with water cooled parts temps. to 1400 F (Impco RC-2, RC-2/140 Resins)

    Good chemical resistance and outdoor stability (Leguval E60, E81, E90, F30, F30S, F33, F35, K25R, K26, K70, N30, N50, N50S, W16, W16T, W20)

    Improved chemical resistance; good outdoor stability (Leguval K41, W35, W41)

    Excellent resistance to uv; good heat stability and resistance to gasoline and water (Paraplex RG-10)

    Properly cured grades are resistant at R.T. to weathering, water and aq. salt sol'ns., and generally also to dilute acids and aliphatic hydrocarbons (Vestopal 110, 120L, 120T, 121L, 123L, 140L, 141L, 145, 145T, 150, 150B, 150T, 151, 152, 160, 160T, 221L, 250, 251, 252, 310, 320, 330, 400, 510, 511, 530, 710, 730, 810, 811, 830)

    3 wks catalyzed life @ 25 C (with 1 phr t-butyl perbenzoate) (Stypol 83-2000)

Storage Stability:

    3 mo. shelf life (Aropol 7241T-15, 7242T-15, 7362, 7530T-16, 7532, 8339T-12, 8420-09, 8343T-11, 8343T-12, 8343T-21, 8343T-22; Hetron 32A, 325FS, 470P, 27196)

    6 mo. shelf life (Aropol 7241-12, 7320, 7329, 7420, 8110, 8310, 8319, 8321, 8329, 8329-25, 8349-15)

    2 mos. in the dark @ 20 C (Crystic Fireguard 75PA)

    3 mos. in the dark @ 20 C (Crystic 2-406PA, 2-489PA, 345PALV, 346PA, 385PA, 387, 406PA, 489PA; Crystic Gelcoat 68PA; Vestopal 120T, 145T, 150T, 160T,

221L, 250, 251, 252, 510, 511, 530)

3 mos. in the dark @ 25 C (Crystic Gelcoat 68E)

4 mos. in the dark @ 25 C (Crystic Gelcoat 33E, 65E)

6 mos. stability when stored under cool and dark conditions (Leguval E60, E81, E90, F30, F30S, F33, F35, K25R, K26, K41, K70, N30, N50, N50S, W16, W16T, W20, W35, W41)

6 mos. in the dark @ 20 C (Crystic 189LV, 189MV, 191LV, 193HR, 195, 196, 196E, 196PALV, 198, 2-491PA, 272, 272E, 302, 306, 323PA, 328PALV, 405PA, 471PALV, 474PA, 491, 491PA, 625MV, 17449; Crystic Pregel 17, 27; Vestopal 110, 120L, 121L, 123L, 140L, 141L, 145, 150, 150B, 151, 152, 160, 310, 320, 330, 400, 710, 730, 810, 811, 830)

6 mos. in the dark @ 25 C (Crystic 385E, 392E, 491E, 625E)

12 mos. in the dark @ 20 C (Crystic 392)

12 mos min. shelf life @ −18 C (Magnamite AS4/1655)

One-year shelf life @ < 120 F (Aropol 7280)

Shelf life of no less than 12 mos. (Crystic 57)

Ref. Index:

1.5315 (Neolyn 20)

1.5467 (Neolyn 35D)

1.548 (cured cast) (Crystic 195)

1.550 (cured cast) (Crystic 306)

1.552 (cured cast) (Crystic 193HR)

1.553 (cured cast) (Crystic 385E, 385PA)

1.556 (cured cast) (Crystic 198, 625E, 625MV)

1.557 (cured cast) (Crystic 392, 392E)

1.560 (cured cast) (Crystic 191LV, 196, 196E)

1.5629 (Neolyn 40)

1.566 (cured cast) (Crystic 189MV)

1.568 (cured cast) (Crystic 302)

1.569 (cured cast) (Crystic 189LV)

1.570 (cured cast) (Crystic 272, 272E, 387)

MECHANICAL PROPERTIES:

Tens. Str.:

380 kg/cm² (ANE-D30)

410 kg/cm² (ANE)

450 kg/cm² (ANE-K10)

520 kg/cm² (ANE-K20)

4 MPa (break) (Vestopal 330)

10 MPa (break) (Vestopal 710, 810, 811)

12 MPa (break) (Vestopal 320)

25 MPa (break) (Vestopal 310)

30 MPa (Crystic 345PALV)

34 MPa (Durez 27962, 29276)

Polyester, thermoset *(cont'd.)*

35 MPa (Crystic 328PALV, 346PA)

38 MPa (Durez 23410)

40 MPa (break) (Vestopal 120T, 221L)

41 MPa (Durez 24337, 31572)

43 MPa (Crystic 323PA)

45 MPa (Durez 31375)

48 MPa (Crystic 306; Durez 24668)

50 MPa (Crystic 387)

55 MPa (Crystic 195, 302, 385PA; Vestopal 120L); (break) (Vestopal 152)

60 MPa (Crystic 191LV, 198, 385E, 474PA, 17449; Crystic Gelcoat 33E; Vestopal 123L)

62 MPa (Durez 24150)

64 MPa (Crystic 2-406PA)

65 MPa (Crystic 193HR, 405PA, 406PA, 471PALV, 625E, 625MV; Vestopal 121L); (break) (Vestopal 141L, 530)

69 MPa (Crystic 196, 196E)

70 MPa (Crystic 189LV, 189MV, 392, 392E; Crystic Gelcoat 68E); (break) (Vestopal 140L, 145T, 150T, 160T)

75 MPa (Crystic 2-489PA, 2-491PA, 272, 272E, 489PA, 491E, 491PA; Crystic Gelcoat 65E); (break) (Vestopal 160)

80 MPa (break) (Vestopal 145, 400, 510, 511)

85 MPa (break) (Vestopal 150, 150B, 151, 250, 730, 830)

90 MPa (break) (Vestopal 110, 251, 252)

3200 psi (inj. molded) (Plenco 1528)

4500 psi (Stypol 40-5732)

4800 psi (Aropol 7020, 7030)

5300 psi (Aropol 8110)

5500 psi (Aropol 7021, 7229, 7362; Plenco 5123 Black Pellets)

5700 psi (Aropol 7329)

6000 psi (Stypol 40-4033, 40-8006)

6895 psi (Stypol 40-4074)

7000 psi (Plenco 1529 Black)

7500 psi (Stypol 40-7106)

7900 psi (Aropol 7131, 7141)

8000 psi (Plenco 1530)

8200 psi (Aropol 8420-09)

8600 psi (Aropol 7240)

8700 psi (Aropol 8321)

8760 psi (Paraplex RG-10 film with 65% nitrocellulose)

9000 psi (Aropol 7240T-15, 7240WT-12, 7280, 8420)

9100 psi (Aropol 7241T-15, 7242T-15)

9075 psi (Aropol 7221, 7241-12)

9300 psi (Aropol 7530T-16; Stypol 40-5727)

9400 psi (Aropol 8319, 8329)

10,000 psi (Aropol 8310, 8329-25, 8339T-12, 8349-15, 8343T-11, 8343T-12, 8343T-21, 8343T-22)

10,900 psi (Aropol 7320)

11,000 psi (Aropol 7420)

11,300 psi (Aropol 7532)

11,500 psi (Hetron 197AT)

13,043 psi (Hetron 325FS)

Tens. Elong.:

0.5% (break) (Crystic 346PA)

0.7% (break) (Crystic 345PALV)

0.8% (Aropol 8110)

0.9% (Aropol 7020, 7030)

1.1% (Aropol 7021)

1.2% (Stypol 40-4033, 40-8006)

1.3% (break) (Crystic 328PALV)

1.4% (Hetron 197AT; Stypol 40-4074); (break) (Crystic 323PA)

1.5% (Aropol 7221, 7229, 7241-12, 7241T-15, 7242T-15, 7362); (break) (Vestopal 120L, 120T, 121L, 123L, 221L)

1.6% (Aropol 8321); (break) (ANE)

1.7% (Aropol 7329; Stypol 40-7016)

1.8% (Aropol 8420-09); (break) (Crystic 387)

1.9% (Aropol 7240, 8310)

1.95% (Aropol 7131, 7141)

2.0% (break) (Crystic 191LV, 302, 405PA, 406PA, 17449; Vestopal 145T, 150T, 152, 160T)

2.1% (Aropol 7240T-15, 7240WT-12, 7280, 7530T-16, 8420); (break) (Crystic 198, 474PA)

2.2% (break) (Crystic 471PALV, 625E, 625MV; Vestopal 160, 530)

2.3% (break) (Crystic 193HR, 196, 196E)

2.4% (break) (Crystic 2-406PA)

2.5% (break) (Crystic 195; Vestopal 141L, 250, 400, 510, 511)

2.9% (break) (Crystic 2-491PA, 385PA, 491E, 491PA)

3.0% (Aropol 8329-25, 8339T-12, 8349-15, 8343T-11, 8343T-12, 8343T-21, 8343T-22); (break) (ANE-D30; Crystic 385E, 392, 392E; Crystic Gelcoat 65E; Vestopal 140L, 145, 251, 252)

3.2% (Stypol 40-5727)

3.5% (Aropol 8319, 8329); (break) (ANE-K20; Crystic 2-489PA, 272E, 489PA; Vestopal 150, 150B, 151, 730, 830)

3.8% (Aropol 7320); (break) (Crystic 272)

4.0% (break) (Crystic 189LV, 189MV, 306; Crystic Gelcoat 33E, 68E, 68PA; Vestopal 110)

4.1% (Aropol 7532)

4.2% (Dion-Iso 6246)
4.4% (break) (ANE-K10)
4.5% (break) (Paraplex RG-10 film with 65% nitrocellulose)
5.5% (Aropol 7420)
20% (break) (Vestopal 310)
25% (break) (Vestopal 320)
30–35% (Stypol 40-5732)
45% (break) (Vestopal 710, 810, 811)
80% (break) (Vestopal 330)

Tens. Mod.:

2.0 GPa (Vestopal 310)
3.0 GPa (Crystic 306, 385E; Crystic Gelcoat 33E)
3.1 GPa (Crystic 387)
3.2 GPa (Crystic 189LV, 189MV, 385PA)
3.4 GPa (Crystic 2-406PA, 392, 392E)
3.5 GPa (Crystic 2-489PA, 2-491PA, 272, 272E, 302, 489PA, 491E, 491PA; Crystic Gelcoat 65E, 68E, 68PA)
3.6 GPa (Crystic 191LV, 471PALV; Vestopal 141L)
3.7 GPa (Crystic 328PALV; Vestopal 110)
3.75 GPa (Vestopal 123L)
3.8 GPa (Crystic 193HR, 196, 196E, 198, 323PA, 405PA, 474PA, 625E, 17449; Vestopal 140L, 251, 252)
3.9 GPa (Vestopal 400)
4.0 GPa (Crystic 406PA, 625MV); Vestopal 120L, 121L, 145, 150, 150B, 151, 152, 510, 511, 530, 730, 830)
4.1 GPa (Vestopal 120T, 145T, 150T, 250)
4.2 GPa (Vestopal 221L)
4.3 GPa (Vestopal 160, 160T)
4.5 GPa (Crystic 195)
5.2 GPa (Crystic 345PALV)
7.3 GPa (Crystic 346PA)
13.8 GPa (Durez 31375)
15.8 GPa (Durez 31572)
460,000 psi (Stypol 40-4033, 40-8006)
520,000 psi (Stypol 40-4074)
2.2×10^6 psi (Hetron 325FS)

Flex. Str.:

800 kg/cm² (Leguval F30S)
850 kg/cm² (ANE)
1000 kg/cm² (ANE-D30, ANE-K10; Leguval W20)
1100 kg/cm² (Leguval F35, K26, N50, N50S, W16)
1150 kg/cm² (ANE-K20)
1200 kg/cm² (Leguval K25R)

1300 kg/cm² (Leguval K41, W35, W41)
1500 kg/cm² (Leguval F33, N30)
30 MPa (Vestopal 710, 810, 811)
58 MPa (Crystic 346PA)
59 MPa (Durez 27962)
62 MPa (Durez 24337)
66 MPa (Durez 31572)
69 MPa (Durez 23410, 29276)
83 MPa (Durez 24668, 31375)
85 MPa (Vestopal 120T)
90 MPa (Vestopal 221L)
100 MPa (Vestopal 120L, 121L)
103 MPa (Durez 24150)
110 MPa (Vestopal 123L)
120 MPa (Vestopal 141L, 150T, 152, 160, 160T, 310, 400, 510, 511, 530)
130 MPa (Vestopal 140L, 250)
140 MPa (Vestopal 251, 252)
150 MPa (Vestopal 145T, 150, 150B, 151, 730, 830)
155 MPa (Vestopal 145)
160 MPa (Vestopal 110)
6500 psi (Stypol 40-5732)
7500 psi (Plenco 1528)
8800 psi (Plenco 1523 Black Pellets)
9000 psi (Stypol 40-4033, 40-8006)
11,000 psi (Aropol 7362; Plenco 1529 Black)
12,500 psi (Plenco 1530)
12,812 psi (Stypol 40-4074)
13,000 psi (Aropol 7229, 8420)
13,200 psi (Aropol 7020, 7030)
13,300 psi (Aropol 7329)
14,000 psi (Aropol 7021, 8321)
14,300 psi (Aropol 8110)
16,000 psi (Aropol 8420-09)
16,800 psi (Stypol 40-5727)
17,000 psi (Aropol 8310, 8329-25, 8339T-12, 8349-15, 8343T-11, 8343T-12, 8343T-21, 8343T-22)
17,800 psi (Aropol 7131, 7141, 8319, 8329)
18,000 psi (Hetron 197AT)
19,000 psi (Aropol 7532)
20,500 psi (Aropol 7240)
21,000 psi (Aropol 7320, 7530T-16)
21,500 psi (Aropol 7240T-15, 7240WT-12, 7280)
22,000 psi (Aropol 7420; Stypol 40-7016)

Polyester, thermoset *(cont'd.)*

23,235 psi (Hetron 325FS)
23,400 psi (Aropol 7221, 7241-12, 7241T-15, 7242T-15)

Flex. Mod.:
33,000 kg/cm² (ANE-K20)
34,000 kg/cm² (ANE-D30)
34,300 kg/cm² (ANE-K10)
36,000 kg/cm² (ANE)
420,000 psi (Aropol 7329)
470,000 psi (Aropol 7020, 7030)
480,000 psi (Aropol 7229)
500,000 psi (Aropol 8329-25, 8339T-12, 8349-15, 8343T-11, 8343T-12, 8343T-21, 8343T-22)
510,000 psi (Aropol 7532)
520,000 psi (Aropol 7131, 7141, 8319, 8329)
530,000 psi (Aropol 7021)
540,000 psi (Aropol 7240T-15, 7240WT-12, 7280)
550,000 psi (Aropol 7240, 7320; Stypol 40-4033, 40-4074, 40-8006)
560,000 psi (Aropol 7221, 7241-12, 7241T-15, 7242T-15, 7420)
580,000 psi (Aropol 7530T-16)
600,000 psi (Aropol 8420)
610,000 psi (Aropol 8310)
620,000 psi (Aropol 8420-09)
640,000 psi (Aropol 8321)
690,000 psi (Aropol 8110)
860,000 psi (Hetron 197AT)
1.5×10^6 psi (Aropol 7362; Hetron 325FS)

Compr. Str.:
120 MPa (Vestopal 310)
140 MPa (Vestopal 221L)
150 MPa (Vestopal 110, 145, 145T, 150T)
152 MPa (Durez 31375)
155 MPa (Vestopal 141L, 150, 150B, 151, 160, 160T, 730, 830)
160 MPa (Vestopal 120L, 140L, 400, 510, 511, 530, 710)
165 MPa (Vestopal 152, 251, 252)
170 MPa (Vestopal 123L)
172 MPa (Durez 31572)
175 MPa (Vestopal 250)
180 MPa (Vestopal 120T, 121L)
22,000 psi (Hetron 197AT)
25,151 psi (Hetron 325FS)

Impact Str. (Izod):
3 cm kp/cm² (Leguval F30S)
9 cm kp/cm² (Leguval W20)

10 cm kp/cm² (Leguval F35, K25R, K26, N50, N50S, W16)
15 cm kp/cm² (Leguval W35)
18 cm kp/cm² (Leguval K41, W41)
20 cm kp/cm² (Leguval F33, N30)
16 J/m (Durez 24337)
17 J/m (Durez 23410)
18 J/m (Durez 24668)
19 J/m (Durez 24150)
21 J/m (Durez 27962)
24 J/m (Durez 29276)
32 J/m (Durez 31572)
42 J/m (Durez 31375)
0.29 ft lb/in. (inj. molded) (Plenco 1528)
0.45 ft lb/in. (inj. molded) (Plenco 1530)
0.47 ft lb/in. (inj. molded) (Plenco 1523 Black Pellets)
0.50 ft lb/in. (transfer molded) (Plenco 1529 Black)
2.34 ft lb/in. unnotched (Stypol 40-4033, 40-4074, 40-8006)
7.0 ft lb/in. (Hetron 197AT)
8.0 ft lb/in. (Hetron 325FS)
Impact Str. (Charpy):
5 mJ/mm² (Vestopal 221L)
6 mJ/mm² (Vestopal 120T)
8 mJ/mm² (Vestopal 121L)
9 mJ/mm² (Vestopal 123L)
10 mJ/mm² (Vestopal 120L, 141L, 150T, 152, 160, 160T)
11 mJ/mm² (Vestopal 140L, 250, 510, 511)
12 mJ/mm² (Vestopal 145T, 251, 252, 400, 530)
14 mJ/mm² (Vestopal 145)
15 mJ/mm² (Vestopal 110, 150, 150B, 151)
20 mJ/mm² (Vestopal 310)
Hardness:
Ball Indentation 140 MPa (Vestopal 310)
Ball Indentation 210 MPa (Vestopal 221L)
Ball Indentation 215 MPa (Vestopal 145T, 510, 511)
Ball Indentation 220 MPa (Vestopal 110, 140L, 141L, 145, 152, 160, 160T)
Ball Indentation 225 MPa (Vestopal 150, 150B, 151, 400, 730, 830)
Ball Indentation 230 MPa (Vestopal 150T, 530)
Ball Indentation 235 MPa (Vestopal 251, 252)
Ball Indentation 240 MPa (Vestopal 123L, 250)
Ball Indentation 245 MPa (Vestopal 120L, 120T, 121L)
Barcol 20 (Stypol 40-5732)
Barcol 35 (Crystic 387)
Barcol 38 (ANE-K10; Crystic 306; Crystic Gelcoat 33E)

Polyester, thermoset (cont'd.)

Barcol 39 (Aropol 7320)
Barcol 40 (Aropol 7229, 7329; Crystic 2-489PA; Crystic Fireguard 75PA; Crystic Gelcoat 68E, 68PA)
Barcol 40–42 (Aropol 7420)
Barcol 42 (Aropol 7532; Crystic 328PALV, 489PA; Crystic Gelcoat 65E)
Barcol 43 (Crystic 2-491PA, 302, 345PALV)
Barcol 44 (ANE-D30)
Barcol 45 (ANE-K20; Aropol 7131, 7141, 7221, 7241-12, 7362; Crystic 189LV, 189MV, 196, 196E, 272, 272E, 385E, 385PA, 392, 392E, 406PA, 471PALV, 491E, 491PA; Hetron 197ATStypol 40-7016)
Barcol 46 (Aropol 7020, 7030, 7241T-15, 7242T-15, 8319, 8329; Crystic 2-406PA)
Barcol 47 (ANE; Crystic 195; Stypol 40-4033, 40-4074)
Barcol 48 (Aropol 7021; Crystic 191LV, 193HR, 323PA, 17449)
Barcol 49 (Aropol 7240, 7240T-15, 7240WT-12, 7280, 7530T-16)
Barcol 50 (Aropol 8110, 8310, 8321, 8329-25, 8339T-12, 8349-15, 8420-09, 8343T-11, 8343T-12, 8343T-21, 8343T-22; Crystic 198, 405PA, 474PA, 625E, 625MV)
Barcol 51 (Aropol 8420; Crystic 346PA)
Barcol 55–60 (Hetron 325FS)

Mold Shrinkage:
0.004 in./in. (Durez 31375)
0.005 in./in. (Durez 31572)

Water Absorp.:
0.10% (Durez 31375)
0.15% (Durez 31572)

THERMAL PROPERTIES:

Soften. Pt. (Vicat):
100 C (ANE-K10)
108 C (ANE-K20)
110 C (ANE-D30)

Conduct.:
0.16 W/m K (Vestopal 110, 120L, 120T, 121L, 123L, 140L, 141L, 145, 145T, 150, 150B, 150T, 151, 152, 310, 320, 400, 510, 511, 530, 730, 830)
0.17 W/m K (Vestopal 160, 160T, 221L, 250, 251, 252)

Distort. Temp.:
45–50 C (Vestopal 310)
52 C (1.8 MPa) (Crystic 306; Crystic Gelcoat 33EStypol 40-5732)
55 C (1.8 MPa) (Crystic 345PALV, 387)
56 C (Aropol 8420-09); (1.8 MPa) (Crystic 189MV)
60 C (Aropol 8319, 8329); (1.8 MPa) (Crystic 189LV, 195, 2-406PA)
62 C (1.8 MPa) (Crystic 406PA)
64 C (1.8 MPa) (Crystic 193HR, 346PA)
65 C (Aropol 8329-25, 8339T-12, 8349-15, 8343T-11, 8343T-12, 8343T-21, 8343T-22); (1.8 MPa) (Crystic 302)

65–70 C (Vestopal 110, 120L, 121L, 140L, 141L)
66 C (1.8 MPa) (Crystic 191LV)
68 C (Aropol 7532)
69 C (1.8 MPa) (Crystic 405PA; Stypol 40-4033, 40-4074, 40-8006)
70 C (1.8 MPa) (Crystic 323PA, 385E, 385PA; Crystic Gelcoat 68E, 68PA)
70–75 C (Vestopal 120T, 123L, 145)
72 C (1.8 MPa) (Crystic 196E, 328PALV)
73 C (Aropol 7420, 8110, 8310); (1.8 MPa) (Crystic 196)
75 C (1.8 MPa) (Crystic 2-489PA, 2-491PA, 272, 272E, 471PALV, 489PA, 491E, 491PA; Crystic Gelcoat 65E)
75–80 C (Vestopal 250)
80 C (Aropol 7329, 7530T-16)
80–85 C (Vestopal 145T)
85–90 C (Vestopal 152, 251, 252)
87 C (Aropol 7320)
89 C (Aropol 8321)
90 C (Stypol 40-5727); (1.8 MPa) (Crystic 392, 392E)
90–95 C (Vestopal 221L)
94 C (1.8 MPa) (Crystic 625E, 625MV)
95 C (Stypol 40-7016)
95–100 C (Vestopal 160, 160T)
100–105 C (Vestopal 400)
103 C (Aropol 7229)
105 C (Vestopal 730, 830)
105–110 C (Vestopal 150, 150B, 151)
105–115 C (Vestopal 510, 511)
110 C (1.8 MPa) (Crystic 198, 474PA)
110–115 C (Vestopal 150T)
111 C (Aropol 7020, 7030)
115 C (1.8 MPa) (Crystic 17449)
121 C (Aropol 7240)
123 C (Aropol 7240T-15, 7240WT-12, 7280)
124 C (Aropol 7021)
125–130 C (Vestopal 530)
126 C (Aropol 7221, 7241-12, 7241T-15, 7242T-15)
133 C (Aropol 7131, 7141)
185 C (Aropol 7362)
191 C (264 psi) (Durez 23410, 24337)
204 C (264 psi) (Durez 29276)
218 C (264 psi) (Durez 24668, 27962)
232 C (264 psi) (Durez 24150, 31572)
246 C (264 psi) (Durez 31375)
375 F (264 psi) (Plenco 1528)

Polyester, thermoset (cont'd.)

500 F (264 psi) (Plenco 1529 Black, 1530)
530 F (264 psi) (Plenco 1523 Black Pellets)
Coeff. of Linear Exp.:
60×10^{-6} K^{-1} (Vestopal 530)
70×10^{-6} K^{-1} (Vestopal 400, 510, 511)
80×10^{-6} K^{-1} (Vestopal 150, 150B, 150T, 151, 221L, 730, 830)
90×10^{-6} K^{-1} (Vestopal 110, 120T, 121L, 123L, 145, 160, 160T, 310, 320)
94×10^{-6} K^{-1} (Vestopal 145T)
100×10^{-6} K^{-1} (Vestopal 120L, 152, 250, 251, 252)
110×10^{-6} K^{-1} (Vestopal 141L)
115×10^{-6} K^{-1} (Vestopal 140L)
Sp. Heat:
1.09 kJ/kgK (Vestopal 250)
1.13 kJ/kgK (Vestopal 221L)
1.17 kJ/kgK (Vestopal 150T, 251, 252)
1.20 kJ/kgK (Vestopal 120T, 160T, 510, 511)
1.26 kJ/kgK (Vestopal 152, 160)
1.30 kJ/kgK (Vestopal 121L, 145T, 150, 150B, 151, 730, 830)
1.35 kJ/kgK (Vestopal 123L)
1.38 kJ/kgK (Vestopal 141L, 145, 530)
1.45 kJ/kgK (Vestopal 110, 120L, 140L, 310, 320, 400)
Flamm.:
V-0 (Durez 24150, 27962, 31375)
V-0/V-1 (Durez 31572)
V-1 (Durez 29276)
HB (Durez 24337)
DOT rating—flammable liq. (Aropol 7221, 7280)
DOT rating 73–100 F—flammable liq. (Aropol 7131, 7240, 7240WT-12, 7242T-15, 7530T-16, 7532, 8339T-12, 8420, 8420-09, 8343T-11, 8343T-12, 8343T-21, 8343T-22; Hetron 197AT, 470P, 700DMA)
DOT rating 73–160 F—flammable liq. (Aropol 7240T-15)

ELECTRICAL PROPERTIES:
Dissip. Factor:
0.002 (800 Hz) (Vestopal 400)
0.003 (800 Hz) (Vestopal 120L)
0.0037 (800 Hz) (Vestopal 123L)
0.0038 (800 Hz) (Vestopal 121L)
0.0042 (800 Hz) (Vestopal 510, 511)
0.0048 (800 Hz) (Vestopal 120T, 152)
0.005 (800 Hz) (Vestopal 110, 140L, 141L, 145, 160, 160T, 250, 251, 252)
0.0052 (800 Hz) (Vestopal 150T)
0.006 (1000 Hz) (Crystic 196, 196E)
0.0064 (800 Hz) (Vestopal 150, 150B, 151)

0.007 (800 Hz) (Vestopal 310)
0.0085 (1 kHz) (Stypol 83-2000)
0.0094 (800 Hz) (Vestopal 221L)
0.010 (1 kHz) (Durez 31375, 31572); (1 mHz) (Vestopal 530); (1 MC) (Aropol 7362)
0.0107 (Aropol 7021)
0.018 (1 MHz) (Crystic 17449)

Dielec. Str.:
13.8 MV/m (Durez 23410, 24150, 24337, 24668, 31572)
15.7 MV/m (Durez 27962, 31375)
35 kV/mm (Vestopal 110, 120L, 120T, 121L, 123L, 145, 145T, 152, 160, 160T, 221L, 250, 251, 252, 530)
40 kV/mm (Vestopal 140L, 141L, 400, 510, 511)
45 kV/mm (Vestopal 150, 150B, 150T, 151)
300 V/mil (Plenco 1523 Black Pellets, 1528)
320 V/mil (Plenco 1529 Black)
330 V/mil (Plenco 1530)
350 V/mil (Impco RC-2, RC-2/140 Resins)
429 V/mil (Aropol 7362)
443 V/mil (Aropol 7021)

Dielec. Const.:
2.75 (1 MC) (Aropol 7362)
3.0 (800 Hz) (Vestopal 120L, 510, 511); (1000 Hz) (Crystic 196, 196E)
3.1 (800 Hz) (Vestopal 400, 530)
3.12 (1 MHz) (Crystic 17449)
3.2 (800 Hz) (Vestopal 120T)
3.3 (800 Hz) (Vestopal 121L, 123L, 141L, 152)
3.4 (50 Hz) (Vestopal 145); (800 Hz) (Vestopal 140L, 150T, 250, 251, 252)
3.5 (800 Hz) (Vestopal 150, 150B, 151, 160, 160T, 221L)
3.6 (1 kHz) (Stypol 83-2000)
3.7 (800 Hz) (Vestopal 110)
4.29 (Aropol 7021)
4.3 (800 Hz) (Vestopal 310)
5.3 (1 kHz) (Durez 31375)
5.6 (1 kHz) (Durez 31572)

Vol. Resist.:
10^{12} ohm-m (Durez 31572)
10^{13} ohm-m (Durez 31375)
1×10^{14} ohm-cm (Durez 23410, 24337, 27962, 29276)
1×10^{15} ohm-cm (Durez 24150, 24668)
> 10^{15} ohm-cm (ANE, -K10, -K20; Vestopal 110, 120L, 120T, 121L, 123L, 140L, 141L, 145, 145T, 150, 150B, 150T, 151, 152, 160, 221L, 250, 251, 252, 310, 320, 400, 510, 511, 530)

Polyester, thermoset *(cont'd.)*

Surf. Resist.:
> 10^{13} ohm (Vestopal 110, 120L, 120T, 121L, 123L, 140L, 141L, 145, 145T, 150, 150B, 150T, 151, 152, 160, 221L, 250, 251, 252, 310, 320, 400, 510, 511, 530)
> 10^{15} ohm (ANE, -K10, -K20)

Arc Resist.:
126 s (Aropol 7362)
190 s (Durez 31572)
205 s (Durez 31375)

CURING CHARACTERISTICS:

Promoted:
Dion Cor-Res 6693FR, 6694, 7000; Dion FR 6114, FR 6124, FR 6125, FR 6127, FR 6604T, FR 6657, FR 6692T; Dion-Iso 6631T; Hetron 470P; Polylite 32-050, 32-180, 32-300, 32-345, 32-356, 32-357, 32-358, 32-367, 32-739, 32-740, 32-773, 33-031, 33-049, 33-067, 33-071, 33-072, 33-081, 33-084, 33-086, 33-089, 33-092, 33-401, 33-402, 33-404, 33-441, 33-450, 33-770, 33-773

Prepromoted:
Aropol 7240T-15, 7240WT-12, 7241-12, 7241T-15, 7242T-15, 7530T-16, 8329-25, 8339T-12, 8349-15, 8420-09, 8343T-11, 8343T-12, 8343T-21, 8343T-22; Stypol 40-4033, 40-4074, 40-5732, 40-8006

Preaccelerated:
Crystic 113PA, 196PALV, 2-406PA, 2-489PA, 2-491PA, 323PA, 328PA, 328PALV, 345PA, 345PALV, 346PA, 385PA, 405PA, 406PA, 471PALV; Leguval K25R (amine accelerator), K26 (amine accelerator), K41 (amine accelerator), K70 (amine acclerator), 474PA, 489PA, 491PA; Crystic Gelcoat 39PA, 46PA, 47PA, 65PA, 68PA; Leguval K25R (amine accelerator), K26 (amine accelerator), K41 (amine accelerator), K70 (amine accelerator); Vestopal 150B

Typical Cure:
Ambient temp. gel and cure (Polylite 32-050)
Cured at either elevated or room temps. (Hetron 32A)
Cured by exposure to uv radiation with addition of suitable photoinitiatior (e.g., alkyl benzoin ethers) (Chempol 19-4807)
Filled, with Lupersol DDM (1 phr) (Hetron 325FS)
Std. 180 F bath—1% benzoyl peroxide (Aropol 7020, 7021, 7030, 7141, 7221, 7229, 7240, 7280, 7320, 7329, 7362, 7420, 8110, 8310, 8319, 8321, 8329, 8420)
Std. 180 F bath—1% benzoyl peroxide, 20% styrene (Aropol 8101)
With MEK peroxide (1%) (Hetron 470P)
R.T. cure—1% MEK peroxide (Hetron 27196; Stypol 40-4033, 40-4074, 40-8006)
R.T. cure—1% MEK peroxide (60%) (Aropol 7240T-15, 7240WT-12, 7241-12, 7241T-15, 7242T-15, 7530T-16, 8329-25, 8339T-12, 8349-15, 8420-09, 8343T-11, 8343T-12, 8343T-21, 8343T-22)
R.T. cure—1% MEK peroxide (60%), 0.5% cobalt octoate (6%) (Aropol 7131)
R.T. cure—1% Lupersol DDM (11% act.), 0.5% of 6% cobalt octoate (Aropol 7532)
R.T. cure @ 100 pbw with Catalyst M (1 pbw) (Crystic 2-406PA, 2-489PA, 489PA)

R.T. cure @ 100 pbw with Catalyst M (1 pbw), Accelerator E (3 pbw) (Crystic 385E)

R.T. cure @ 100 pbw with Catalyst M (2 pbw) (Crystic 196PALV, 2-491PA, 323PA, 328PALV, 345PALV, 346PA, 385PA, 405PA, 406PA, 471PALV, 474PA, 491PA; Crystic Fireguard 75PA; Crystic Gelcoat 68PA)

R.T. cure @ 100 pbw with Catalyst M (2 pbw), Accelerator E (1 pbw) (Crystic 405E)

R.T. cure @ 100 pbw with Catalyst M (2 pbw), Accelerator E (4 pbw) (Crystic 195, 196E, 272E, 392, 625MV; Crystic Gelcoat 33E, 65E, 68E)

R.T. cure @ 100 pbw with Catalyst O (4 pbw), Accelerator G (3 pbw) (Crystic 17449)

R.T. cure @ 100 pbw with Catalyst Paste H (4 pbw), Accelerator E (4 pbw) (Crystic 189LV, 189MV, 191LV, 193HR, 196, 198, 272, 302, 306, 387, 491)

35 C @ 100 pbw, Catalyst M (1 pbw), Accelerator E (3 pbw) (Crystic 491E, 625E)

Pot Life:

10 min (Crystic Fireguard 75PA)

Gel Time:

< 1 min (149 C) (Magnamite AS4/1655)

2 min (180 F) (Dion FR 6114)

3 min (Aropol 7420; Dion-Iso 8200); (150–190 F) (Aropol 7280); (180 F) (Dion FR 6124)

3.25 min (150–190 F) (Aropol 7021)

3.5 min (Aropol 7221)

4 min (Dion FR 8300; Dion-Iso 8101); (150–190 F) (Aropol 8321)

4–5 min (180 F) (Dion FR 6125, FR 6127)

4–6 min (Polylite 31-006, 31-039); (150–190 F) (Aropol 7020)

5 min (Aropol 7240, 7320, 8420); (150–190 F) (Aropol 7030, 8310, 8319, 8329)

5–7 min (Polylite 31-447)

5.5 min (150–190 F) (Aropol 8110)

5.5–7.5 min (Polylite 31-451)

6 min (Aropol 7141, 7362; Hetron 32A)

6–9 min (Polylite 31-571)

7 min (Aropol 7131; Stypol 40-8006)

7–9 min (Polylite 31-546, 31-830)

7–10 min (Polylite 31-585)

8 min (Aropol 7532; Crystic 196, 272); (180 F) (Dion FR 6308)

8–10 min (Polylite 31-402, 32-120)

8–12 min (Polylite 32-039, 32-128, 32-129)

8.5 min (Crystic 323PA)

9 min (Aropol 8420-09; Crystic 189LV, 189MV, 392, 491); (180 F) (Dion FR 6399)

10 min (Aropol 7229; Crystic 385PA, 471PALV, 625MV; Crystic Gelcoat 68E, 68PA); (150–190 F) (Aropol 8101); (R.T.) (Dion-Iso 6246, 6315HV)

10–15 min (Polylite 31-450, 32-037, 32-122, 32-125)

10–19 min (Stypol 40-4074)

10.7 min (Stypol 83-2000)

11 min (Aropol 8343T-11; Crystic 193HR, 387)

11–13 min (Aropol 7241-12)

11–14 min (Polylite 31-587)

12 min (Aropol 7240WT-12, 8339T-12, 8343T-12; Crystic 198, 2-491PA, 474PA, 491PA); (180 F) (Dion Cor-Res 6693HV; Dion-Iso 6314)

12–18 min (Polylite 32-112)

12–20 min (Polylite 34-308)

14 min (Crystic 196PALV; Crystic Gelcoat 33E, 65E)

14–18 min (Polylite 32-111)

15 min (Aropol 7240T-15, 7241T-15, 7242T-15, 7329, 8349-15; Crystic 196E; Hetron 470P); (R.T.) (Dion Cor-Res 6694; Dion-Iso 6631T)

15–18 min (Polylite 32-127)

15–19 min (Hetron 325FS; Stypol 40-4033)

15–20 min (Polylite 32-050)

16 min (Aropol 7530T-16; Crystic 2-406PA, 406PA); (@ 180 F) (Dion Cor-Res 6694HV); (R.T.) (Dion FR 6657)

17 min (Crystic 191LV, 302)

18 min (Crystic 272E, 328PALV, 345PALV, 346PA, 405PA; Hetron 27196)

20 min (@ R.T.) (Dion Cor-Res 6693FR; Dion FR 6604T, FR 6655T, FR 6692T)

20–25 min (Polylite 32-033, 32-035)

21 min (Aropol 8343T-21)

22 min (Aropol 8343T-22)

23 min (Crystic 2-489PA, 306, 489PA)

25 min (Aropol 8329-25; Crystic 195); (@ R.T.) (Dion Cor-Res 7000)

25–35 min (Polylite 32-032)

30 min (Crystic 385E, 392E, 491E, 625E)

30–150 min (55% in styrene sol'n.) (ANE-D30, -K20)

30–180 min (55% in styrene sol'n.) (ANE, -K10)

38 min (Crystic 17449)

45 min (Crystic 405E)

Cure Time:

3–5 min (Polylite 31-402)

5.5–8.5 min (Polylite 31-006)

5.5–9.0 min (Polylite 31-039)

6–9 min (Polylite 31-447)

6.5–9 min (Polylite 31-451)

7–11 min (Polylite 31-571)

8–11 min (Polylite 31-585)

9–12 min (Polylite 31-546)

10–13 min (Polylite 31-830)

12–16 min (Polylite 31-587)

12–18 min (Polylite 31-450)

Total time to peak exotherm:

4 min (Aropol 7280, 7420)

4.3 min (Aropol 7021)
4.75 min (Aropol 7221)
5.0–7.5 min (Aropol 7020)
6.0 min (Aropol 7030, 7320, 8321)
7 min (Aropol 7141, 7240, 8310)
7.5 min (Aropol 8110, 8319, 8329, 8420)
8.3 min (Hetron 32A)
8.5 min (Aropol 7362)
13 min (Aropol 7229)
14.5 min (Stypol 83-2000)
15 min (Aropol 7131)
17–23 min (Aropol 7241-12)
17.5 min (Aropol 7329)
18 min (Aropol 7532)
19 min (Aropol 8101)
20 min (Aropol 8420-09)
21 min (Aropol 7240WT-12, 8339T-12, 8343T-12)
23 min (Aropol 8349-15, 8343T-11)
28 min (Aropol 7240T-15, 7241T-15, 7242T-15)
29 min (Aropol 7530T-16)
31 min (Aropol 8343T-22)
33 min (Aropol 8343T-21)
35 min (Aropol 8329-25)
Peak Exotherm:
210 F (Aropol 8101)
310–330 F (Polylite 31-830)
317 F (Aropol 7530T-16, 7532)
318 F (Aropol 8420-09)
330 F (Aropol 8343T-22)
332 F (Aropol 8339T-12, 8343T-12, 8343T-21)
335 F (Aropol 8319, 8343T-11)
348 F (Aropol 8349-15)
355 F (Aropol 8420)
356 F (Aropol 8329-25)
360–390 F (Aropol 7241-12)
365 F (Aropol 7362)
365–385 F (Polylite 31-546)
375 F (Aropol 8110)
380 F (Aropol 7242T-15)
380–400 F (Polylite 31-451)
383 F (Aropol 7240T-15, 7241T-15)
385 F (Aropol 8329)
390 F (Aropol 8321; Hetron 32A)

Polyester, thermoset *(cont'd.)*

390–410 F (Polylite 31-587)
392 F (Aropol 7240WT-12)
395 F (Aropol 7329)
395–415 F (Polylite 31-006)
400 F (Aropol 8310)
400–420 F (Polylite 31-585)
400–430 F (Polylite 31-402)
405 F (Aropol 7131)
408 F (Stypol 83-2000)
415–435 F (Polylite 31-039)
420 F (Aropol 7221)
420–440 F (Polylite 31-447)
425 F (Aropol 7420)
425–450 F (Polylite 31-571)
426 F (Aropol 7229)
430 F (Aropol 7320)
430–460 F (Aropol 7020)
435–455 F (Polylite 31-450)
445 F (Aropol 7030, 7240)
455 F (Aropol 7141)
460 F (Aropol 7021)
475 F (Aropol 7280)

TOXICITY/HANDLING:

Dust from machining laminates may be harmful if inhaled (Crystic 323PA, 328PALV, 345PALV, 346PA, 387; Crystic Prefil F)

Nontoxic in liquid form when used in ventilated areas; inert when cured (Impco RC-2, RC-2/140 Resins)

Observe normal safety precautions for handling polyester resin; use in adequately ventilated areas; avoid prolonged/repeated skin contact (Magnamite AS4/1655)

STORAGE/HANDLING:

Store and use at R.T. (Impco RC-2, RC-2/140 Resins)

Avoid exposure to heat and moisture in storage (Aropol 7020, 7021, 7030, 7131)

Store < 80 F; avoid exposure to heat and moisture (Aropol 7141, 7221, 7229, 7240, 7241-12, 7241T-15, 7320, 7329, 7362, 7420, 7532, 8110, 8310, 8319, 8321, 8329, 8329-25, 8349-15, 8420-09; Hetron 32A, 197AT, 27196)

Store < 80 F; avoid exposure to heat and moisture; settling may occur (Aropol 7240T-15, 7240WT-12, 7242T-15, 7530T-16, 8339T-12, 8343T-11, 8343T-12, 8343T-21, 8343T-22)

Store < 100 F; avoid exposure to heat and moisture (Aropol 8101)

Use caution in opening drums due to possible pressure build-up from acetone (Aropol 7280)

Store in closed containers ≤ 38 C; avoid exposure to direct sunlight (Chempol 19-4807)

Keep away from flames (Crystic 57)

Polyester, thermoset *(cont'd.)*

Highly flammable; keep away from naked flames; recommended storage ≤ 20 C (Crystic 189LV, 189MV, 191LV, 193HR, 195, 196, 198, 2-406PA, 2-489PA, 2-491PA, 272, 302, 306, 323PA, 328PALV, 385PA, 392, 387, 405PA, 406PA, 471PALV, 474PA, 489PA, 491, 491PA, 625MV, 17449; Crystic Gelcoat 68PA; Crystic Prefil F)

Flammable; store in the dark in closed containers ≤ 25 C (Crystic 196E, 272E, 385E, 392E, 405E, 491E, 625E; Crystic Gelcoat 33E, 65E, 68E)

Flammable; keep away from naked flames; store ≤ 20 C; this filled resin is prone to separation and settling on storage; mix thoroughly before use (Crystic 345PALV, 346PA)

Flammable; keep away from naked flames; store ≤ 20 C in tightly closed containers (Crystic Pregel 17, 27)

Flammable; keep away from naked flames; store ≤ 20 C; resin must be stirred well before adding catalyst (Crystic Fireguard 75PA)

Liquid base is flammable; keep away from naked flames; store ≤ 20 C (Crystic Stopper)

Store < 77 F; avoid exposure to heat and moisture; store in stainless steel or epoxy- or phenolic-lined tanks for bulk quantities (Hetron 325FS)

Store < 80 F; avoid exposure to heat and moisture; some thixotrope settling may occur; store bulk in stainless steel or epoxy- or phenolic-lined tanks (Hetron 470P)

May become cloudy and crystallize when stored for several weeks at 15 C or below—stir and warm before use to restore reasonable homogeneity (Paraplex G-60)

Allow material to reach ambient temps. when removing from 40 F storage; protect from contamination by grease, dust, and dirt; dry-ice refrigeration not required for shipping (Magnamite AS4/1655)

STD. PKGS.:

5-gal (50 lb net) pails, 55-gal (550 lb net) drums (Hetron 32A)

55-gal steel drums (Hetron 325FS)

55-gal (450 lb net) drums (Hetron 700DMA)

55 gal (500 lb net) drums (Aropol 7020, 7021, 7030, 7131, 7141, 7221, 7229, 7240, 7240T-15, 7240WT-12, 7241-12, 7241T-15, 7242T-15, 7280, 7320, 7329, 7362, 7420, 7530T-16, 7532, 8101, 8110, 8310, 8319, 8321, 8329, 8329-25, 8339T-12, 8349-15, 8420, 8420-09, 8343T-11, 8343T-12, 8343T-21, 8343T-22)

55-gal (550 lb net) drums (Hetron 197AT)

55-gal (575 lb net6) drums (Hetron 27196)

25-kg containers (Crystic Stopper)

25-kg and 220-kg steel containers (Crystic Gelcoat 68E)

25-kg and 225-kg steel containers (Crystic Gelcoat 33E, 65E, 68PA; Crystic Prefil F; Crystic Pregel 17, 27)

25-kg and 225-kg steel containers and bulk (Crystic 191LV, 193HR, 195, 196, 198, 272, 302, 306, 323PA, 328PALV, 345PALV, 346PA, 405PA, 406PA, 471PALV, 474PA, 491PA, 625MV)

200-kg containers (Crystic 57)

200-kg steel containers and bulk (Crystic 2-406PA, 2-489PA, 2-491PA, 489PA)

Polyester, thermoset *(cont'd.)*

225-kg steel containers (Crystic 196E, 272E, 385E, 405E, 491E, 625E; Crystic Fireguard 75PA)

22.7-kg rolls in sealed polyethylene bags and cardboard containers (Magnamite AS4/1655)

Polyimide resin

STRUCTURE:
Thermoplastic polyimide:

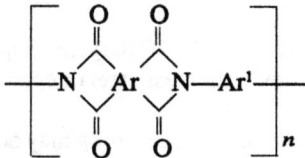

Thermoset polyimide

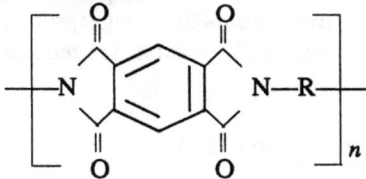

TRADENAME EQUIVALENTS:
Thermoplastic polyimide:
Envex 1000, 1115, 1228, 1315, 1330, 3540 [Rogers Corp.]
Glass-reinforced:
Envex 1620 [Rogers Corp.]
Thermoset polyimide:
Kapton Type F, H, V [DuPont]
PI-730 [Fiberite]
Thermid AL-600, LR-600, MC-600 [Nat'l. Starch & Chem.]
Glass-reinforced:
PI-740 (chopped glass fibers), -750 ($^{1}/_{2}$ in. chopped glass fibers) [Fiberite]

MODIFICATIONS/SPECIALTY GRADES:
Film:
Kapton Type H, V

Film coated with Teflon FEP:
  Kapton Type F
Acetylenic-capped:
  Thermid AL-600, LR-600, MC-600
Flame-retardant:
  Kapton Type H
Graphite-lubricated:
  Envex 1315, 1330, 3540
MoS₂-lubricated:
  Envex 1115
PTFE-lubricated:
  Envex 1228
CATEGORY:
  Thermoplastic resin, thermosetting resin
PROCESSING:
Injection molding:
  PI-730
Compression molding:
  PI-740, -750; Thermid MC-600
Lamination:
  Kapton Type F, H, V; Thermid AL-600
Can be metalized, punched, formed, or adhesive coated:
  Kapton Type F, H, V
Transfer molding:
  PI-740
APPLICATIONS:
  Automobile industry: engine components (PI-730)
  Aviation industry: jet engine components (PI-730)
  Electrical/electronic industry: coil insulation (Kapton Type F, H, V); connectors (PI-730); magnet wire insulation (Kapton Type F, H, V); magnetic and pressure-sensitive tapes and tubing (Kapton Type F, H, V); motor slot liner (Kapton Type F, H, V); printed circuit substrate (Kapton Type F, H, V); transformer and capacitor insulation (Kapton Type F, H, V); wire and cable tapes (Kapton Type F, H, V)
  Industrial applications: adhesives (Thermid LR-600); bearings (Envex 1228, 1315, 1330; PI-730); bushings (Envex 1228, 1315); carriage bearings (Envex 1228); coatings (Thermid AL-600, LR-600); composite prepregs (Thermid AL-600, LR-600); gaskets (Envex 1000); insulators (Envex 1000); laminates (Thermid AL-600); neat and composite molding (Thermid MC-600); piston rings (Envex 1228); sealing rings (Envex 1228); seals (Envex 1000, 1115, 1315, 1330; PI-730); self-lubricating applications: (Envex 1115, 1315); structural parts/applications (Envex 1000, 1115, 1315, 1620, 3540); vacuum bearings (Envex 1115); valve seats (Envex 1000; PI-730); welding torch components (PI-730)

Polyimide resin *(cont'd.)*

Nuclear industry: nuclear components (Envex 1115); nuclear reactor parts (PI-730)
PROPERTIES:
Form:
Liquid (Thermid AL-600, LR-600)
Powder (Thermid MC-600
Solid (Envex 1115, 1228, 1315, 1330, 3540)
Solid; avail. in assorted stock shapes and as molded and/or machined parts (Envex
    1000)
Film avail. in a variety of constructions, e.g., 30-μm film with 2.5-μm Teflon FEP
    coatings around 25-μm Kapton Type H (Kapton Type F)
Film avail. 1-, 2-, 3-, and 5-mil widths (Kapton Type H)
Film avail. in 2-, 3-, and 5-mil widths (Kapton Type V)
Color:
Amber (Thermid AL-600, LR-600)
Tan (Thermid MC-600)
Composition:
50% solids in N-methyl pyrrolidone (Thermid LR-600)
75% solids in ethanol (Thermid AL-600)
80% polyimide and 20% FEP (Kapton Type F)
100% polyimide (Kapton Type H, V)
GENERAL PROPERTIES:
Solubility:
Sol. in ethanol (Thermid AL-600)
Sol. in ketones (Thermid AL-600, LR-600)
Sol. in methylene chloride (Thermid AL-600, LR-600)
Sol. in N-methyl pyrrolidone (Thermid LR-600, MC-600)
Sol. in THF (Thermid AL-600, LR-600)
M.W.:
1099 (Thermid MC-600)
1171 (Thermid LR-600)
1355 (Thermid AL-600)
Sp. Gr.:
1.19 (Thermid AL-600, LR-600)
1.33 (Envex 1000)
1.37 (Thermid MC-600)
1.4 (Envex 1315)
1.5 (Envex 1115)
1.51 (Envex 1330)
1.52 (Envex 1228)
1.85 (PI-730)
1.90 (PI-740)
1.95 (PI-750)

Density:
0.00153 kg/m³ (Kapton Type F)
1.42 g/cm³ (Kapton Type H)

Visc.:
50,000 cps (Thermid AL-600)
100,000 cps (Thermid LR-600)

M.P.:
None (Kapton Type H)
374–410 F (Thermid MC-600)

Stability:
High degree of chemical resistance to a wide variety of chemicals incl. solvents and oils, e.g., benzene, toluene, methanol, acetone, transformer oil, p-cresol (Kapton Type F, H, V)

Resistant to oils and solvents, e.g,, jet fuels, kerosene, toluene, freon, and trichloroethylene; unaffected by dilute acid sol'ns.; good thermal oxidative stability (PI-730, –740, -750)

Storage Stability:
6+ mos. shelf life @ 40 F (Thermid AL-600, LR-600)
> 1 yr shelf life at ambient temps. (PI-730, -740, -750)
Indefinite shelf life (Thermid MC-600)

Ref. Index:
1.78 (Becke Line) (Kapton Type H)

MECHANICAL PROPERTIES:

Tens. Str.:
22.8 MPa (Envex 1228)
55.2 MPa (Envex 1330)
55.3 MPa (Envex 1115)
65.5 MPa (Envex 1315)
69.1 MPa (Envex 3540)
88.3 MPa (Envex 1000)
165 MPa (Kapton Type V); (break) (Kapton Type F)
172 MPa (break) (Kapton Type H)
8000 psi (PI-730)
12,000 psi (neat resin) (Thermid AL-600, LR-600, MC-600)
15,000 psi (PI-740)
21,000 psi (PI-750)

Tens. Elong.:
1.3% (break) (Envex 3540)
1.7% (break) (Envex 1228)
2% (break) (neat resin) (Thermid AL-600, LR-600, MC-600)
2.4% (break) (Envex 1115)
3.0% (break) (Envex 1315)
3.6% (break) (Envex 1330)

Polyimide resin *(cont'd.)*

   5.1% (break) (Envex 1000)
   65% (break) (Kapton Type F)
   70% (Kapton Type V); (break) (Kapton Type H)
Tens. Mod.:
   2069 MPa (Envex 1228)
   2593 MPa (Envex 1000)
   2855 MPa (Envex 1330)
   3276 MPa (Envex 1315)
   3448 MPa (Envex 1115)
   13,861 MPa (Envex 3540)
   2.86 GPa (Kapton Type F)
   3.0 GPa (Kapton Type H)
   570,000 psi (neat resin) (Thermid AL-600, LR-600, MC-600)
Flex. Str.:
   51.7 MPa (Envex 1228)
   101.4 MPa (Envex 1330)
   110 MPa (Envex 1115)
   118.6 MPa (Envex 1315)
   124 MPa (Envex 3540)
   164.8 MPa (Envex 1000)
   10,000 psi (PI-730)
   19,000 psi (neat resin) (Thermid AL-600, LR-600, MC-600)
   36,000 psi (PI-740)
   36,000 psi (PI-750)
Flex. Mod.:
   2965 MPa (Envex 1228)
   3055 MPa (Envex 1000)
   4000 MPa (Envex 1315)
   4448 MPa (Envex 1330)
   4634 MPa (Envex 1115)
   14,895 MPa (Envex 3540)
   650,000 psi (neat resin) (Thermid AL-600, LR-600, MC-600)
   2×10^6 psi (PI-730)
   2.8×10^6 psi (PI-740)
   3.1×10^6 psi (PI-750)
Compr. Str.:
   90.3 MPa (Envex 1228)
   165.5 MPa (Envex 1330)
   193 MPa (Envex 1115)
   204 MPa (Envex 1000)
   206.9 MPa (Envex 1315)
   244.1 MPa (Envex 3540)
   25,000 psi (neat resin) (Thermid AL-600, LR-600, MC-600)

32,000 psi (PI-750)
34,000 psi (PI-740)
35,000 psi (PI-730)
Compr. Mod.:
2400 MPa (Envex 1228)
3227 MPa (Envex 1000)
3310 MPa (Envex 1330)
3647 MPa (Envex 1115)
3751 MPa (Envex 1315)
7034 MPa (Envex 3540)
Tear Str. (Elmendorf):
0.39 g/μm (Kapton Type F)
8 g (Kapton Type H)
Impact Str. (Izod):
70 g • m (Kapton Type F)
23 J/mm (Kapton Type H)
0.4 ft lb/in. notched (Izod) (PI-730)
7 ft lb/in. notched (PI-740)
22 ft lb/in. notched (PI-750)
Hardness:
Rockwell M93 (Envex 1228)
Rockwell M103 (Envex 1330)
Rockwell M111 (Envex 1115, 1315)
Rockwell M117 (PI-730)
Rockwell M118 (PI-740, -750)
Rockwell M122 (Envex 1000)
Shore D91 (neat resin) (Thermid AL-600, LR-600, MC-600)
Water Absorp.:
1% wt. gain (1000 h, 50 C, 95% r.h.) (neat resin) (Thermid AL-600, LR-600, MC-600)

THERMAL PROPERTIES:
Conduct.:
0.155 w/m • K (Kapton Type H)
9.0×10^{-4} Cal/s/cm²/C/cm (PI-730, -740)
9.1×10^{-4} Cal/s/cm²C/cm (PI-750)
Distort. Temp.:
260 C (Envex 3540)
288 C (Envex 1000, 1115, 1228, 1315, 1330)
600+ F (264 psi) (PI-730, -740, -750)
Coeff. of Linear Exp.:
1.0×10^{-5} in./in./C (PI-750)
1.5×10^{-5} in./in./C (PI-740)
2.0×10^{-5} in./in./C (Kapton Type H)
2.7×10^{-5} in./in./C (PI-730)

Polyimide resin *(cont'd.)*

    13×10^{-6} in./in.C (Envex 3540)
    45×10^{-6} in./in./C (Envex 1330)
    46.8×10^{-6} in./in./C (Envex 1315)
    54×10^{-6} in./in./C (Envex 1000)
    59.4×10^{-6} in./in./C (Envex 1228)
    64.8×10^{-6} in./in./C (Envex 1115)
    4.42×10^{-5} in./in./F (neat resin) (Thermid AL-600, LR-600, MC-600)

Sp. Heat:
    0.261 Cal/g/C (Kapton Type H)

Flamm.:
    VTM-0 (Kapton Type H)
    VEO (PI-730, -740, -750)

ELECTRICAL PROPERTIES:

Dissip. Factor:
    0.0022 (Kapton Type F)
    0.0025 (1 kHz) (Kapton Type H)
    0.0025 (1 kHz) (Kapton Type V)
    0.003 (1 MC) (PI-730, -740)
    0.004 (1 MC) (PI-750)

Dielec. Str.:
    12.3 MV/m (Envex 1000)
    12.9 V/mil (Envex 1228)
    213 V/μm (Kapton Type V)
    267 V/μm (Kapton Type F)
    276 V/μm (Kapton Type H)
    400 V/M (PI-730)
    450 V/mil (PI-740, -750)

Dielec. Const.:
    2.8 (Kapton Type F)
    3.5 (1 kHz) (Kapton Type H)
    3.6 (1 kHz) (Kapton Type V)
    4.7 (1 MC) (PI-730)
    4.8 (1 MC) (PI-740)
    5.1 (1 MC) (PI-750)
    5.38 (10 MHz) (neat resin) (Thermid AL-600, LR-600, MC-600)

Vol. Resist.:
    1.5×10^{14} ohm-m (Kapton Type F)
    8×10^{15} ohm-m (Kapton Type V)
    10^{16} ohm-cm (Kapton Type H)
    1.4×10^{16} ohm-cm (Envex 1000)
    1.5×10^{16} ohm-cm (Envex 1228)

Arc Resist.:
    125 s (PI-730, -740, -750)

Polyimide resin (cont'd.)

CURING CHARACTERISTICS:
Suggested/typical cure:
  Post cure 24 h @ 480 F (PI-730, -740, -750)
STORAGE/HANDLING:
  Hygroscopic—keep containers closed when not in use (PI-730, -740, -750)

Polyphenylene sulfide

SYNONYMS:
  PPS
STRUCTURE:

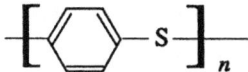

TRADENAME EQUIVALENTS:
  Fortron 0214P1, 0205P4 [Hoechst-Celanese]
  Liquinite PPS [LNP]
  RTP 1378 [Fiberite]
  Ryton R-9 901, R-11 70, V-1 [Phillips]
Carbon-reinforced:
  EMI-X OC-1008 (40% carbon fiber), OC-100-10 (50% carbon fiber) [LNP]
  RTP 1383 (20% carbon fiber), 1385 (30% carbon fiber), 1387 (40% carbon fiber) [Fiberite]
  Stat-Kon OC-1003 (15% carbon fiber), OC-1006 (30% carbon fiber), OCL-4036 (30% carbon fiber) [LNP]
  Thermocomp OC-1006 (30% carbon fiber), OCL-4036 (30% carbon fiber) [LNP]
Glass-reinforced:
  Fortron 1140B1 (40% fiberglass), 1140B4 (40% fiberglass) [Hoechst-Celanese]
  RTP 1301 (10% glass fibers), 1303 (20% glass fibers), 1305 (30% glass fibers), 1307 (40% glass fibers) [Fiberite]
  Ryton R-3 02 Black, R-4 (40% glass fiber), R-5 [Phillips]
  SF-30GR (30% glass fiber), SF-40GF (40% glass) [Compounding Technology]
  Thermocomp OF-1006 (30% fiber glass), OF-1008 (40% fiber glass), OFL-4036 (30% fiber glass) [LNP]
Glass/calcium carbonate-filled:
  Fortron 6165B4 (65% $CaCO_3$/fiberglass) [Hoechst-Celanese]
Glass/mineral-reinforced:
  RTP 1379, 1379S [Fiberite]

269

Polyphenylene sulfide (cont'd.)

Ryton R-8, R-10 Avocado R-4004B, R-10 Black 5002C and 5004A, R-10 Copper R-2005B, R-10 Gold R-1001B, R-10 Gray R-5000B, R-10 Green R-4000B, R-10 Lt. Green R-7002B, R-10 Natural 7006A and 7007A, R-10 Tan R-7002B [Phillips]

Glass/talc-filled:

Fortron 6265B4 (65% talc/fiberglass) [Hoechst-Celanese]

Nickel-reinforced:

PDX-83238 (40% nickel) [LNP]

MODIFICATIONS/SPECIALTY GRADES:

Conductive grade (EMI shielding):

EMI-X OC-1008, OC-100-10; PDX-83238; RTP 1383, 1385, 1387

Statically dissipative:

Stat-Kon OC-1003, OC-1006, OCL-4036

Flame-retardant:

EMI-X OC-1008, OC-100-10; Fortron 1140B1; PDX-83238; RTP 1301, 1303, 1305, 1307, 1378, 1383, 1385, 1387; Ryton R-3 02 Black, R-4, R-5, R-8, R-9 901, R-10 Avocado R-4004B, R-10 Black 5002C and 5004A, R-10 Copper R-2005B, R-10 Gold R-1001B, R-10 Gray R-5000B, R-10 Green R-4000B, R-10 Lt. Green R-7002B, R-10 Natural 7006A and 7007A, R-10 Tan R-7002B, R-11 70; SF-30GR, SF-40GF; Liquinite PPS; Stat-Kon OC-1003, OC-1006, OCL-4036; Thermocomp OC-1006, OCL-4036, OF-1006, OF-1008, OFL-4036

PTFE/TFE-lubricated:

Stat-Kon OCL-4036 (15% PTFE); Thermocomp OCL-4036 (15% PTFE), OFL-4036 (15% TFE)

CATEGORY:

Engineering thermoplastic resin

PROCESSING:

Compression molding:

Ryton V-1

Injection molding:

Fortron 0205P4, 0214P1, 1140B1, 1140B4, 6165B4, 6265B4; RTP 1301, 1303, 1305, 1307, 1378, 1379, 1379S, 1383, 1385, 1387; Ryton R-3 02 Black, R-4, R-5, R-8, R-10 Avocado R-4004B, R-10 Black 5002C and 5004A, R-10 Copper R-2005B, R-10 Gold R-1001B, R-10 Gray R-5000B, R-10 Green R-4000B, R-10 Lt. Green R-7002B, R-10 Natural 7006A and 7007A, R-10 Tan R-7002B, R-11 70

Potting/encapsulation:

Ryton R-3 02 Black, R-9 901

APPLICATIONS:

Automotive applications: (Ryton R-3 02 Black); fuel line systems (Fortron 0205P4, 0214P1, 1140B1, 1140B4, 6165B4, 6265B4); under-the-hood parts (Fortron 0205P4, 0214P1, 1140B1, 1140B4, 6165B4, 6265B4; Ryton R-4, R-5, R-8)

Aviation industry: avionics housings (EMI-X OC-1008, OC-100-10; PDX-83238)

Consumer products: appliances (Ryton R-3 02 Black, R-4, R-5, R-8, R-10 Avocado R-

4004B, R-10 Black 5002C and 5004A, R-10 Copper R-2005B, R-10 Gold R-1001B, R-10 Gray R-5000B, R-10 Green R-4000B, R-10 Lt. Green R-7002B, R-10 Natural 7006A and 7007A, R-10 Tan R-7002B); microwave ovenware (Ryton R-11 70)

Electrical/electronic industry: (Fortron 0205P4, 0214P1, 1140B1, 1140B4, 6165B4, 6265B4; EMI-X OC-1008, OC-100-10; PDX-83238; Ryton R-3 02 Black, R-4, R-5, R-8, R-10 Avocado R-4004B, R-10 Black 5002C and 5004A, R-10 Copper R-2005B, R-10 Gold R-1001B, R-10 Gray R-5000B, R-10 Green R-4000B, R-10 Lt. Green R-7002B, R-10 Natural 7006A and 7007A, R-10 Tan R-7002B); business machines and office equipment (EMI-X OC-1008, OC-100-10; PDX-83238); coils (Ryton R-10 Avocado R-4004B, R-10 Black 5002C and 5004A, R-10 Copper R-2005B, R-10 Gold R-1001B, R-10 Gray R-5000B, R-10 Green R-4000B, R-10 Lt. Green R-7002B, R-10 Natural 7006A and 7007A, R-10 Tan R-7002B); components (Stat-Kon OC-1003, OC-1006, OCL-4036); connectors (Fortron 0205P4, 0214P1, 1140B1, 1140B4, 6165B4, 6265B4; RTP 1301, 1303, 1305, 1307; Ryton R-3 02 Black, R-10 Avocado R-4004B, R-10 Black 5002C and 5004A, R-10 Copper R-2005B, R-10 Gold R-1001B, R-10 Gray R-5000B, R-10 Green R-4000B, R-10 Lt. Green R-7002B, R-10 Natural 7006A and 7007A, R-10 Tan R-7002B); electronic packaging (Stat-Kon OC-1003, OC-1006, OCL-4036); housings (EMI-X OC-1008, OC-100-10; PDX-83238; Ryton R-10 Avocado R-4004B, R-10 Black 5002C and 5004A, R-10 Copper R-2005B, R-10 Gold R-1001B, R-10 Gray R-5000B, R-10 Green R-4000B, R-10 Lt. Green R-7002B, R-10 Natural 7006A and 7007A, R-10 Tan R-7002B); potting/encapsulation (Ryton R-3 02 Black, R-9 901); structural electrical parts (Ryton R-8); switchgear (Fortron 0205P4, 0214P1, 1140B1, 1140B4, 6165B4, 6265B4; Ryton R-10 Avocado R-4004B, R-10 Black 5002C and 5004A, R-10 Copper R-2005B, R-10 Gold R-1001B, R-10 Gray R-5000B, R-10 Green R-4000B, R-10 Lt. Green R-7002B, R-10 Natural 7006A and 7007A, R-10 Tan R-7002B)

FDA-approved applications: (Ryton V-1)

Food-contact applications: cookware coatings (Ryton V-1)

Functional additives: modifier (Ryton V-1)

Industrial applications: cams (Ryton R-4, R-5); caps (Fortron 0205P4, 0214P1, 1140B1, 1140B4, 6165B4, 6265B4); coatings (Ryton V-1); fittings (Fortron 0205P4, 0214P1, 1140B1, 1140B4, 6165B4, 6265B4); gears (Fortron 0205P4, 0214P1, 1140B1, 1140B4, 6165B4, 6265B4); housings (Ryton R-4, R-5); impellers (Ryton R-4, R-5); load applications (RTP 1379, 1379S); low load applications (RTP 1301, 1303, 1305, 1307); mechanical handling (Fortron 0205P4, 0214P1, 1140B1, 1140B4, 6165B4, 6265B4); molded parts (Fortron 0205P4, 0214P1, 1140B1, 1140B4, 6165B4, 6265B4); potable water pipe fittings (Ryton R-4); powder coatings (Liquinite PPS); pumps (RTP 1301, 1303, 1305, 1307; Ryton R-4, R-5); release coatings (Ryton V-1); thin sections (Ryton R-3 02 Black)

Polyphenylene sulfide *(cont'd.)*

PROPERTIES:
Form:
Powder (Liquinite PPS)
Fine powder (Ryton V-1)
Pellets (Fortron 0205P4, 0214P1)

Color:
Natural (Ryton R-10 Natural 7006A and 7007A)
Natural, black (RTP 1301, 1303, 1305, 1307, 1378, 1379, 1379S)
Avocado (Ryton R-10 Avocado R-4004B)
Copper (Ryton R-10 Copper R-2005B)
Gold (Ryton R-10 Gold R-1001B)
Gray (Ryton R-10 Gray R-5000B)
Light green (Ryton R-10 Lt. Green R-7002B)
Green (Ryton R-10 Green R-4000B)
Light tan (Ryton V-1)
Tan (Ryton R-10 Tan R-7002B)
Black (Liquinite PPS; RTP 1383, 1385, 1387; Ryton R-10 Black 5002C and 5004A)

Composition:
0.4% volatiles (Ryton V-1)

Fineness:
Passes through 60 mesh screen (Ryton V-1)

GENERAL PROPERTIES:
Melt Flow:
30 g/10 min (316 C) (Ryton R-9 901)
4000–6000 g/10 min (Ryton V-1)

Sp. Gr.:
1.34–1.35 (Liquinite PPS)
1.37 (RTP 1301)
1.38 (RTP 1383)
1.39 (Stat-Kon OC-1003)
1.42 (RTP 1385)
1.44 (RTP 1303)
1.45 (Stat-Kon OC-1006; Thermocomp OC-1006)
1.46 (RTP 1387)
1.49 (EMI-X OC-1008)
1.53 (EMI-X OC-100-10; RTP 1305)
1.55 (SF-30GR)
1.57 (Stat-Kon OCL-4036)
1.62 (RTP 1307)
1.64 (RTP 1378)
1.65 (PDX-83238; RTP 1379S; SF-40GF)
1.66 (RTP 1379)

Polyphenylene sulfide (cont'd.)

Density:
1.35 g/cm³ (Fortron 0205P4, 0214P1)
1.57 g/cm³ (Ryton R-3 02 Black)
1.6 g/cm³ (Ryton R-4)
1.64 g/cm³ (Fortron 1140B1, 1140B4)
1.7 g/cm³ (Ryton R-5)
1.8 g/cm³ (Ryton R-8)
1.9 g/cm³ (Ryton R-9 901)
1.96 g/cm³ (Ryton R-10 Black 5002C and 5004A)
1.97 g/cm³ (Ryton R-10 Natural 7007A)
1.98 g/cm³ (Ryton R-10 Natural 7006A, R-11 70)
2.03 g/cm³ (Ryton R-10 Green R-4000B)
2.07 g/cm³ (Ryton R-10 Gray R-5000B, R-10 Tan R-7002B)
2.08 g/cm³ (Fortron 6165B4, 6265B4)
2.1 g/cm³ (Ryton R-10 Avocado R-4004B, R-10 Gold R-1001B, R-10 Lt. Green R-7002B)
2.11 g/cm³ (Ryton R-10 Copper R-2005B)
550–585 g/l (Liquinite PPS)

Visc.:
Low (Fortron 0205P4, 1140B4, 6165B4, 6265B4)
Medium (Fortron 0214P1, 1140B1

M.P.:
285 C (Ryton V-1)
525–530 F (Liquinite PPS)
600–650 F (SF-30GR, -40GF)

Stability:
Excellent resistance to acids, bases, solvents (Thermocomp OCL-4036, OF-1008)

MECHANICAL PROPERTIES:

Tens. Str.:
68.9 MPa (Ryton R-10 Black 5004A, R-10 Natural 7007A)
69 MPa (Ryton R-10 Avocado R-4004B, R-10 Copper R-2005B, R-10 Gold R-1001B, R-10 Gray R-5000B, R-10 Green R-4000B, R-10 Lt. Green R-7002B, R-10 Tan R-7002B)
75.6 MPa (Ryton R-11 70)
75.9 MPa (Ryton R-9 901)
79.2 MPa (Ryton R-10 Black 5002C, R-10 Natural 7006A)
91 MPa (Ryton R-8)
107 MPa (Ryton R-3 02 Black)
128 MPa (Ryton R-5)
130 MPa (Ryton R-4)
10,800 psi (Liquinite PPS)
12,000 psi (RTP 1301, 1379S)
12,400 psi (Fortron 0214P1)

Polyphenylene sulfide (cont'd.)

12,800 psi (Fortron 0205P4)
13,000 psi (RTP 1379)
14,000 psi (RTP 1303)
14,100 psi (Fortron 6265B4)
16,500 psi (Fortron 6165B4)
17,000 psi (RTP 1305)
18,000 psi (Stat-Kon OC-1003)
19,000 psi (Thermocomp OFL-4036)
19,500 psi (SF-30GR)
20,000 psi (RTP 1307, 1378; Thermocomp OF-1006)
21,000 psi (PDX-83238)
22,000 psi (RTP 1383; SF-40GF)
23,000 psi (Fortron 1140B4; Thermocomp OF-1008)
24,000 psi (Fortron 1140B1)
25,000 psi (RTP 1385)
25,500 psi (Stat-Kon OCL-4036; Thermocomp OCL-4036)
26,500 psi (RTP 1387)
27,000 psi (Stat-Kon OC-1006; Thermocomp OC-1006)
28,000 psi (EMI-X OC-100-10)
30,000 psi (EMI-X OC-1008)

Tens. Elong.:
0.5% (RTP 1385, 1387; Ryton R-9 901, R-10 Avocado R-4004B, R-10 Black 5004A, R-10 Copper R-2005B, R-10 Gold R-1001B, R-10 Gray R-5000B, R-10 Green R-4000B, R-10 Lt. Green R-7002B, R-10 Natural 7007A, R-10 Tan R-7002B)
0.54% (38 C) (Ryton R-8)
0.6% (Ryton R-10 Black 5002C, R-10 Natural 7006A, R-11 70)
0.7% (Fortron 6265B4)
0.75% (RTP 1383)
0.8% (Ryton R-5)
1.0% (EMI-X OC-100-10; Fortron 6165B4; RTP 1378)
1–2% (Liquinite PPS)
1.1% (Ryton R-3 02 Black)
1.25% (Ryton R-4)
1.3% (RTP 1305, 1307)
1.4% (RTP 1303, 1379, 1379S)
1.5% (EMI-X OC-1008; Fortron 1140B1, 1140B4; RTP 1301)
2.0% (PDX-83238; Stat-Kon OC-1003)
2–3% (Thermocomp OC-1006)
2.6% (Stat-Kon OCL-4036)
2.8% (Stat-Kon OC-1006)
3% (Fortron 0214P1)
3–4% (SF-30GR, SF-40GF; Thermocomp OF-1006, OF-1008, OFL-4036)
4.1% (Fortron 0205P4)

Tens. Mod.:
850,000 psi (RTP 1301)
1.2×10^6 psi (RTP 1303)
1.6×10^6 psi (RTP 1305; SF-30GR)
1.7×10^6 psi (RTP 1378, 1379, 1379S)
2.0×10^6 psi (RTP 1307; SF-40GF)
2.5×10^6 psi (RTP 1383)
3.7×10^6 psi (RTP 1385)
4.5×10^6 psi (RTP 1387)

Flex. Str.:
113.7 MPa (Ryton R-10 Black 5004A)
117.1 MPa (Ryton R-10 Avocado R-4004B, R-10 Copper R-2005B, R-10 Gold R-1001B, R-10 Gray R-5000B, R-10 Green R-4000B, R-10 Lt. Green R-7002B)
120.6 MPa (Ryton R-9 901, R-10 Natural 7007A, R-10 Tan R-7002B)
123.8 MPa (Ryton R-11 70)
127.5 MPa (Ryton R-10 Black 5002C, R-10 Natural 7006A)
134 MPa (Ryton R-3 02 Black)
150 MPa (38 C) (Ryton R-8)
165 MPa (Ryton R-5)
180 MPa (Ryton R-4)
17,000 psi (RTP 1301, 1379)
18,000 psi (RTP 1379S)
20,000 psi (RTP 1303)
21,200 psi (Fortron 0205P4)
21,300 psi (Fortron 0214P1)
21,900 psi (Fortron 6265B4)
24,000 psi (Thermocomp OFL-4036)
25,000 psi (RTP 1378)
25,500 psi (Stat-Kon OC-1003)
26,000 psi (PDX-83238)
26,300 psi (Fortron 6165B4)
27,000 psi (RTP 1383)
28,000 psi (RTP 1305; SF-30GR; Thermocomp OCL-4036)
29,000 psi (Thermocomp OF-1006)
30,000 psi (RTP 1307)
31,000 psi (RTP 1385)
31,500 psi (SF-40GF)
32,000 psi (Fortron 1140B4; Thermocomp OF-1008)
34,000 psi (RTP 1387; Thermocomp OC-1006)
35,000 psi (Fortron 1140B1)
36,000 psi (Stat-Kon OC-1006)
40,000 psi (EMI-X OC-1008, OC-100-10)

Polyphenylene sulfide (cont'd.)

Flex. Mod.:
 9646 MPa (Ryton R-3 02 Black)
 11,000 MPa (Ryton R-4)
 12,400 MPa (Ryton R-10 Avocado R-4004B, R-10 Black 5002 C and 5004A, R-10
   Copper R-2005B, R-10 Gold R-1001B, R-10 Gray R-5000B, R-10 Green R-4000B,
   R-10 Lt. Green R-7002B, R-10 Natural 7006A and 7007A, R-10 Tan R-7002B)
 13,780 MPa (Ryton R-5)
 13,800 MPa (Ryton R-11 70)
 14,000 MPa (38 C) (Ryton R-8)
 14,469 MPa (Ryton R-9 901)
 540,000 psi (Fortron 0214P1)
 550,000 psi (Fortron 0205P4)
 750,000 psi (RTP 1301)
 1.1×10^6 psi (RTP 1303)
 1.3×10^6 psi (Thermocomp OFL-4036)
 1.4×10^6 psi (RTP 1305, 1378, 1379, 1379S)
 1.5×10^6 psi (SF-30GR)
 1.6×10^6 psi (RTP 1307; Thermocomp OF-1006)
 1.7×10^6 psi (SF-40GF)
 1.8×10^6 psi (Thermocomp OF-1008)
 1.9×10^6 psi (Fortron 1140B1)
 2.0×10^6 psi (Fortron 1140B4; SF-40GF; Stat-Kon OC-1003; Thermocomp OCL-
   4036)
 2.1×10^6 psi (PDX-83238; RTP 1383)
 2.4×10^6 psi (Fortron 6165B4)
 2.45×10^6 psi (Thermocomp OC-1006)
 2.5×10^6 psi (RTP 1385)
 2.8×10^6 psi (Fortron 6265B4)
 3.0×10^6 psi (Stat-Kon OCL-4036)
 3.1×10^6 psi (Stat-Kon OC-1006)
 3.5×10^6 psi (RTP 1387)
 4.0×10^6 psi (EMI-X OC-1008)
 4.5×10^6 psi (EMI-X OC-100-10)
Compr. Str.:
 75.9 MPa (Ryton R-9 901)
 110 MPa (Ryton R-8)
 113.7 MPa (Ryton R-10 Avocado R-4004B, R-10 Copper R-2005B, R-10 Gold R-
   1001B, R-10 Gray R-5000B, R-10 Green R-4000B, R-10 Lt. Green R-7002B, R-
   10 Natural 7007A, R-10 Tan R-7002B)
 120.6 MPa (Ryton R-10 Black 5004A)
 127.5 MPa (Ryton R-10 Black 5002C, R-10 Natural 7006A)
 145 MPa (Ryton R-4)
 15,000 psi (RTP 1378)

18,000 psi (RTP 1301)
21,000 psi (RTP 1379, 1379S)
22,500 psi (RTP 1303)
24,000 psi (RTP 1305, 1383)
25,000 psi (RTP 1307)
26,000 psi (RTP 1385)
27,000 psi (RTP 1387)

Shear Str.:

8000 psi (Thermocomp OCL-4036)

Impact Str. (Izod):

31 J/m notched (38 C) (Ryton R-8)

32 J/m notched (Ryton R-10 Black 5004A)

37 J/m notched (Ryton R-9 901, R-10 Avocado R-4004B, R-10 Copper R-2005B, R-10 Gold R-1001B, R-10 Gray R-5000B, R-10 Green R-4000B, R-10 Lt. Green R-7002B, R-10 Natural 7007A, R-10 Tan R-7002B)

42.3 J/m notched (Ryton R-11 70)

43 J/m notched (Ryton R-10 Black 5002C)

53 J/m notched (Ryton R-3 02 Black, R-10 Natural 7006A)

69 J/m notched (Ryton R-5)

74 J/m notched (Ryton R-4)

0.5 ft lb/in. notched (Fortron 0205P4, 0214P1)

0.6 ft lb/in. notched (RTP 1379S)

0.7 ft lb/in. notched (Fortron 6265B4; RTP 1301, 1379)

0.8 ft lb/in. notched (Fortron 6165B4; PDX-83238; RTP 1383; Thermocomp OCL-4036)

0.9 ft lb/in. notched (Stat-Kon OC-1003)

1.0 ft lb/in. notched (Stat-Kon OCL-4036)

1.1 ft lb/in. notched (EMI-X OC-1008; Stat-Kon OC-1006; Thermocomp OC-1006, OFL-4036)

1.2 ft lb/in. notched (EMI-X OC-100-10; RTP 1303, 1378, 1385, 1387)

1.4 ft lb/in. notched (Fortron 1140B4; RTP 1305, 1307; Thermocomp OF-1006)

1.5 ft lb/in. notched (SF-30GR; Thermocomp OF-1008)

1.6 ft lb/in. notched (Fortron 1140B1; SF-40GF)

Hardness:

Rockwell M85 (Fortron 6265B4)

Rockwell M93 (Fortron 0205P4, 0214P1)

Rockwell M98 (Fortron 1140B4, 6165B4)

Rockwell M100 (Fortron 1140B1)

Rockwell R118 (RTP 1378)

Rockwell R120 (RTP 1379, 1379S)

Rockwell R121 (RTP 1301, 1303; Ryton R-8; Thermocomp OF-1006)

Rockwell R122 (RTP 1305, 1383)

Rockwell R123 (RTP 1307, 1385, 1387; Ryton R-4; Thermocomp OF-1008)

Polyphenylene sulfide *(cont'd.)*

Rockwell R124 (Liquinite PPS)

Mold Shrinkage:

1–3 mils/in. (FD) (Fortron 1140B1, 1140B4)

0.001 in./in. (Stat-Kon OC-1006)

0.0015 in./in. (EMI-X OC-1008, OC-100-10; PDX-83238; Stat-Kon OCL-4036)

0.004 in./in. (Stat-Kon OC-1003)

Water Absorp.:

0.01% (Fortron 0205P4, 0214P1, 6265B4)

0.015% (Fortron 1140B1, 1140B4)

0.02% (EMI-X OC-100-10)

0.03% (EMI-X OC-1008; PDX-83238; Stat-Kon OCL-4036)

0.04% (Stat-Kon OC-1003, OC-1006)

THERMAL PROPERTIES:

Soften. Pt. (Vicat):

530 F (SF-30GR, 40GF)

Conduct.:

2.1 Btu/h/ft²/F/in. (RTP 1301, 1303, 1305, 1383)

2.2 Btu/h/ft²/F/in. (RTP 1307, 1379, 1379S)

2.5 Btu/h/ft²/F/in. (RTP 1385)

2.7 Btu/h/ft²/F/in. (RTP 1378; Thermocomp OFL-4036)

2.8 Btu/h/ft²/F/in. (Thermocomp OF-1006)

2.9 Btu/h/ft²/F/in. (SF-30GR)

3.1 Btu/h/ft²/F/in. (Thermocomp OF-1008)

3.3 Btu/h/ft²/F/in. (RTP 1387; SF-40GF)

4.2 Btu/h/ft²/F/in. (Stat-Kon OC-1003)

5.2 Btu/h/ft²/F/in. (Stat-Kon OC-1006; Thermocomp OC-1006)

5.4 Btu/h/ft²/F/in. (Stat-Kon OCL-4036; Thermocomp OCL-4036)

6.0 Btu/h/ft²/F/in. (EMI-X OC-1008)

6.5 Btu/h/ft²/F/in. (EMI-X OC-100-10)

Distort. Temp.:

243 C (1.8 MPa) (Ryton R-4)

244 C (1.8 MPa) (Ryton R-8)

260 C (1.8 MPA) (Ryton R-3 02 Black, R-5, R-9 901, R-11 70)

> 260 C (1.8 MPa) (Ryton R-10 Avocado R-4004B, R-10 Black 5002C and 5004A, R-10 Copper R-2005B, R-10 Gold R-1001B, R-10 Gray R-5000B, R-10 Green R-4000B, R-10 Lt. Green R-7002B, R-10 Natural 7006A and 7007A, R-10 Tan R-7002B)

221 F (264 psi) (Fortron 0205P4, 0214P1)

450 F (264 psi) (RTP 1301)

495 F (264 psi) (SF-30GR; Stat-Kon OC-1003)

498 F (264 psi) (Fortron 6165B4)

500 F (264 psi) (PDX-83238; RTP 1303, 1305, 1307, 1378, 1379, 1379S, 1383, 1385, 1387; SF-40GF; Stat-Kon OCL-4036; Thermocomp OCL-4036, OF-1006, OFL-

278

4036)

502 F (264 psi) (Fortron 1140B4)

505 F (264 psi) (EMI-X OC-1008, OC-100-10; Fortron 1140B1; Stat-Kon OC-1006; Thermocomp OC-1006, OF-1008)

514 F (264 psi) (Fortron 6265B4)

Coeff. of Linear Exp.:

2.0×10^{-5} cm/cm/C (Ryton R-9 901)

2.2×10^{-5} cm/cm/C (Ryton R-4)

2.8×10^{-5} cm/cm/C (Ryton R-8)

0.4×10^{-5} in./in./F (EMI-X OC-100-10)

0.5×10^{-5} in./in./F (EMI-X OC-1008)

0.6×10^{-5} in./in./F (Stat-Kon OC-1006; Thermocomp OC-1006)

0.8×10^{-5} in./in./F (PDX-83238; Stat-Kon OCL-4036; Thermocomp OCL-4036)

1.3×10^{-5} in./in./F (SF-40GF; Stat-Kon OC-1003; Thermocomp OF-1006)

1.4×10^{-5} in./in./F (SF-30GR)

1.5×10^{-5} in./in./F (Thermocomp OF-1008, OFL-4036)

Flamm.:

5V/V-0 (Ryton R-4, R-8)

V-0 (EMI-X OC-1008, OC-100-10; Fortron 1140B1; PDX-83238; RTP 1301, 1303, 1305, 1307, 1378, 1383, 1385, 1387; Ryton R-3 02 Black, R-5, R-9 901, R-10 Avocado R-4004B, R-10 Black 5002C and 5004A, R-10 Copper R-2005B, R-10 Gold R-1001B, R-10 Gray R-5000B, R-10 Green R-4000B, R-10 Lt. Green R-7002B, R-10 Natural 7006A and 7007A, R-10 Tan R-7002B, R-11 70; SF-30GR, SF-40GF; Liquinite PPS; Stat-Kon OC-1003, OC-1006, OCL-4036; Thermocomp OC-1006, OCL-4036, OF-1006, OF-1008, OFL-4036)

VEO (RTP 1379, 1379S)

ELECTRICAL PROPERTIES:

Dissip. Factor:

0.0004 (1 kHz) (Thermocomp OF-1006)

0.0010 (1 kHz) (Fortron 6165B4); (1 MHz) (RTP 1301, 1303, 1305, 1307, 1379, 1379S; SF-30GR, SF-40GF)

0.0011 (1 kHz) (Thermocomp OF-1008)

0.0012 (1 MHz) (RTP 1378)

0.0013 (1 kHz) (Fortron 0205P4, 0214P1, 1140B4)

0.0020 (1 kHz) (Fortron 1140B1)

0.0032 (2450 MHz) (Ryton R-11 70)

0.0049 (1 kHz) (Ryton R-10 Tan R-7002B)

0.0060 (1 kHz) (Fortron 6265B4)

0.0096 (1 kHz) (Ryton R-10 Gray R-5000B)

0.011 (1 kHz) (Ryton R-10 Avocado R-4004B, R-10 Lt. Green R-7002B)

0.012 (1 kHz) (Ryton R-10 Gold R-1001B)

0.014 (1 kHz) (Ryton R-9 901, R-10 Copper R-2005B)

0.017 (1 kHz) (Ryton R-8)

Polyphenylene sulfide *(cont'd.)*

0.021 (1 kHz) (Ryton R-10 Green R-4000B)
Dielec. Str.:
12,600 V/mm (Ryton R-10 Black 5002C)
12,800 V/mm (Ryton R-10 Black 5004A)
15,200 V/mm (Ryton R-10 Green R-4000B)
15,700 V/mm (R-10 Natural 7006A and 7007A)
15,800 V/mm (Ryton R-10 Tan R-7002B)
16,400 V/mm (Ryton R-10 Gray R-5000B)
16,800 V/mm (Ryton R-10 Copper R-2005B)
18,300 V/mm (Ryton R-10 Gold R-1001B)
18,900 V/mm (Ryton R-10 Lt. Green R-7002B)
20,500 V/mm (Ryton R-10 Avocado R-4004B)
13,000 V/cm (Ryton R-8)
17,200 V/cm (Ryton R-4)
340 V/mil (RTP 1379, 1379S)
350 V/mil (RTP 1303, 1305, 1307, 1378; SF-30GR, SF-40GF)
360 V/mil (RTP 1301)
375 V/mil (Thermocomp OF-1006, OF-1008)
400 V/mil (Fortron 1140B1; Ryton R-9 901)
410 V/mil (Fortron 0205P4)
420 V/mil (Fortron 6165B4)
450 V/mil (Fortron 6265B4)
500 V/mil (Fortron 1140B4)
520 V/mil (Fortron 0214P1)
4500 V/mil (Liquinite PPS)
Dielec. Const.:
2.9 (1 kHz) (Fortron 0205P4)
3.0 (1 kHz) (Fortron 0214P1); (1 MHz) (RTP 1307)
3.1 (1 MHz) (SF-40GF)
3.3 (1 MHz) (RTP 1301)
3.4 (1 MHz) (SF-30GR)
3.5 (1 kHz) (Fortron 1140B4); (1 MHz) (RTP 1303, 1305)
3.6 (1 kHz) (Fortron 1140B1)
3.8 (1 kHz) (Thermocomp OF-1006); (1 MHz) (RTP 1378, 1379, 1379S)
3.9 (1 kHz) (Ryton R-4; Thermocomp OF-1008)
4.0 (1 kHz) (Fortron 6265B4)
4.5 (1 kHz) (Fortron 6165B4)
4.6 (1 kHz) (Ryton R-8, R-9 901)
4.86 (2450 MHz) (Ryton R-11 70)
5.4 (1 kHz) (Ryton R-10 Gold R-1001B, R-10 Lt. Green R-7002B)
5.5 (1 kHz) (Ryton R-10 Avocado R-4004B)
5.8 (1 kHz) (Ryton R-10 Tan R-7002B)
5.9 (1 kHz) (Ryton R-10 Gray R-5000B)

6.1 (1 kHz) (Ryton R-10 Copper R-2005B, R-10 Green R-4000B)

Vol. Resist.:

10 ohm-cm (EMI-X OC-100-10; PDX-83238)

30 ohm-cm (RTP 1387)

40 ohm-cm (RTP 1385)

75 ohm-cm (RTP 1383)

100 ohm-cm (EMI-X OC-1008; Stat-Kon OC-1006, OCL-4036)

1000 ohm-cm (Stat-Kon OC-1003)

2.7×10^{13} ohm-cm (Ryton R-10 Gold R-1001B)

5.8×10^{13} ohm-cm (Ryton R-10 Avocado R-4004B)

10^{15} ohm-cm (Ryton R-10 Green R-4000B)

1.2×10^{15} ohm-cm (Ryton R-10 Copper R-2005B)

2×10^{15} ohm-cm (Ryton R-8)

2.7×10^{15} ohm-cm (Ryton R-10 Tan R-7002B)

3×10^{15} ohm-cm (Ryton R-10 Gray R-5000B)

3.1×10^{14} ohm-cm (Ryton R-10 Lt. Green R-7002B)

3.1×10^{15} ohm-cm (Ryton R-9 901)

10^{16} ohm-cm (Fortron 0205P4, 0214P1, 1140B1, 1140B4, 6165B4, 6265B4; Liquinite
   PPS; RTP 1301, 1303, 1305, 1307, 1378, 1379, 1379S)

4.5×10^{16} ohm-cm (Ryton R-4)

4.7×10^{16} ohm-cm (Thermocomp OF-1006)

5.0×10^{16} ohm-cm (Thermocomp OF-1008)

Surf. Resist.:

0 ohm/sq. (PDX-83238)

10 ohm/sq. (EMI-X OC-100-10)

100 ohm/sq. (EMI-X OC-1008; Stat-Kon OC-1006, OCL-4036)

150 ohm/sq. (Thermocomp OCL-4036)

1000 ohm/sq. (Stat-Kon OC-1003)

Arc Resist.:

34 s (Ryton R-4)

50 s (RTP 1378)

75 s (RTP 1301)

116 s (Ryton R-10 Black 5002C, R-10 Natural 7006A)

120 s (Fortron 0205P4; RTP 1303, 1305, 1307, 1379, 1379S)

124 s (Fortron 0214P1)

131 s (Ryton R-10 Gold R-1001B)

134 s (Fortron 1140B4)

136 s (Fortron 1140B1)

170 s (Ryton R-10 Lt. Green R-7002B)

180 s (Ryton R-9 901)

182 s (Fortron 6165B4; Ryton R-8, R-10 Black 5004A, R-10 Natural 7007A)

191 s (Ryton R-10 Avocado R-4004B)

195 s (Fortron 6265B4)

Polyphenylene sulfide (cont'd.)

202 s (Ryton R-10 Copper R-2005B)
211 s (Ryton R-10 Gray R-5000B, R-10 Green R-4000B)
212 s (Ryton R-10 Tan R-7002B)

TOXICITY/HANDLING:

Off-gas products produced during molding can be irritating to the mucous membranes; use with adequate ventilation when injection molding resins (Ryton R-3 02 Black, R-4, R-5, R-8, R-9 901, R-10 Avocado R-4004B, Ryton R-10 Black 5002C and 5004A, R-10 Copper R-2005B, R-10 Gold R-1001B, R-10 Gray R-5000B, R-10 Green R-4000B, R-10 Lt. Green R-7002B, R-10 Natural 7006A and 7007A, R-10

Polystyrene

SYNONYMS:

Benzene, ethenyl-, homopolymer
Ethenylbenzene, homopolymer
PS

EMPIRICAL FORMULA:

$[CH(C_6H_5)CH_2]_n$

STRUCTURE:

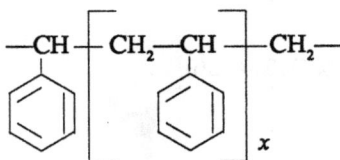

CAS No.:

9003-53-6

TRADENAME EQUIVALENTS:

Amoco G1, G2, G3, G18, H2R, H3E, H4, H4E, H4R, H5M, H5M, M2R, M9, M9R, R1, R2, R3, R5, R7, R12 [Amoco]
Bapolan 6050, 6060, 6145, 6270, 6445, 6450 [Bamberger]
Cosden Polystyrene 500, 525, 550, 625, 710, 825, 825E [Cosden]
Fostalite [Hocehst-Celanese]
Fostarene 50, 57, 58, 817 [Hoechst-Celanese]
Fosta Tuf-Flex 326, 329, 702, 707, 717, 730, 782, 840, 880, 929 [Hoechst-Celanese]
Fyrid KS1 [Borg-Warner]
Kama CS, OPS [Kama]
Lustrex 2220, 2400, 3300, 3350, 4220, 4300 [Monsanto]

Polypenco Q200.5 [Polymer Corp.] (cross-linked)

PS 110R, 110S, 110SL, 224EP, 224EV, 224M, 310S, 344/E, 314/E, 410S, 444, B44HH, 510TG, 8120, 8720 [Mobil]

Styron 412D, 420D, 430U, 470, 475U, 492U, 495, 666U, 678U, 685D [Dow]

Superdense 700, 710 [Hammond Plastics]

Superflex 405, 432, 505, 550, FR-1 [Hammond Plastics]

Superflow 390 [Hammond Plastics]

Vestyron 114, 214, 214A, 314, 325, 512, 512A, 512 TA, 516, 616, 620, 620A, 620TA, 628, 638, 640, TSG 502, TSG 504, TSG 602, TSG 604 [Huls AG]

Calcium carbonate-filled:

PS-H63CC-1 (10% calcium carbonate) [Washington Penn]

Glass-reinforced:

RTP 401 (10% glass fiber), 403 (20% glass fiber), 405 (30% glass fiber), 407 (40% glass fiber) [Fiberite]

Thermocomp CF-1004 (20% glass fiber), CF-1006 (30% glass fiber), CF-1008 (40% glass fiber) [LNP]

Mica-filled:

PS-H23MF-1 (10% mica) [Washington Penn]

Rubber-modified:

Fosta Tuf-Flex 240, 271, 283, 326, 474, 721-M, 742 [Hoechst-Celanese]

Expandable beads:

Fostafoam Type 3775, 4775, 5775 [Hoechst-Celanese]

Vestypor Series, 01N, 01S, 10N, 10S, 20N, 20S, 30N, 30S [Huls AG]

MODIFICATIONS/SPECIALTY GRADES:

Crystal grade:

Amoco G1, G2, G3, G18, R1, R2, R3, R5, R7, R12; PS 110R, 110S, 110SL, 224EP, 224EV, 224M

Impact grade:

Amoco H2R, H3E, H5M, M2R; Styron 475U, 495; Vestyron 616, 620, 620A, 620TA, 628

High-impact:

Amoco H4, H4E, H4R; Bapolan 6445, 6450; Cosden Polystyrene 825, 825E; Fosta Tuf-Flex 240, 283, 329, 474, 702, 707, 717, 721-M, 730, 742, 782, 840, 880; Fyrid KS1; Lustrex 3300, 3350, 4220, 4300; PS 410S, 444, 444HH, 510TG; PS-H23MF-1; PS 8120, 8720; Superdense 700; Superflex 505, 550, FR-1; Thermocomp CF-1008; Vestyron 638, 640

Medium-impact:

Amoco M9, M9R; Bapolan 6050, 6270; Cosden Polystyrene 625, 710; Fosta Tuf-Flex 271, 326, 929; Lustrex 2400, 2220; PS 310S, 314/E, 344/E; PS-H63CC-1; Styron 412D, 420D, 430U, 470, 492U; Superflex 405, 432; Thermocomp CF-1004, CF-1006; Vestyron 512, 512A, 512TA, 516, TSG502, TSG504, TSG602, TSG604

Flame-retardant:

Fosta Tuf-Flex 702, 707; Fyrid KS1; PS 8120, 8720; FR-1; Vestypor 01S, 10S, 20S,

Polystyrene (cont'd.)

30S
Heat-resistant:
Amoco R2, R3, R5; Bapolan 6145; Cosden Polystyrene 550; Fosta Tuf-Flex 283, 474; Fostarene 50, 57, 58; Lustrex 2400, 3300, 3350; PS 310S, 444HH, 685D; Vestyron 114, 516, 616, 638

UV-stabilized:
Fostalite; PS 110SL

High-flow:
Bapolan 6060, 6270; Cosden Polystyrene 500, 625; Fosta Tuf-Flex 326, 717, 929; Fostarene 817; Fyrid KS1; Lustrex 4220; PS 224EP, 224EV, 314/E, 344/E; Styron 412D, 678U; Superdense 710; Superflex 405, 505; Vestyron 314, 325, 512, 512A, 512TA

Antistatic grade:
Vestyron 214A, 512A, 512TA, 620A, 620TA

Externally lubricated:
PS 224EP, 224EV, 224M, 314/E, 344/E, 444, 444HH

CATEGORY:
Thermoplastic resin

PROCESSING:
Extrusion:
Amoco R12; Bapolan 6145, 6445; Cosden Polystyrene 710, 825E; Fostalite; Fostarene 57, 58; Fosta Tuf-Flex 329, 730, 840, 880; Lustrex 3300, 3350, 4300; Polypenco Q200.5; PS 110S, 110SL, 310S, 410S, 510TG, 8120, 8720; Styron 420D, 430U, 470, 475U, 492U, 495, 666U, 685D; Superdense 700, 710; Superflex 405, 550; Superflow 390; Vestyron 114, 638, 640

Injection molding:
Amoco G1, G2, G3, G18, H2R, H3E, H4, H4E, H4R, H5M, M2R, M9, M9R, M9R, R1, R2, R3, R5, R7, R12; Bapolan 6050, 6060, 6270, 6450; Cosden Polystyrene 500, 525, 550, 625, 825; Fostarene 50, 57, 58, 817; Fosta Tuf-Flex 240, 271, 283, 326, 329, 474, 702, 707, 717, 721-M, 742, 782, 929; Fyrid KS1; Lustrex 2220, 2400, 3300, 3350, 4220, 4300; PS 110R, 224EP, 224EV, 224M, 314/E, 344/E, 444, 444HH, 510TG, 8120, 8720; PS-H23MF-1, -H63CC-1; RTP 401, 403, 405, 407; Styron 412D, 420D, 430U, 470, 475U, 492U, 495, 666U, 678U, 685U; Superdense 700, 710; Superflex 405, 432, 505, 550, FR-1; Superflow 390; Thermocomp CF-1004, CF-1006, CF-1008; Vestyron 114, 214, 214A, 314, 325, 512, 512A, 512TA, 516, 616, 620, 620A, 620TA, 628, 640, TSG502, TSG504, TSG602, TSG604

Compression molding:
Styron 412D, 470, 475U, 666U; Superdense 700, 710; Superflex 405; Superflow 390

Blow molding:
Amoco R3, R5; Cosden Polystyrene 825; Styron 420D, 430U, 475U, 666U, 685D; Superflex 405; Superflow 390

Thermoforming:
Amoco H2R; Bapolan 6145; Fosta Tuf-Flex 730, 840, 880; Kama CS, OPS; Lustrex

Polystyrene *(cont'd.)*

3300, 3350, 4300; PS 410S, 8120, 8720; Styron 420D, 470, 475U, 492U; Superflex 432; Vestyron 638, 640

Vacuum forming:

Styron 495

Structural foam processing:

Amoco M9, M9R; Cosden Polystyrene 550; Fostafoam Type 3775, 4775, 5775; Fostarene 57, 58; Fosta Tuf-Flex 717, 929; Fyrid KS1; Lustrex 4220; PS 8120; Styron 430U; Vestypor Series, 01N, 01S, 20N, 20S, 30N, 30S; Vestyron 114, TSG502, TSG504, TSG602, TSG604

APPLICATIONS:

Architectural applications: bathroom cabinets (Vestyron 214A, 512A, 620, 620A); insulation (Vestypor 01N, 01S, 10N, 10S, 20N, 20S, 30N, 30S); shutters (PS-H63CC-1); wall tile (Fostarene 817); window frames (Lustrex 4220)

Automotive applications: (Amoco H4R; Fosta Tuf-Flex 721-M; PS-H23MF-1; RTP 401, 403, 405, 407); ducts (PS 444HH); grills (Vestyron 616); heating ducts (Styron 492U); interior parts (Fosta Tuf-Flex 474; Styron 492U)

Consumer products: air conditioner/fan grills (Fosta Tuf-Flex 742); appliances (Amoco M9R; Bapolan 6445; Fosta Tuf-Flex 240, 474, 702, 707; Lustrex 3350; PS 444, 444HH, 510TG, 8120); appliance housings (Fosta Tuf-Flex 283, 721-M; Lustrex 2400); appliance parts (Bapolan 6450; Cosden Polystyrene 825; PS 444HH; Styron 412D, 430U, 475U, 666U, 685D; Superflex 505, 550; Superflow 390); battery cases (Fostarene 50); billiard balls (Superdense 700, 710); cameras (Fosta Tuf-Flex 326); camping articles (Vestyron TSG 502, TSG 504, TSG 602, TSG 604); carrying cases (Cosden Polystyrene 825); containers (Amoco G1, G18, H4E, R7; Bapolan 6270, 6450; Cosden Polystyrene 625, 825; Fosatrene 817; Fosta Tuf-Flex 730; Kama OPS; Lustrex 2220, 2400, 4220; PS 224EP, 224EV, 224M, 310S, 314/E, 344/E, 410S, 444, 510TG; Styron 412D, 430U, 470, 495, 666U, 678U; Superflex 405, 505, 550; Superflow 390; Vestyron 114, 214, 620, 620A, 640); cosmetic packaging (Lustrex 2220; PS 224EP; Superflex 405); cutlery (Bapolan 6050, 6450; Cosden Polystyrene 550; Fostarene 58; Fosta Tuf-Flex 271; Lustrex 2400, 3300; PS 110R, 224M; 325); dinnerware (Styron 420D); drinkware (Amoco G1; Bapolan 6050, 6060; Cosden Polystyrene 825E; Fostarene 817; Lustrex 2400, 3300, 4300; PS 110R, 110S, 224EP, 224EV, 224M, 310S, 410S; Styron 420D, 470, 678U; Vestyron 114, 325, 620); furniture (Amoco H4, H4R, M9; Cosden Polysty-rene 825; Fosta Tuf-Flex 271, 329, 721-M, 782; Lustrex 4220; PS 314/E, 344/E, 444; Superflex 505; Vestypor 01N, 01S, 10N, 10S, 20N, 20S, 30N, 30S; Vestyron 512, TSG 502, TSG 504, TSG 602, TSG 604); greeting cards (Kama OPS); frames (PS 224EP; Vestyron 114); hair dryers (Superflex FR-1); hangers (Bapolan 6050, 6060, 6450; PS 314/E, 344/E; Vestyron 214, 512, 620); hobby crafts (Lustrex 2220); housewares (Amoco G1, G18, H4, H5M, M2R, M9, M9R; Bapolan 6050, 6060, 6450; Fostarene 50, 817; Fosta Tuf-Flex 326, 474, 929; Lustrex 2220, 2400, 4220; PS 444; PS-H63CC-1; Styron 666U, 685D; Superflex 405, 505; Superflow 390; Vestyron 214, 314, 512, 620, 628, TSG 502, TSG 504, TSG 602, TSG 604);

Polystyrene *(cont'd.)*

ice buckets (Fostafoam Type 4775); jars (PS 224EP, 224EV, 224M); jewelry (Vestyron 214); novelties (Bapolan 6050, 6060, 6270, 6450; Fostarene 817); photographic equipment (Lustrex 3350; Vestyron 516, 616); photographic films and transparencies (Kama OPS); refrigeration (Bapolan 6445; Cosden Polystyrene 825E; Fosta Tuf-Flex 240, 730; Lustrex 3350; PS 410S; Vestyron 620, 638); refrigeration insulation (Vestypor 01N, 01S, 10N, 10S, 20N, 20S, 30N, 30S); report covers (Kama OPS); shoe soles (Vestypor 01N, 01S, 10N, 10S, 20N, 20S, 30N, 30S); tackle boxes (PS 444); toilet articles (Vestyron 512); toys (Amoco G1, H4, H5M, M9; Bapolan 6050, 6060, 6270, 6450; Cosden Polystyrene 625, 825, 329; Fostafoam Type 4775; Fosta Tuf-Flex 742; Lustrex 4220; PS 224EP, 314/E, 344/E, 444; Styron 412D, 430U, 475U, 495, 666U, 678U; Superflex 505, 550; Superflow 390; Vestyron 214, 314, 512, 620); window envelopes (Kama OPS)

Electrical/electronic industry: (Fosta Tuf-Flex 702, 707; Polypenco Q200.5; PS 8120; Vestyron TSG 502, TSG 504, TSG 602, TSG 604); audio/video cassettes (Lustrex 2400, 4220; PS 110R, 110S, 224M, 314/E, 344/E; Styron 412D, 430U; Superflex 432; Vestyron 516); business machines and office equipment (Fosta Tuf-Flex 702, 707; Lustrex 4220; PS 8120, 8720; Vestyron 620, TSG 502, TSG 504, TSG 602, TSG 604); cassette housings (Amoco M2R, M9R; Vestyron 114); computers (PS 8120; Superflow 390); electrical components (Superflex 432, FR-1); housings (Fosta Tuf-Flex 702, 707; Lustrex 4220; PS 8120; Superflex FR-1; Vestyron TSG 502, TSG 504, TSG 602, TSG 604); insulators (Polypenco Q200.5); lighting equipment (Fostalite; PS 110S, 110SL); speaker cabinets (PS 8120); stereo/TV cabinets (Amoco H4R; Fostarene 50; Fosta Tuf-Flex 474, 702, 707; Lustrex 4220; PS 444HH, 8720; Styron 492U; Superflex FR-1; Vestyron 616); telecommunications (Fyrid KS1; Polypenco Q200.5); TV (PS 8120)

FDA-approved applications: (Amoco G1, G2, G3, G18, H2R, H3E, H4, H4E, H4R, H5M, M2R, M9, M9R, M9R, R1, R2, R3, R5, R7, R12; Bapolan 6050, 6145, 6270, 6445, 6450; Kama OPS; PS 110S, 224EP, 224EV, 224M, 310S, 410S, 444, 444HH, 510TG; Styron 412D, 420D, 430U, 470, 475U, 492U, 495, 666U, 678U, 685D; Superflex 405; Superflow 390)

Food-contact applications: food packaging/containers (Amoco H2R, H3E, H4, R1, R2; Bapolan 6050, 6145, 6270, 6445, 6450; Fostarene 57, 58; Fosta Tuf-Flex 326, 730, 880; Kama OPS; Lustrex 4300; Vestyron 314, 640)

Industrial applications: beanbag filler (Fostafoam Type 3775); blown bottles and containers (Amoco M2R, R3, R5); caps (Superflex 432; Vestyron 214); casting molds (Vestypor 01N, 01S, 10N, 10S, 20N, 20S, 30N, 30S); closures (Amoco M2R; Lustrex 2220, 2400; PS 314/E, 344/E; Superflex 432; Vestyron 314, 620); coating (Amoco R12); films (Fostarene 50, 58; Styron 685D); flotation (Fostafoam Type 3775; Vestypor 01N, 01S, 10N, 10S, 20N, 20S, 30N, 30S); foam sheet (Amoco R1, R3; Cosden Polystyrene 550; Fosatrene 57, 58; PS 110S; Styron 685D; Superflow 390); foils (638, 640); high-clarity containers (Amoco G2, G3); housings (RTP 401, 403, 405, 407; Styron 492U); intricate/hard-to-fill moldings (Fostafoam Type 5775; Fosatrene 817); large parts (Bapolan 6145); monofilament (Fostarene 50);

multiple-cavity molds (Bapolan 6060, 6270, 6450); packaging/displays (Amoco H4E, R7; Bapolan 6050, 6060, 6145, 6445; Cosden Polystyrene 710; Fostafoam Type 4775; Fostarene 58, 817; Fosta Tuf-Flex 283, 474, 730, 880, 929; Lustrex 3350, 4300; Styron 420D, 470; Superflex 432; Vestypor 01N, 01S, 10N, 10S, 20N, 20S, 30N, 30S; Vestyron 114, 214, 214A, 314. 512, 512A, 620, 620A, 638, 640); packaging film (Vestyron 512); pipe (PS 410S); plated parts (Superdense 700, 710); profiles (Bapolan 6445; Fosta Tuf-Flex 730; PS 310S, 410S); sheet (Bapolan 6145, 6445; Cosden Polystyrene 825E; Fosta Tuf-Flex 840, 880; PS 110S; Styron 475U; Superflex 550; Vestypor 10N, 10S, 20N, 20S; Vestyron 638); signs (Bapolan 6145; Fosta Tuf-Flex 283; Kama OPS; PS 410S); spools (Lustrex 4220); structural foam (Amoco H4R, H5M, M9, M9R; Lustrex 4220; Vestyron TSG 502, TSG 504, TSG 602, TSG 604); technical parts (Amoco H5M; Bapolan 6270; Cosden Polystyrene 625; Fosta Tuf-Flex 271, 283; PS 314/E, 344/E; Vestyron 114, 214, 314, 512, 512TA, 516, 616, 620, 620TA, 628, 640); thick sections (Cosden Polystyrene 525; Fostafoam Type 5775; Styron 685D; Vestypor 30N, 30S); thin sections (Amoco G2, G3, H3E, H4E; Bapolan 6060, 6270, 6445; Cosden Polystyrene 500, 625; Fostafoam Type 5775; Fostarene 817; Fosta Tuf-Flex 717, 929; Lustrex 2220; PS 224EP, 224EV, 224M, 310S, 314/E, 344/E, 410S, 8120, 8720; Styron 470, 495, 678U); trays (PS 310S, 410S; Vestyron 214A, 512, 512A, 620A); wood replacement (Styron 475U)

Medical applications: (Fostarene 50, 817); laboratory items (Amoco G1; Vestyron 114. 214, 325)

PROPERTIES:

Form:

Solid (Fostarene 50, 58, 817; Fosta Tuf-Flex 240, 271, 283, 326, 329, 474, 702, 707, 717, 721-M, 730, 742, 782, 840, 880, 929)

Crystal (Amoco G1, G2, G3, G18, R1, R2, R3, R5, R7, R12; PS 110R, 110S, 110SL, 224EP, 224EV, 224M)

Beads (Fostafoam Type 3775, 4775, 5775)

Beads (0.5–0.9 mm) (Vestypor 30N, 30S)

Beads (0.8–1.25 mm) (Vestypor 20N, 20S)

Beads (0.8–1.8 mm) (Vestypor 10N, 10S)

Beads (1.6–2.5 mm) (Vestypor 01N, 01S)

Pellets (Fostalite; Superflex 405, 432, 505, 550, FR-1; Superflow 390; Thermocomp CF-1004, CF-1006, CF-1008)

1/8 in. pellets (PS 110R, 110S, 110SL, 310S, 314/E, 410S, 510TG)

1/16 in. pellets (PS 224EP, 224EV, 224M, 344/E, 444, 444HH)

Cylindrical pellets (Vestyron 114, 214, 214A, 314, 325, 512, 512A, 512TA, 516, 616, 620, 620A, 620TA, 628, 638, 640, TSG 502, TSG 504, TSG 602, TSG 604)

Crystalline (Fostarene 57)

Biaxially oriented sheet (2–20 mil) (Kama CS)

Biaxially oriented sheet (1.25–20 mil) (Kama OPS)

Rod, plate (Polypenco Q200.5)

Polystyrene (cont'd.)

Color:
Transparent (Polypenco Q200.5)
Natural (Vestyron 325)
Natural and opaque colors (Superdense 700, 710; Superflex 405)
Natural and white (Vestyron 638, 640)
Natural and custom colors (PS 8120, 8720; Vestyron 114, 214, 214A, 314, 512, 512A, 512TA, 516, 616, 620, 620A, 620TA, 628, TSG 502, TSG 504, TSG 602, TSG 604)
Natural and pigmented (Thermocomp CF-1004, CF-1006, CF-1008)
Natural, black (RTP 401, 403, 405, 407)
Clear (Styron 666U, 678U, 685D)
Clear and colors (Fostarene 50)
White (Vestypor 01N, 01S, 10N, 10S, 20N, 20S, 30N, 30S)
Semiopaque and opaque colors (Superflex 550)
Opaque (Styron 412D, 420D, 430U, 470, 475U, 492U, 495)
Opaque colors (Superflex 432, 505; Superflow 390)
Opaque colors in wide range with improved fade resistance (Superflex FR-1)
Opaque clear, white, brown, black; also avail. in transparent or opaque red, blue, green, yellow, amber (Kama OPS)
Avail. in precompounded colors (PS 110R, 110S, 110SL, 224EP, 224EV, 224M, 310S, 314/E, 344/E, 410S, 444, 444HH, 510TG)

Odor:
None (PS 110S, 224EP, 224EV, 224M)

Taste:
None (PS 110S, 224EP, 224EV, 224M)

GENERAL PROPERTIES:

Solubility:
Sol. in aromatic hydrocarbons (Kama OPS)
Sol. in chlorinated hydrocarbons (Kama OPS)

Melt Flow:
1.1 g/10 min (PS 110S, 110SL)
1.6 g/10 min (Styron 685D)
1.7 g/10 min (PS 410S)
1.8 g/10 min (Amoco R1)
2.0 g/10 min (Amoco R2; PS 110R, 310S, 444HH)
2.2 g/10 min (Amoco H2R)
2.4 g/10 min (PS 510TG)
2.5 g/10 min (Vestyron 114, 616)
2.7 g/10 min (Amoco M2R; Styron 420D)
2.9 g/10 min (Styron 495)
3.0 g/10 min (Amoco H3E, H4E, R3; Bapolan 6445; PS 224M, 444; Styron 470, 492U; Vestyron 638)
3.5 g/10 min (PS 8720)
4.0 g/10 min (Bapolan 6145)

288

4.5 g/10 min (Amoco H4, H4R)

5.0 g/10 min (Amoco R5; PS 8120; Superflex 432; Vestyron 628)

5.6 g/10 min (Styron 430U)

6.0 g/10 min (PS 224EP; Vestyron 516)

6–8 g/10 min (Superdense 700)

7.0 g/10 min (Amoco R7; Superflex FR-1; Vestyron 214, 640)

7.5 g/10 min (PS-H63CC-1; Styron 475U, 666U; Superflow 390)

8.0 g/10 min (Amoco G1, H5M; Bapolan 6050, 6450; PS 314/E, 344/E; Superflex 405; Vestyron 214A, 620)

8–10 g/10 min (Superdense 710)

9.0 g/10 min (Amoco M9R; Superflex 505; Vestyron 620TA)

10.0 g/10 min (Amoco M9; PS-H23MF-1; Vestyron 620A)

12.0 g/10 min (Amoco G2, R12; PS 224EV; Styron 678U)

14.0 g/10 min (Bapolan 6060; Styron 412D)

16.0 g/10 min (Vestyron 314, 512)

18.0 g/10 min (Amoco G3, G18; Bapolan 6270; Vestyron 512TA)

19 g/10 min (Vestyron 512A)

25 g/10 min (Vestyron 325)

Sp. Gr.:

1.04 (Cosden Polystyrene 625, 710, 825, 825E; Fosta Tuf-Flex 240, 271, 283, 326, 329, 474, 717, 721-M, 730, 742, 782, 840, 880, 929; PS 310S, 314/E, 344/E, 410S, 444, 444HH, 510TG; Styron 666U, 678U, 685D)

1.05 (Bapolan 6050, 6145, 6270, 6445, 6450; Cosden Polystyrene 500, 525, 550; Fostalite; Fostarene 50, 57, 58, 817; Kama CS, OPS; Lustrex 2220, 2400; Polypenco Q200.5; PS 110R, 110S, 110SL, 224EP, 224EV, 224M; Styron 412D, 420D, 430U, 470, 475U, 492U, 495)

1.06 (Lustrex 3350, 4220, 4300)

1.13 (RTP 401)

1.14 (Superflex FR-1)

1.15 (PS 8120, 8720)

1.20 (RTP 403; Thermocomp CF-1004)

1.28 (RTP 405; Thermocomp CF-1006)

1.38 (RTP 407; Thermocomp CF-1008)

Sp. Vol.:

20.1 in.^{3}/lb (Thermocomp CF-1008)

21.6 in.^{3}/lb (Thermocomp CF-1006)

23.1 in.^{3}/lb (Thermocomp CF-1004)

Density:

1.03 g/cm^3 (Amoco H2R, H3E, H4, H4E, H4R, H5M, M2R, M9, M9R; Vestypor 01N, 01S, 10N, 10S, 20N, 20S, 30N, 30S)

1.05 g/cm^3 (Amoco G1, G2, G3, G18, R1, R2, R3, R5, R7, R12; Vestyron 114, 214, 214A, 314, 325, 512, 512A, 512TA, 516, 616, 620, 620A, 620TA, 628, 638, 640)

1.08 g/cm^3 (PS-H23MF-1)

Polystyrene *(cont'd.)*

1.10 g/cm³ (PS-H63CC-1)

M.P.:

242 F (Thermocomp CF-1004, CF-1006, CF-1008)

Stability:

May discolor when exposed to uv light (Fosta Tuf-Flex 702, 707)

Excellent dimensional stability; excellent chemical resistance to acids, weak and strong alkalis; good resistance to alcohols, grease, and oils; attacked by oxidizing agents (Kama OPS)

High radiation resistance; good resistance to load deformation (Polypenco Q200.5)

Excellent thermal stability; excellent resistance to organic acids, aliphatic amines, bases, polyglycols, salts, pharmaceuticals, foodstuffs, and beverages; good resistance to weak and strong inorganic acids, alcohols, veg. oils, condiments (Styron 412D, 420D, 430U, 470, 475U, 492U, 495)

Excellent thermal stability; excellent resistance to weak and strong inorganic acids, weak organic acids, alcohols, aliphatic amines, bases, polyglycols, salts, and to pharmaceuticals, foodstuffs, and beverages; good resistance to strong organic acids and veg. oils (Styron 666U, 678U, 685D)

Good resistance to bases; fair resistance to acids; poor resistance to solvents (Thermocomp CF-1008)

Expanded parts are resistant to water, acids (except strongly oxidizing acids), sol'ns. of alkalis and common alcohols; not resistant to ketones, ethers, esters, petrols, chlorinated hydrocarbons, and mineral oils; prolonged exposure to weathering produces yellowing of the surface but no substantial change in mechanical strength (Vestypor 01N, 01S, 10N, 10S, 20N, 20S, 30N, 30S)

Good resistance to aq. sol'ns. of salts, acids, and alkalis (except strong oxidizing agents); highly resistant to foodstuffs and beverages; many organic substances, e.g., aromatic hydrocarbons, ethers, esters, aldehydes, ketones, and halogenated hydrocarbons cause swelling, dissolution, or embrittlement (Vestyron 114, 214, 214A, 314, 325, 512, 512A, 512TA, 516, 616, 620, 620A, 620TA, 628, 638, 640)

Ref. Index:

1.586–1.606 (Styron 666U, 678U, 685D)

1.59 (Vestyron 114, 214, 314, 325)

Opaque (Vestyron 214A)

MECHANICAL PROPERTIES:

Tens. Str.:

197 kg/cm² (break) (Styron 495)

230 kg/cm² (yield) (Styron 475U)

232 kg/cm² (yield) (Styron 495)

240 kg/cm² (break) (Styron 470)

245 kg/cm² (yield) (Styron 470)

246 kg/cm² (yield and break) (Styron 412D)

260 kg/cm² (break) (Styron 475U)

267 kg/cm² (yield) (Styron 492U)

281 kg/cm² (yield) (Styron 420D)
288 kg/cm² (break) (Styron 492U)
295 kg/cm² (break) (Styron 420D)
316 kg/cm² (break) (Styron 430U)
323 kg/cm² (yield) (Styron 430U)
415 kg/cm² (yield and break) (Styron 678U)
429 kg/cm² (yield and break) (Styron 666U)
575 kg/cm² (yield and break) (Styron 687D)
0.15–0.25 MPa (15 kg/m³ density) (Vestypor 01N, 01S, 10N, 10S, 20N, 20S, 30N, 30S)
20 MPa (break) (Amoco H4; Vestyron 640)
21 MPa (Fosta Tuf-Flex 880); (break) (Amoco H5M)
22 MPa (break) (Amoco H4E, M9; Vestyron 512A, 512TA, 620A, 620TA)
23 MPa (break) (Amoco M9R)
24 MPa (Fosta Tuf-Flex 717)
24.1 MPa (yield) (PS 510TG)
25 MPa (Fosta Tuf-Flex 840); (break) (Vestyron 512, 620, 638)
26 MPa (Fosta Tuf-Flex 329, 742); (break) (Amoco H4R)
26.9 MPa (yield) (PS 314/E, 344/E, 410S)
27 MPa (break) (Amoco H3E)
28 MPa (Fosta Tuf-Flex 240, 730, 929)
28 MPa (break) (Amoco H2R)
29 MPa (Fosta Tuf-Flex 782); (yield) (PS 444)
30 MPa (break) (Vestyron 628)
31 MPa (Fosta Tuf-Flex 283)
33.1 MPa (yield) (PS 444HH)
34.5 MPa (yield) (PS 310S)
35 MPa (Fosta Tuf-Flex 326, 474, 721-M); (break) (Vestyron 516)
37 MPa (break) (Amoco G3)
39 MPa (Fosatrene 817); (break) (Amoco G2, M2R)
40 MPa (break) (Amoco G18; Vestyron 616)
41 MPa (Fosta Tuf-Flex 271); (break) (Amoco G1)
41.4 MPa (break) (PS 224EV)
42 MPa (break) (Amoco R12)
43 MPa (break) (Vestyron 325)
45 MPa (break) (Vestyron 214A, 314)
46 MPa (break) (Amoco R7)
46.2 MPa (break) (PS 224M)
47 MPa (break) (Amoco R5)
50 MPa (break) (Amoco R3; Vestyron 214)
51.7 MPa (break) (PS 110R, 224EP)
52 MPa (Fostarene 50); (break) (Amoco R1, R2; Fostalite)
53 MPa (Fosatrene 57, 58)
56.5 MPa (break) (PS 110S); (yield) (PS 110SL)

Polystyrene *(cont'd.)*

60 MPa (break) (Vestyron 114)
25–35 psi (1.0 pcf) (Fostafoam Type 3775)
40–60 psi (1.5 pcf) (Fostafoam Type 4775, 5775)
2700 psi (Cosden Polystyrene 825)
2800 psi (yield) (Lustrex 4220); (break) (PS 8720)
2900 psi (yield) (Lustrex 4300)
3000–3300 psi (yield) (Superflex 550)
3050 psi (break) (Superdense 710)
3100 psi (Cosden Polystyrene 625); (yield) (Fosta Tuf-Flex 707)
3200 psi (yield) (PS 8720)
3300 psi (yield) (Superflex 505)
3400 psi (Bapolan 6450)
3500 psi (Bapolan 6270); (yield) (PS-H63CC-1)
3500–4100 (break) (Superflex 550)
3600 psi (yield) (Lustrex 3350)
3700 psi (yield) (Lustrex 2220, 3300; Superflex 432); (break) (PS 8120)
3800 psi (Cosden Polystyrene 825E; Kama CS); (yield) (PS-H23MF-1)
3900 psi (yield) (PS 8120)
4000 psi (yield) (Fosta Tuf-Flex 702)
4118 psi (yield) (Superflex FR-1)
4200 psi (Cosden Polystyrene 710)
4400 psi (Bapolan 6445); (break) (Superdense 700)
5200 psi (yield) (Lustrex 2400); (break) (Superflex 405)
5600 psi (Cosden Polystyrene 500)
5800 psi (Cosden Polystyrene 525)
7000 psi (Superflow 390)
7500 psi (Cosden Polystyrene 550)
7800 psi (RTP 401)
8000 psi (Bapolan 6145)
8000–10,000 psi (Polypenco Q200.5)
8000–12,000 psi (Kama OPS)
8100 psi (Bapolan 6050)
11,000 psi (RTP 403)
11,500 psi (Thermocomp CF-1004)
12,000 psi (RTP 405)
13,500 psi (Thermocomp CF-1006)
14,000 psi (RTP 407)
15,000 psi (Thermocomp CF-1008)
Tens. Elong.:
1.0% (RTP 403, 405, 407)
1.2% (break) (Amoco G3)
1.3% (RTP 401)
1.4% (break) (Amoco G18)

1.5% (break) (Amoco G1, G2, R12; Fostarene 817)

1.7% (break) (Amoco R7)

2.0% (Styron 678U; Superflow 390); (yield and break) (Styron 666U); (break) (Amoco R3, R5; Vestyron 214, 214A, 314, 325)

2–3% (Thermocomp CF-1004, CF-1006, CF-1008)

2.2% (break) (Superdense 710)

2.4% (yield and break) (Styron 685D)

2.5% (break) (Fostalite; Fostarene 50, 57, 58)

3.0% (Polypenco Q200.5; PS 110R, 110S, 110SL, 224EP, 224EV, 224M); (break) (Amoco R1, R2; Superdense 700; Vestyron 114)

3–40% (Kama OPS)

8% (yield) (PS-H23MF-1)

10% (break) (Superflex 405)

17.5% (Kama CS)

20% (Fosta Tuf-Flex 271); (break) (Vestyron 616)

25% (Fosta Tuf-Flex 326, 929; PS 310S); (break) (Amoco M2R)

30% (Cosden Polystyrene 625, 710; PS 314/E, 344/E, 444, 444HH); (break) (Fosta Tuf-Flex 702; Lustrex 2220, 2400; Superflex 432; Vestyron 512, 512A, 512TA)

35% (Fosta Tuf-Flex 283, 474, 721-M; PS 510TG); (break) (Styron 430U)

40% (Fosta Tuf-Flex 240, 717; PS 410S); (break) (Amoco M9, M9R; Lustrex 3350; PS 8120; Styron 412D, 420D; Superflex 505, FR-1; Vestyron 620, 620A, 620TA, 628); (yield) (PS-H63CC-1)

45% (Fosta Tuf-Flex 782); (break) (Amoco H4E, H4R; PS 8720; Styron 492U, 495; Superflex 550; Vestyron 640)

50% (Cosden Polystyrene 825; Fosta Tuf-Flex 329, 730, 742, 880); (break) (Amoco H2R; Fosta Tuf-Flex 707, 840; Lustrex 3300; Styron 470, 475U; Vestyron 638)

55% (Cosden Polystyrene 825E)

56% (break) (Amoco H4)

65% (break) (Amoco H5M; Lustrex 4220, 4300)

70% (break) (Amoco H3E)

Tens. Mod.:

18,980 kg/cm² (Styron 492U)

19,500 kg/cm² (Styron 475U)

21,000 kg/cm² (Styron 470)

21,090 kg/cm² (Styron 495)

23,200 kg/cm² (Styron 420D)

23,900 kg/cm² (Styron 430U)

26,000 kg/cm² (Styron 412D)

32,700 kg/cm² (Styron 678U)

33,400 kg/cm² (Styron 666U)

34,100 kg/cm² (Styron 685D)

1200 MPa (0.6 g/cc density) (Vestyron TSG 602, TSG 604)

1500 MPa (0.6 g/cc density) (Vestyron TSG 502, TSG 504)

Polystyrene *(cont'd.)*

1800 MPa (Fosta Tuf-Flex 730, 880)

1850 MPa (Vestyron 640)

1930 MPa (Fosta Tuf-Flex 840)

1960 MPa (Fosta Tuf-Flex 717)

2000 MPa (Fosta Tuf-Flex 329; Vestyron 638)

2070 MPa (Fosta Tuf-Flex 742)

2140 MPa (Fosta Tuf-Flex 240, 474, 721-M, 782)

2150 MPa (Vestyron 620A)

2200 MPa (Vestyron 620, 620TA)

2280 MPa (Fosta Tuf-Flex 283)

2300 MPa (Vestyron 628)

2346 MPa (Fosta Tuf-Flex 929)

2400 MPa (Vestyron 616)

2415 MPa (Fosta Tuf-Flex 271, 326)

2600 MPa (Vestyron 512A)

2650 MPa (Vestyron 512, 512TA)

2760 MPa (Fostarene 817)

2800 MPa (Vestyron 516)

3000 psi (Superdense 710)

3100 MPa (Fostalite; Fostarene 50, 57, 58)

3200 MPa (Vestyron 214A, 314, 325)

3250 MPa (Vestyron 214)

3350 MPa (Vestyron 114)

3.2 GPa (Amoco G1, G2, G3, G18, R1, R2, R3, R5, R7, R12)

4000 psi (Superdense 700)

230,000 psi (Superflex 550)

240,000 psi (Cosden Polystyrene 825; Fosta Tuf-Flex 707; Polypenco Q200.5; Superflex 505)

250,000 psi (Cosden Polystyrene 825E; Superflex 432)

260,000 psi (PS 8120)

270,000 psi (Lustrex 4220; PS 8720)

275,000 psi (Lustrex 4300)

290,000 psi (Cosden Polystyrene 625)

300,000 psi (Cosden Polystyrene 710; Fosta Tuf-Flex 702)

320,000 psi (Lustrex 3350)

330,000 psi (Cosden Polystyrene 500; Superflex FR-1)

340,000 psi (Lustrex 3300)

350,000 psi (Cosden Polystyrene 525)

360,000 psi (Cosden Polystyrene 550; Lustrex 2220)

400,000 psi (Lustrex 2400)

420,000 psi (Superflow 390)

440,000 psi (Superflex 405)

450,000 psi (Kama OPS)

900,000 psi (RTP 401)
1.05×10^6 psi (Thermocomp CF-1004)
1.2×10^6 psi (RTP 403)
1.3×10^6 psi (RTP 405; Thermocomp CF-1006)
1.65×10^6 psi (Thermocomp CF-1008)
2.0×10^6 psi (RTP 407)

Flex. Str.:
0.15–0.23 MPa (15 kg/m³ density) (Vestypor 01N, 01S, 10N, 10S, 20N, 20S, 30N, 30S)
20 MPa (break, 0.6 g/cc density) (Vestyron TSG 602, TSG 604)
21 MPa (break, 0.6 g/cc density) (Vestyron TSG 502, TSG 504)
33.1 MPa (yield) (PS 510TG)
34.5 MPa (yield) (PS 314/E, 344/E, 410S)
35 MPa (Vestyron 640)
37.9 MPa (yield) (PS 444)
38 MPa (Vestyron 620A, 620TA)
40 MPa (Vestyron 620)
42 MPa (Vestyron 512A, 512TA)
45 MPa (Vestyron 512, 638)
48.3 MPa (yield) (PS 310S, 444HH)
50 MPa (Vestyron 628)
53.8 MPa (yield) (PS 224EV)
54 MPa (Amoco G3)
55 MPa (Amoco G2, R12)
56 MPa (Amoco G18)
56.5 MPa (yield) (PS 224EP)
59 MPa (Amoco G1)
60.7 MPa (yield) (PS 224M)
65 MPa (Vestyron 616)
68.9 MPa (yield) (PS 110R)
70 MPa (Vestyron 516)
75 MPa (break) (Vestyron 325)
76 MPa (Amoco R5)
79 MPa (Amoco R7)
80 MPa (break) (Vestyron 314)
83 MPa (Amoco R3)
85 MPa (break) (Vestyron 214A)
86 MPa (Fostalite)
87 MPa (Amoco R1, R2)
89.6 MPa (yield) (PS 110S, 110SL)
90 MPa (break) (Vestyron 214)
110 MPa (Vestyron 114)
25–35 psi (1.0 pcf) (Fostafoam Type 3775)
40–60 psi (1.5 pcf) (Fostafoam Type 4775, 5775)

Polystyrene *(cont'd.)*

    3200–4000 psi (yield) (Superflex 550)
    4500 psi (break) (PS 8720)
    5300 psi (Cosden Polystyrene 825)
    5500 psi (Bapolan 6450)
    5600 psi (Cosden Polystyrene 625)
    5700 psi (PS-H23MF-1)
    5704 psi (Superflex FR-1)
    5800 psi (break) (PS 8120)
    5900 psi (Fosta Tuf-Flex 702)
    6100 psi (PS-H63CC-1)
    6300 psi (Bapolan 6445)
    6500 psi (Cosden Polystyrene 825E)
    7700 psi (Cosden Polystyrene 710)
    9800 psi (Cosden Polystyrene 525)
    10,000 psi (Cosden Polystyrene 500)
    10,000–15,000 psi (Polypenco Q200.5)
    13,000 psi (Bapolan 6050, 6145)
    14,200 psi (RTP 401)
    14,600 psi (Cosden Polystyrene 550)
    15,000 psi (Thermocomp CF-1004)
    16,200 psi (Thermocomp CF-1006)
    16,500 psi (RTP 403)
    16,800 psi (RTP 405)
    17,500 psi (Thermocomp CF-1008)
    19,000 psi (RTP 407)
    360,000 psi (Bapolan 6270)

Flex. Mod.:
    1.7 GPa (Amoco H3E)
    1.8 GPa (Amoco H5M)
    1.9 GPa (Amoco H4E, H4R)
    2.0 GPa (Amoco H4)
    2.1 GPa (Amoco M9R)
    2.2 GPa (Amoco H2R, M9)
    2.8 GPa (Amoco M2R)
    3.1 GPa (Amoco G1, G2, G3, G18, R1, R2, R3, R5, R7, R12)
    319 psi (Superflex FR-1)
    145,000 psi (PS-H63CC-1)
    180,000 psi (PS-H23MF-1)
    270,000 psi (Fyrid KS1)
    280,000 psi (Cosden Polystyrene 710; PS 8720)
    300,000 psi (Cosden Polystyrene 825, 825E; PS 8120)
    360,000 psi (Cosden Polystyrene 625)
    380,000 psi (Cosden Polystyrene 525)

390,000 psi (Cosden Polystyrene 500)
430,000 psi (Cosden Polystyrene 550)
850,000 psi (RTP 401)
950,000 psi (Thermocomp CF-1004)
1.1×10^6 psi (RTP 403)
1.2×10^6 psi (RTP 405; Thermocomp CF-1006)
1.5×10^6 psi (Thermocomp CF-1008)
1.6×10^6 psi (RTP 407)
Compr. Str.:
0.08–0.12 MPa (15 kg/m³ density) (Vestypor 01N, 01S, 10N, 10S, 20N, 20S, 30N, 30S)
8–13 psi (1.0 pcf) (Fostafoam Type 3775)
18–22 psi (1.5 pcf) (Fostafoam Type 4775, 5775)
14,500 psi (RTP 401)
16,000 psi (RTP 403)
16,700 psi (Thermocomp CF-1004)
17,000 psi (RTP 405)
17,300 psi (Thermocomp CF-1006)
17,500 psi (RTP 407)
17,800 psi (Thermocomp CF-1008)
Tear Str.:
1.1 lb/mil (Kama OPS)
Shear Str.:
0.45–0.55 MPa (15 kg/m³ density) (Vestypor 01N, 01S, 10N, 10S, 20N, 20S, 30N, 30S)
7500–8000 psi (Polypenco Q200.5)
Impact Str. (Izod):
2.4 cm kg/cm notched (Styron 666U, 678U, 687D)
6.5 cm kg/cm notched (Styron 412D)
9.2 cm kg/cm notched (Styron 420D)
10.9 cm kg/cm notched (Styron 492U)
13.0 cm kg/cm notched (Styron 430U)
15.8 cm kg/cm notched (Styron 470)
25.5 cm kg/cm notched (Styron 495)
28.7 cm kg/cm notched (Styron 475U)
16 J/m notched (PS 110R, 110S, 110SL, 224EP, 224EV, 224M)
19 J/m notched (Fosatrene 817)
21 J/m (Fostalite; Fostarene 57, 58); notched (Fostarene 50)
30 J/m notched (Amoco M2R)
40 J/m notched (Amoco M9R)
43 J/m notched (Fosta Tuf-Flex 326)
50 J/m notched (Amoco M9)
51 J/m notched (Fosta Tuf-Flex 929)
59 J/m notched (Fosta Tuf-Flex 271)
69 J/m notched (Fosta Tuf-Flex 782; PS 310S)

Polystyrene *(cont'd.)*

70 J/m notched (Amoco H4R)
75 J/m notched (Amoco H2R; Fosta Tuf-Flex 717; PS 314/E, 344/E)
78 J/m notched (Fosta Tuf-Flex 283)
80 J/m notched (Amoco H5M; Fosta Tuf-Flex 474, 721-M)
90 J/m notched (Amoco H4E)
92 J/m notched (Fosta Tuf-Flex 730)
95 J/m notched (Amoco H4)
96 J/m notched (PS 444HH)
105 J/m notched (Amoco H3E)
107 J/m notched (Fosta Tuf-Flex 240, 329, 742)
110 J/m notched (PS 410S, 444)
133 J/m notched (PS 510TG)
161 J/m (Fosta Tuf-Flex 840, 880)
0.25 ft lb/in. notched (Superflow 390)
0.35–0.50 ft lb/in. (Polypenco Q200.5)
0.44 ft lb/in. notched (Superdense 710)
0.46 ft lb/in. notched (Superflex 405)
0.5 ft lb/in. (Bapolan 6145)
0.6 ft lb/in. (Bapolan 6050)
0.7 ft lb/in. notched (RTP 401)
0.9 ft lb/in. notched (Cosden Polystyrene 625; Lustrex 2220, 2400; Superflex 432; Thermocomp CF-1004)
0.95 ft lb/in. notched (PS-H63CC-1)
1.0 ft lb/in. (Bapolan 6270); notched (RTP 403; Superflex 505, FR-1; Thermocomp CF-1006)
1.09 ft lb/in. notched (PS-H23MF-1)
1.1 ft lb/in. notched (RTP 405, 407)
1.2 ft lb/in. notched (Cosden Polystyrene 710; Thermocomp CF-1008)
1.3 ft lb/in. notched (Lustrex 3300, 3350, 4220; Superflex 550)
1.6 ft lb/in. (Bapolan 6450)
1.7 ft lb/in. (Bapolan 6445)
1.8 ft lb/in. notched (Lustrex 4300)
1.9 ft lb/in. (Fosta Tuf-Flex 702)
2.0 ft lb/in. notched (Cosden Polystyrene 825, 825E; PS 8120)
2.2 ft lb/in. unnotched (Fyrid KS1)
3.2 ft lb/in. (Fosta Tuf-Flex 707)
3.3 ft lb/in. notched (PS 8720)
4.0 ft lb/in. notched (Superdense 700)
Impact Str. (Charpy):
6 mJ/mm² (0.6 g/cc density) (Vestyron TSG 502, TSG 504)
10 mJ/mm²) (0.6 g/cc density) (Vestyron TSG 602, TSG 604)
12 mJ/mm² (Vestyron 325)
15 mJ/mm² (Vestyron 314)

18 mJ/mm² (Vestyron 214, 214A)
20 mJ/mm² (Vestyron 114)
50 mJ/mm² (Vestyron 512, 512A, 512TA, 516)
70 mJ/mm² (Vestyron 616)
75 mJ/mm² (Vestyron 628)
80 mJ/mm² (Vestyron 620, 620A, 620TA)
No break (Vestyron 638, 640)

Hardness:

Ball Indentation 75 MPa (Vestyron 640)
Ball Indentation 85 MPa (Vestyron 638)
Ball Indentation 90 MPa (Vestyron 620A)
Ball Indentation 95 MPa (Vestyron 620TA)
Ball Indentation 100 MPa (Vestyron 620, 628)
Ball Indentation 105 MPa (Vestyron 512A, 512TA, 616)
Ball Indentation 120 MPa (Vestyron 512)
Ball Indentation 130 MPa (Vestyron 516)
Ball Indentation 150 MPa (Vestyron 114, 214, 214A, 314, 325)
Rockwell L38 (Fosta Tuf-Flex 707)
Rockwell L44 (Styron 495)
Rockwell L50 (Fosta Tuf-Flex 840, 880; PS 510TG)
Rockwell L55 (Fosta Tuf-Flex 240, 702; PS 410S, 8720)
Rockwell L56 (Fosta Tuf-Flex 730)
Rockwell L60 (Fosta Tuf-Flex 742)
Rockwell L62 (PS 8120; Styron 470)
Rockwell L65 (Fosta Tuf-Flex 329, 717; PS 314/E, 344/E, 444)
Rockwell L67 (PS 444HH)
Rockwell L74 (PS 310S)
Rockwell L75 (Fosta Tuf-Flex 283, 474, 721-M, 782)
Rockwell M10 (Cosden Polystyrene 825)
Rockwell M15 (Bapolan 6445)
Rockwell M20 (Bapolan 6450; Cosden Polystyrene 825E)
Rockwell M30 (Bapolan 6270)
Rockwell M35 (Cosden Polystyrene 625)
Rockwell M40 (Cosden Polystyrene 710; Fosta Tuf-Flex 929; Styron 420D)
Rockwell M45 (Fosta Tuf-Flex 271, 326)
Rockwell M48 (Superflex FR-1)
Rockwell M60 (Cosden Polystyrene 550)
Rockwell M65 (Cosden Polystyrene 525)
Rockwell M70 (PS 110R, 110S, 110SL, 224EP, 224EV)
Rockwell M71 (Styron 678U)
Rockwell M74 (Styron 666U)
Rockwell M75 (Cosden Polystyrene 500; Fostalite; Fostarene 50, 57, 58, 817; PS 224M)

Polystyrene *(cont'd.)*

Rockwell M76 (Styron 685D)
Rockwell M80 (Bapolan 6050, 6145)
Rockwell M85 (Kama OPS)
Rockwell M90 (Thermocomp CF-1004)
Rockwell M92 (Thermocomp CF-1006)
Rockwell M93 (Thermocomp CF-1008)
Rockwell R110–120 (Polypenco Q200.5)
Rockwell R117 (RTP 401)
Rockwell R119 (RTP 403)
Rockwell R121 (RTP 405, 407)
Rockwell R140 (Kama CS)

THERMAL PROPERTIES:

Soften. Pt. (Vicat):

76 C (Vestyron 512TA)
77 C (Vestyron 620TA)
78 C (Vestyron 512A, 620A)
80 C (Vestyron 314, 512, 620, 640)
83 C (Vestyron 325)
85 C (Amoco G3)
86 C (Vestyron 214A)
88 C (Amoco M9; Vestyron 214)
90 C (Styron 678U; Vestyron 638)
91 C (Amoco G2; Fosta Tuf-Flex 326; Vestyron 628)
92 C (Fostarene 817; Fosta Tuf-Flex 929)
93 C (Amoco G1, G18; PS 224M)
95 C (Fosta Tuf-Flex 707)
96 C (Fosta Tuf-Flex 742; Lustrex 4220; PS 224EP; Vestyron 616)
97 C (Fosta Tuf-Flex 717, 782, 880; Lustrex 2220, 4300; Styron 495; Vestyron 516)
98 C (PS 314/E, 344/E, 444, 510TG)
99 C (Amoco H4, H4E, H5M, M9R; Fosta Tuf-Flex 702)
100 C (Fosta Tuf-Flex 283, 329; PS 410S; Styron 666U)
101 C (Fosta Tuf-Flex 240, 721-M, 730, 840; Vestyron 114)
102 C (Amoco H4R, M2R, R3, R5, R7, R12; Lustrex 3300; PS 310S)
103 C (Amoco H3E; Fosta Tuf-Flex 474; Lustrex 3350; PS 444HH)
104 C (Amoco H2R; Fosta Tuf-Flex 271; Styron 492U)
105 C (Lustrex 2400)
106 C (Amoco R2; Fostalite)
107 C (Fostarene 50, 57, 58; PS 110SL)
108 C (Amoco R1; Styron 685D)
167 F (Superdense 710)
176 F (Superdense 700)
190 F (Bapolan 6270; Superflex 405)
192 F (Cosden Polystyrene 625)

195 F (Superflex 550)
196 F (Cosden Polystyrene 500)
200 F (Cosden Polystyrene 525, 825; Superflex 432)
202 F (Bapolan 6450)
205 F (Superflex 505)
208 F (PS-H23MF-1)
210 F (Bapolan 6445; Cosden Polystyrene 710, 825E; PS 8720)
212 F (Bapolan 6050; PS-H63CC-1)
213 F (Superflow 390)
215 F (PS 8120)
220 F (Bapolan 6145)
224 F (Cosden Polystyrene 550)

Conduct.:

0.040 W/m (15 kg/m^3 density) (Vestypor 01N, 01S, 10N, 10S, 20N, 20S, 30N, 30S)
0.16 W/m K (Vestyron 114, 214, 214A, 314, 325)
0.17 W/m K (Vestyron 512, 512A, 512TA, 516, 616, 620, 620A, 620TA, 628, 638, 640)
0.0002 Cal/s/cm^2/C/cm (Kama OPS)
0.25 Btu/h/ft^2/F/in. (1.5 pcf) (Fostafoam Type 4775, 5775)
0.26 Btu/h/ft^{2}F/in. (1.0 pcf) (Fostafoam Type 3775)
1.3 Btu/h/ft^2/F/in. (RTP 401)
1.7 Btu/h/ft^2/F/in. (RTP 403)
1.8 Btu/h/ft^2/F/in. (Thermocomp CF-1004)
2.0 Btu/h/ft^2/F/in. (Thermocomp CF-1006)
2.1 Btu/h/ft^2/F/in. (RTP 405)
2.2 Btu/h/ft^2/F/in. (RTP 407; Thermocomp CF-1008)

Distort. Temp.:

67 C (Vestyron 620A, 620TA)
68 C (Amoco G3; Vestyron 512TA)
70 C (Vestyron 512A, 640)
71 C (Vestyron 314)
72 C (Vestyron 512, 620)
74 C (Amoco M9; Fosta Tuf-Flex 326; Vestyron 325); (264 psi) (Fosta Tuf-Flex 929)
76 C (Vestyron 214A)
77 C (Amoco G1; Vestyron 214)
78 C (18.6 kg/cm^2) (Styron 475U)
78–80 C (15 kg/m^3 density) (Vestypor 01N, 01S, 10N, 10S, 20N, 20S, 30N, 30S)
79 C (Amoco G2, H4E, H5M, M9R; Fostarene 817)
80 C (Vestyron 638); (264 psi) (Fosta Tuf-Flex 707)
81 C (Fosta Tuf-Flex 283, 742); (18.6 kg/cm^2) (Styron 495, 678U)
82 C (Amoco G18, H2R, H3E, H4, H4R); (264 psi) (Fosta Tuf-Flex 240, 880; Lustrex 4220)
83 C (264 psi) (Fosta Tuf-Flex 702)

Polystyrene *(cont'd.)*

84 C (Vestyron 628); (264 psi) (Fosta Tuf-Flex 730; Lustrex 2220, 4300)

85 C (Amoco M2R, R3, R5, R7, R12; Vestyron 616); (264 psi) (Fosta Tuf-Flex 329, 717, 782)

87 C (Vestyron 516)

88 C (264 psi) (Fosta Tuf-Flex 721-M, 840; Lustrex 3300)

89 C (Amoco R2; Vestyron 114); (264 psi) (Lustrex 3350)

90 C (18.6 kg/cm²) (Styron 470)

91 C (264 psi) (Fostalite; Fosta Tuf-Flex 474; Lustrex 2400)

92 C (Amoco R1); (18.6 kg/cm²) (Styron 430U, 666U)

93 C (264 psi) (Fostarene 50, 57, 58; Fosta Tuf-Flex 271)

95 C (18.6 kg/cm²) (Styron 420D)

96 C (18.6 kg/cm²) (Styron 492U)

103 C (18.6 kg/cm²) (Styron 685D)

145 F (264 psi) (Superdense 710)

150 F (Cosden Polystyrene 625)

155 F (Cosden Polystyrene 500, 825); (264 psi) (Superdense 700)

160 F (Cosden Polystyrene 525); (264 psi) (Fyrid KS1)

162 F (Cosden Polystyrene 825E; Superflex FR-1)

162–172 F (264 psi) (Superflex 550)

165 F (Cosden Polystyrene 710); (264 psi) (PS-H63CC-1)

170 F (264 psi) (PS-H23MF-1; Superflex 405)

174 F (Cosden Polystyrene 550)

180 F (264 psi) (PS 8720; RTP 401)

185 F (264 psi) (Superflex 505; Superflow 390)

188 F (264 psi) (PS 8120)

200 F (264 psi) (RTP 403; Superflex 432; Thermocomp CF-1004)

200–225 F (264 psi) (Polypenco Q200.5)

210 F (264 psi) (RTP 405, 407)

215 F (264 psi) (Thermocomp CF-1006)

220 F (264 psi) (Thermocomp CF-1008)

Coeff. of Linear Exp.:

6.0×10^{-5} K^{-1} (Vestypor 01N, 01S, 10N, 10S, 20N, 20S, 30N, 30S; Vestyron 638, 640)

7.0×10^{-5} K^{-1} (Vestyron 114, 214, 214A, 314, 325)

9.0×10^{-5} K^{-1} (Vestyron 512, 512A, 512TA, 516, 616, 620, 620A, 620TA, 628)

6.8×10^{-5} in./in./C (Kama OPS)

1.6×10^{-5} in./in./F (Thermocomp CF-1008)

1.9×10^{-5} in./in./F (Thermocomp CF-1006)

2.2×10^{-5} in./in./F (Thermocomp CF-1004)

3.5×10^{-5} in./in./F (1.0 pcf) (Fostafoam Type 3775); (1.5 pcf) (Fostafoam Type 4775, 5775)

5.0×10^{-5} in./in./F (Polypenco Q200.5)

Sp. Heat:

0.24 Btu/lb/F (Thermocomp CF-1008)

0.26 Btu/lb/F (Thermocomp CF-1006)
0.28 Btu/lb/F (Thermocomp CF-1004)
Flamm.:
V-0/5V (Fyrid KS1)
V-0 (Fosta Tuf-Flex 702; PS 8120, 8720; Superflex FR-1)
HB (Amoco G1, G2, G3, H3E, H4, H4E, H4R, M2R, M9, M9R, R1, R2, R3, R5; RTP
401, 403, 405, 407)
Slow burning (Kama CS, OPS)
0.8 in./min (Thermocomp CF-1008)
0.9 in./min (Thermocomp CF-1006)
1.0 in./min (Thermocomp CF-1004)

ELECTRICAL PROPERTIES:
Dissip. Factor:
0.00003–0.0005 (Vestypor 01N, 01S, 10N, 10S, 20N, 20S, 30N, 30S)
0.0001–0.0003 (60 cycles) (Kama OPS)
0.0002 (1 MHz) (Polypenco Q200.5; Vestyron 214); (1000 cps) (Styron 666U, 678U,
685D)
0.0003 (60×10^6 Hz) (Fostalite); (1 MHz) (Vestyron 325)
0.0004 (1 MHz) (Vestyron 314)
0.0004–0.0008 (1000 cps) (Styron 412D, 420D, 430D, 470, 475U, 492U, 495)
0.0005 (1 MHz) (Vestyron 516)
0.0006 (1 MHz) (Vestyron 515, 628, 638)
0.0007 (1 MHz) (Vestyron 640)
0.0007–0.008 (60 to 10^6 Hz) (Thermocomp CF-1006)
0.00075 (1 MHz) (Vestyron 512, 620)
0.001 (1 MHz) (Lustrex 2220, 3350, 4220; Vestyron 114); (1 MC) (RTP 401, 403)
0.002 (1 MC) (RTP 405)
0.003 (1 MC) (RTP 407)
0.005 (1 MHz) (Vestyron 214A)
0.010 (1 MHz) (Vestyron 512A, 512TA, 620A, 620TA)
Dielec. Str.:
2 kV/mm (Vestypor 01N, 01S, 10N, 10S, 20N, 20S, 30N, 30S)
22.6 kV/mm (Fostalite)
> 100 kV/mm (Vestyron 114, 214, 214A, 314, 325, 512, 512A, 512TA, 516, 616, 620,
620A, 620TA, 628, 638, 640)
300–500 V/mil (Styron 412D, 420D, 430D, 470, 475U, 492U, 495)
425 V/mil (RTP 401, 403)
450 V/mil (RTP 405, 407)
500 V/mil (Styron 666U, 678U, 685D)
500–700 V/mil (Kama OPS)
550 V/mil (Thermocomp CF-1006)
700–1200 V/mil (Polypenco Q200.5)

Polystyrene *(cont'd.)*

Dielec. Const.:
1.02–1.104 (100–400 MHz) (Vestypor 01N, 01S, 10N, 10S, 20N, 20S, 30N, 30S)
2.5 (Kama OPS); (1 MHz) (Vestyron 214, 314, 325, 512, 516)
2.50–2.53 (1000 cps) (Styron 412D, 420D, 430D, 470, 475U, 492U, 495)
2.54 (1000 cps) (Styron 666U, 678U, 685D)
2.55 (1 MHz) (Lustrex 2220; Polypenco Q200.5)
2.6 (60×10^6 Hz) (Fostalite); (1 MHz) (Lustrex 3350, 4220; Vestyron 616, 620, 628, 638, 640)
2.7 (1 MHz) (Vestyron 214A, 512A, 512TA, 620A, 620TA)
2.81 (60 to 10^6 Hz) (Thermocomp CF-1006)
2.9 (1 MC) (RTP 401)
3.0 (1 MC) (RTP 403)
3.2 (1 MC) (RTP 405)
3.5 (1 MC) (RTP 407)

Vol. Resist.:
> 10^{14} ohm-cm (Fostalite; Vestyron 214A, 512A, 512TA, 620A, 620TA)
10^{15} ohm-cm (Polypenco Q200.5)
10^{16} ohm-cm (Lustrex 2220, 3350, 4220; RTP 401, 403, 405, 407; Vestyron 616)
> 10^{16} ohm-cm (Styron 412D, 420D, 430D, 470, 475U, 492U, 495, 666U, 678U, 685D; Vestyron 114, 214, 314, 325, 512, 516, 620, 628, 638, 640)
10^{17}–10^{19} ohm-cm (Kama OPS)

Surf. Resist.:
10^{10} ohm (Vestyron 214A, 512A, 512TA, 620A, 620TA)
> 10^{13} ohm (Vestyron 114, 214, 314, 325, 512, 516, 616, 620, 628, 638, 640)
4×10^{13} ohm (Vestypor 01N, 01S, 10N, 10S, 20N, 20S, 30N, 30S)

Arc Resist.:
20–135 s (Styron 412D, 420D, 430D, 470, 475U, 492U, 495)
40 s (RTP 407)
50 s (RTP 405)
60 s (RTP 403)
60–135 s (Styron 666U, 678U, 685D)
75 s (RTP 401)

TOXICITY/HANDLING:
May release hazardous or toxic vapors on prolonged exposure to high heat—use with adequate ventilation (Fosta Tuf-Flex 702, 707)
Low degree of toxicity; inhalation of dust or vapors may be hazardous; avoid contact with molten resin during fabrication (Styron 412D, 420D, 430D, 470, 475U, 492U, 495, 666U, 678U, 685D)

STORAGE/HANDLING:
Avoid melt temps. > 500 F to prevent resin degradation and possible machine damage through corrosion and/or high internal pressure (Fosta Tuf-Flex 702)
Avoid melt temps. > 475 F to prevent resin degradation and possible machine damage through corrosion and/or high internal pressure (Fosta Tuf-Flex 707)

Product does not absorb moisture; however, store in a cool, dry place; should moisture be found it may be removed by drying in an air-circulating oven for 1–2 h @ 160 F (Fostalite)

Dispose of waste product as hazardous material (PS 8120, 8720)

Combustible (Styron 412D, 420D, 430D, 470, 475U, 492U, 495, 666U, 678U, 685D)

Polysulfone

SYNONYMS:
PSO

STRUCTURE:

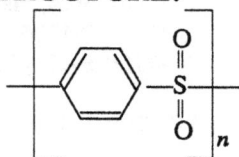

TRADENAME EQUIVALENTS:
Udel P-1700, P-1700FR, P-1710, P-1720, P-3500 [Amoco]

Carbon-reinforced:
Stat-Kon GC-1003 (15% carbon fiber), GC-1006 (30% carbon fiber) [LNP]
Thermocomp GC-1006 (30% carbon fiber) [LNP]

Glass-reinforced:
PF-10GF/15T (10% glass fiber), -20GF/15T (20% glass fiber), -30GF/15T (30% glass fiber) [Compounding Technology]
RTP 901 (10% glass fiber), 903 (20% glass fiber), 905 (30% glass fiber), 907 (40% glass fiber) [Fiberite]
Thermocomp GF-1004 (20% glass fiber), GF-1006 (30% glass fiber), GF-1006FR (30% glass fiber), GF-1008 (40% glass fiber) [LNP]
Udel GF-110, GF-120, GF-130 [Amoco]

MODIFICATIONS/SPECIALTY GRADES:
Statically dissipative:
Stat-Kon GC-1003, GC-1006

Flame-retardant:
RTP 905, 907; Stat-Kon GC-1003, GC-1006; Thermocomp GC-1006, GF-1004, GF-1006, GF-1006FR, GF-1008; Udel GF-110, GF-120, GF-130, P-1700, P-1700FR, P-1710, P-1720

Polysulfone *(cont'd.)*

Low smoke generation grade:
Thermocomp GF-1006FR
PTFE/TFE-lubricated:
PF-10GF/15T (15% PTFE), -20GF/15T (15% PTFE), -30GF/15T (15% PTFE)

CATEGORY:
Engineering thermoplastic resin

PROCESSING:
Extrusion:
Udel P-1700, P-1700FR, P-1710, P-1720, P-3500
Injection molding:
RTP 901, 903, 905, 907; Thermocomp GC-1006, GF-1004, GF-1006, GF-1006FR, GF-1008; Udel GF-110, GF-120, GF-130, P-1700, P-1700FR, P-1710, P-1720

APPLICATIONS:
Automotive applications: (Udel P-1700, P-1700FR, P-1710, P-1720, P-3500); fuses (Udel P-1700, P-1700FR, P-1710, P-1720, P-3500); sensors (Udel P-1700, P-1700FR, P-1710, P-1720, P-3500); under-the-hood parts (Udel P-1700, P-1700FR, P-1710, P-1720, P-3500)

Aviation industry: (Thermocomp GF-1006FR); aerospace (Udel P-1700, P-1700FR, P-1710, P-1720, P-3500)

Consumer products: appliances (Udel P-1700, P-1700FR, P-1710, P-1720, P-3500); microwave cookware (Udel P-1700, P-1700FR, P-1710, P-1720, P-3500)

Electrical/electronic industry: (RTP 901, 903, 905, 907; Udel P-1700, P-1700FR, P-1710, P-1720, P-3500); connectors (Udel P-1700, P-1700FR, P-1710, P-1720, P-3500); electrical components (Stat-Kon GC-1003, GC-1006); electrical enclosures/packaging (Stat-Kon GC-1003, GC-1006); housings (Udel P-1700, P-1700FR, P-1710, P-1720, P-3500); printed circuit boards (Udel P-1700, P-1700FR, P-1710, P-1720, P-3500); TV/stereo components (Udel P-1700, P-1700FR, P-1710, P-1720, P-3500)

FDA-approved applications: (Udel P-1700, P-1710)

Food service applications: (RTP 901, 903, 905, 907); milking machine parts (Udel P-1700, P-1700FR, P-1710, P-1720, P-3500)

Industrial applications: insulation (Thermocomp GF-1004, GF-1006, GF-1008); process and sanitary pipe (Udel P-1700, P-1700FR, P-1710, P-1720, P-3500); plumbing (RTP 901, 903, 905, 907; Udel P-1700, P-1700FR, P-1710, P-1720, P-3500); pollution control equipment (Udel P-1700, P-1700FR, P-1710, P-1720, P-3500); pumps (RTP 901, 903, 905, 907); structural components (Thermocomp GF-1004, GF-1006, GF-1008)

Medical applications: dental and surgical instruments (Udel P-1700, P-1700FR, 1710, P-1720, P-3500); instrument housings (Udel P-1700, P-1700FR, P-1710, P-1720, P-3500); pacemakers (Udel P-1700, P-1700FR, P-1710, P-1720, P-3500); surgical tool trays (Udel P-1700, P-1700FR, P-1710, P-1720, P-3500)

PROPERTIES:
Form:
    Transparent pellets (Udel P-1700, P-3500)
    Translucent pellets (Udel P-1700FR, P-1720)
    Opaque pellets (Udel P-1710)
Color:
    Natural, black (RTP 901, 903, 905, 907)
    Light beige (Udel P-1710)
    Light amber; avail. opaque and transparent colors (Udel P-1700, P-3500)

GENERAL PROPERTIES:
Melt Flow:
    4 dgm/min (Udel GF-110)
    7–8 dgm/min (Udel GF-120, GF-130)
    3.5 g/10 min (Udel P-3500)
    6.5 g/10 min (Udel P-1700, P-1700FR, P-1710, P-1720)
Sp. Gr.:
    1.30 (Stat-Kon GC-1003)
    1.32 (RTP 901)
    1.36 (PF-10GF/15T)
    1.37 (Stat-Kon GC-1006; Thermocomp GC-1006)
    1.38 (RTP 903; Thermocomp GF-1004)
    1.45 (Thermocomp GF-1006)
    1.46 (RTP 905; Thermocomp GF-1006FR)
    1.49 (PF-20GF/15T)
    1.55 (Thermocomp GF-1008)
    1.56 (RTP 907)
    1.59 (PF-30GF/15T)
Sp. Vol.:
    17.9 in.^{3}/lb (Thermocomp GF-1008)
    19.0 in.^{3}/lb (Thermocomp GF-1006)
Density:
    1.24 mg/m^3 (Udel P-1700, P-1700FR, P-3500)
    1.25 mg/m^3 (Udel P-1710, P-1720)
    1.33 mg/m^3 (Udel GF-110)
    1.40 mg/m^3 (Udel GF-120)
    1.49 mg/m^3 (Udel GF-130)

MECHANICAL PROPERTIES:
Tens. Str.:
    68.9 MPa (yield) (Udel P-1720)
    70.3 MPa (yield) (Udel P-1700, P-1700FR, P-1710, P-3500)
    78.0 MPa (Udel GF-110)
    96.6 MPa (Udel GF-120)

Polysulfone *(cont'd.)*

107.6 MPa (Udel GF-130)
11,000 psi (RTP 901)
11,800 psi (PF-10GF/15T)
13,000 psi (PF-20GF/15T)
15,000 psi (RTP 903; Thermocomp Gf-1004)
16,000 psi (PF-30GF/15T)
17,000 psi (RTP 905)
17,500 psi (Thermocomp Gf-1006FR)
18,000 psi (Stat-Kon GC-1003; Thermocomp GF-1006)
19,000 psi (RTP 907)
20,000 psi (Thermocomp GF-1008)
23,000 psi (Stat-Kon GC-1006; Thermocomp GC-1006)

Tens. Elong.:
1.7% (RTP 907)
2% (PF-20GF/15T; Stat-Kon GC-1006; Thermocomp GF-1008; Udel GF-130)
2–3% (PF-30GF/15T; Thermocomp GC-1006)
2.1% (RTP 905)
2.7% (Stat-Kon GC-1003)
2.8% (RTP 903)
3% (RTP 901; Thermocomp GF-1004, GF-1006; Udel GF-120)
4% (Udel GF-110)
50–100% (break) (Udel P-1700, P-1700FR, P-1710, P-1720, P-3500)

Tens. Mod.:
2482 MPa (Udel P-1700, P-1700FR, P-1710, P-1720, P-3500)
3655 MPa (Udel GF-110)
5172 MPa (Udel GF-120)
7379 MPa (Udel GF-130)
700,000 psi (RTP 901)
880,000 psi (RTP 903)
1.05×10^6 psi (PF-30GF/15T)
1.2×10^6 psi (RTP 905)
1.7×10^6 psi (RTP 907)

Flex. Str.:
106.2 MPa (Udel P-1700, P-1700FR, P-1710, P-1720, P-3500)
127.6 MPa (Udel GF-110)
148.3 MPa (Udel GF-120)
154.5 MPa (Udel GF-130)
18,000 psi (RTP 901)
20,000 psi (RTP 903)
21,000 psi (Thermocomp GF-1004)
23,000 psi (RTP 905; Stat-Kon GC-1006)
24,000 psi (Thermocomp GF-1006, GF-1006FR)
25,000 psi (RTP 907)

26,500 psi (Stat-Kon GC-1003)
27,000 psi (Thermocomp GF-1008)
32,000 psi (Thermocomp GC-1006)

Flex. Mod.:
2689 MPa (Udel P-1700, P-1700FR, P-1710, P-1720, P-3500)
3793 MPa (Udel GF-110)
5517 MPa (Udel GF-120)
7586 MPa (Udel GF-130)
550,000 psi (PF-10GF/15T)
600,000 psi (RTP 901)
700,000 psi (RTP 903)
850,000 psi (Thermocomp GF-1004)
10^6 psi (PF-30GF/15T; RTP 905)
1.2×10^6 psi (RTP 907; Thermocomp GF-1006, GF-1006FR)
1.4×10^6 psi (Stat-Kon GC-1003)
1.6×10^6 psi (Thermocomp GF-1008)
2.05×10^6 psi (Stat-Kon GC-1006; Thermocomp GC-1006)

Compr. Str.:
14,500 psi (RTP 901)
19,000 psi (RTP 903)
21,000 psi (Thermocomp GF-1004)
22,000 psi (RTP 905)
24,000 psi (RTP 907; Thermocomp GF-1006)
25,500 psi (Thermocomp GF-1008)

Shear Str.:
9000 psi (Thermocomp GF-1004)
9500 psi (Thermocomp GC-1006, GF-1006)
10,000 psi (Thermocomp GF-1008)

Impact Str. (Izod):
64 J/m notched (Udel GF-110)
69 J/m notched (Udel GF-120, P-1700, P-1700FR, P-1710, P-1720, P-3500)
75 J/m notched (Udel GF-130)
1.0 ft lb/in. notched (PF-10GF/15T, -20GF/15T; RTP 901; Stat-Kon GC-1003)
1.1 ft lb/in. notched (RTP 903)
1.2 ft lb/in. notched (Stat-Kon GC-1006; Thermocomp GC-1006)
1.3 ft lb/in. notched (RTP 905; Thermocomp GF-1004, GF-1006FR)
1.6 ft lb/in. notched (RTP 907)
1.7 ft lb/in. notched (PF-30GF/15T)
1.8 ft lb/in. notched (Thermocomp GF-1006)
2.0 ft lb/in. notched (Thermocomp GF-1008)

Tens. Impact:
48 ft lb/in.^{2} (Thermocomp GF-1004)
63 ft lb/in.^{2} (Thermocomp GF-1006)

Polysulfone *(cont'd.)*

82 ft lb/in.² (Thermocomp GF-1008)

Hardness:

Rockwell M69 (Udel P-1700, P-1700FR, P-1710, P-1720, P-3500)

Rockwell M92 (Thermocomp GF-1004, GF-1006, GF-1008)

Rockwell R122 (RTP 901)

Rockwell R123 (RTP 903, 905, 907)

Mold Shrinkage:

0.2% (Udel GF-130)

0.3% (Udel GF-120)

0.4% (Udel GF-110)

0.007 mm/mm (Udel P-1700, P-1700FR, P-1710, P-1720, P-3500)

0.0015 in./in. (Stat-Kon GC-1006)

0.002 in./in. (Stat-Kon GC-1003)

Water Absorp.:

0.15% (Stat-Kon GC-1006)

0.20% (Stat-Kon GC-1003)

0.3% (Udel P-1700, P-1700FR, P-1710, P-1720, P-3500)

THERMAL PROPERTIES:

Conduct.:

0.26 W/mC (Udel P-1700, P-1700FR, P-1710, P-1720, P-3500)

1.9 Btu/h/ft²/F/in. (RTP 901)

2.0 Btu/h/ft²/F./in. (Thermocomp GF-1004)

2.1 Btu/h/ft²/F/in. (RTP 903)

2.2 Btu/h/ft²/F/in. (RTP 905; Thermocomp GF-1006)

2.5 Btu/h/ft²/F/in. (Thermocomp GF-1008)

2.6 Btu/h/ft²/F/in. (RTP 907)

3.7 Btu/h/ft²/F/in. (Stat-Kon GC-1003)

5.5 Btu/h/ft²/F/in. (Stat-Kon GC-1006; Thermocomp GC-1006)

Distort. Temp.:

174 C (1.8 MPa) (Udel P-1700, P-1700FR, P-1710, P-1720, P-3500)

179 C (1.8 MPa) (Udel GF-110)

180 C (1.8 MPa) (Udel GF-120)

181 C (1.8 MPa) (Udel GF-130)

345 F (66 psi) (PF-10GF/15T)

350 F (264 psi) (PF-20GF/15T; RTP 901)

360 F (264 psi) (RTP 903; Stat-Kon GC-1003; Thermocomp GF-1004)

362 F (264 psi) (RTP 905)

365 F (264 psi) (RTP 907; Stat-Kon GC-1006; Thermocomp GC-1006, GF-1006, GF-1006FR)

370 F (264 psi) (PF-30GF/15T; Thermocomp GF-1008)

Coeff. of Linear Exp.:

5.6×10^{-5} mm/mm/C (Udel P-1700, P-1700FR, P-1710, P-1720, P-3500)

0.6×10^{-5} in./in./F (Stat-Kon GC-1006; Thermocomp GC-1006)

1.2 × 10⁻⁵ in./in./F (Thermocomp GF-1008)

1.4 × 10⁻⁵ in./in./F (Stat-Kon GC-1003; Thermocomp GF-1006, GF-1006FR)

1.6 × 10⁻⁵ in./in./F (PF-30GF/15T)

1.7 × 10⁻⁵ in./in./F (PF-20GF/15T; Thermocomp GF-1004)

Flamm.:

V-0 (RTP 905, 907; Stat-Kon GC-1003, GC-1006; Thermocomp GC-1006, GF-1004, GF-1006, GF-1006FR, GF-1008; Udel GF-110, GF-120, GF-130, P-1700, P-1700FR, P-1710)

V-0/V-1 (Udel P-1720)

V-1 (PF-20GF/15T, -30GF/15T; RTP 901, 903)

10 mm AEB (Udel P-3500)

ELECTRICAL PROPERTIES:

Dissip. Factor:

0.0008 (60 Hz) (Udel P-1700, P-1700FR, P-1710, P-1720, P-3500)

0.001 (60 Hz) (Udel GF-110, GF-120, GF-130)

0.0015 (1 kHz) (Thermocomp GF-1006)

0.0020–0.0050 (60 to 10⁶ Hz) (Thermocomp GF-1006FR)

0.005 (1 MHz) (RTP 901, 903, 905, 907)

Dielec. Str.:

370 V/mil (Udel P-1720)

425 V/mil (Udel P-1700, P-1700FR, P-1710, P-3500)

430 V/mil (RTP 901, 903, 905, 907)

475 V/mil (Udel GF-110, GF-120, GF-130)

480 V/mil (Thermocomp GF-1006, GF-1006FR)

Dielec. Const.:

3.07 (60 Hz) (Udel P-1700, P-1700FR, P-1710, P-3500)

3.15 (60 Hz) (Udel P-1720)

3.3 (60 Hz) (Udel GF-110)

3.4 (60 Hz) (Udel GF-120)

3.5 (60 Hz) (Udel GF-130); (1 MHz) (RTP 901)

3.55 (1 kHz) (Thermocomp GF-1006)

3.6 (1 MHz) (RTP 903)

3.65 (1 MHz) (RTP 905)

3.65–3.52 (60 to 10⁶ Hz) (Thermocomp GF-1006FR)

3.75 (1 MHz) (RTP 907)

Vol. Resist.:

100 ohm-cm (Stat-Kon GC-1003, GC-1006)

10¹⁵ ohm-cm (RTP 901, 903, 905, 907)

> 10¹⁶ ohm-cm (Udel GF-110, GF-120, GF-130)

5 × 10¹⁶ ohm-cm (Udel P-1700, P-1700FR, P-1710, P-1720, P-3500)

10¹⁷ ohm-cm (Thermocomp GF-1006)

Surf. Resist.:

100 ohm/sq. (Stat-Kon GC-1003, GC-1006)

Polysulfone (cont'd.)

Arc Resist.:
  60 s (Udel P-1720)
  100 s (RTP 901, 903, 905)
  115 s (Thermocomp GF-1006)
  120 s (RTP 907)
  122 s (Udel P-1700, P-1700FR, P-1710, P-3500)

STD. PKGS.:
  22.7-kg bags (Udel P-1700, P-1700FR, P-1710, P-1720, P-3500)
  50-lb multiwall bags (Udel GF-110, GF-120, GF-130)

Polytetrafluoroethylene

SYNONYMS:
  PTFE
  TFE

EMPIRICAL FORMULA:
  $(C_2F_4)_n$

STRUCTURE:
  $[—CF_2—CF_2—]_n$

CAS No.:
  9002-84-0

TRADENAME EQUIVALENTS:
  Hostaflon TF 1100, TF 1400, TF 1620, TF 1620 F, TF 1640, TF 1665, TF 1740, TF 1760, TF 2024, TF 2026, TF 2051, TF 2053, TF 2071, TF 5032, TF 5033, TF 5034, TF 5444, TF 5515, TF 5537 [Hoechst Celanese]
  Klingerflon PTFE (Virgin) [Klinger Engineering Plastics]
  LNP Packing Blend B [LNP]
  LNP Reprocessed TFE #1 Grade, TFE #2 Grade, TFE #3 Grade, TFE F-2080, TFE OX-60, TFE Utility Grade [LNP]
  Polypenco TFE [Polymer Corp.]
  Teflon 6A, 6C, 6H, 6L, 7A, 7C, 8, 8A, 9A, 9B, 64 [DuPont]
  TL-101, -102, -113, -115A, -115AH, -126, -126H, -127 [LNP]
  Bronze-filled:
    FC-446CS [LNP]
    Fluorocomp 144 (40% bronze), 145 (50% bronze), 146 (60% bronze), 147 (70% bronze), 148 (80% bronze) [LNP]
    Hostaflon TF 4406 (60% bronze 90/10) [Hoechst Celanese]
    Klingerflon PTFE (60% bronze filled) [Klinger Engineering Plastics]

Polytetrafluoroethylene (cont'd.)

Bronze/MoS₂-filled:
Fluorocomp 181 (40% bronze, 5% MoS₂), 182 (55% bronze, 5% MoS₂), 183 (65% bronze, 5% MoS₂) [LNP]

Carbon-filled:
Fluorocomp 193 (15% carbon), 195 (25% carbon) [LNP]
Hostaflon TF 4215 (25% electrographitized carbon) [Hoechst Celanese]
Thermocomp FC-102CF (10% Pitch carbon fiber) [LNP]

Carbon/graphite-filled:
Fluorocomp 190 (10% carbon/graphite), 191 (25% carbon/graphite), 192 (35% carbon/graphite), 192HE (35% carbon/graphite) [LNP]

Coke flour-filled:
Fluorocomp 131 (5% coke flour), 133 (15% coke flour), 135 (25% coke flour) [LNP]

Glass-filled:
Fluorocomp 101 (5% glass fiber), 102 (10% glass fiber), 103 (15% glass fiber), 104 (20% glass fiber), 105 (25% glass fiber), 106 (30% glass fiber) [LNP]
Hostaflon TF 2103 (15% glass fiber), TF 4103 (15% milled glass fiber), TF 4105 (25% milled glass fiber) [Hoechst Celanese]
Klingerflon PTFE (25% glass filled) [Klinger Engineering Plastics]

Glass/carbon-filled:
Fluorocomp 197 (10% glass fiber, 5% carbon), 199 (5% glass fiber, 10% carbon) [LNP]

Glass/graphite-filled:
Fluorocomp 169 (10% glass fiber, 10% graphite), 170 (5% glass fiber, 5% graphite), 171 (20% glass fiber, 5% graphite) [LNP]

Glass/MoS₂-filled:
Fluorocomp 174 (15% glass fiber, 5% MoS₂), 175 (20% glass fiber, 20% MoS₂), 176 (23% glass fiber, 2% MoS₂), 177 (12.5% glass fiber, 12.5% MoS₂) [LNP]

Graphite-filled:
Fluorocomp 122 (10% graphite), 123 (15% graphite), 124 (20% graphite) [LNP]
Hostaflon TF 4303 (15% graphite) [Hoechst Celanese]
Klingerflon PTFE (15% graphite filled) [Klinger Engineering Plastics]

Inorganic-filled:
FC-477 SM [LNP]

Polymer-filled:
PDX-79625FF [LNP]

Poly-p-oxybenzoate-filled:
Polycomp 160 (10% Ekonol), 161 (25% Ekonol) [LNP]

Polyphenylene sulfide-filled:
Polycomp 139, 142, 148, 149, 158, 184, 185 [LNP]

Stainless steel-filled:
FC 60% SS [LNP]

Polytetrafluoroethylene (cont'd.)

CATEGORY:
Thermoplastic resin
PROCESSING:
Calendering:
Teflon 6A
Compression molding:
Fluorocomp 101, 102, 103, 104, 105, 106, 122, 123, 124, 131, 133, 135, 144, 145, 146,
147, 148, 169, 170, 171, , 174, 175, 176, 177, 181, 182, 183, 190, 191, 192, 192HE,
193, 195, 197, 199; Hostaflon TF 1400, TF 1620, TF 1620F, TF 1640, TF 1665, TF
1740, TF 1760, TF 4103, TF 4105, TF 4215, TF 4303, TF 4406; LNP Reprocessed
TFE #1 Grade, #2 Grade, TFE Utility Grade; Polycomp 139, 142, 148, 149, 158,
184, 185; Teflon 7A, 7C, 8, 8A
Extrusion:
Hostaflon TF 1100, TF 1665, TF 2024, TF 2026, TF 2051, TF 2053, TF 2071, TF 2103;
LNP Reprocessed TFE #1 Grade, #2 Grade, TFE Utility Grade; Polycomp 139, 142,
148, 149, 158, 184, 185; Teflon 6A, 6C, 6H, 6L, 8, 8A, 9A, 9B, 64
APPLICATIONS:
Electrical/electronic industry: (Fluorocomp 101, 102, 103, 104, 105, 106; Hostaflon
TF 1740, TF 1760); cable and wire jacketing (Teflon 6C, 6H, 64); cable insulation
(Hostaflon TF 2026, TF 2051); electrical tapes (Hostaflon TF 2026, TF 2051;
Teflon 6A)

FDA-approved applications: TL-101

Food-contact applications: TL-101

Functional additives: extreme pressure additive (TL-101, -102, -115A, -115AH);
grease/oil thickeners (LNP Reprocessed TFE #3 Grade, TFE F-2080, TFE OX-60;
TL-101, -102, -115A, -115AH); lubricant (FC 60% SS, -446CS, -447SM; LNP
Reprocessed TFE #3 Grade; TL-101, -102, -113, -115A, -115AH, -126, -126H,
-127)

Industrial applications: aerosols (TL-101, -102); bearing applications (Fluorocomp
122, 123, 124, 131, 133, 135, 169, 170, 171, 190, 191, 192, 193, 195, 197, 199);
bearings and bushings (Fluorocomp 101, 102, 103, 104, 105, 106, 144, 145, 146,
147, 148, 174, 175, 176, 177, 181, 182, 183; Klingerflon PTFE; Polypenco TFE);
ceramics coatings (Hostaflon TF 5032, TF 5034, TF 5444); chemical industry
(Fluorocomp 101, 102, 103, 104, 105, 106, 174, 175, 176, 177); coatings (Hostaflon
TF 5032, TF 5033, TF 5444, TF 5515, TF 5537; Teflon 6L); complicated shapes
(Teflon 9A, 9B); diaphragms (Hostaflon TF 1620 F); films (Hostaflon TF 1620, TF
1740, TF 1760); gaskets/seals (Fluorocomp 101, 102, 103, 104, 105, 106, 122, 123,
124, 131, 133, 135, 144, 145, 146, 147, 148, 169, 170, 171, 174, 175, 176, 177, 181,
182, 183, 190, 191, 192, 193, 195, 197, 199; LNP Reprocessed TFE #1 Grade, TFE
Utility Grade; Polypenco TFE); glass coatings (Hostaflon TF 5444); impregnation
of absorbent materials (Hostaflon TF 5032); insulators (Fluorocomp 101, 102, 103,
104, 105, 106, 174, 175, 176, 177; Hostaflon TF 2053; Polypenco TFE); lubricating

linings for bearings (Hostaflon TF 5033); metal coatings (Hostaflon TF 5034, TF 5444); molded parts (Teflon 7A, 7C, 8A); packings (Fluorocomp 122, 123, 124, 131, 133, 135, 169, 170, 171, 190, 191, 192, 193, 195, 197, 199; LNP Reprocessed TFE #3 Grade, TFE F-2080, TFE OX-60, TFE Utility Grade); pots/pans nonstick coatings (Hostaflon TF 5032); pump parts (Klingerflon PTFE; Polypenco TFE); rings/o-rings (Fluorocomp 101, 102, 103, 104, 105, 106, 122, 123, 124, 131, 133, 135, 144, 145, 146, 147, 148, 169, 170, 171, 174, 175, 176, 177, 181, 182, 183, 190, 191, 192, 193, 195, 197, 199; Klingerflon PTFE; Polypenco TFE); rollers (Polypenco TFE); sheet (Teflon 8A); tapes (Hostaflon TF 2024; Teflon 6C, 6H, 6L); thin sections (Hostaflon TF 2103; Teflon 9A, 9B); thread sealants (LNP Reprocessed TFE F-2080, TFE #3 Grade, TFE OX-60); tubing/hoses (Hostaflon TF 2026, TF 2051, TF 2053; LNP Reprocessed TFE #1 Grade; Teflon 6C, 6H, 6L, 9A, 9B, 64); valves/valve seats (Fluorocomp 101, 102, 103, 104, 105, 106, 144, 145, 146, 147, 148, 174, 175, 176, 177, 181, 182, 183; LNP Packing Blend B; LNP Reprocessed TFE #1 Grade; Polypenco TFE); wear applications (Fluorocomp 101, 102, 103, 104, 105, 106, 122, 123, 124, 131, 133, 135, 169, 170, 171, 174, 175, 176, 177, 190, 191, 192, 193, 195, 197, 199)

PROPERTIES:

Form:

Dispersion (Hostaflon TF 5032, TF 5033), TF 5034)

Aq. dispersion (Hostaflon TF 5444, TF 5515, TF 5537)

Solid (Klingerflon PTFE (60% bronze filled), PTFE (25% glass filled))

Paste powder (Hostaflon TF 2024, TF 2026, TF 2051, TF 2053, TF 2071, TF 2103)

Powder (LNP Reprocessed TFE #2 Grade, TFE Utility Grade; Teflon 6L)

Free-flowing powder (Hostaflon TF 4103, TF 4105, TF 4215, TF 4303, TF 4406; Teflon 6A, 6C, 6H, 64)

Powder (avg. 40–60 mesh) (LNP Reprocessed TFE F-2080)

Powder (avg. 60 mesh) (LNP Reprocessed TFE #1 Grade, TFE OX-60)

Powder (avg. 100 mesh) (LNP Reprocessed TFE #3 Grade)

Powder (3–8 μ spherical particles) (TL-101, -102, -127)

Powder (irregular granules) (Hostaflon TF 1100, TF 1400, TF 1740, TF 1760)

Powder (4–35 μ large, irregular particles) (TL-113)

Powder (6–25 μ) (TL-115AH)

Powder (6–25 μ irregular particles) TL-115A)

Powder (6–26 μ irregular particles) (TL-126)

Powder (spherical granules) (Hostaflon TF 1620, TF 1620 F, TF 1640, TF 1665)

Granular (Teflon 7A, 7C, 8A, 9A, 9B)

Free-flowing granular (Teflon 8)

Nonpelletized and pelletized (Polycomp 139, 142, 148, 149, 158, 160, 161, 184, 185; PDX-79625 FF)

Avail. in sheet, rod, or tube (Klingerflon PTFE (15% graphite filled))

Avail. in rod, sheet, tubing, tape, bar (Polypenco TFE)

Polytetrafluoroethylene (cont'd.)

Color:
Unpigmented (Hostaflon TF 5515)
White (Teflon 6A, 6C, 6H, 6L, 64; TL-101, -102, -126, -126H, -127)
White or very slightly off-white (LNP Reprocessed TFE #1 Grade)
White to light gray (TL-115A, -115AH)
Off-white to slight gray (LNP Reprocessed TFE #2 Grade)
Off-white to gray (LNP Reprocessed TFE #3 Grade, TFE F-2080, TFE OX-60)
Gray or slightly tan; also avail. pigmented with darker colors to cover spotting (LNP
Reprocessed TFE Utility Grade)
Avail. in wide range of colors (Hostaflon TF 5444, Tf 5537)

Composition:
35% conc. (Hostaflon TF 5033)
60% conc. (FC 60% SS; FC-446CS; Hostaflon TF 5032)
79.5% conc. (FC-477SM)

GENERAL PROPERTIES:

Sp. Gr.:
1.25 (Hostaflon TF 5033)
1.5 (Hostaflon TF 2024, TF 5032)
1.52 (Hostaflon TF 5034)
1.79–1.87 (Polycomp 149)
1.84–1.92 (Polycomp 184)
1.87 (Polycomp 161)
1.88–1.96 (Polycomp 185)
1.97–2.03 (Polycomp 148)
1.98–2.04 (Polycomp 158)
2.02 (Fluorocomp 192HE)
2.02 ± 0.04 (PDX-79625 FF)
2.03 (Polycomp 160)
2.03–2.099 (Polycomp 142)
2.04 (Fluorocomp 135, 195)
2.05 (Fluorocomp 192)
2.08 (Fluorocomp 122, 123, 191, 193)
2.09 (Fluorocomp 124; Hostaflon TF 4215)
2.10 (Fluorocomp 133, 190; Hostaflon TF 2071, TF4303)
2.1–2.3 (Polypenco TFE)
2.11 (Fluorocomp 197)
2.12 (Fluorocomp 131; Thermocomp FC-102CF)
2.12–2.16 (Polycomp 139)
2.13 (Fluorocomp 170)
2.13–2.23 (Klingerflon PTFE (Virgin))
2.14 (Fluorocomp 101)
2.14–2.16 (Teflon 64)
2.144 (Teflon 9A)

2.15 (Fluorocomp 199; Hostaflon TF 1100; Teflon 8A, 9B)
2.15–2.20 (LNP Reprocessed TFE #1 Grade, #2 Grade, TFE Utility Grade; Teflon 6C, 6L)
2.155 (Hostaflon TF 1760)
2.16 (Fluorocomp 102, 169; Hostaflon TF 2026, TF 2051, TF 2053)
2.16–2.28 (TL-126, -126H)
2.17 (Hostaflon TF 1400, TF 1620, TF 1665)
2.18 (Fluorocomp 171; Hostaflon TF 1640, TF 1740, TF 2103)
2.18–2.23 (Teflon 6H)
2.18–2.24 (Teflon 6A)
2.18–2.28 (TL-101, -102, -113, -115A, -115AH, -127)
2.19 (Fluorocomp 103; Hostaflon TF 4103)
2.19–2.21 (Klingerflon PTFE (15% graphite filled))
2.20 (Hostaflon TF 4105)
2.21 (Fluorocomp 104, 106, 176)
2.22 (Fluorocomp 105, 174)
2.22–2.25 (Klingerflon PTFE (25% glass filled))
2.24 (FC-477SM)
2.30 (Fluorocomp 177)
2.47 (Fluorocomp 175)
3.05 (Fluorocomp 144)
3.14 (Fluorocomp 181)
3.35 (Fluorocomp 145)
3.67 (Fluorocomp 182)
3.78 (FC 60% SS; Fluorocomp 146)
3.8–3.97 (Klingerflon PTFE (60% bronze filled))
3.90 (FC-446CS; Hostaflon TF 4406)
4.16 (Fluorocomp 183)
4.25 (Fluorocomp 147)
4.70 (Fluorocomp 148)

Bulk Density:
275 g/l (Teflon 7C)
380 g/l (Hostaflon TF 1760)
400 g/l (TL-126, -126H)
425 g/l (TL-102)
440 g/l (TL-101)
450 g/l (Hostaflon TF 2026, TF 2051, TF 2053, TF 2071)
475 g/l (Teflon 7A)
475 ± 100 g/l (Teflon 6A, 6C, 6H, 6L, 64)
480 g/l (Hostaflon TF 2103)
500–700 g/l (LNP Reprocessed TFE F-2080)
525 g/l (TL-113)
560 g/l (Hostaflon TF 1400)

Polytetrafluoroethylene *(cont'd.)*

   575 g/l (Teflon 9B)
   600 g/l (Hostaflon TF 4215; Teflon 9A)
   600–900 g/l (LNP Reprocessed TFE #3 Grade)
   630 g/l (TL-115A, -115AH)
   650 g/l (Hostaflon TF 1100, TF 4303)
   680 g/l (Teflon 8A)
   700 g/l (Hostaflon TF 4103, TF 4105)
   720 g/l (Teflon 8)
   800 g/l (Hostaflon TF 1665, TF 1740)
   800–1000 g/l (LNP Reprocessed TFE OX-60)
   840 g/l (Hostaflon TF 1620)
   850 g/l (Hostaflon TF 1640)
   890 g/l (Hostaflon TF 1620 F)
   900–1100 g/l (LNP Reprocessed TFE #1 Grade)
   1350 g/l (Hostaflon TF 4406)

Visc.:
   0.002–0.004 Pa•s (Hostaflon TF 5033)
   0.010–0.015 Pa•s (Hostaflon TF 5034)
   0.012–0.020 Pa•s (Hostaflon TF 5032)

Melt Visc.:
   1.5×10^8 poise (TL-101)

M.P.:
   327 ± 10 C (Teflon 6A, 6C, 6H, 6L, 64)
   621 ± 9 F (Polypenco TFE)

Stability:
   Excellent resistance to acids, alkalis, and solvents (Klingerflon PTFE (15% graphite filled), PTFE (25% glass filled), PTFE (Virgin); Thermocomp FC-102CF)
   Excellent resistance to solvents; filler is attacked by acids and alkalis (Klingerflon PTFE (60% bronze filled))
   Excellent corrosion resistance and excellent weathering; high heat and low temp. impact resistance; almost totally unaffected by chemicals encountered in commercial use (Polypenco TFE)
   Inert to almost all chemicals and solvents; excellent weather resistance (Teflon 6A, 6C, 6H, 6L, 64)

pH:
   8–9 (Hostaflon TF 5032, TF 5034)
   9 (Hostaflon TF 5033)

Surface Tension:
   20.2 dynes/cm (TL-126, -126H)
   20.7 dynes/cm (TL-113)
   20.9 dynes/cm (TL-102)

MECHANICAL PROPERTIES:
Tens. Str.:
12 MPa (Hostaflon TF 4105)
14 MPa (Hostaflon TF 2024, TF 4215, TF 4406)
16 MPa (Hostaflon TF 4103, TF 4303)
17.24–27.4 MPa (Teflon 6C, 64)
22 MPa (Hostaflon TF 1100)
27 MPa (Hostaflon TF 2103)
28 MPa (Hostaflon TF 2053, TF 2071)
30 MPa (Hostaflon TF 1400, TF 2051)
31 MPa (Hostaflon TF 1665)
32 MPa (Hostaflon TF 1620, TF 1640)
33 MPa (Hostaflon TF 1620 F, TF 1740, TF 1760)
34 MPa (Hostaflon TF 2026)
70–200 kg/cm² (Klingerflon PTFE (25% glass filled))
91–189 kg/cm² (Klingerflon PTFE (15% graphite filled))
105-140 kg/cm² (Klingerflon PTFE (60% bronze filled))
140–350 kg/cm² (Klingerflon PTFE (Virgin))
620 psi (Polycomp 149)
1000 psi (Polycomp 185)
1220 psi (Polycomp 184)
1400–2200 psi (LNP Reprocessed TFE Utility Grade)
1500 psi (Fluorocomp 192HE; Polycomp 161)
1500–2200 psi (LNP Reprocessed TFE #2 Grade)
1500–5000 psi (Polypenco TFE)
1600 psi (Polycomp 148)
1600–2400 psi (LNP Reprocessed TFE #1 Grade)
1700 psi (Fluorocomp 147, 148)
1800 psi (Polycomp 142; PDX-79625 FF)
2000 psi (Fluorocomp 192)
2050 psi (Teflon 8)
2200 psi (Fluorocomp 183, 191; Polycomp 158)
2300 psi (Fluorocomp 106, 135, 182, 195)
2400 psi (FC-477SM; Fluorocomp 175; Polycomp 160)
2500 psi (FC-446CS; Fluorocomp 171)
2500–4000 psi (Teflon 6A, 6H, 6L)
2550 psi (Fluorocomp 177)
2600 psi (Fluorocomp 124, 146, 176)
2700 psi (Fluorocomp 105, 145, 169, 181)
2800 psi (Fluorocomp 174)
2850 psi (Fluorocomp 144)
2900 psi (FC 60% SS; Fluorocomp 104, 123, 199; Teflon 8A)
3000 psi (Fluorocomp 133, 193, 197)

Polytetrafluoroethylene *(cont'd.)*

    3100 psi (Fluorocomp 103)
    3300 psi (Fluorocomp 122)
    3400 psi (Fluorocomp 102, 170; 9A, 9B)
    3450 psi (Fluorocomp 190)
    3600 psi (Fluorocomp 131; Polycomp 139)
    3700 psi (Fluorocomp 101)
    4400 psi (Thermocomp FC-102CF)
    5300 psi (Teflon 7A)
    5900 psi (Teflon 7C)

Tens. Elong.:
    25% (Fluorocomp 148)
    50% (break) (Polycomp 149)
    50–100% (LNP Reprocessed TFE Utility Grade)
    50–150% (LNP Reprocessed TFE #2 Grade)
    55% (Fluorocomp 192)
    65% (Fluorocomp 183, 191)
    70% (Fluorocomp 175)
    75–350% (Polypenco TFE)
    80% (Fluorocomp 192HE)
    80–160% (Klingerflon PTFE (60% bronze filled))
    85% (break) (Polycomp 184)
    100% (Fluorocomp 135, 147, 195); (break) (Hostaflon TF 4406; Polycomp 185)
    100–200% (LNP Reprocessed TFE #1 Grade)
    100–300% (Klingerflon PTFE (25% glass filled))
    100–400% (Klingerflon PTFE (Virgin))
    125% (Teflon 8)
    130% (FC-446CS; Fluorocomp 182); (break) (Polycomp 148, 161)
    140% (Fluorocomp 146); (break) (PDX-79625 FF)
    150% (break) (Hostaflon TF 4303)
    170% (Fluorocomp 124, 145)
    180–300% (Klingerflon PTFE (15% graphite filled))
    200% (Fluorocomp 123, 181); (break) (Hostaflon TF 4215; Polycomp 142)
    210% (FC 60% SS)
    220% (Fluorocomp 144)
    225–450% (Teflon 6C, 6L, 64)
    230% (Fluorocomp 106, 133, 171, 190); (break) (Hostaflon TF 2024)
    235% (Fluorocomp 199); (break) (Polycomp 160)
    240% (Fluorocomp 122, 169, 197)
    245% (Fluorocomp 176)
    240% (Fluorocomp 177)
    250% (Fluorocomp 105, 193; Teflon 8A)
    250–500% (Teflon 6H)
    260% (break) (Hostaflon TF 2051; Polycomp 158)

265% (Fluorocomp 174)
251% (FC-477SM)
270% (Fluorocomp 104, 170)
280% (Fluorocomp 131)
290% (Fluorocomp 103)
300% (break) (Hostaflon TF 2026, TF 2053, TF 2071, TF 2103, TF 4103, TF 4105)
300–600% (Teflon 6A)
310% (Fluorocomp 102); (break) (Polycomp 139)
320% (Fluorocomp 101)
325% (Teflon 9A, 9B)
350% (Teflon 7A, 7C); (break) (Hostaflon TF 1100)
400% (break) (Hostaflon TF 1620, TF 1665, TF 1760)
450% (break) (Hostaflon TF 1400)
480% (break) (Hostaflon TF 1620 F, TF 1640)
500% (break) (Hostaflon TF 1740)
Tens. Mod.:
50,000–90,000 psi (Polypenco TFE)
Flex. Str.:
No break (Polypenco TFE)
800 psi (1% strain) (Fluorocomp 170)
825 psi (1% strain) (Fluorocomp 131)
850 psi (1% strain) (Fluorocomp 122)
875 psi (1% strain) (Fluorocomp 169)
900 psi (1% strain) (Fluorocomp 123)
950 psi (1% strain) (Fluorocomp 124, 190)
1050 psi (1% strain) (Fluorocomp 171, 197)
1100 psi (1% strain) (Fluorocomp 133, 193)
1125 psi (1% strain) (Fluorocomp 101)
1170 psi (1% strain) (Fluorocomp 174)
1175 psi (1% strain) (Fluorocomp 102, 135, 195, 199)
1200 psi (1% strain) (Fluorocomp 106)
1225 psi (1% strain) (Fluorocomp 103)
1250 psi (1% strain) (Fluorocomp 104)
1275 psi (1% strain) (Fluorocomp 105)
1350 psi (1% strain) (Fluorocomp 176, 177)
1375 psi (1% strain) (Fluorocomp 191)
1400 psi (1% strain) (Fluorocomp 144)
1425 psi (1% strain) (Fluorocomp 192)
1550 psi (1% strain) (Fluorocomp 175)
1600 psi (1% strain) (Fluorocomp 181)
1700 psi (1% strain) (Fluorocomp 145)
2000 psi (1% strain) (Fluorocomp 146)
2100 psi (1% strain) (Fluorocomp 182)

Polytetrafluoroethylene *(cont'd.)*

2400 psi (1% strain) (Fluorocomp 147)
2500 psi (1% strain) (Fluorocomp 183)
2700 psi (1% strain) (Fluorocomp 148)
3000 psi (Thermocomp FC-102CF)

Flex. Mod.:

3500–6300 kg/cm² (Klingerflon PTFE (Virgin))
13,862 kg/cm² (Klingerflon PTFE (60% bronze filled))
14,284 kg/cm² (Klingerflon PTFE (15% graphite filled))
16,900 kg/cm² (Klingerflon PTFE (25% glass filled))
90,000–115,000 psi (Polypenco TFE)
105,000 psi (Fluorocomp 131)
115,000 psi (Fluorocomp 170)
120,000 psi (Fluorocomp 101, 122)
125,000 psi (Fluorocomp 123, 124)
130,000 psi (Fluorocomp 190)
132,000 psi (Fluorocomp 174)
135,000 psi (Fluorocomp 102, 199)
140,000 psi (Fluorocomp 133, 144, 169, 193, 197)
153,000 psi (Fluorocomp 177)
155,000 psi (Fluorocomp 103, 171)
160,000 psi (Fluorocomp 135, 181, 191, 195)
170,000 psi (Fluorocomp 104, 145)
172,000 psi (Fluorocomp 176)
180,000 psi (Fluorocomp 192)
190,000 psi (Fluorocomp 105, 106)
200,000 psi (Fluorocomp 146)
210,000 psi (Fluorocomp 175, 182)
220,000 psi (Thermocomp FC-102CF)
240,000 psi (Fluorocomp 147)
250,000 psi (Fluorocomp 183)
270,000 psi (Fluorocomp 148)

Compr. Str.:

480 psi (1% strain) (PDX-79625 FF)
650 psi (1% strain) (Polycomp 158)
680 psi (1% strain) (Polycomp 139)
700 psi (1% strain) (Fluorocomp 193)
750 psi (1% strain) (Fluorocomp 122, 169, 170, 190, 197, 199)
790 psi (1% strain) (Polycomp 142)
800 psi (1% strain) (Fluorocomp 131, 195)
810 psi (1% strain) (Fluorocomp 101)
840 psi (1% strain) (Fluorocomp 102)
850 psi (1% strain) (Fluorocomp 171)
875 psi (1% strain) (Fluorocomp 123)

Polytetrafluoroethylene *(cont'd.)*

880 psi (1% strain) (Fluorocomp 103)
900 psi (1% strain) (Fluorocomp 133, 191)
910 psi (1% strain) (Fluorocomp 104)
950 psi (1% strain) (Fluorocomp 105)
970 psi (1% strain) (Polycomp 148)
980 psi (1% strain) (Fluorocomp 106)
990 psi (1% strain) (Fluorocomp 174)
1000 psi (1% strain) (Fluorocomp 124, 135, 144, 192; Polycomp 184)
1100 psi (1% strain) (Polycomp 185)
1150 psi (1% strain) (Fluorocomp 145, 181)
1160 psi (1% strain) (Fluorocomp 176)
1170 psi (1% strain) (Fluorocomp 177)
1350 psi (1% strain) (Fluorocomp 175)
1390 psi (1% strain) (Polycomp 149)
1450 psi (1% strain) (Fluorocomp 182)
1500 psi (1% strain) (Fluorocomp 146)
1700 psi (1% strain) (Fluorocomp 183)
1800 psi (1% strain) (Fluorocomp 147)
2000 psi (1% strain) (Fluorocomp 148)
Compr. Mod.:
10,000 psi (Fluorocomp 144)
11,000 psi (Fluorocomp 145, 181)
15,000 psi (Fluorocomp 182)
16,000 psi (Fluorocomp 146)
18,000 psi (Fluorocomp 183)
20,000 psi (Fluorocomp 147)
22,000 psi (Fluorocomp 148)
48,000 psi (PDX-79625 FF)
65,000 psi (Polycomp 158)
68,000 psi (Fluorocomp 197; Polycomp 139)
70,000 psi (Fluorocomp 170, 199)
73,000 psi (Fluorocomp 169)
74,000 psi (Fluorocomp 190)
76,000 psi (Fluorocomp 122)
79,000 psi (Fluorocomp 123; Polycomp 142)
80,000 psi (Fluorocomp 131)
82,000 psi (Fluorocomp 124)
83,000 psi (Fluorocomp 101)
88,000 psi (Fluorocomp 102)
90,000 psi (Fluorocomp 133, 171, 193)
94,000 psi (Fluorocomp 103)
95,000–115,000 psi (Polypenco TFE)
96,000 psi (Fluorocomp 191)

323

Polytetrafluoroethylene *(cont'd.)*

  97,000 psi (Polycomp 148)
  98,000 psi (Fluorocomp 104)
  100,000 psi (Fluorocomp 135, 195; Polycomp 184)
  101,000 psi (Fluorocomp 174)
  103,000 psi (Fluorocomp 105)
  108,000 psi (Fluorocomp 192)
  109,000 psi (Fluorocomp 106)
  110,000 psi (Polycomp 185)
  118,000 psi (Fluorocomp 176)
  120,000 psi (Fluorocomp 177)
  138,000 psi (Polycomp 149)
  141,000 psi (Fluorocomp 175)

Shear Str.:
  3500 psi (Thermocomp FC-102CF)

Impact Str. (Izod):
  10.60 kg cm/cm notched (Klingerflon PTFE (60% bronze filled))
  11.90 kg cm/cm notched (Klingerflon PTFE (25% glass filled))
  16.00 kg cm/cm notched (Klingerflon PTFE (Virgin))
  5.6 ft lb/in. notched (Thermocomp FC-102CF)

Tens. Impact:
  30–200 ft lb/in.^{2} (Polypenco TFE)

Hardness:
  Ball Indentation 30.0 MPa (Hostaflon TF 4103, TF 4105)
  Ball Indentation 32.0 MPa (Hostaflon TF 4303)
  Ball Indentation 38.0 MPa (Hostaflon TF 4215)
  Ball Indentation 40.0 MPa 9Hostaflon TF 4406)
  Rockwell R10–20 (Polypenco TFE)
  Shore D50–60 (Klingerflon PTFE (Virgin))
  Shore D50–65 (Teflon 6A, 6C, 6H, 6L, 64)
  Shore D55 (Fluorocomp 131; Klingerflon PTFE (15% graphite filled), PTFE (25% glass filled))
  Shore D56 (Fluorocomp 170; Polycomp 139)
  Shore D57 (Fluorocomp 122, 169, 190, 197)
  Shore D58 (Fluorocomp 171)
  Shore D59 (Fluorocomp 101, 123, 133, 193, 199)
  Shore D60 (Fluorocomp 102, 124, 144, 181, 191; Polycomp 142, 158)
  Shore D61 (Fluorocomp 103, 135, 195; Polycomp 148)
  Shore D62 (Fluorocomp 104, 174, 192; Polycomp 184)
  Shore D63 (Fluorocomp 105; Polycomp 149, 185)
  Shore D64 (Fluorocomp 106, 145)
  Shore D65 (Fluorocomp 176; Klingerflon PTFE (60% bronze filled))
  Shore D66 (Fluorocomp 177, 182)
  Shore D68 (Fluorocomp 146)

Shore D69 (Fluorocomp 175)
Shore D70 (Fluorocomp 183)
Shore D73 (Fluorocomp 147)
Shore D76 (Fluorocomp 148)

THERMAL PROPERTIES:

Conduct.:

0.35 W/m•K (Hostaflon TF 4103, TF 4105)
0.7 W/m•K (Hostaflon TF 4215, TF 4406)
0.93 W/m•K (Hostaflon TF 4303)
2.1 Btu/h/ft^2/F/in. (Fluorocomp 101)
2.4 Btu/h/ft^2/F/in. (Fluorocomp 102)
2.7 Btu/h/ft^2/F/in. (Fluorocomp 103)
2.8 Btu/h/ft^2/F/in. (Fluorocomp 170)
2.9 Btu/h/ft^2/F/in. (Fluorocomp 104)
3.1 Btu/h/ft^2/F/in. (Fluorocomp 105, 131)
3.2 Btu/h/ft^2/F/in. (Fluorocomp 174)
3.3 Btu/h/ft^2/F/in. (Fluorocomp 106, 190)
3.4 Btu/h/ft^2/F/in. (Fluorocomp 176)
3.6 Btu/h/ft^2/F/in. (Fluorocomp 177)
3.8 Btu/h/ft^2/F/in. (Fluorocomp 171)
3.9 Btu/h/ft^2/F/in. (Fluorocomp 122; Thermocomp FC-102CF)
4.0 Btu/h/ft^2/F/in. (Fluorocomp 197)
4.1 Btu/h/ft^2/F/in. (Fluorocomp 175)
4.2 Btu/h/ft^2/F/in. (Fluorocomp 199)
4.3 Btu/h/ft^2/F/in. (Fluorocomp 133, 144, 181, 193)
4.5 Btu/h/ft^2/F/in. (Fluorocomp 135, 195)
4.7 Btu/h/ft^2/F/in. (Fluorocomp 169)
4.8 Btu/h/ft^2/F/in. (Fluorocomp 123)
5.0 Btu/h/ft^2/F/in. (Fluorocomp 145, 182, 191)
5.8 Btu/h/ft^2/F/in. (Fluorocomp 146)
5.9 Btu/h/ft^2/F/in. (Fluorocomp 124)
6.0 Btu/h/ft^2/F/in. (Fluorocomp 183)
6.3 Btu/h/ft^2/F/in. (Fluorocomp 192)
6.6 Btu/h/ft^2/F/in. (Fluorocomp 147)
7.5 Btu/h/ft^2/F/in. (Fluorocomp 148)

Distort. Temp.:

100–140 F (264 psi) (Polypenco TFE)
215 F (Thermocomp FC-102CF)

Coeff. of Linear Exp.:

7.54 × 10^{-5} cm/cm/C (Klingerflon PTFE (25% glass filled))
7.88 × 10^{-5} cm/cm/C (Klingerflon PTFE (60% bronze filled))
10.0 × 10^{-5} cm/cm/C (Klingerflon PTFE (Virgin))
0.7 × 10^{-4} K^{-1} (Hostaflon TF 4406)

Polytetrafluoroethylene *(cont'd.)*

1.0×10^{-4} K^{-1} (Hostaflon TF 4105, TF 4215)
1.1×10^{-4} K^{-1} (Hostaflon TF 4103, TF 4303)
2.9×10^{-5} in./in./F (Fluorocomp 175)
3.7×10^{-5} in./in./F (Fluorocomp 177)
3.8×10^{-5} in./in./F (Fluorocomp 106)
3.9×10^{-5} in./in./F (Fluorocomp 176)
4.0×10^{-5} in./in./F (Fluorocomp 192)
4.1×10^{-5} in./in./F (Fluorocomp 105)
4.5×10^{-5} in./in./F (Fluorocomp 104, 135, 171, 195)
4.6×10^{-5} in./in./F (Fluorocomp 174)
4.7×10^{-5} in./in./F (Fluorocomp 133, 169, 191, 193)
4.8×10^{-5} in./in./F (Fluorocomp 124, 199)
5.0×10^{-5} in./in./F (Fluorocomp 103)
5.2×10^{-5} in./in./F (Fluorocomp 197)
5.4×10^{-5} in./in./F (Fluorocomp 144)
5.5×10^{-5} in./in./F (Fluorocomp 102)
$5.5–7.5 \times 10^{-5}$ in./in./F (Polypenco TFE)
5.8×10^{-5} in./in./F (Fluorocomp 123, 170)
6.0×10^{-5} in./in./F (Fluorocomp 101)
6.1×10^{-5} in./in./F (Fluorocomp 131)
6.3×10^{-5} in./in./F (Fluorocomp 190)
6.7×10^{-5} in./in./F (Fluorocomp 122)

Flamm.:
V-0 (Thermocomp FC-102CF)
Nonflammable (Teflon 6H, 6L)
AEB 5 mm (Teflon 6A, 6C, 64)
ATB < 5 s (Teflon 6A, 6C, 64)

ELECTRICAL PROPERTIES:

Dissip. Factor:
< 0.0002 (60–10^8 cps) (Teflon 6A, 6H, 6L); (100 Hz–60 MHz) (Teflon 6C, 64); (1 MHz) (Klingerflon PTFE (Virgin))
0.0004 (1 MHz) (LNP Reprocessed TFE #1 Grade)
0.0006 (1 MHz) (LNP Reprocessed TFE #2 Grade)
0.0025 (1 MHz) (LNP Reprocessed TFE Utility Grade)
< 0.003 (1 MHz) (Klingerflon PTFE (25% glass filled))

Dielec. Str.:
> 13 kV/mm (Klingerflon PTFE (25% glass filled))
23.6 kV/mm (Teflon 6C, 64)
> 24 kV/mm (Klingerflon PTFE (Virgin))
50 kV/mm (Hostaflon TF 1400)
60 kV/mm (Hostaflon TF 1640, TF 1665)
70 kV/mm (Hostaflon TF 1620)
80–90 kV/mm (Hostaflon TF 1740)

90 kV/mm (Hostaflon TF 1760)

500–650 V/mil (Polypenco TFE)

600 V/mil (LNP Reprocessed TFE Utility Grade; Teflon 6A, 6H, 6L)

1000 V/mil (LNP Reprocessed TFE #2 Grade)

1200 V/mil (LNP Reprocessed TFE #1 Grade)

3400 V/mil (Teflon 7A, 7C)

Dielec. Const.:

2.0–2.1 (60 Hz) (Polypenco TFE)

2.1 (1 MHz) (Klingerflon PTFE (Virgin))

2.2 ($60-10^8$ cps) (Teflon 6A, 6H, 6L); (100 Hz–60 MHz) (Teflon 6C, 64)

2.26 (1 MHz) (LNP Reprocessed TFE #1 Grade)

2.28 (1 MHz) (LNP Reprocessed TFE #2 Grade)

2.29 (1 MHz) (LNP Reprocessed TFE Utility Grade)

2.35 (1 MHz) (Klingerflon PTFE (25% glass filled))

Vol. Resist.:

> 10^6 ohm-cm (Klingerflon PTFE (15% graphite filled))

> 10^7 ohm-cm (Klingerflon PTFE (60% bronze filled))

> 10^{16} ohm-cm (Klingerflon PTFE (25% glass filled))

10^{17} ohm-cm (LNP Reprocessed TFE #1 Grade, #2 Grade, Utility Grade)

> 10^{17} ohm-cm (Polypenco TFE)

> 10^{18} ohm-cm (Klingerflon PTFE (Virgin); Teflon 6A, 6C, 6H, 6L, 64)

Surf. Resist.:

500 ohm/sq. (Thermocomp FC-102CF)

> 10^9 ohm (Klingerflon PTFE (60% bronze filled))

> 10^{14} ohm (Klingerflon PTFE (15% graphite filled))

> 10^{15} ohm (Klingerflon PTFE (25% glass filled))

10^{16} ohm (LNP Reprocessed TFE #1 Grade, #2 Grade, Utility Grade)

> 10^{16} ohm (Klingerflon PTFE (Virgin))

TOXICITY/HANDLING:

At elevated temps. vapors are liberated which may be harmful—use with adequate ventilation (Teflon 6A, 6C, 64)

Adequate ventilation should be provided above 500 F to prevent inhalation of decomposition products (TL-101, -102, -113, 115A, -115AH, -126, -126H, -127)

STD. PKGS.:

50-lb Leverpak containers (Teflon 6A, 6C, 64)

4- and 18-lb Fiberpak and 50-lb Leverpak containers (Teflon 6H)

3- and 16-lb Fiberpak and 45-lb Leverpak containers (Teflon 6L)

25-lb Fiberpak and 100-lb Leverpak containers (Teflon 7A, 7C)

25-lb Fiberpak and 100-lb Leverpak containers with polyethylene liners (Teflon 8A, 9A, 9B)

100-lb Leverpak containers with polyethylene liners (Teflon 8)

Polyvinyl acetate

SYNONYMS:
Acetic acid, ethenyl ester, homopolymer
Ethenyl acetate, homopolymer
PVAc

EMPIRICAL FORMULA:
$(C_4H_6O_2)_x$

STRUCTURE:
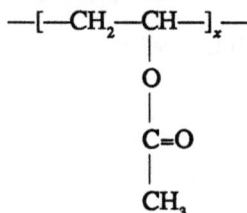

CAS No.:
9003-20-7

TRADENAME EQUIVALENTS:
Daratak 17-200, 17-230, 17-300, 52L, 55L, 56L, 61L, 62L, 65L, 71L, 78L, SP1004,
    SP1006, SP1011, SP1041, SP1047, SP1065, SP1066, SP1079 [W.R. Grace]
Texicote 03-001, 03-020, 03-021, 03-040 [Scott Bader]

MODIFICATIONS/SPECIALTY GRADES:
Carboxylated:
Daratak 78L
Stabilized with polyvinyl alcohol:
Texicote 03-001
Plasticized:
Daratak SP1004, SP1047, SP1066
Externally plasticized:
Daratak SP1065; Texicote 03-020, 03-021

CATEGORY:
Thermoplastic resin

APPLICATIONS:
Industrial applications: adhesives (Daratak 17-200, 17-230, 17-300, 52L, 55L, 56L,
    61L, 62L, 65L, 71L, 78L, SP1004, SP1006, SP1011, SP1041, SP1047, SP1065,
    SP1066; Texicote 03-001, 03-040); building industry (Texicote 03-001); cement
    (Daratak SP1065; Texicote 03-001); coatings (Daratak 52L, 55L, 78L, SP1065,
    SP1079; Texicote 03-020, 03-021); films (Daratak 17-200, 17-230, 17-300, 52L,
    55L, 56L, 61L, 62L, 65L, 71L, 78L, SP1004, SP1006, SP1011, SP1041, SP1047,
    SP1065, SP1066, SP1066); fire-retardant applications (Daratak 61L, SP1011,
    SP1066); heat sealing applications (Daratak 52L, 61L, SP1011); mastics (Daratak
    61L, SP1065, SP1066); packaging applications (Daratak 71L); paints (Daratak

Polyvinyl acetate (cont'd.)

SP1011, SP1066); paper industry (Daratak 56L, 61L, 65L); patching compounds (Daratak SP1065); primer sealers (Daratak 56L); sealing applications (Daratak 55L, SP1006, SP1041, SP1066); textile applications (Daratak 52L, 56L, SP1066); wallpaper (Texicote 03-020, 03-021)

PROPERTIES:

Form:

Emulsion; 0.5 μ particle size (Texicote 03-020, 03-021)

Emulsion; 1.0 μ particle size (Daratak 56L, 61L, 62L, 78L, SP1011, SP1065, SP1066, SP1066)

Emulsion; 1–2 μ particle size (Texicote 03-001)

Emulsion; 1–5 μ particle size (Texicote 03-040)

Emulsion; 1.5 μ particle size (Daratak 17-200, 17-230, 17-300, 52L, 55L, 65L, 71L, SP1004, SP1006, SP1041, SP1047)

Color:

White (Daratak 17-200, 17-230, 17-300, 52L, 55L, 56L, 61L, 62L, 65L, 71L, 78L, SP1004, SP1006, SP1011, SP1041, SP1047, SP1065, SP1066, SP1066)

Odor:

Slight (Daratak SP1004, SP1066)

Slight, characteristic (Daratak 17-200, 17-230, 17-300, 52L, 55L, 56L, 61L, 62L, 65L, 71L, 78L, SP1006, SP1011, SP1041, SP1047, SP1065, SP1066)

Composition:

50 ± 1% solids (Texicote 03-001)

54–56% solids (Daratak 17-200, 17-230, 17-300, 52L, 55L, 56L, 61L, 62L, 65L, 71L, 78L, SP1004, SP1006, SP1011, SP1041, SP1047, SP1066)

55 ± 1% solids (Texicote 03-020, 03-021, 03-040)

55–57% solids (Daratak SP1065, SP1066, SP1079)

6.5% film plasticizer (Texicote 03-020)

10% film plasticizer (Texicote 03-021)

GENERAL PROPERTIES:

Solubility:

Sol. in alkalis (Daratak 78L)

Sp. Gr.:

1.08 (Texicote 03-001)

1.09 (Texicote 03-020, 03-021)

1.10 (Texicote 03-040)

1.19 (of film) (Daratak 55L, SP1006, SP1011, SP1066)

Density:

9.1 lb/gal (emulsion) (Daratak 55L, SP1004, SP1006, SP1011, SP1041, SP1047, SP1066)

9.3 lb/gal (emulsion) (Daratak SP1066)

9.9 lb/gal (solids) (Daratak SP1066)

9.93 lb/gal (Daratak 17-200, 17-230, 17-300, 52L, 55L, 56L, 61L, 62L, 65L, 71L, 78L, SP1006, SP1011, SP1065, SP1066)

329

Polyvinyl acetate *(cont'd.)*

Visc.:
0.25–1.0 poise (Texicote 03-020, 03-021)
12–16 poise (Texicote 03-040)
40–60 poise (Texicote 03-001)
300–600 cps (Daratak 52L, 56L)
500–900 cps (Daratak SP1066)
900–1500 cps (Daratak SP1066)
1200–1500 cps (Daratak SP1004)
1200–1800 cps (Daratak SP1006)
1400–2000 cps (Daratak 65L)
1500–2000 cps (Daratak 17-200)
1500–2200 cps (Daratak 71L)
1500–3000 cps (Daratak SP1065)
1600–1800 cps (Daratak SP1047)
1800–3800 cps (Daratak SP1011)
2000–4000 cps (Daratak 61L)
2000–6000 cps (Daratak 78L)
2300–2700 cps (Daratak 62L)
2400–2800 cps (Daratak 17-230)
2500–2900 cps (Daratak SP1041)
3000–4000 cps (Daratak 17-300, 55L)

Stability:
Emulsions exhibit 20+ min mechanical stability; films exhibit excellent light stability and aging characteristics (Daratak 17-200, 17-230, 17-300, 52L, 56L, 61L, 62L, 65L, 71L, 78L, SP1065)

Emulsions exhibit 20+ min mechanical stability; films exhibit excellent light stability and aging characteristics, and very good water resistance (Daratak SP1011, SP1066)

Emulsions exhibit freeze-thaw stability and 20+ min mechanical stability; films have excellent light stability and aging characteristics and fair water resistance (Daratak 55L, SP1006)

Emulsions exhibit excellent freeze-thaw stability, 20+ min mechanical stability, and borax compatibility; films have excellent light stability and poor water resistance (Daratak SP1079)

Emulsions are not freeze-thaw stable nor borax compatible; films have excellent aging characteristics and good water resistance (Daratak SP1047)

Films have excellent light stability and water resistance (Daratak SP1041)

Borax-tolerant (Daratak SP1011)

Borax-incompatible (Daratak SP1004, SP1006, SP1041)

Freeze-thaw stable (Daratak SP1004; Texicote 03-001, 03-040)

Freeze-thaw unstable (Texicote 03-020, 03-021)

pH:
4.0–5.0 (Daratak SP1079)

4.0–5.5 (Texicote 03-001, 03-020, 03-021, 03-040)
4.0–6.0 (Daratak 78L)
4.5–5.5 (Daratak 61L, 65L, SP1065, SP1066)
4.5–6.0 (Daratak 17-200, 17-230, 17-300, 55L, SP1006)
4.5–6.5 (Daratak SP1004)
4.6–5.5 (Daratak SP1011)
5.0–6.0 (Daratak 71L, SP1041, SP1047)
5.0–7.0 (Daratak 62L)
5.5–7.0 (Daratak 52L, 56L)

STD. PKGS.:
200-kg lacquer-lined metal drums; bulk road tanker (Texicote 03-001, 03-020, 03-021, 03-040)

Polyvinyl butyral (CTFA)

SYNONYMS:
Polyvinyl butyral resin
PVB
Vinyl acetyl polymers, butyrals

STRUCTURE:

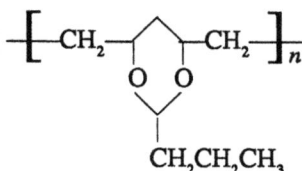

CAS No.:
63148-65-2

TRADENAME EQUIVALENTS:
Butacite 10, 14, 106, 125, 140 [DuPont]
Butvar [Amer. Cyanamid]

CATEGORY:
Thermoplastic resin, adhesive interlayer

APPLICATIONS:
Automotive applications: safety glass/windshields (Butacite 10, 14, 106, 125, 140; Butvar)
Industrial applications: adhesives (Butvar); architectural glazing (Butacite 10, 14); coatings (metal, textile, wood, etc.) (Butvar); films (Butvar); insulation (Butvar); molded parts (Butvar)

Polyvinyl butyral *(cont'd.)*

PROPERTIES:
Form:
15- and 30-mil sheeting in various widths (Butacite 10, 14, 106, 125, 140)
Color:
Clear, standard solid colors, and with gradient bands on clear or colors (Butacite)

GENERAL PROPERTIES:
Solubility:
Sol. in alcohols (Butacite 10, 14, 106, 125, 140)
Swells when subjected to aromatic solvents (Butacite 10, 14, 106, 125, 140)
Limited sol. in most chlorinated hydrocarbons (Butacite 10, 14, 106, 125, 140)
Sol. in esters (Butacite 10, 14, 106, 125, 140); limited sol. in higher esters (Butacite 10, 14, 106, 125, 140)
Sol. in ketones (Butacite 10, 14, 106, 125, 140)
Sp. Gr.:
1.07–1.08 (Butacite 10, 14, 106, 125, 140)
Stability:
Relatively stable under alkaline conditions; unstable in the presence of acids (Butacite 10, 14, 106, 125, 140)
Ref. Index:
1.48 (Butacite 10, 14, 106, 125, 140)

MECHANICAL PROPERTIES:
Tens. Str.:
3500 psi (@ 0.3% water) (Butacite 10, 14, 106, 125, 140)

THERMAL PROPERTIES:
Conduct.:
0.10 PCU/h-ft^2-C/ft (Butacite 10, 14, 106, 125, 140)
Sp. Heat:
0.45 PCU/lb-C (Butacite 10, 14, 106, 125, 140)

Polyvinyl methyl ether (CTFA)

SYNONYMS:
Ethene, methoxy-, homopolymer
Methoxyethene, homopolymer
Poly (methyl vinyl ether)
PVM

EMPIRICAL FORMULA:
$(C_2H_6O)_x$

STRUCTURE:

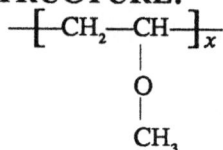

CAS No.:

9003-09-2

TRADENAME EQUIVALENTS:

Gantrez M-154, M-555, M-556, M-574 [GAF]

APPLICATIONS:

Functional additives: binder (Gantrez M-154, M-555, M-556, M-574); plasticizer (Gantrez M-154, M-555, M-556, M-574); tackifier (Gantrez M-154, M-555, M-556, M-574)

Industrial applications: latex modification (Gantrez M-154, M-555, M-556, M-574); printing inks (Gantrez M-154, M-555, M-556, M-574); textile sizes and finishes (Gantrez M-154, M-555, M-556, M-574)

PROPERTIES:

Form:

Liquid (Gantrez M-154, M-555, M-556, M-574)

Composition:

50% active in water (Gantrez M-154)

50% active in toluene (Gantrez M-555, M-556)

70% active in toluene (Gantrez M-574)

GENERAL PROPERTIES:

Ionic Nature:

Nonionic (Gantrez M-154, M-555, M-556, M-574)

Solubility:

Sol. in diverse organic solvents (Gantrez M-154, M-555, M-556, M-574)

Sol. in water; thermally reversible sol. in aq. systems (Gantrez M-154, M-555, M-556, M-574)

Styrene-acrylonitrile resin

SYNONYMS:
SAN
STRUCTURE:

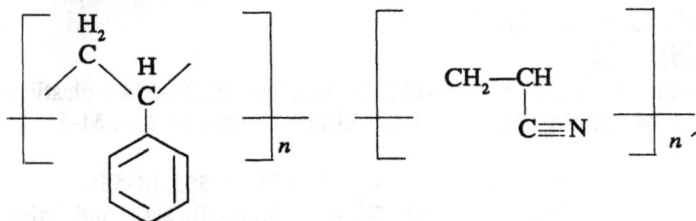

TRADENAME EQUIVALENTS:
Darex 165L [W.R. Grace]
Lustran SAN 31, 33, 35, 41, 43, 47, 49, 51 [Monsanto]
Tyril 860B, 867B, 870C, 880B, 1000, 1000B, 1020, 1021, 1030, 1031 [Dow]
Glass-reinforced:
RTP 501 (10% glass fiber), 503 (20% glass fiber), 505 (30% glass fiber), 506 (35% glass fiber), 507 (40% glass fiber) [Fiberite]
SN-20GF (20% glass fiber), -30GF (30% glass fiber), -40GF (40% glass fiber) [Compounding Technology]
Thermocomp BF-1006 (30% glass fiber), BF-1006FR (30% glass fiber), BF-1008 (40% glass fiber) [LNP]

TYPES/MODIFICATIONS/SPECIALTY GRADES:
Latex:
Darex 165L
Antistatic grade:
Lustran SAN 41, 43; Tyril 1030
High-flow:
Lustran SAN 33; Tyril 860B
Flame-retardant:
Thermocomp BF-1006FR
UV-stabilized:
Lustran SAN 47, 49; Tyril 870C, 1020, 1030
CATEGORY:
Thermoplastic resin

Styrene-acrylonitrile resin (cont'd.)

PROCESSING:
Injection blow molding:
Tyril 880B, 1000, 1000B, 1020, 1021, 1030, 1031
Compression molding:
Tyril 1000, 1000B, 1020, 1021, 1030, 1031
Extrusion:
Lustran SAN 49; Tyril 867B, 1000, 1000B
Injection molding:
Lustran SAN 31, 33, 35, 41, 43, 47; RTP 501, 503, 505, 506, 507; Thermocomp BF-1006, BF-1006FR, BF-1008; Tyril 860B, 867B, 870C, 1000, 1000B, 1020, 1021, 1030, 1031

APPLICATIONS:
Architectural applications: light louvers (Lustran SAN 47, 49); screen door panels (Lustran SAN 47, 49)

Automotive applications: (Lustran SAN; Tyril 1000, 1000B, 1020, 1021, 1030, 1031); light lenses (Lustran SAN 41, 43

Consumer products: appliances (Thermocomp BF-1006FR; Tyril 1000, 1000B, 1020, 1021, 1030, 1031); blender jars (Lustran SAN 31, 35); cameras (Thermocomp BF-1006FR); cosmetic packaging (Lustran SAN 31, 33; Tyril 1000, 1000B, 1020, 1021, 1030, 1031); drinkware (Lustran SAN 31); fan blades (Lustran SAN 31; RTP 501, 503, 505, 506, 507); furniture (Tyril 1000, 1000B, 1020, 1021, 1030, 1031); housewares (Tyril 1000, 1000B, 1020, 1021, 1030, 1031); refrigerator parts (Lustran SAN 31, 33); tableware (Lustran SAN 31); toys (Lustran SAN 31; Tyril 1000, 1000B, 1020, 1021, 1030, 1031)

Electrical/electronic industry: audio/video cassettes (Lustran SAN 41, 43; Tyril 1000, 1000B, 1020, 1021, 1030, 1031); business machines and office equipment (Lustran SAN 31; Thermocomp BF-1006FR); electrical components (Lustran SAN); tape hubs (RTP 501, 503, 505, 506, 507)

Food service applications: food containers (Lustran SAN 31)

Industrial applications: batteries (Lustran SAN 35); bonding/impregnating fibers and webs (Darex 165L); complex parts (Tyril 860B); films (Darex 165L); glazing (Tyril 1000, 1000B, 1020, 1021, 1030, 1031); housings (RTP 501, 503, 505, 506, 507; Thermocomp BF-1006FR); interior applications (Tyril 870C); intricate/hard-to-fill moldings (Lustran SAN 33); large parts (Tyril 860B); molded parts (Thermocomp BF-1006FR); pumps (RTP 501, 503, 505, 506, 507); structural components (Lustran SAN 35); thin sections (Lustran SAN 33)

Medical applications: parts (Lustran SAN 31; Tyril 1000, 1000B, 1020, 1021, 1030, 1031)

PROPERTIES:
Form:
Latex; particle size 0.3 μ (Darex 165L)
Cylindrical pellets (Tyril 860B, 867B, 870C, 880B)

Styrene-acrylonitrile resin (cont'd.)

Color:
Clear (Lustran SAN 35)
Natural, black (RTP 501, 503, 505, 506, 507)
Crystone color (Tyril 1020, 1030, 1031)
Avail. clear, water-white, transparent and opaque colors (Lustran SAN 31)
Avail. in wide range of transparent, translucent, and opaque colors (Tyril 867B, 880B)

Composition:
23.8% acrylonitrile (Lustran SAN 31)
32% acrylonitrile (Lustran SAN 33)
32.1% acrylonitrile (Lustran SAN 35)
50% total solids in water; 71% styrene, 29% acrylonitrile (Darex 165L)

GENERAL PROPERTIES:

Melt Flow:
3.3 g/10 min (Tyril 880B)
4.5 g/10 min (Tyril 867B, 870C, 1020)
5.5 g/10 min (Tyril 1021, 1031)
7.4 g/10 min (Lustran SAN 35)
7.5 g/10 min (Tyril 1000B, 1030)
7.9 g/10 min (Lustran SAN 31)
8.0 g/10 min (Tyril 1000)
9.5 g/10 min (Tyril 860B)
14 g/10 min (Lustran SAN 33)

Sp. Gr.:
1.07 (Lustran SAN 31, 1020)
1.08 (Tyril 860B, 867B, 870C, 880B, 1000, 1000B, 1021, 1030, 1031)
1.15 (RTP 501)
1.22 (RTP 503; SN-20GF)
1.31 (RTP 505; Thermocomp BF-1006)
1.32 (SN-40GF)
1.35 (RTP 506)
1.40 (RTP 507; SN-30GF; Thermocomp BF-1008)
1.50 (Thermocomp BF-1006FR)

Density:
8.7 lb/gal (Darex 165L)

Visc.:
50 cps (Darex 165L)

Stability:
Excellent mechanical stability (Darex 165L)
Good resistance to bases; fair resistance to acids; poor resistance to solvents (Thermocomp BF-1008)
Excellent thermal stability; excellent resistance to 25% acetic acid, ammonium hydroxide, corn oil, formaldehyde, gasoline, halogen salts, hydrochloric acid, mineral oil, sulfate salts, and 40% sulfuric acid (Tyril 860B, 867B, 870C, 880B)

High heat resistance (Tyril 1000, 1000B, 1021)
Excellent chemical resistance (Tyril 1000B)
Good chemical resistance (Tyril 1020, 1030, 1031)
Excellent weathering characterstics (Tyril 1020, 1021, 1030, 1031)

Ref. Index:
1.565 (Tyril 880B)
1.569 (Tyril 860B, 867B, 870C)

pH:
9 (Darex 165L)

Surface Tension:
47 dynes/cm (42% solids) (Darex 165L)

Haze:
1.0% (Lustran SAN 35)
1.2% (Lustran SAN 33)
1.4% (Lustran SAN 31)

Transmittance:
90% (Lustran SAN 31)

MECHANICAL PROPERTIES:

Tens. Str.:
64 MPa (yield) (Tyril 1030, 1031)
69 MPa (yield) (Tyril 1000B, 1020, 1021)
70 MPa (yield) (Tyril 1000)
75.9 MPa (yield) (Lustran SAN 31)
78.0 MPa (yield) (Lustran SAN 33)
79.4 MPa (yield) (Lustran SAN 35)
10,400 psi (Tyril 860B)
11,000 psi (Tyril 867B, 870C)
11,500 psi (RTP 501)
11,900 psi (Tyril 880B)
14,000 psi (Thermocomp BF-1006FR)
15,000 psi (SN-20GF)
17,000 psi (SN-40GF)
17,400 psi (Thermocomp BF-1006)
18,000 psi (RTP 503)
18,100 psi (SN-30GF)
18,600 psi (Thermocomp BF-1008)
19,000 psi (RTP 505)
19,500 psi (RTP 506)
20,000 psi (RTP 507)

Tens. Elong.:
1–2% (SN-20GF, -30GF, -40GF)
1.4% (RTP 507)
1.5% (RTP 506)

Styrene-acrylonitrile resin *(cont'd.)*

    1.6% (RTP 505)
    1.8% (RTP 503)
    2.0% (RTP 501)
    2.5% (break) (Tyril 860B, 1000B, 1020, 1021, 1030, 1031)
    2.6% (break) (Tyril 1000)
    2.7% (break) (Lustran SAN 31)
    3.0% (break) (Lustran SAN 35; Tyril 867B, 870C, 880B)
    3.2% (break) (Lustran SAN 33)

Tens. Mod.:
    3.23 GPa (Lustran SAN 33)
    3.33 GPa (Lustran SAN 31)
    3.4 GPa (Lustran SAN 35)
    3448 MPa (Tyril 1030, 1031)
    3552 MPa (Tyril 1000B)
    3586 MPa (Tyril 1000)
    3793 MPa (Tyril 1020, 1021)
    500,000 psi (Tyril 860B)
    535,000 psi (Tyril 867B, 870C)
    560,000 psi (Tyril 880B)
    800,000 psi (RTP 501)
    1.2×10^6 psi (RTP 503)
    1.6×10^6 psi (RTP 505)
    1.9×10^6 psi (RTP 506)
    2.0×10^6 psi (RTP 507)

Flex. Str.:
    103 MPa (Tyril 1020, 1021)
    112 MPa (Tyril 1030, 1031)
    112.8 MPa (Lustran SAN 31)
    121 MPa (Tyril 1000)
    128 MPa (Tyril 1000B)
    18,000 psi (RTP 501)
    19,100 psi (SN-20GF)
    20,000 psi (RTP 503; Thermocomp BF-1006FR)
    21,000 psi (SN-40GF)
    22,000 psi (RTP 505)
    22,500 psi (SN-30GF)
    23,000 psi (RTP 506)
    24,000 psi (RTP 507)

Flex. Mod.:
    3.12 GPa (Lustran SAN 33)
    3.6 GPa (Lustran SAN 31)
    3.62 GPa (Lustran SAN 35)
    3586 MPa (Tyril 1000, 1030, 1031)

3724 MPa (Tyril 1000B)
4138 MPa (Tyril 1020, 1021)
750,000 psi (RTP 501)
10^6 psi (RTP 503; SN-20GF)
1.3×10^6 psi (SN-40GF)
1.4×10^6 psi (RTP 505)
1.5×10^6 psi (Thermocomp BF-1006)
1.6×10^6 psi (Thermocomp BF-1006FR)
1.7×10^6 psi (RTP 506; SN-30GF)
1.8×10^6 psi (RTP 507)
1.85×10^6 psi (Thermocomp BF-1008)

Compr. Str.:
16,500 psi (RTP 501)
19,000 psi (RTP 503)
20,500 psi (SN-40GF)
21,000 psi (RTP 505)
21,500 psi (RTP 506)
22,000 psi (RTP 507)
23,000 psi (SN-30GF)

Shear Str.:
8500 psi (SN-20GF)
9000 psi (SN-40GF)
9500 psi (SN-30GF)

Impact Str. (Izod):
16 J/m (Tyril 1000, 1000B, 1020, 1021, 1030, 1031)
24.0 J/m (Lustran SAN 35)
24.6 J/m (Lustran SAN 33)
26.7 J/m (Lustran SAN 31)
0.4 ft lb/in. (Tyril 860B)
0.45 ft lb/in. (Tyril 867B, 870C)
0.5 ft lb/in. (Tyril 880B)
0.7 ft lb/in. notched (RTP 501)
0.9 ft lb/in. notched (SN-20GF)
1.0 ft lb/in. notched (RTP 503; SN-40GF; Thermocomp BF-1006, BF-1006FR)
1.1 ft lb/in. notched (RTP 505, 506, 507; SN-30GF; Thermocomp BF-1008)

Hardness:
Rockwell M79 (Tyril 1000)
Rockwell M80 (Tyril 860B, 867B, 870C, 880B, 1030, 1031)
Rockwell M82 (Tyril 1021)
Rockwell M83 (Lustran SAN 31; Tyril 1020)
Rockwell M86 (Tyril 1000B)
Rockwell M94 (Thermocomp BF-1006)
Rockwell M97 (Thermocomp BF-1008)

Styrene-acrylonitrile resin *(cont'd.)*

Rockwell R121 (RTP 501)
Rockwell R122 (RTP 503)
Rockwell R123 (RTP 505, 506, 507)
Mold Shrinkage:
0.003–0.007 cm/cm (Tyril 1000, 1000B, 1020, 1021, 1030, 1031)
THERMAL PROPERTIES:
Soften. Pt. (Vicat):
104 C (Tyril 1030, 1031)
110 C (Tyril 1000B)
112 C (Tyril 1000, 1020, 1021)
112.8 C (Lustran SAN 31)
229 F (Tyril 860B)
230 F (Tyril 867B, 870C)
231 F (Tyril 880B)
Conduct.:
1.4 Btu/h/ft^2/F/in. (RTP 501)
1.9 Btu/h/ft^2/F/in. (RTP 503)
1.95 Btu/h/ft^2/F/in. (RTP 505)
2.0 Btu/h/ft^2/F/in. (RTP 506)
2.1 Btu/h/ft^2/F/in. (RTP 507)
Distort. Temp.:
93.9 C (Lustran SAN 33)
95.6 C (Lustran SAN 31)
96.6 C (Lustran SAN 35)
97 C (264 psi) (Tyril 1030)
102 C (264 psi) (Tyril 1020, 1021, 1031)
103 C (264 psi) (Tyril 1000B)
108 C (264 psi) (Tyril 1000)
190 F (264 psi) (Thermocomp BF-1006FR)
200 F (264 psi) (SN-20GF)
205 F (264 psi) (RTP 501)
210 F (264 psi) (RTP 503; SN-40GF)
212 F (264 psi) (RTP 505)
214 F (Tyril 860B); (264 psi) (RTP 506)
215 F (264 psi) (SN-30GF; Thermocomp BF-1006)
217 F (Tyril 867B, 870C, 880B); (264 psi) (RTP 507)
220 F (264 psi) (Thermocomp BF-1008)
Coeff. of Linear Exp.:
1.5 × 10^{-5} in./in./F (Thermocomp BF-1008)
1.7 × 10^{-5} in./in./F (SN-30GF)
1.8 × 10^{-5} in./in./F (Thermocomp BF-1006)
2.0 × 10^{-5} in./in./F (SN-40GF)
2.1 × 10^{-5} in./in./F (Thermocomp BF-1006FR)

2.3×10^{-5} in./in./F (SN-20GF)

3.6×10^{-5} in./in./F (Tyril 860B, 867B, 870C)

3.7×10^{-5} in./in./F (Tyril 880B, 1000, 1000B, 1020, 1021, 1030, 1031)

Sp. Heat:

0.31 Btu/lb/F (Tyril 860B, 867B, 870C, 880B)

Flamm.:

V-0 (Thermocomp BF-1006FR)

HB (RTP 501, 503, 505, 506, 507; SN-20GF, -30GF, -40GF; Thermocomp BF-1006, BF-1008)

Resins will burn and once ignited may burn rapidly under certain conditions of heat and oxygen supply (Tyril 860B, 867B, 870C, 880B)

ELECTRICAL PROPERTIES:

Dissip. Factor:

0.0020–0.0300 (60 to 10^6 Hz) (Thermocomp BF-1006FR)

0.006 (1 kHz) (Tyril 860B, 867B, 870C)

0.007 (1 kHz) (Tyril 880B)

0.008 (1 MHz) (RTP 501, 503, 505, 506, 507)

Dielec. Str.:

450 V/mil (Thermocomp BF-1006FR)

475 V/mil (SN-20GF)

490 V/mil (SN-40GF)

500 V/mil (RTP 501, 503, 505, 506, 507; SN-30GF)

Dielec. Const.:

2.83–2.77 (60 to 10^6 Hz) (Thermocomp BF-1006FR)

3.0 (1 kHz) (Tyril 860B, 867B, 870C)

3.18 (1 kHz) (Tyril 880B)

3.3 (1 MHz) (RTP 501)

3.5 (1 MHz) (RTP 503)

3.6 (1 MHz) (RTP 505)

3.7 (1 MHz) (RTP 506)

3.8 (1 MHz) (RTP 507)

Vol. Resist.:

4×10^{15} ohm-cm (SN-20GF, -30GF, -40GF)

10^{16} ohm-cm (RTP 501, 503, 505, 506, 507)

Arc Resist.:

60 s (RTP 505, 506, 507)

70 s (RTP 503)

75 s (RTP 501)

TOXICITY/HANDLING:

Low degree of toxicity; avoid contact with molten resins during fabrication and exposure to dusts during machining; during heat fabricating, some vapors (monomers, carbon monoxide) may be released—use with adequate ventilation (Tyril 860B, 867B, 870C, 880B)

Styrene-acrylonitrile resin *(cont'd.)*

Low degree of toxicity; inhalation of vapors may be irritation; avoid breathing dusts or vapors; spills can be hazardous (Tyril 1000, 1000B, 1020, 1021, 1030, 1031)

STORAGE/HANDLING:

Dusts and fines in air may be combustible—keep from contact with open flame (Tyril 860B, 867B, 870C, 880B)

Resin will burn, and once ignited may burn rapidly under the right conditions of heat and oxygen feed; in burning, they may emit a dense black smoke (Tyril 1000, 1000B, 1020, 1021, 1030, 1031)

Terpene/polyterpene resin

Derived from dipentene (monocyclic):
SYNONYMS:
    Cajeputene
    Cinene
    Cyclohexene, 1-methyl-4-(1-methylethenyl)-
    DL-Limonene
    Limonene, inactive
    dl-p-Mentha-1,8-diene
    1-Methyl-4-(1-methylethenyl) cyclohexene
EMPIRICAL FORMULA:
    $C_{10}H_{16}$
CAS No.: 138-86-3; 7705-14-8
STRUCTURE:

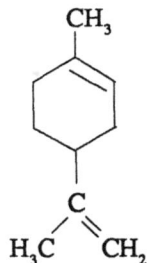

Derived from limonene (monocyclic):
EMPIRICAL FORMULA:
    $C_{10}H_{16}$
CAS No.: 138-86-3
STRUCTURE:

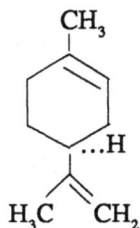

Terpene/polyterpene resin (*cont'd.*)

Derived from α-pinene (dicyclic):
EMPIRICAL FORMULA:

$C_{10}H_{16}$

STRUCTURE:

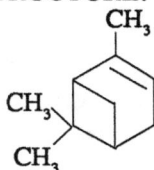

Derived from β-pinene (dicyclic):
SYNONYMS:

Nopinene

EMPIRICAL FORMULA:

$C_{10}H_{16}$

STRUCTURE:

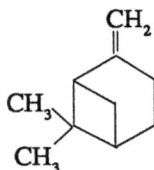

Derived from myrcene (acyclic):
SYNONYMS:

7-Methyl-3-methylene-1,6-octadiene

EMPIRICAL FORMULA:

$C_{10}H_{16}$

TRADENAME EQUIVALENTS:

Croturez [Crosby]

Nevpene 9500 [Neville]

Nevtac 80, 99 Super, 100, 115, 130 [Neville]

Nirez 1000 Series [Reichhold]

Piccofyn A100, A115, A135, D125 [Hercules]

Piccolyte A115, A125, A135, C100, C115, C125, C135, D100, D115, D135, HM-110
    Resin, S100, S115, S115SF, S125, S125SF, S135 [Hercules]

Picconol A200, A201 [Hercules]

Plusplice 100 [C.P. Hall]

Zonarez 7010, 7025, 7040, 7055, 7070, 7085, 7100, 7115, 7125, B-10, B-25, B-40, B-
    55, B-70, B-85, B-115, B-125 [Arizona]

Zonatec 105, 115 [Arizona]

Terpene/polyterpene resin (cont'd.)

TYPES/MODIFICATIONS/SPECIALTY GRADES:
Alkylated, phenolic-modified:
Piccofyn A100, A115, A135, D125
Dipentene-derived:
Piccolyte D100, D115, D135; Zonarez 7010, 7025, 7040, 7055, 7070, 7085, 7100, 7115, 7125
d-Limonene-derived:
Piccolyte C100, C115, C125, C135
α-Pinene-derived:
Piccolyte A115, A125, A135
β-Pinene-derived:
Piccolyte S100, S115, S115SF, S125, S125SF, S135; Zonarez B-10, B-25, B-40, B-55, B-70, B-85, B-115, B-125
Antioxidant additive:
Piccolyte S115SF, S125SF

CATEGORY:
Thermoplastic resin

APPLICATIONS:
Architectural applications: construction applications (Piccolyte D100, D115, D135)
Consumer products: toys (Piccolyte S100, S115, S115SF, S125, S125SF, S135)
Electrical/electronic industry:
FDA-approved applications: (Piccofyn A100, A115, A135, D125; Piccolyte A115, A125, A135, C100, C115, C125, C135, D100, D115, D135, HM-110 Resin, S100, S115, S115SF, S125, S125SF, S135; Picconol A200, A201; Zonarez 7010, 7025, 7040, 7055, 7070, 7085, 7100, 7115, 7125, B-10, B-25, B-40, B-55, B-70, B-85, B-115, B-125)
Food-contact applications: (Piccolyte A115, A125, A135, C100, C115, C125, C135); chewing gums (Piccolyte 115, C125, C135; Zonarez 7010, 7025, 7040, 7055, 7070, 7085, 7100, 7115, 7125, B-10, B-25, B-40, B-55, B-70, B-85, B-115, B-125); food container closures (Piccolyte S115, S115SF, S125, S125SF, S135); food packaging (Piccofyn A100, A115, A135, D125; Piccolyte C100, C115, C125, C135, D100, D115, D135, HM-110 Resin, S100, S115, S115SF, S125, S125SF, S135; Picconol A200, A201); food processing (Piccolyte C100, C115, C125, C135, D100, D115, D135, HM-110 Resin, S100, S115, S115SF, S125, S125SF, S135)
Functional additives: gloss promoter (Piccolyte A115, A125, A135); masticatory agent (Piccolyte C115, C125, C135); modifier (Piccofyn A100, A115, A135; Piccolyte A115, A125, A135, C100, C115, C125, C135, D100, D115, D135, S100, S115, S115SF, S125, S125SF, S135); processing aid (Piccofyn A100, A115, A135); reinforcing agent (Piccofyn A100, A115, A135; Piccolyte A115, A125, A135, D100, D115, D135); tackifier (Croturez; Nevpene 9500; Nevtac 80, 99 Super, 100, 115, 130; Nirez 1000 Series; Piccofyn A100, A115, A135; Piccolyte A115, A125, A135, C100, C115, C125, C135, D100, D115, D135, HM-110 Resin,

Terpene/polyterpene resin (cont'd.)

S100, S115, S115SF, S125, S125SF, S135; Picconol A200, A201; Plusplice 100; Zonarez 7000 Series, B Series; Zonatec 105, 115)

Industrial applications: adhesives/sealants (Nevpene 9500; Nevtac 80, 99 Super, 100, 115, 130; Piccofyn A100, A115, A135, D125; Piccolyte A115, A125, A135, C100, C115, C125, C135, D100, D115, D135, HM-110 Resin, S100, S115, S115SF, S125, S125SF, S135; Picconol A200, A201; Plusplice 100; Zonarez 7010, 7025, 7040, 7055, 7070, 7085, 7100, 7115, 7125, B-10, B-25, B-40, B-55, B-70, B-85, B-115, B-125); caulking compounds (Nevpene 9500; Nevtac 80, 99 Super, 100, 115, 130; Piccolyte A115, A125, A135; Zonarez 7010, 7025, 7040, 7055, 7070, 7085, 7100, 7115, 7125, B-10, B-25, B-40, B-55, B-70, B-85, B-115, B-125); coatings (Nevpene 9500; Nevtac 80, 99 Super, 100, 115, 130; Piccofyn A100, A115, A135, D125; Piccolyte A115, A125, A135, C100, C115, C125, C135, D100, D115, D135; Plusplice 100; Zonarez 7010, 7025, 7040, 7055, 7070, 7085, 7100, 7115, 7125, B-10, B-25, B-40, B-55, B-70, B-85, B-115, B-125); concrete (Nevtac 80, 99 Super, 100, 115, 130; Zonarez 7010, 7025, 7040, 7055, 7070, 7085, 7100, 7115, 7125, B-10, B-25, B-40, B-55, B-70, B-85, B-115, B-125); elastomers (Piccolyte D100, D115, D135); floor waxes (Piccofyn A100, A115, A135); laminates (Piccolyte C100, C115, C125, C135, D100, D115, D135, S100, S115, S115SF, S125, S125SF, S135; Picconol A200, A201; Plusplice 100); lost-wax investment castings (Piccolyte A115, A125, A135, C100, C115, C125, C135, D100, D115, D135, S100, S115, S115SF, S125, S125SF, S135; Zonarez 7010, 7025, 7040, 7055, 7070, 7085, 7100, 7115, 7125, B-10, B-25, B-40, B-55, B-70, B-85, B-115, B-125); paints, varnishes (Piccolyte A115, A125, A135; Zonarez 7010, 7025, 7040, 7055, 7070, 7085, 7100, 7115, 7125, B-10, B-25, B-40, B-55, B-70, B-85, B-115, B-125); paper coatings/ laminations (Nevpene 9500; Nevtac 80, 99 Super, 100, 115, 130; Piccolyte C100, C115, C125, C135); printing inks (Piccolyte S100, S115, S115SF, S125, S125SF, S135; Zonarez 7010, 7025, 7040, 7055, 7070, 7085, 7100, 7115, 7125, B-10, B-25, B-40, B-55, B-70, B-85, B-115, B-125); rubber (Croturez; Nevpene 9500; Nevtac 80, 99 Super, 100, 115, 130; Nirez 1000 Series; Piccofyn A100, A115, A135; Piccolyte A115, A125, A135, C100, C115, C125, C135, S100, S115, S115SF, S125, S125SF, S135; Picconol A200, A201; Plusplice 100; Zonarez 7010, 7025, 7040, 7055, 7070, 7085, 7100, 7115, 7125, B-10, B-25, B-40, B-55, B-70, B-85, B-115, B-125; Zonatec 105, 115); textile coatings/sizing (Nevtac 80, 99 Super, 100, 115, 130; Piccolyte S100, S115, S115SF, S125, S125SF, S135); tires (Nevpene 9500; Nevtac 80, 99 Super, 100, 115, 130); waterproofing (Piccolyte A115, A125, A135, D100, D115, D135, S100, S115, S115SF, S125, S125SF, S135; Zonarez 7010, 7025, 7040, 7055, 7070, 7085, 7100, 7115, 7125, B-10, B-25, B-40, B-55, B-70, B-85, B-115, B-125)

Marine applications: coatings (Nevtac 80, 99 Super, 100, 115, 130)

Medical applications: (Piccolyte S100, S115, S115SF, S125, S125SF, S135); gelatin capsules (Zonarez 7010, 7025, 7040, 7055, 7070, 7085, 7100, 7115, 7125, B-10, B-25, B-40, B-55, B-70, B-85, B-115, B-125)

Transportation industry: traffic coatings (Nevtac 80, 99 Super, 100, 115, 130)

Terpene/polyterpene resin (cont'd.)

PROPERTIES:
Form:
Liquid emulsion (Picconol A200, A201)

Solid (Nevpene 9500; Nevtac 80, 99 Super, 100, 115, 130; Piccofyn A100, A115, A135, D125; Piccolyte A115, A125, A135, C100, C115, C125, C135, D100, D115, D135, HM-110 Resin, S100, S115, S115SF, S125, S125SF, S135)

Crushed (Piccofyn A100, A115, A135)

Flakes (Nevpene 9500; Nevtac 99 Super, 100, 115, 130; Piccofyn D125; Piccolyte C100, C115, C125, C135, D100, D115, D135, HM-110 Resin, S100, S115, S115SF, S125, S125SF, S135)

Color:
Gardner 2 (50% in toluene) (Piccolyte S100, S115, S115SF, S125, S125SF, S135); (base resin, 50% in toluene) (Picconol A200, A201); (50% in heptane) (Zonarez 7010, 7025, 7040, 7055)

Gardner 2+ (50% in heptane) (Zonarez B-10, B-25, B-40)

Gardner 3– (50% in heptane) (Zonarez B-55, B-70)

Gardner 3 (50% in toluene) (Piccolyte D115, D135); (50% in heptane) (Zonarez 7070, 7085, 7100, 7115, 7125, B-85, B-115, B-125; Zonatec 105, 115)

Gardner 4 (50% in toluene) (Nirez 1000 Series); Piccolyte A115, A125, C100, C115, C125, C135, D100)

Gardner 5 (50% in toluene) (Nevtac 80; Piccolyte A135)

Gardner 5–8 (Plusplice 100)

Gardner 6 (50% in toluene) (Nevtac 100)

Gardner 7 (50% in toluene) (Nevtac 99 Super, 115; Piccofyn A100, A115, A135)

Gardner 8 (50% in toluene) (Nevpene 9500; Nevtac 130; Piccofyn D125; Piccolyte HM-110 Resin)

GENERAL PROPERTIES:
Solubility:
Sol. in alcohols (Piccofyn D125); sol. in long-chain alcohols (Piccofyn A100, A115, A135; Piccolyte S100, S115, S115SF, S125, S125SF, S135); sol. in long-chain aliphatic alcohols (Piccolyte C100, C115, C125, C135, D100, D115, D135); insol. in lower m.w. alcohols (Piccofyn A100, A115, A135; Piccolyte C100, C115, C125, C135, D100, D115, D135, S100, S115, S115SF, S125, S125SF, S135); insol. in alcohols (Piccolyte A115, A125, A135)

Sol. in aliphatic hydrocarbons (Piccofyn A100, A115, A135; Piccolyte A115, A125, A135, C100, C115, C125, C135, D100, D115, D135, HM-110 Resin, S100, S115, S115SF, S125, S125SF, S135; Zonarez 7085, 7115, 7125, B-85, B-115, B-125); insol. (Piccofyn D125)

Sol. in amyl acetate (Piccolyte C100, C115, C125, C135, D100, D115, D135; Zonarez 7085, 7115, 7125)

Sol. in aromatic hydrocarbons (Nevpene 9500; Nevtac 80, 99 Super, 100, 115, 130; Piccofyn A100, A115, A135, D125; Piccolyte A115, A125, A135, C100, C115, C125, C135, D100, D115, D135, HM-110 Resin, S100, S115, S115SF, S125,

Terpene/polyterpene resin (cont'd.)

S125SF, S135; Zonarez 7085, 7115, 7125, B-85, B-115, B-125)

Sol. in butyl acetate (Piccolyte C100, C115, C125, C135, D100, D115, D135; Zonarez 7085, 7115, 7125)

Sol. in chlorinated hydrocarbons (Nevpene 9500; Nevtac 80, 99 Super, 100, 115, 130; Piccofyn A100, A115, A135, D125; Piccolyte A115, A125, A135, C100, C115, C125, C135, D100, D115, D135, S100, S115, S115SF, S125, S125SF, S135; Zonarez 7085, 7115, 7125, B-85, B-115, B-125)

Sol. in esters (Nevpene 9500; Nevtac 80, 99 Super, 100, 115, 130; Piccofyn D125; Piccolyte S100, S115, S115SF, S125, S125SF, S135; Zonarez B-85, B-115, B-125)

Sol. in ethers (Nevpene 9500; Nevtac 80, 99 Super, 100, 115, 130; Piccofyn D125; Piccolyte C100, C115, C125, C135, D100, D115, D135, S100, S115, S115SF, S125, S125SF, S135)

Sol. in ethyl acetate (Piccolyte A115, A125, A135)

Sol. in ethyl ether (Zonarez 7085, 7115, 7125, B-85, B-115, B-125)

Sol. in 2-ethylhexyl alcohol (Zonarez 7085, 7115, 7125, B-85, B-115, B-125)

Insol. in glycols (Piccofyn A100, A115, A135, D125; Piccolyte A115, A125, A135, C100, C115, C125, C135, D100, D115, D135, S100, S115, S115SF, S125, S125SF, S135)

Sol. in ketones (Piccolyte A115, A125, A135); sol. in ketones (except acetone) (Nevpene 9500; Nevtac 80, 99 Super, 100, 115, 130); sol. in higher ketones (S100, S115, S115SF, S125, S125SF, S135); insol. in ketones (Piccolyte C100, C115, C125, C135, D100, D115, D135)

Sol. in MIBK (Piccofyn A100, A115, A135)

Sol. in mineral oil (Piccolyte C100, C115, C125, C135)

Sol. in naphthenic hydrocarbons (Nevpene 9500; Nevtac 80, 99 Super, 100, 115, 130)

Partly sol. in paraffinic hydrocarbons (Nevpene 9500; Nevtac 80, 99 Super, 100, 115, 130)

Sol. in petroleum ether (Zonarez 7085, 7115, 7125, B-85, B-115, B-125)

Sol. in terpene hydrocarbons (Nevpene 9500; Nevtac 80, 99 Super, 100, 115, 130; Zonarez 7085, 7115, 7125, B-85, B-115, B-125)

Sol. in turpentine (Piccolyte C100, C115, C125, C135)

Sol. in VM&P naphtha (Piccolyte A115, A125, A135, C100, C115, C125, C135, D100, D115, D135)

Insol. in water (Piccofyn D125; Piccolyte S100, S115, S115SF, S125, S125SF, S135)

M.W.:

1040 (Nevpene 9500; Nevtac 99 Super)

1070 (Nevtac 80)

1280 (Nevtac 100)

1410 (Nevtac 130)

1415 (Nevtac 115)

Sp. Gr.:

0.920 (Zonarez B-10

0.930 (Zonarez B-25)

Terpene/polyterpene resin (cont'd.)

0.940 (Zonarez B-40)
0.950 (Nevtac 80)
0.960 (Nevtac 100; Plusplice 100; Zonarez B-55, B-70)
0.970 (Nevtac 115; Zonarez 7010, 7025, 7040, 7085, 7100, B-125)
0.980 (Nevtac 99 Super, 130; Zonarez 7055, 7070, B-115)
0.980–0.995 (Nirez 1000 Series)
0.99 (Zonarez 7115, 7125, B-85
1.050 (Nevpene 9500)

Density:
0.98 kg/l (Piccolyte A115, A125, A135, D100, D115, D135; Picconol A200, A201)
0.99 kg/l (Piccolyte C100, C115, C125, C135, S100, S115, S115SF, S125, S125SF, S135)
1.02 kg/l (Piccolyte HM-110 Resin)
1.03 kg/l (Piccofyn A100, A115, A135)
1.04 kg/l (Piccofyn D125)

Visc.:
105 stokes (70% solids in toluene) (Piccofyn D125)
1000 cps (Picconol A201)
1500 cps (Picconol A200)
L (70% solids in toluene) (Piccofyn A100)
P (70% solids in toluene) (Piccofyn A115)
V (70% solids in toluene) (Piccofyn A135)

Melt Visc.:
189 C (1 poise) (Piccolyte C100)
192 C (1 poise) (Piccolyte HM-110 Resin)
200 C (1 poise) (Piccolyte A115)
201 C (1 poise) (Piccolyte D100)
202 C (1 poise) (Piccolyte S100)
204 C (1 poise) (Piccolyte C115)
214 C (1 poise) (Piccolyte C125)
210 C (1 poise) (Piccolyte A125)
211 C (1 poise) (Piccolyte D115)
220 C (1 poise) (Piccolyte A135, S115, S115SF)
224 C (1 poise) (Piccolyte C135)
225 C (1 poise) (Piccolyte S125, S125SF)
232 C (1 poise) (Piccolyte D135, S135)

Softening Pt. (Ring & Ball):
10 C (Zonarez 7010, B-10)
10–135 C (Croturez; Nirez 1000 Series)
25 C (Zonarez 7025, B-25)
40 C (Zonarez 7040, B-40)
55 C (Zonarez 7055, B-55)
57 C (emulsion solids) (Picconol A200)

Terpene/polyterpene resin *(cont'd.)*

70 C (Zonarez 7070, B-70)
80 C (emulsion solids) (Picconol A201)
81 C (Nevtac 80)
85 C (Zonarez 7085, B-85)
95 C (Nevpene 9500)
99 C (Nevtac 99 Super)
100 C (Piccofyn A100; Piccolyte C100, D100, S100; Plusplice 100; Zonarez 7100)
102 C (Nevtac 100)
105 C (Zonatec 105)
109 C (Piccolyte HM-110 Resin)
114 C (Nevtac 115)
115 C (Piccofyn A115; Piccolyte A115, C115, D115, S115, S115SF; Zonarez 7115, B-115; Zonatec 115)
125 C (Piccofyn D125; Piccolyte A125, C125, S125, S125SF; Zonarez 7125, B-125)
131 C (Nevtac 130)
135 C (Piccofyn A135; Piccolyte A135, C135, D135, S135)
Flash Pt.:
216 C (COC) (Piccolyte A115)
219 C (COC) (Piccolyte A125)
224 C (COC) (Piccofyn A100)
229 C (COC) (Piccofyn A115)
230 C (COC) (Piccolyte C100)
232 C (COC) (Piccolyte D100, S100)
234 C (COC) (Piccolyte S115, S115SF)
235 C (COC) (Piccolyte C115)
238 C (COC) (Piccolyte A135, D115, S125, S125SF)
241 C (COC) (Piccolyte C125)
254 C (COC) (Piccoalyte D135)
259 C (COC) (Piccolyte S135)
260 C (COC) (Piccolyte C135)
268 C (COC) (Piccofyn A135)
440 F (COC) (Nevtac 80)
450 F (COC) (Nevpene 9500; Nevtac 99 Super, 100)
460 F (COC) (Nevtac 115)
475 F (COC) (Nevtac 130)
525 F (Piccolyte HM-110 Resin)
Fire Pt.:
570 F (Piccolyte HM-110 Resin)
Acid No.:
Nil (Plusplice 100)
< 1 (Piccofyn D125; Zonarez 7010, 7025, 7040, 7055, 7070, 7085, 7100, 7115, 7125, B-10, B-25, B-40, B-55, B-70, B-85, B-115, B-125)
> 1 (Piccolyte HM-110 Resin)

Terpene/polyterpene resin (cont'd.)

Iodine No.:

84 (Plusplice 100)

97 (Nevpene 9500)

110 (Nevtac 80)

135 (Nevtac 99 Super)

145 (Nevtac 100)

150 (Nevtac 115)

173 (Nevtac 130)

Saponification No.:

< 1 (Zonarez 7010, 7025, 7040, 7055, 7070, 7085, 7100, 7115, 7125, B-10, B-25, B-40, B-55, B-70, B-85, B-115, B-125)

> 1 (Piccolyte HM-110 Resin)

< 2 (Piccofyn D125)

Bromine No.:

22.0 (Piccolyte HM-110 Resin)

27.0 (Piccolyte A135, C100, C135, D135, S125, S125SF, S135)

28.0 (Piccolyte C115, C125, D100, D115)

30.0 (Piccolyte S100, S115, S115SF)

31.0 (Piccolyte A125)

31.5 (Piccolyte A115)

77 (Piccofyn D125)

Stability:

Excellent resistance to alkalis, dilute acids, and moisture (Nevtac 80, 99 Super, 100, 115, 130)

Good color stability and resistance to uv light (Piccofyn A100, A115, A135)

Good color, heat, and uv light stability; good resistance to acids, alkalis, and water (Piccolyte A115, A125, A135)

Good color retention, thermal stability; inert to acids, alkalis, and salts (Piccolyte D100, D115, D135)

Good color retention and thermal stability; good resistance to aging, heat, and uv light; inert to acids, alkalis, and salts (Piccolyte C100, C115, C125, C135)

Excellent uv- and heat-stability; good color stability and resistance to acids, alkalis, and water (Piccolyte HM-110 Resin)

High thermal stability; good color stability and resistance to acids, alkalis, and water (Piccolyte S100, S115, S115SF, S125, S125SF, S135)

Fair mechanical stability; dried films have good color stability and are highly resistant to acids, alkalis, and water (Picconol A200, A201)

Storage Stability:

Excellent storage stability (Picconol A200, A201)

Ref. Index:

1.53 (Piccolyte A115, A125, A135)

pH:

8.5 (Picconol A201)

Terpene/polyterpene resin *(cont'd.)*

9 (Picconol A200)
Surface Tension:
32 dynes/cm (Picconol A200)
35 dynes/cm (Picconol A201)
STORAGE/HANDLING:
Crushed forms prone to gradual oxidation; maintain strict inventory control (Piccofyn A100, A115, A135)
Flaked forms prone to gradual oxidation; maintain strict inventory control (Piccofyn D125; Piccolyte C100, C115, C125, C135, D100, D115, D135, S100, S115, S115SF, S125, S125SF, S135)
STD. PKGS.:
Bag, drum, tankcar (Croturez)
175-kg net lightweight metal drums (Piccolyte A115, A125, A135)
175-kg net lightweight metal drums (solid); 22.7-kg net multiwall paper bags (flakes) (Piccolyte C100, C115, C125, C135, D100, D115, D135, S100, S115, S115SF, S125, S125SF, S135)
182-kg net lightweight metal drums (solid); 22.7-kg net multiwall paper bags (crushed) (Piccofyn A100, A115, A135)
182-kg net lightweight metal drums (solid); 22.7-kg net multiwall paper bags (flakes) (Piccofyn D125)
209-kg net fiber drums (Picconol A200, A201)

Toluenesulfonamide/formaldehyde resin

SYNONYMS:
Benzenesulfonamide, 4-methyl-, polymer with formaldehyde
EMPIRICAL FORMULA:
$(C_7H_9NO_2S \cdot CH_2O)_x$
STRUCTURE:

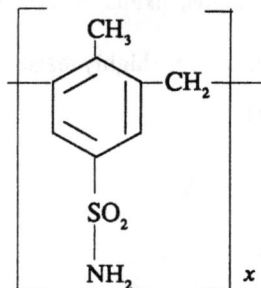

Toluenesulfonamide/formaldehyde resin (cont'd.)

CAS No.:
25035-71-6
TRADENAME EQUIVALENTS:
Santolite MHP, MS80% [Monsanto]
CATEGORY:
Plasticizing resin, flexibilizer, modifier, adhesion promoter for film-forming resins, e.g., nitrocellulose, cellulose acetate, ethyl cellulose, polyvinyl acetate, vinyl, polyamide formulations
APPLICATIONS:
Food applications: food packaging (Santolite MHP, MS-80%)

Industrial applications: adhesives (Santolite MHP, MS-80%); cellophane coatings (Santolite MHP, MS-80%); lacquers (Santolite MS-80%); molding or solution coating (Santolite MHP, MS-80%); spray lacquers (Santolite MHP, MS-80%); vinyl lacquers (Santolite MHP, MS-80%)

Personal products: fingernail polish (Santolite MHP, MS-80%)
PROPERTIES:
Form:
Hard resin (Santolite MHP)

Viscous liquid, free flowing above 20 C (Santolite MS-80%)
Color:
Practically colorless (Santolite MHP)

APHA 80 max. (Santolite MS-80%)
Odor:
Faint formaldehyde odor that disappears after heating (Santolite MHP)
Composition:
80% sol'n. in butyl acetate (Santolite MS-80%)
GENERAL PROPERTIES:
Solubility:
Sol. in alcohols (Santolite MHP)

Sol. in aromatic hydrocarbons (Santolite MHP)

Miscible with all common organic solvents and thinners except petroleum hydrocarbons (Santolite MS-80%)

Sol. in esters (Santolite MHP)

Sol. in ethers (Santolite MHP)

Sol. in ketones (Santolite MHP)

Insol. in min. oil (Santolite MHP)

Immiscible with paint and varnish oils (Santolite MS-80%)

Insol. in veg. oil (Santolite MHP)

Practically insol. in water (Santolite MHP, MS-80%)
Sp.gr.:
1.256 (Santolite MS-80%)

1.35 (Santolite MHP)

Toluenesulfonamide/formaldehyde resin *(cont'd.)*

Density:
   10.5 lb/gal (Santolite MS-80%)
   11.2 lb/gal (Santolite MHP)
Softening Pt.:
   62 C (Santolite MHP)
Flash Pt.:
   90–95 F (COC) (Santolite MS-80%)
Ref. Index:
   1.4275–1.4325 (25 g/75 g butyl acetate) (Santolite MHP)
pH:
   1.0 (Santolite MHP)

Urea-formaldehyde resin

SYNONYMS:
Urea, polymer with formaldehyde

EMPIRICAL FORMULA:
$(CH_4N_2O \cdot CH_2O)_x$

STRUCTURE:

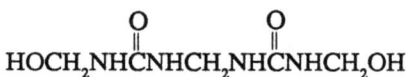

CAS No.:
9011-05-6

TRADENAME EQUIVALENTS:
Hercules Resin 917 [Hercules]
Kymene 435, 760, 760-35, 835, 917 [Hercules]
Uformite F-200E, F-210, F-222, F-240, F-240N [Rohm & Haas]

CATEGORY:
Thermosetting amino resin

APPLICATIONS:
Automotive applications: primers (Uformite F-240, F-240N)

Functional additives: modifier (Hercules Resin 917; Uformite F-240); pigment-grinding medium (Uformite F-240)

Industrial applications: coatings (Hercules Resin 917; Uformite F-240); enamels (Uformite F-200E, F-210, F-222, F-240, F-240N); films (Hercules Resin 917; Uformite F-200E, F-222); furniture finishes (Uformite F-200E); metal coatings (Uformite F-210, F-222, F-240, F-240N); paper applications (Kymene 435, 760, 760-35, 835, 917)

Medical applications: Hospital equipment finishes (Uformite F-210)

PROPERTIES:
Form:
Liquid (Hercules Resin 917)
Clear solution (Uformite F-200E, F-240, F-240N)
Clear to slightly hazy solution (Uformite F-210, F-222)

Color:
Colorless (Uformite F-200E, F-210, F-222, F-240, F-240N)
Gardner 3–4 (Hercules Resin 917; Kymene 435, 760, 760-35, 835, 917)

Urea-formaldehyde resin *(cont'd.)*

Composition:
   25% solids (Kymene 917); 25% solids in water (Hercules Resin 917)
   27.5 ± 1% solids (Kymene 760)
   35% solids (Kymene 435)
   35 ± 1% solids (Kymene 760-35, 835)
   50 ± 2% solids in xylene:butanol (1:1) (Uformite F-200E, F-210)
   50 ± 2% solids in xylene:butanol (2:3) (Uformite F-222)
   60 ± 2% solids in high-flash naphtha (Uformite F-240N)
   60 ± 2% solids in xylene:butanol (1:1.5) (Uformite F-240)

GENERAL PROPERTIES:

Ionic Nature:
   Cationic (Hercules Resin 917; Kymene 435, 760, 760-35, 835, 917)

Sp. Gr.:
   1.01 (Uformite F-200E, F-210, F-222)
   1.02 (Uformite F-240)
   1.03 (Uformite F-240N)
   1.106 (Hercules Resin 917; Kymene 917)
   1.150 (Kymene 435)

Density:
   1.15 kg/l (Kymene 760-35, 835)
   8.4 lb/gal (Uformite F-200E, F-210, F-222)
   8.5 lb/gal (Uformite F-240)
   8.6 lb/gal (Uformite F-240N)
   9.3 lb/gal (Kymene 760)

Visc.:
   14–19 cps (Kymene 760)
   15–25 cps (Hercules Resin 917; Kymene 917)
   38–45 cps (Kymene 435, 760-35, 835)
   K–P (Uformite F-240)
   T–W (Uformite F-222)
   W–Z (Uformite F-200E, F-210)
   Z_1–Z_5 (Uformite F-240N)

F.P.:
   –4 C (Kymene 435)
   –3.9 C (Kymene 760-35, 835)
   –1.7 C (Hercules Resin 917; Kymene 760, 917)

Acid No.:
   2–5 (Uformite F-240N)
   3–6 (Uformite F-240)
   4–7 (Uformite F-210)
   6–9 (Uformite F-200E)
   14–18 (Uformite F-222)

Urea-formaldehyde resin *(cont'd.)*

Storage Stability:
>3 mo. shelf life (stored < 32 C) (Hercules Resin 917; Kymene 435)
>3 mo shelf life (stored < 32 C and not allowed to freeze) (Kymene 917)
>3 mo shelf life (stored ≤ 90 F) (Kymene 760, 760-35, 835)

pH:
>7.1–7.3 (Hercules Resin 917; Kymene 760, 760-35, 835)
>7.2 (Kymene 435, 917)

TOXICITY/HANDLING:
>Evolves fumes of free formaldehyde which may be irritating to skin, eyes, and respiratory tract (Hercules Resin 917)

STORAGE/HANDLING:
>Store < 32 C; do not allow to freeze; if frozen, thoroughly agitate immediately after thawing (Hercules Resin 917)
>Store between 4.4 and 32 C for optimum shelf life (Kymene 435)
>Store as cool as possible but > 40 F for optimum shelf life (Kymene 760, 760-35, 835)
>Frozen resin can usually be salvaged if agitated immediately after thawing (Kymene 917)

STD. PKGS.:
>208-1 (227-kg net) fiber or heavy-weight metal drums (Hercules Resin 917)

357

TRADENAME PRODUCTS AND GENERIC EQUIVALENTS

ABS 124ESG, 124ESGE, 125M, 224M, 236F, 236MA, 236R, 301K, 323MA, 400P, 410ESG, 414MA, 420A, 424ESM, 500FR-1, 504ESG, 506ESG, 510EC, 522ESM, 550ES, 606ED, 707H, 707K, 808K, 910K [Mobil]—ABS resin

ABS-G1FG-2 [Washington Penn]—ABS resin, glass-reinforced

Abson 010, 042, 110, 120, 130, 135, 140, 161, 171 [Mobay]—ABS resin

A-C Copolymer 400A [Allied-Signal]—Ethylene/vinyl acetate copolymer

A-C Polyethylene 400 [Allied-Signal]—Ethylene/vinyl acetate copolymer

Acrylite GP, MS-2 Sheet [Cyro Industries]—Acrylic resin

Acryloid A-10, A-11, A-21, A-21LV, A-30, A-101, AT-50, AT-51, AT-56, AT-63, AT-64, AT-70, AT-71, AT-75, B-44, B-48N, B-50, B-82, B-99, C-10LV, NAD-10, WR-97 (crosslinkable) [Rohm & Haas]—Acrylic resin

Acrysol I-94 (crosslinkable), WS-12, WS-24, WS-32, WS-50 [Rohm & Haas]—Acrylic resin

Allabond Twenty/twenty Adhesive, Twenty/twenty NM [Bacon Industries]—Epoxy resin

Amoco G1, G2, G3, G18, H2R, H3E, H4, H4E, H4R, H5M, H5M, M2R, M9, M9R, R1, R2, R3, R5, R7, R12 [Amoco]—Polystyrene

ANE, ANE-D30, ANE-K10, ANE-K20, ANE-P10, ANE-P20 [Ciech Plastofarb]—Polyester, thermoset

Aropol 7020, 7021, 7030, 7131, 7141, 7221, 7229, 7240, 7240T-15, 7240WT-12, 7241-12, 7241T-15, 7242T-15, 7280, 7320, 7329, 7362, 7420, 7530T-16, 7532, 8101, 8110, 8310, 8319, 8321, 8329, 8329-25, 8339T-12, 8349-15, 8420, 8420-09, 8343T-11, 8343T-12, 8343T-21, 8343T-22 [Ashland]—Polyester, thermoset

Arylef U 100 [Solvay]—Polyester, thermoplastic (polyarylate)

AS-10GF, -20GF, -30GF, -40GF [Compounding Technology]—ABS resin, glass-reinforced

AT-20GF, -30GF, -40GF [Compounding Technology]—Acetal resin, glass-reinforced

Bapolan 6050, 6060, 6145, 6270, 6445, 6450 [Bamberger]—Polystyrene

Bapolan 8445, 8635, 8640 [Bamberger]—ABS resin

Barco Bond MB-14X, MB-127, MB-165, MB-175 [Astro]—Epoxy resin

BASF Wax EVA 1 [BASF AG]—Ethylene/vinyl acetate copolymer

Butacite 10, 14, 106, 125, 140 [DuPont]—Polyvinyl butyral

Butvar [Amer. Cyanamid]—Polyvinyl butyral

TRADENAME PRODUCTS AND GENERIC EQUIVALENTS

C-84 [Bacon Industries]—Epoxy resin

CA-394-60, -398-3, -398-6, -398-10, -398-30 [Eastman]—Cellulose acetate

Calibre 200-3, 200-6, 200-10, 200-15, 300-3, 300-6, 300-10, 300-15, 700-3, 700-6, 700-10, 700-15, 800-3, 800-6, 800-10, 800-15, 7070 [Dow]—Polycarbonate resin

Calibre 510, 550 [Dow]—Polycarbonate resin, glass-reinforced

Carboset 514A, 514H, 515, 515, 525, 531, XL-11, XL-19 [BF Goodrich]—Acrylic resin

Cast Cellulose Acetate Film 904 [Georgia-Pacific]—Cellulose acetate

Cast Cellulose Acetate Sheet 867 [Xcel]—Cellulose acetate

Celanex 2000, 2002, 2012 [Hoechst-Celanese]—Polyester, thermoplastic (PBT)

Celanex 3200, 3210, 3211, 3300, 3310, 3311, 3400 [Hoechst-Celanese]—Polyester, thermoplastic (PBT), glass-reinforced

Celcon AS270, AS450, C-400, C-401, EP90, LW90, LW90-S2, LW90-SC, M25, M25-01, M25-04, M50, M90, M90-04, M90-08, M140, M270, M270-04, M450, U10, U10-11, U V25, UV90, WR25 Black, WR90 Black [Hoechst-Celanese]—Acetal resin

Celcon GC-25A, LWGC-S2 [Hoechst-Celanese]—Acetal resin, glass-reinforced

Celcon MC90, MC90-HM, MC270, MC270-HM [Hoechst-Celanese]—Acetal resin, mineral-filled

Cellidor A, S 100-25, S 100-30, S 100-33, S200-17, S 200-22, S 200-27, S 200-32 [Bayer AG]—Cellulose acetate

Chempol 13-1905, 13-2211 [Freeman]—Polyester, thermoset (alkyd)

Chempol 19-4807 [Freeman]—Polyester, thermoset

CI-2, CI-3, CI-6 [Bacon Industries]—Epoxy resin

Cosden Polystyrene 500, 525, 550, 625, 710, 825, 825E [Cosden]—Polystyrene

Coumin [St. Lawrence Resin Products]—Coumarone-indene resin

CP-51, CP-61, CP-80, CP-81, CP-82 [Continental Polymers]—Acrylic resin

Croturez [Crosby]—Terpene/polyterpene resin

Crystic 2-406 PA, 2-489PA, 2-491 PA, 57, 113PA, 189LV, 189MV, 191LV, 193 HR, 195, 196, 196E, 196 PALV, 198, 272, 272E, 302, 306, 323 PA, 328PA, 345PA, 328PALV, 345PALV, 346PA, 385E, 385PA, 387, 392, 392E, 405E, 405PA, 406PA, 471PALV, 474PA, 489PA, 491, 491E, 491PA, 625E, 625MV, 17449 [Scott Bader]—Polyester, thermoset

Crystic Fireguard 75PA [Scott Bader]—Polyester, thermoset

Crystic Fireguard Gelcoat 70PA [Scott Bader]—Polyester, thermoset

Crystic Gelcoat 33E, 39PA, 46PA, 47PA, 65E, 65PA, 68E, 68PA [Scott Bader]—Polyester, thermoset

Crystic LS7 [Scott Bader]—Acrylic resin

Crystic Prefil F [Scott Bader]—Polyester, thermoset

Crystic Pregel 17, 27 [Scott Bader]—Polyester, thermoset

Crystic Release Agent No. 1 [Scott Bader]—Cellulose acetate

Crystic Stopper [Scott Bader]—Polyester, thermoset

Cumar LX-509, P-10, P-25, R-1, R-3 (W-1), R-5 (W-1½), R-6 (W-2, 2½), R-7, R-9, R-

10 (V-1), R-11 (V-1¹/₂), R-12 (V-2, V-2¹/₂), R-12A (V-3), R-13, R-14 (MH-1), R-15 (MH-1¹/₂), R-16 (MH-2, 2¹/₂), R-16A (MH-3), R-17 (RH-17), R-19, R-21, R-27, R-28, R-29 (25°) [Neville]—Coumarone-indene resin

Cyanacryl C, L, R [Amer. Cyanamid]—Acrylic resin

Cycolac AR, CG, CGA, CTB, DFA-R, DH, EP-3510, EPB-3570, ETS-3540, FBK, GSE, GSM, GT, HM, KGA, KJB, KJM, KJT, KJU, KJW, L, LS, LXB, SF, T, TA, X-11, X-15, X-17, X-37, Z36, Z48, Z86, ZA2, ZA4, ZA7, ZB2, ZB6, ZC7 [Borg-Warner]—ABS resin

Cycolac KGA [Borg-Warner]—ABS resin, glass-reinforced

Cymel 303, 370, 373, 380, 1141 [Amer. Cyanamid]—Melamine-formaldehyde resin

Daratak 17-200, 17-230, 17-300, 52L, 55L, 56L, 61L, 62L, 65L, 71L, 78L, SP1004, SP1006, SP1011, SP1041, SP1047, SP1065, SP1066, SP1079 [W.R. Grace]—Polyvinyl acetate

Daratak 74L [W.R. Grace]—Acrylic resin

Darex 165L [W.R. Grace]—Styrene-acrylonitrile resin

DC 2900 HW [Hardman]—Epoxy resin

Delrin 100, 100 AF, 500, 500 AF, 500 CL, 500 F, 570, 900, 900 F, 907 F [DuPont]—Acetal resin

D.E.N. 431, 438, 439, 485 [Dow]—Epoxy resin

D.E.R. 317, 324, 325, 332, 337, 330, 331, 383, 642U, 672U, 652-A75, 661, 661-A80, 662, 663U, 664, 664U, 667, 671-EE75, 671-EEA75, 671-EEK75, 671-MK75, 671-T75, 671-X75, 671-XM75, 673MF, 684EK40, 732, 736 [Dow]—Epoxy resin

Dion Cor-Res 6693FR, 6693HV, 6694, 6694HV, 7000 [Koppers]—Polyester, thermoset

Dion FR 6114, 6124, 6125, 6127, 6308, 6399, 6604T, 6655T, 6657, 6692T, 8300 [Koppers]—Polyester, thermoset

Dion-Iso 6246, 6314, 6315HV, 6631T, 8101, 8200 [Koppers]—Polyester, thermoset

Dow ABS 213, 350, 500, PG912, PG940 Resin [Dow]—ABS resin

Durel 400, P400, P410, P430 [Occidental]—Polyester, thermoplastic (polyarylate)

Durez 23410, 24150, 24337, 24668, 27962, 29276 [Occidental]—Polyester, thermoset (alkyd)

Durez 31375, 31572 [Occidental]—Polyester, thermoset

E-231, -3810, -3938, -8405 [Fiberite]—Epoxy resin

E-260, -260H, -261H, -264H, -2748, -3800, -3824, -8353, -8354, -8354J, -9405 [Fiberite]—Epoxy resin, glass-reinforced

E-2068, -9451 [Fiberite]—Epoxy resin, mineral-filled

E-Form 7-51, 7-61, 7-61M, 7-61MM, 7-67 [Acme Chem. & Insulation]—Epoxy resin, mineral/glass-filled

E-Form 7-72 [Acme Chem. & Insulation]—Epoxy resin, mineral-filled

Elvax 40-P, 150, 210, 220, 230, 240, 250, 260, 265, 310, 350, 360, 410, 420, 450, 460, 470, 550, 560, 565, 650, 660, 670, 750, 760, 770, 3120, 3128, 3135SB, 3135X [DuPont]—Ethylene/vinyl acetate copolymer

EMI-X DA-30, DA-35, DA-40 [LNP]—Polycarbonate resin, aluminum flake-filled

EMI-X DC-1008 [LNP]—Polycarbonate resin, carbon-reinforced

EMI-X OC-1008 , OC-100-10 (50% [LNP]—Polyphenylene sulfide, carbon-reinforced

EMI-X WA-40 [LNP]—Polyester, thermoplastic (PBT), aluminum flake-filled

EMI-X WC-1008 [LNP]—Polyester, thermoplastic (PBT), carbon-reinforced

Envex 1000, 1115, 1228, 1315, 1330, 3540 [Rogers Corp.]—Polyimide resin, thermoplastic

Envex 1620 [Rogers Corp.]—Polyimide resin, thermoplastic, glass-reinforced

EP 1760 LV, 1765 TS, 2220 TC, 2225 MV, 2230 TS, 2305 LK, 2306 LK, 2400 TC, 2404 GE, 2405 FR, 2408 TS, 2415 TC, 2420 TC, 2740 HT, 2770 TC, 2800 GU, 2805 TC, 2810 TS, 2820 TS, 2825 TS, 2840 TS [Hardman]—Epoxy resin

Epiceram [Delta Plastics]—Epoxy resin

Epolite 1301, 1302, 1350, 1353, 1354, 2300, 2302, 2353, 2354, 2360, 3300, 3301, 3302, 3306, 3353, 3354, 3357 [Hexcel/Rezolin]—Epoxy resin

Epotuf Resin 37-127, 37-128, 37-130, 37-134, 37-135, 37-139, 37-140, 37-141, 37-151, 37-170, 37-200, 37-250 [Reichhold]—Epoxy resin

EVA 3121, 3134, 3152, 3159, 3167, 3174, 3200, PE-4928 [DuPont]—Ethylene/vinyl acetate copolymer

FA-1, -8, -13, -14, -21 [Bacon Industries]—Epoxy resin

FC-60% SS [LNP]—Polytetrafluoroethylene, stainless steel-filled

FC-446CS [LNP]—Polytetrafluoroethylene, bronze-filled

FC-477 SM [LNP]—Polytetrafluoroethylene, inorganic-filled

FFA-2, -5, -9 [Bacon Industries]—Epoxy resin

Fluorocomp 122, 123, 124 [LNP]—Polytetrafluoroethylene, graphite-filled

Fluorocomp 101, 102, 103, 104, 105, 106 [LNP]—Polytetrafluoroethylene, glass-filled

Fluorocomp 131, 133), 135 [LNP]—Polytetrafluoroethylene, coke flour-filled

Fluorocomp 144, 145, 146, 147, 148 [LNP]—Polytetrafluoroethylene, bronze-filled

Fluorocomp 169, 170, 171 [LNP]—Polytetrafluoroethylene, glass/graphite-filled

Fluorocomp 174, 175, 176, 177 [LNP]—Polytetrafluoroethylene, glass/MoS$_2$-filled

Fluorocomp 181, 182, 183 [LNP]—Polytetrafluoroethylene, bronze/MoS$_2$-filled

Fluorocomp 190, 191, 192, 192HE [LNP]—Polytetrafluoroethylene, carbon/graphite-filled

Fluorocomp 193, 195 [LNP]—Polytetrafluoroethylene, carbon-filled

Fluorocomp 197, 199 [LNP]—Polytetrafluoroethylene, glass/carbon-filled

Fortron 0214P1, 0205P4 [Hoechst-Celanese]—Polyphenylene sulfide

Fortron 1140B1, 1140B4 [Hoechst-Celanese]—Polyphenylene sulfide, glass-reinforced

Fortron 6165B4 [Hoechst-Celanese]—Polyphenylene sulfide, glass/calcium carbonate-filled

Fortron 6265B4 [Hoechst-Celanese]—Polyphenylene sulfide, glass/talc-filled

Fosta Tuf-Flex 240, 271, 283, 326, 474, 721-M, 742 [Hoechst-Celanese]—Polystyrene, rubber-modified

Fosta Tuf-Flex 326, 329, 702, 707, 717, 730, 782, 840, 880, 929 [Hoechst-Celanese]—Polystyrene

Fostafoam Type 3775, 4775, 5775 [Hoechst-Celanese]—Polystyrene

Fostalite [Hocehst-Celanese]—Polystyrene

Fostarene 50, 57, 58, 817 [Hoechst-Celanese]—Polystyrene

Fulton 441, 441D [LNP]—Acetal resin

Fyrid KS1 [Borg-Warner]—Polystyrene

Gafite 1432F, 1432Z, 1452F, 1462Z, 1465F, 1482Z, 1614G, 1632F, 1632Z, 1642Z, 1662Z, LW 4424R , LW X-4424R, LW X-4424S [GAF]—Polyester, thermoplastic (PBT), glass-reinforced

Gafite 1600A, 1602F, 1602Z, 1700A, LW X-4612R [GAF]—Polyester, thermoplastic (PBT)

Gafite LW X-5632Z [GAF]—Polyester, thermoplastic (PBT), glass/mica-reinforced

Gafite LW6362R, LW 6443B, LW 6443F, LW 6443R, LW 7342R, LW X-6443R, LW X-7443R [GAF]—Polyester, thermoplastic (PBT), mica-reinforced

Gaflex 355ZS, 372, 372ZS, 540ZS, 547, 547ZS, 555, 555ZS, 572ZS [GAF]—Polyester, thermoplastic

Gantrez M-154, M-555, M-556, M-574 [GAF]—Polyivnyl methyl ether

Gillfab 1045, 1053 [M.C. Gill]—Epoxy/glass laminate

Gillfab 1109 [M.C. Gill]—Epoxy/nylon laminate

Heloxy WC-8002, WC-8004 [Wilmington Chem. Corp.]—Epoxy resin

Hercules Resin 917 [Hercules]—Urea-formaldehyde resin

Hetron 32A, 197AT, 325FS, 470P, 700 Series, 700DMA, 27196 [Ashland]—Polyester, thermoset

Hostaflon TF 1100, TF 1400, TF 1620, TF 1620 F, TF 1640, TF 1665, TF 1740, TF 1760, TF 2024, TF 2026, TF 2051, TF 2053, TF 2071, TF 5032, TF 5033, TF 5034, TF 5444, TF 5515, TF 5537 [Hoechst Celanese]—Polytetrafluoroethylene

Hostaflon TF 2103, TF 4103, TF 4105 [Hoechst Celanese]—Polytetrafluoroethylene, glass-filled

Hostaflon TF 4215 [Hoechst Celanese]—Polytetrafluoroethylene, carbon-filled

Hostaflon TF 4303 [Hoechst Celanese]—Polytetrafluoroethylene, graphite-filled

Hostaflon TF 4406 [Hoechst Celanese]—Polytetrafluoroethylene, bronze-filled

Hostaform C 2521, C 2541, C 9021, C 9021 M, C 9021 TF, C 13021, C 13031, C 27021, C 32021, C 32021 AST, C 52021, S 9063, S 9064, S 27063, S 27064, S 27073, S 27076, T 1020 [Hoechst-Celanese]—Acetal resin

Hostaform C 9021 GV1/30, C 9021 GV1/40, C 9021 GV3/10, C 9021 GV3/20, C 9021 GV3/30 [Hoechst-Celanese]—Acetal resin, glass-reinforced

Hostaform C 9021 K [Hoechst-Celanese]—Acetal resin, mineral-filled

HTG-4275, -4450 [DuPont]—Polyester, thermoplastic

Hysol 1C, 3X, 0151, 309, 608, 615, 907, 1105, 9340 Epoxi-Patch Kit [Hysol Div./Dexter]—Epoxy resin

Hysol 6C Epoxi-Patch Kit [Hysol Div./Dexter]—Epoxy resin, aluminum-filled

Hysol C8-4143/HD3404, C8-4143/HD3475, C9-4183/HD3404, C9-4183/HD3469, C9-4183/HD3485, C9-4183/HD3537, C9-4183/HD3561, C9-4183/HD3615, C9-

4183/HD3719 [Hysol Div./Dexter]—Epoxy resin, silica-filled
Hysol C18F, C60, C61, EO1016, EO1017, ES4212, ES4312, ES4412, MG15F, MG15F-
01, MG15F-02, MG18, MH19F, MH19F-01, MH19F-02, PC12-007, R8-2038/
HD3404, R8-2038/HD3475, R9-2039/HD3404, R9-2039/HD3469, R9-2039/
HD3561, R9-2039/HD3615, R9-2039/HD3719 [Hysol Div./Dexter]—Epoxy resin
Hysol PC20 [Hysol/Dexter Corp.]—Acrylic resin
Hytrel Series, 4056, 5526, 5555HS, 6346, 7246, G-4074, G-4075, G-4766, G-4774, G-
5544 [DuPont]—Polyester, thermoplastic
Impco RC-2, RC-2/140 Resins [Impco]—Polyester, thermoset
Interset Concrete Bonder, Concrete Topping Resin #2, Deep Penetrating SEaler, Epoxy
Coal Tar, Epoxy Gel, Injection Resin (F) [Int'l. Thermoset]—Epoxy resin
Joncryl 61LV, 67, 74, 77, 78, 85, 130, 142, 678 [S.C. Johnson]—Acrylic resin
Kama CS, OPS [Kama]—Polystyrene
Kapton Type F, H, V [DuPont]—Polyimide resin, thermoset
Klingerflon PTFE (15% graphite filled) [Klinger Engineering Plastics]—
Polytetrafluoroethylene, graphite-filled
Klingerflon PTFE (25% glass filled) [Klinger Engineering Plastics]—Polytetrafluoroeth-
ylene, glass-filled
Klingerflon PTFE (Virgin) [Klinger Engineering Plastics]—Polytetrafluoroethylene
Klingerflon PTFE [Klinger Engineering Plastics]—Polytetrafluoroethylene, bronze-
filled
Knightset EM-3 [GAF]—Melamine-formaldehyde resin
Kodapak PET Polyester [Eastman]—Polyester, thermoplastic (PET)
Korad 2, A, Klear [Georgia-Pacific]—Acrylic resin
Kymene 435, 760, 760-35, 835, 917 [Hercules]—Urea-formaldehyde resin
LCA-1, -4, -4LV, -9, -14, -20, -21, -27 [Bacon Industries]—Epoxy resin
Leguval E60, E81, E90, F33, F30, F30S, F35, K25R, K26, K41, K70, N30, N50, N50S,
W16, W16T, W20, W35, W41 [Bayer AG]—Polyester, thermoset
Lekutherm X18, X20, X23, X24, X30S, X201, X227, X256, X257 [Bayer AG]—Epoxy
resin
Levapren 336, 400, 408, 450, 452, 456, KA 8114 [Bayer AG]—Ethylene/vinyl acetate
copolymer
Lexan 8A13, 8A23, 8A33, 8A43, 8B13, 8B23, 8B33, 8B35, 8B36, 8B43, 8B327, 8B328,
8C20, 101, 103, 104, 123, 121, 124, 141, 143, 144, 150, 303, 500, 920, 920A, 940,
940A, 950, 950A, 1500, 2014, 8010-112, 8020, 8030, 8040, 8060, 8800, 8800-112,
CF-20 MB-112, CF-23 MB-112, CF-25 MB-112, CF-26 MB-112, CF-60 MB-112,
FL-900, FL-910, FL-930, FL-1000, FL-1800, FR60-112, FR60SE-112, FR63-112,
FR65-112, FR66-112, FR88-112, S100, S200, SG400, Sheet 9034, Thermoclear,
XL Sheet [General Electric]—Polycarbonate resin
Lexan 3412, 3413, 3414 [General Electric]—Polycarbonate resin, glass-reinforced
Liquinite FEP [LNP]—Fluorinated ethylene-propylene
Liquinite PPS [LNP]—Polyphenylene sulfide

LNP Packing Blend B [LNP]—Polytetrafluoroethylene

LNP Reprocessed TFE #1 Grade, TFE #2 Grade, TFE #3 Grade, TFE F-2080, TFE OX-60, TFE Utility Grade [LNP]—Polytetrafluoroethylene

Lucite L, SAR [DuPont]—Acrylic resin

Lucky ABS HF-350, HF-380, HI-100, HI-121, HI-151, HR-420, HR-450, MP-211, RS-600, SH-610 [Standard Polymers]—ABS resin

Lustran ABS 248, 252, 448, 452, 456, 545, 648, 743, 750-10802, 752, 780-10802, 782-10846, 860, 865, 1152, -HR850, -HR851, PG-298, PG-299, PG-300, FR ABS911, FR ABS914HM, FR ABS921UV, Ultra ABS HX, MX [Monsanto]—ABS resin

Lustran SAN 31, 33, 35, 41, 43, 47, 49, 51 [Monsanto]—Styrene-acrylonitrile resin

Lustrex 2220, 2400, 3300, 3350, 4220, 4300 [Monsanto]—Polystyrene

M-2015 [Fiberite]—Melamine-formaldehyde resin, fabric-filled

M-2037, -2840, -2880, -3882 [Fiberite]—Melamine-formaldehyde resin, glass-filled

M-4536, -7510 [Fiberite]—Melamine-formaldehyde resin, cellulose-filled

M-6204 [Fiberite]—Melamine-formaldehyde resin, mineral-filled

Magnamite AS/3501-5A, AS/3501-6, AS4/1908, AS4/1919, AS4/2220-3, AS4/3501-5A, AS4/3501-6, AS4/3502, HMS/1908, HMS/3501-5A, HMS/3501-6 [Hercules]—Epoxy resi, graphite-reinforced

Magnamite AS4/1655 [Hercules]—Polyester, thermoset, graphite-reinforced

Magnum ABS 213, 341, 343, 545, 941, 4500UV, 9010, 9020, 9030, 9408, 9450P, CLR95, FG960 [Dow]—ABS resin

Makrofol E, G, KG, N, SG, SKG, SN [Bayer AG]—Polycarbonate resin

Makrolon 2400, 2403, 2405, 2600, 2603, 2605, 2800, 2803, 2805, 2807, 2808, 2809, 3100, 3103, 3105, 3108, 3109, 3119, 3200, 3203, 3208, 6030, 6550, 6560, 6553, 6555, 6557, 6603, 6655, 6870 [Bayer AG]—Polycarbonate resin

Makrolon 8020, 8025, 8030, 8035, 8315, 8320, 8324, 8344, 9310, 9410, 9415 [Bayer AG]—Polycarbonate resin, glass-reinforced

Malon 1070, 1080, 1090 [M.A. Industries]—Polyester, thermoplastic

Mamax 2020 [M.A. Industries]—Polyester, thermoplastic

Marc Polycarbonate [MRC Enterprises]—Polycarbonate resin (reprocessed)

Margard Sheet [General Electric]—Polycarbonate resin

Merlon 5300, 6450, 6560, 6870, AL-400, CD-2000, FAR, HMS-3119, M39, M40, M50, MPG-700, MPG-750, SF-600, T70, T85 [Mobay]—Polycarbonate resin

Merlon 8310, 9310, 9510 [Mobay]—Polycarbonate resin, glass-reinforced

Migralube DL-4030, DL-4410, DL-4530 [LNP]—Polycarbonate resin

Modic E-100H, E-200H, E-300K, E-300S, E-310K [Mitsubishi]—Ethylene/vinyl acetate copolymer

Mylar Series, Type A, AB, C, D, EL, HS, KB, M, M461, MO, PB, S, T, VB, WC, WC 11, WC 11 G, WC 22 [DuPont]—Polyester, thermoplastic (PET)

Natro-Rez 10, 15, 25, 50-D [Harwick]—Coumarone-indene resin

NeoCryl A-550, A-601, A-602, A-603A, A-604, A-610, A-620, A-621, A-624, A-630, A-1031, A-1038, A-1044, A-1045, B-700, B-705, B-723, B-725, B-726, B-728, B-

734, B-750, B-1019, B-1041, B-1042, B-1054, BT-7, BT-8, BT-20, BT-24, BT-175, CC-6, CL-300, EX-450, EX-482, EX-484, EX-485, EX-487, P-920A Glaze, SR-270, SR-272, SR-274, SR-276, S-1004 [Polyvinyl Chem. Industries]—Acrylic resin
Neolyn 20, 35D, 40 [Hercules]—Polyester, thermoset (alkyd)
Nevpene 9500 [Neville]—Terpene/polyterpene resin
Nevtac 80, 99 Super, 100, 115, 130 [Neville]—Terpene/polyterpene resin
Nirez 1000 Series [Reichhold]—Terpene/polyterpene resin
Nobestos D-7700 [Rogers Corp.]—Acrylic resin
Norbond 406, 2162, 2162 Clear Amber [R.H. Carlson]—Epoxy resin
Norcast 23, 157, 176X, 424, 971, 3215, 3216, 3220-D, 3222, 3222 LV, Norcast 3231/ Norcure 135 or 135-M, Norcast 3231/Norcure 112, Norcast 3253/Norcure 112, Norcast 3253/Norcure 135M, Norcast 3255, Norcast 3343, Norcast 3370, Norcast 34301 M, Norcast 3420 LC, Norcast 3425 HF, Norcast 3425-M, Norcast 3511, Norcast 3705, Norcast 4200-1, 4200-2, 4200-05, 4915, 5500, 5503, 5509-M, Norcast 5615-10/Norcure 178-M, 7086, 7123, 7124, 7146, 7146-F, 7633, 9100, 9101, 9102, 9311, 9383, 9390, T-30, X-9201, Norcast 141/Catalyst 0.31, Norcast 141/Norcure 122, Norcast 154/Norcure 134, Norcast 154/Norcure 137, Norcast 0364/Norcure 142, Norcast 1750/Norcure 112, Norcast 1750/Norcure 112, Norcast 1750/Norcure 135-M, Norcast 1750/Norcure 138, Norcast 1750/Norcure 170, Norcast 2070-PC, Norcast 2795/Norcure 3416, Norcast 3217M/Norcure 111 & 112, Norcast 3217M/Norcure 122, Norcast 3217M/Norcure 135, Norcast 3230 SP/ Norcure 112 & 170, Norcast 3230 SP/Norcure 3230B & 135M, Norcast 3256/ Norcure 112, Norcast 3256/Norcure 135, Norcast 3257/Norcure 112, Norcast 3257/ Norcure 135, Norcast 3258/Curing Agent D, Norcast 3258/Norcure 112, Norcast 3259 Resin-D-Type, 3424-M, 3510, 3515, 3705-F, 3706-F, 4917 HSS, Norcast 5005/Norcure 150 [R.H. Carlson]—Epoxy resin
Norcast 3230-A/Norcure 112, Norcast 3230-A/Norcure 135, Norcast 3230 LV/Norcure 112 [R.H. Carlson]—Epoxy resin, aluminum oxide-filled
Norcast 4912, 4920 [R.H. Carlson]—Epoxy resin, silver-filled
Norcast 4915 [R.H. Carlson]—Epoxy resin, carbon-filled
Norcast T-27 [R.H. Carlson]—Epoxy resin, aluminum-filled
Novodur HGV, P2T, P2T-AT, PH-AT, PHE, PK, PK-AT, PKT, PL-AT, PLT-AT, PM, PM3C, PM5C, PM-AT, PME, PMT, PMT-AT, PTE, PX [Bayer AG]—ABS resin
Novodur PHGV, PHGV-AT [Bayer AG]—ABS resin, glass-reinforced
P-11, -14, -19, -20, -24, -38, -51, -56, -56A, -70, -76, -78, -80C, -81, -82C, -85, -86 [Bacon Industries]—Epoxy resin
P-74, -75 [Bacon Industries]—Epoxy resin, glass-filled
PA-55-004, -55-011, -55-012, 55-013, 55-024 [Polymer Applications]—Polyester, thermoplastic
Paradene No. 1, No. 2, No 33, No. 35 [Neville]—Coumarone-indene resin
Paraplex AL-111, G-60, RG-10 [Rohm & Haas]—Polyester, thermoset

Parcryl 200,250, 300, 311, 400, 450, 475, 500, 777, 900, 966 [Thibaut & Walker]—Acrylic resin

PC-20CF, -30CF, -40CF [Compounding Technology]—Polycarbonate resin, carbon-reinforced

PDX-79625FF [LNP]—Polytetrafluoroethylene, polymer-filled

PDX-83238 [LNP]—Polyphenylene sulfide, nickel-reinforced

PDX-83393 [LNP]—Polycarbonate resin, nickel-modified

PDX-84356 [LNP]—Polycarbonate resin, stainless steel-modified

PDX-84357 [LNP]—ABS resin

PDX-84357 [LNP]—ABS-stainless steel composite

PDX-84368 [LNP]—Polycarbonate resin

PDX-84440 [LNP]—Ethylene/vinyl acetate copolymer, carbon-reinforced

Petlon 3530, 3550, 4530 [Mobay]—Polyester, thermoplastic (PET), glass-reinforced

PF-10GF/15T, -20GF/15T, -30GF/15T [Compounding Technology]—Polysulfone, glass-reinforced

PI-730 [Fiberite]—Polyimide resin, thermoset

PI-740, -750 [Fiberite]—Polyimide resin, thermoset, glass-reinforced

Piccofyn A100, A115, A135, D125 [Hercules]—Terpene/polyterpene resin

Piccolyte A115, A125, A135, C100, C115, C125, C135, D100, D115, D135, HM-110 Resin, S100, S115, S115SF, S125, S125SF, S135 [Hercules]—Terpene/polyterpene resin

Picconol A200, A201 [Hercules]—Terpene/polyterpene resin

Plenco 801 White [Plastics Engineering]—Melamine-formaldehyde resin, cellulose-filled

Plenco 1523 Black Pellets, 1530 [Plastics Engineering]—Polyester, thermoset (alkyd), glass/mineral-filled

Plenco 1528 [Plastics Engineering]—Polyester, thermoset (alkyd), mineral-filled

Plenco 1529 Black [Plastics Engineering]—Polyester, thermoset (alkyd)

Plexiglas DR, DR G, DR M, HFI-7, HFI-10, MI-7, MI-7G, V-044, V-045, V-052, V-811, V-920, VM, VS [Rohm & Haas]—Acrylic resin

Plusplice 100 [C.P. Hall]—Terpene/polyterpene resin

Pocan B 1305, B 1505 [Bayer AG]—Polyester, thermoplastic (PBT)

Pocan B 3225), B 3235, B 4225, B 4235, B 5335 [Bayer AG]—Polyester, thermoplastic (PBT), glass-reinforced

Polycomp 139, 142, 148, 149, 158, 184, 185 [LNP]—Polytetrafluoroethylene, polyphenylene sulfide-filled

Polycomp 160, 161 [LNP]—Polytetrafluoroethylene, poly-p-oxybenzoate-filled

Polylite 31-000, 31-001, 31-006, 31-039, 31-402, 31-439, 31-447, 31-450, 31-451, 31-546, 31-571, 31-585, 31-587, 31-820, 31-822, 31-830, 31-851, 31-866, 32-032, 32-033, 32-035, 32-037, 32-039, 32-050, 32-111, 32-112, 32-120, 32-122, 32-125, 32-127, 32-128, 32-129, 32-180, 32-300, 32-344, 32-345, 32-356, 32-357, 32-358, 32-367, 32-402, 32-516, 32-528, 32-737, 32-738, 32-739, 32-740, 32-773, 33-089, 33-

092, 33-084, 33-086, 33-401, 33-011, 33-031, 33-049, 33-059, 33-067, 33-071, 33-072, 33-081, 33-404, 33-411, 33-441, 33-450, 33-770, 33-773, 34-308 [Reichhold]—Polyester, thermoset
Polylite 30001-1, 30001-2, 30001-3, 30003-1, 30003-2 [Reichhold]—Polyester, thermoset, glass/mineral-filled
Polylite 30002-1 [Reichhold]—Polyester, thermoset, cellulose-filled
Polylite 30004-3, 30004-4 [Reichhold]—Polyester, thermoset, glass/mineral/cellulose-filled
Polypenco Acetal [Polymer Corp.]—Acetal resin
Polypenco Cast Acrylic Rod [Polymer Corp.]—Acrylic resin
Polypenco Polycarbonate [Polymer Corp.]—Polycarbonate resin
Polypenco Q200.5 [Polymer Corp.]—Polystyrene (cross-linked)
Polypenco TFE [Polymer Corp.]—Polytetrafluoroethylene
Polyset 317, 707-2 [Morton Int'l.]—Epoxy resin, mineral/glass-filled
Polyset 405, 410B, 415-SG [Morton Int'l.]—Epoxy resin, mineral-filled
Polyset EPC50, EPC 68 FR [Morton Int'l.]—Epoxy resin
Presto Adhesive 2125, 2125M88, 2125U88, 2127+, 2127M6 [Presto Mfg.]—Acrylic resin
Protect-A-Glaze Sheet [General Electric]—Polycarbonate resin
PS 110R, 110S, 110SL, 224EP, 224EV, 224M, 310S, 344/E, 314/E, 410S, 444, B44HH, 510TG, 8120, 8720 [Mobil]—Polystyrene
PS-H23MF-1 [Washington Penn]—Polystyrene, mica-filled
PS-H63CC-1 [Washington Penn]—Polystyrene, calcium carbonate-filled
Rhoplex AC-22, AC-34, AC-35, AC-61, AC-73, AC-388, AC-490, AC-507, AC-707, MV-1 [Rohm & Haas]—Acrylic resin
RTP 301, 303, 303TFE10, 305, 305TFE15, 307 [Fiberite]—Polycarbonate resin, glass-reinforced
RTP 401, 403, 405, 407 [Fiberite]—Polystyrene, glass-reinforced
RTP 501, 503, 505, 506, 507 [Fiberite]—Styrene-acrylonitrile resin, glass-reinforced
RTP 601, 603, 605, 607 [Fiberite]—ABS resin, glass-reinforced
RTP 800TFE20, 801, 803, 805), 805TFE15, 807 [Fiberite]—Acetal resin, glass-reinforced
RTP 901, 903, 905, 907 [Fiberite]—Polysulfone, glass-reinforced
RTP 1001, 1003, 1005, 1005FR, 1007 [Fiberite]—Polyester, thermoplastic (PBT), glass-reinforced
RTP 1007 GB 10, 1025, 1026, 1027, 1028 [Fiberite]—Polyester, thermoplastic (PBT), glass/mineral-reinforced
RTP 1105FR [Fiberite]—Polyester, thermoplastic (PET), glass-reinforced
RTP 1301, 1303, 1305, 1307 [Fiberite]—Polyphenylene sulfide, glass-reinforced
RTP 1378 [Fiberite]—Polyphenylene sulfide
RTP 1379, 1379S [Fiberite]—Polyphenylene sulfide, glass/mineral-reinforced
RTP 1383, 1385, 1387 [Fiberite]—Polyphenylene sulfide, carbon-reinforced

367

RTP 1500.5, 1501, 1503, 1505, 1507 [Fiberite]—Polyester, thermoplastic, glass-reinforced

Rynite 430, 530, 545, 555, FR-530 [DuPont]—Polyester, thermoplastic (PET), glass-reinforced

Rynite 935, 940, 940FB [DuPont]—Polyester, thermoplastic (PET), glass/mica-reinforced

Ryton R-3 02 Black, R-4 , R-5 [Phillips]—Polyphenylene sulfide, glass-reinforced

Ryton R-8, R-10 Avocado R-4004B, R-10 Black 5002C and 5004A, R-10 Copper R-2005B, R-10 Gold R-1001B, R-10 Gray R-5000B, R-10 Green R-4000B, R-10 Lt. Green R-7002B, R-10 Natural 7006A and 7007A, R-10 Tan R-7002B [Phillips]—Polyphenylene sulfide, glass/mineral-reinforced

Ryton R-9 901, R-11 70, V-1 [Phillips]—Polyphenylene sulfide

Santolite MHP, MS80% [Monsanto]—Toluenesulfonamide/formaldehyde resin

SF-30GR , SF-40GF [Compounding Technology]—Polyphenylene sulfide, glass-reinforced

SN-20GF, -30GF, -40GF [Compounding Technology]—Styrene-acrylonitrile resin, glass-reinforced

Spauldite FR-4, G-10, G-10-900, G-10CR, G-10-773, G-11, G-11-963, G-11 CR, PEG FR, T-525 [Spaulding Fibre]—Epoxy/glass laminate

Stat-Kon AC-1003 [LNP]—ABS resin, carbon-reinforced

Stat-Kon CDF-1006, DFD-15 [LNP]—Polycarbonate resin, glass/carbon-reinforced

Stat-Kon D, DC-1003FR, DC-1006, DCL-4033 [LNP]—Polycarbonate resin, carbon-reinforced

Stat-Kon FP-P [LNP]—Perfluoroalkoxy resin

Stat-Kon FP-PC-1004 [LNP]—Perfluoroalkoxy resin, carbon-reinforced

Stat-Kon GC-1003, GC-1006 [LNP]—Polysulfone, carbon-reinforced

Stat-Kon KC-1002, KCL-4022 [LNP]—Acetal resin, carbon-reinforced

Stat-Kon OC-1003, OC-1006, OCL-4036 [LNP]—Polyphenylene sulfide, carbon-reinforced

Stat-Kon WC-1002, WC-1006 [LNP]—Polyester, thermoplastic, carbon-reinforced

Stycast 1090, 1090-SI [Emerson & Cuming]—Epoxy resin, glass-reinforced

Stycast 1210, 1263, 1264, 1266, 1269-A, 1467, 1492, 2057, 2651, 2762, 2762-FT, 2850-FT, 2651-40, 2651-MM, 2741, 2850-GT, 2850-KT, 3050, 3180-M [Emerson & Cuming]—Epoxy resin

Stypol 40-4033, 40-4074, 40-5727, 40-5732, 40-7016, 40-8006, 83-2000 [Freeman]—Polyester, thermoset

Styron 412D, 420D, 430U, 470, 475U, 492U, 495, 666U, 678U, 685D [Dow]—Polystyrene

Superdense 700, 710 [Hammond Plastics]—Polystyrene

Superflex 405, 432, 505, 550, FR-1 [Hammond Plastics]—Polystyrene

Superflow 390 [Hammond Plastics]—Polystyrene

Tactix 742 Performance Polymer [Dow]—Epoxy resin

Teflon 6A, 6C, 6H, 6L, 7A, 7C, 8, 8A, 9A, 9B, 64 [DuPont]—Polytetrafluoroethylene

Teflon 100 FEP, 110 FEP, 120 FEP, 140 FEP, 160 FEP [DuPont]—Fluorinated ethylene-propylene

Teflon 340 PFA, 350 PFA [DuPont]—Perfluoroalkoxy resin

Tenite Cellulosic Acetate [Eastman]—Cellulose acetate

Tenite PET Polyester 6857 [Eastman]—Polyester, thermoplastic (PET)

Terlan 6090, 6100, 6300, 6430, 6490, 6530, 6600, 6676 [Terrell]—Polyester, thermoplastic

Texicote 03-001, 03-020, 03-021, 03-040 [Scott Bader]—Polyvinyl acetate

Texicryl 13-003, 13-010, 13-011, 13-100, 13-101, 13-104, 13-203, 13-205, 13-300, 13-302, 13-439, 13-442, 13-206, 13-210 [Scott Bader]—Acrylic resin

Thermid AL-600, LR-600, MC-600 [Nat'l. Starch & Chem.]—Polyimide resin, thermoset

Thermocomp AF-1004, AF-1006, AF-1008 [LNP]—ABS resin, glass-reinforced

Thermocomp BF-1006, BF-1006FR, BF-1008 [LNP]—Styrene-acrylonitrile resin, glass-reinforced

Thermocomp CF-1004, CF-1006, CF-1008 [LNP]—Polystyrene, glass-reinforced

Thermocomp DC-1006, DC-1006PC [LNP]—Polycarbonate resin, carbon-reinforced

Thermocomp DF-1004, DF-1006, DF-1008, DFA-113, DFL-4036 [LNP]—Polycarbonate resin, glass-reinforced

Thermocomp DL-4030 [LNP]—Polycarbonate resin

Thermocomp FC-102CF [LNP]—Polytetrafluoroethylene, carbon-filled

Thermocomp GC-1006 [LNP]—Polysulfone, carbon-reinforced

Thermocomp GF-1004, GF-1006, GF-1006FR, GF-1008 [LNP]—Polysulfone, glass-reinforced

Thermocomp KB-1008, KF-1006, KFX-1002, KFX-1006, KFX-1008, KFX-1008MG [LNP]—Acetal resin, glass-reinforced

Thermocomp KC-1004 [LNP]—Acetal resin, carbon-reinforced

Thermocomp LF-1004, LF-1004M [LNP]—Fluorinated ethylene-propylene, glass-fortified

Thermocomp OC-1006, OCL-4036 [LNP]—Polyphenylene sulfide, carbon-reinforced

Thermocomp OF-1006, OF-1008, OFL-4036 [LNP]—Polyphenylene sulfide, glass-reinforced

Thermocomp WC-1006 [LNP]—Polyester, thermoplastic (PBT), carbon-reinforced

Thermocomp WF-1004, WF-1006, WF-1006FR, WF-1008 [LNP]—Polyester, thermoplastic (PBT), glass-reinforced

Thermocomp YF-1002, YF-1004, YF-1006, YF-1008 [LNP]—Polyester, thermoplastic, glass-reinforced

TL-101, -102, -113, -115A, -115AH, -126, -126H, -127 [LNP]—Polytetrafluoroethylene

TL-120 [LNP]—Fluorinated ethylene-propylene

Tra-Bond 2101, 2112, 2113, 2116, 2126, 2129, 2135D, 2151, 2208, 2211, 2215, 2248 [Tra-Con]—Epoxy resin

Tra-Bond 2122 [Tra-Con]—Epoxy resin, aluminum-filled

Tra-Bond 2123 [Tra-Con]—Epoxy resin, steel-filled

Tra-Bond 2125 [Tra-Con]—Epoxy resin, glass-filled

Tra-Duct 2902, 2924 [Tra-Con]—Epoxy resin, silver-filled

Tyril 860B, 867B, 870C, 880B, 1000, 1000B, 1020, 1021, 1030, 1031 [Dow]—Styrene-acrylonitrile resin

Ucar Acrylic 503, 505, 515, 516, 518 [Union Carbide]—Acrylic resin

Ucar Latex 123, 153, 154, 163, 173, 174, 175, 803, 812, 874 [Union Carbide]—Acrylic resin

Ucar Vehicle 407, 4630, 4358, 4414, 4431, 4550, 4620 [Union Carbide]—Acrylic resin

Udel GF-110, GF-120, GF-130 [Amoco]—Polysulfone, glass-reinforced

Udel P-1700, P-1700FR, P-1710, P-1720, P-3500 [Amoco]—Polysulfone

Uformite F-200E, F-210, F-222, F-240, F-240N [Rohm & Haas]—Urea-formaldehyde resin

Uformite MM-46, MM-47, MM-55, MM-57 [Rohm & Haas]—Melamine-formaldehyde resin

Ultradur B 2550, B 4500, B 4520, B 4550, KR 4015, KR 4036, KR 4070, KR 4071 [Badische]—Polyester, thermoplastic (PBT)

Ultradur B 4300 G2, B 4300 G4, B 4300 G6, B 4300 G10, B 4300 K4, B 4300 K6, B 4305 G2, B 4305 G4, B 4305 G6, KR 4025, KR 4035 [Badische]—Polyester, thermoplastic (PBT), glass-reinforced

Ultradur KR 4001 [Badische]—Polyester, thermoplastic (PBT), mineral-reinforced

Ultradur KR 4011 [Badische]—Polyester, thermoplastic (PBT), glass/mineral-reinforced

Ultraform H 2320, N 2211 PVX, N 2320, N 2320 Black 11001 UV, N 2320 Black 11005 MO, S 2320, W 2320 [Badische]—Acetal resin

Ultraform N 2200 G5 [Badische]—Acetal resin, glass-reinforced

Ultrathene UE 630, UE 631, UE 632, UE 635, UE 637, UE 643, UE 655, UE657 [USI Chem.]—Ethylene/vinyl acetate copolymer

Valox 310-SEO, 325, 340, 357, FV-600, FV-608, FV-609, FV-650, FV-699 [General Electric]—Polyester, thermoplastic (PBT)

Valox 420, 420-SEO, DR-48, DR-51 [General Electric]—Polyester, thermoplastic (PBT), glass-reinforced

Valox 701, 730, 750, 752 [General Electric]—Polyester, thermoplastic (PBT), glass/mineral-reinforced

Valox 744, 745, 760 [General Electric]—Polyester, thermoplastic (PBT), mineral-reinforced

Varcum 6404, 6407, 9413 [Reichhold]—Epoxy resin

Versaflex 1, 2, 4, 5 [W.R. Grace]—Acrylic resin

Vestopal 110, 120L, 120T, 121L, 123L, 140L, 141L, 145, 145T, 150, 150B, 150T, 151, 152, 160, 160T, 221L, 250, 251, 252, 310, 320, 330, 400, 510, 511, 530, 710, 730, 810, 811, 830 [Huls AG]—Polyester, thermoset

Vestypor Series, 01N, 01S, 10N, 10S, 20N, 20S, 30N, 30S [Huls AG]—Polystyrene

TRADENAME PRODUCTS AND GENERIC EQUIVALENTS

Vestyron 114, 214, 214A, 314, 325, 512, 512A, 512 TA, 516, 616, 620, 620A, 620TA, 628, 638, 640, TSG 502, TSG 504, TSG 602, TSG 604 [Huls AG]—Polystyrene

XT Polymer 250, 375 [Cyro Industries]—Acrylic resin

Xydar G-330, G-345, G-430, G-445, M-350, M-450, MG-350, MG-450 [Amoco]—Polyester, thermoplastic (LCP)

Zelux W (various grades) [Westlake Plastics]—Polycarbonate resin

Zonarez 7010, 7025, 7040, 7055, 7070, 7085, 7100, 7115, 7125, B-10, B-25, B-40, B-55, B-70, B-85, B-115, B-125 [Arizona]—Terpene/polyterpene resin

Zonatec 105, 115 [Arizona]—Terpene/polyterpene resin

GENERIC CHEMICAL SYNONYMS
AND CROSS REFERENCES

Acetate fiber. See Cellulose acetate
Acetate film. See Cellulose acetate
Acetic acid, ethenyl ester, homopolymer. See Polyvinyl acetate
Acetic acid, ethenyl ester, polymer with ethene. See Ethylene/vinyl acetate copolymer
Acrylic/acrylate copolymer. See under Acrylic resin
Acrylonitrile-butadiene-styrene resin. See ABS resin
Alkyd resin. See under Polyester, thermoset
Benzene, ethenyl-, homopolymer. See Polystyrene
Benzenesulfonamide, 4-methyl-, polymer with formaldehyde. See Toluenesulfonamide/
    formaldehyde resin
CA. See Cellulose acetate
Cajeputene. See under Terpene/polyterpene resins (dipentene)
Cellulose acetate ester. See Cellulose acetate
Cinene. See under Terpene/polyterpene resins (dipentene)
Coal tar resin. See Coumarone-indene resin
Cyclohexene, 1-methyl-4-(1-methylethenyl)-. See under Terpene/polyterpene resins
    (dipentene)
Cyclomethicone. See Book VI
Dimethicone. See Book VI
Dimethicone copolyol. See Book VI
Epichlorohydrin/bisphenol A resin. See Epoxy resin
Epoxy novolak resin. See Epoxy resin
Ethene, methoxy-, homopolymer. See Polyvinyl methyl ether (CTFA)
Ethenyl acetate, homopolymer. See Polyvinyl acetate
Ethenylbenzene, homopolymer. See Polystyrene
EVA copolymer. See Ethylene/vinyl acetate copolymer
FEP resin. See Fluorinated ethylene-propylene
LCP. See under Polyester, thermoplastic
Limonene. See under Terpene/polyterpene resins (limonene)
DL-Limonene. See under Terpene/polyterpene resins (dipentene)
Limonene, inactive. See under Terpene/polyterpene resins (dipentene)
Liquid crystal polyester. See under Polyester, thermoplastic
Melamine resin. See Melamine-formaldehyde resin

dl-p-Mentha-1,8-diene. See under Terpene/polyterpene resins (dipentene)
Methoxyethene, homopolymer. See Polyvinyl methyl ether (CTFA)
Methyl methacrylate resin. See under Acrylic resin
1-Methyl-4-(1-methylethenyl) cyclohexene. See Terpene/polyterpene resins (dipentene)
Nopinene. See under Terpene/polyterpene resins (β-pinene)
Nylon resin. See Book VI
PBT. See under Polyester, thermoplastic
PC resin. See Polycarbonate resin
PET. See under Polyester, thermoplastic
PFA resin. See Perfluoroalkoxy resin
Phenol-formaldehyde resin. See Book VI
Phenylmethylpolysiloxane. See Book VI
α-Pinene. See under Terpene/polyterpene resins (α-pinene)
β-Pinene. See under Terpene/polyterpene resins (β-pinene)
Polyacetal. See Acetal resin
Polyarylate. See under Polyester, thermoplastic
Polybutylene terephthalate. See under Polyester, thermoplastic
Polyester, saturated. See under Polyester, thermoplastic
Polyester, unsaturated. See under Polyester, thermoset
Polyethylene resin. See Book VI
Polyethylene terephthalate. See under Polyester, thermoplastic
Polymethacrylate resin. See under Acrylic resin
Poly (methyl vinyl ether). See Polyvinyl methyl ether (CTFA)
Polypropylene resin. See Book VI
Polysiloxane-polyether copolymer. See Book VI
Polyurethane. See Book VI
Polyvinyl butyral resin. See Polyvinyl butyral (CTFA)
Polyvinyl chloride. See Book VI
Polyvinyl isobutyl ether. See Book VI
PPS. See Polyphenylene sulfide
PS. See Polystyrene
PSO. See Polysulfone
PTFE. See Polytetrafluoroethylene
PVAc. See Polyvinyl acetate
PVB. See Polyvinyl butyral (CTFA)
SAN. See Styrene-acrylonitrile resin
Saturated polyester. See under Polyester, thermoplastic
Secondary acetate. See Cellulose acetate
Sodium polyacrylate. See Book VI
Sodium polymethacrylate. See Book VI
Tetrafluoroethylene/hexafluoropropylene copolymer. See Fluorinated ethylene-propyl-
    ene

TFE. See Polytetrafluoroethylene

1,3,5-Triazine-2,4,6-triamine, polymer with formaldehyde. See Melamine-formaldehyde resin

Unsaturated polyester. See under Polyester, thermoset

Urea, polymer with formaldehyde. See Urea-formaldehyde resin

Vinylidene chloride copolymer. See Book VI

Vinyl acetyl polymers, butyrals. See Polyvinyl butyral (CTFA)

TRADENAME PRODUCT MANUFACTURERS

Acme Chemical & Insulation Co./ Div. of Allied Prod. Corp.
PO Box 1404
New Haven, CT 06505

Allied-Signal
PO Box 2332R
Columbia Rd. and Park Ave.
Morristown, NJ 07960

Allied Corporation International NV-SA
Haasrode Research Park
B-3030 Heverlee, Belgium

Allied Corporation International NV-SA
International House
Bickenhill Lane
Birmingham B37 7HQ
England

Allied Chemical International Corporation
P.O. Box 99067, Tsimshatsui Post Office
Hong Kong

American Cyanamid Co.
Industrial Chem. Div.
Berdan Ave.
Wayne, NJ 07470

American Cyanamid Co./
Polymer & Chem. Dept.
Berdan Ave.
Wayne, NJ 07470

Cyanamid B.V.
Postbus 1523, 3000 BM
Rotterdam, The Netherlands

Cyanamid India Ltd.
Nyloc House, 254-D2 Dr. Annie Besant Rd.
Bombay 400 025 India

Cyanamid Quimica do Brasil Ltda.
Av. Imperatriz Leopoldina, 86
Sao Paulo, Brazil

Cyanamid Taiwan Corp.
8/F Union Commercial Bldg.
137, Nanking E. Rd., Sec. 2
Taipei, Taiwan, R.O.C.

Amoco Chemicals Corp.
200 East Randolph Dr., PO Box 8640A
Chicago, IL 60680

Amoco Chemical Europe S.A.
15, Rue Rothschild
CH-1211 Geneva 21,
Switzerland

Amoco Performance Products,
Japan Ltd.
10th Floor, Tonichi Building
2-31 Roppongi 6-Chome
Minato Ku, Tokyo 106, Japan

Arizona Chemical Co.
16-00 Route 208
Fairlawn, NJ 07410

Ashland Chemical Co./ Div. Ashland Oil, Inc.
Box 2219
Columbus, OH 43216

Astro Chemical Co., Inc.
1205 Godfrey Lane
Schenectady, NY 12309

Bacon Industries Inc.
192 Pleasant St.
Watertown, MA 02172

Badische Corp.
PO Drawer D
Williamsburg, VA 23187

Bamberger Polymers, Inc.
3003 New Hyde Park Rd.
New Hyde Park, NY 11042

BASF Corp.
100 Cherry Hill Rd.
Parsippany, NJ 07054

BASF Canada Ltd.
PO Box 430
Montreal, Quebec H4L 4V8, Canada

BASF (UK) Ltd.
PO Box 4, Earl Rd., Cheadle Hulme
Cheadle, Cheshire 5K8 60QG, UK

BASF Belgium S.A.
avenue Hamoir-Iaan 14
B-1180 Bruxelles/Brussel
Belgium

BASF AG
ESA/WA-H 201
D-6700 Ludwigshafen, West Germany

BASF Espanola S.A.
Apartado 762
Barcelona 8, Spain

BASF S.A., Compagnie Francaise
MC-NT, 140, Rue Jules Guesde
92303 Levallois-Perret, France

BASF India, Ltd.
Maybaker House, S.K. Ahire Marg.
PO Box 19108
Bombay 400 025 India

BASF Japan Ltd.
C.P.O. Box 1757
Toyko 100-91
Japan

Bayer AG
Sitz der Gesellschaft, Leverkusen
Eintragung, Amtsgericht
Leverkusen HRB 1122, Germany

Bayer UK Ltd.
Bayer House, Strawberry Hill
Newbury, Berkshire RG13 1JA, UK

Borg-Warner Chemicals Inc.
International Center
Parkersburg, WV 26102

Borg Warner Chemicals
20 Coventry Rd., Cubbington
Leamington Spa, Warks CV32 7JW, UK

Borg-Warner Chemical Europe B.V.
Cyprusweg 2, P.O. Box 8122
1005 AC Amsterdam, Havens West
The Netherlands

Ube Cycon, Limited
Daito Building 3F
7-1 Kasumigaseki 3-chome
Chiyoda-ku, Tokyo 100
Japan

R.H. Carlson Co.
41 Chestnut St.
Greenwich, CT 06830

Ciech Plastofarb Div.
ul. Jasna 12
00950 Warsaw, Poland

Compounding Technology, Inc.
3140 Pullman St.
Costa Mesa, CA 92626

Continental Polymers Inc.
2225 East Del Amo Blvd.
Compton, CA 90220

Cosden Oil & Chemical Co.
PO Box 2159, 3350 N. Central Expressway
Dallas, TX 75221

Crosby Chemical Inc.
600 Whitney Bldg.
New Orleans, LA 70130

Cyro Industries
697 Route 46
Clifton, NJ 07015

TRADENAME PRODUCT MANUFACTURERS

Delta Plastics Co.
Santa Fe Springs, CA

Dow Chemical Co.
1703 S. Saginaw Rd.
Midland, MI 48640

Dow Chemical Co. Ltd.
Stana Place, Fairfield Ave.
Staines, Middlesex TW18 4SX, UK

Dow Chemical Europe S.A.
Bachtobelstrasse 3, CH-8810
Horgen, Switzerland

Dow Chemical Pacific Limited
P.O. Box 711
Hong Kong

E.I. DuPont de Nemours & Co.
Nemours Bldg.
Wilmington, DE 19898

Du Pont Co./Polymer Prod. Dept.
15990 N. Barkers Landing Rd., POB 19029
Houston, TX 77224

DuPont Canada
PO Box 26, Toronto Dominion Center
Toronto, Ont. M5K 1B6 Canada

DuPont (UK) Ltd.
Wedgwood Way, Stevenage
Herts SG1 4QN, UK

DuPont (UK) Ltd./Polymer Prods. Dept.
Maylands Ave., Hemel Hempstead
Herts HP2 7DP, UK

DuPont France
0137, rue de l'Universite
F-75334 Paris Cedex 07
France

DuPont de Nemours International S.A.
Case Postale CH-1211
Geneva 24, Switzerland

DuPont Japan Ltd.
P.O. Box 37, Akasaka
Tokyo, 107-91 Japan

Eastman Chemical Products, Inc./ Subsid. of Eastman Kodak Co.
PO Box 431
Kingsport, TN 37662

Eastman Chemical International A.G.
Hemel Hempstead
P.O. Box 66
Kodak House, Station Road
Herts, HP1 1JU England

Eastman Japan Ltd.
Nishi-Shinbashi Mitsui Bldg.
1-24-14 Nishi-Shinbashi
Minato-Ku, Tokyo 105
Japan

Emerson & Cuming—*See W.R. Grace & Co.*

Fiberite Corp.
501-559 W. Third St.
Winona, MN 55987

Freeman Chemical Corp./Subsid. of H.H. Robertson Co.
PO Box 247
Port Washington, WI 53074

GAF Corp./Chemical Products
1361 Alps Rd.
Wayne, NJ 07470

GAF Europe
40 Alan Turing Rd., Surrey Research Park
Guildford, Surrey, UK

General Electric Co.
General Electric Co./Plastics Div.
One Plastics Ave.
Pittsfield, MA 01201

General Electric Co./Silicone Prod. Div.
Mechanicsville Rd.
Waterford, NY 12188

General Electric Plastics B.V.
Plasticslaan 1, P.O. Box 117
4600 AC Bergen op Zoom
The Netherlands

377

GE Plastics Ltd.
Birchwood Park
Risley, Warrington
Cheshire WA3 6DA
England

General Electric (USA) Plastics Japan Ltd.
No. 35 Kowa Building 5F
14-14 Akasaka 1-chome
Minato-ku, Tokyo 107
Japan

Georgia-Pacific Corp.
300 W. Laurel St.
Bellingham, WA 98225

M.C. Gill Corp.
4056 Easy St.
El Monte, CA 91731

BFGoodrich Co.
6100 Oak Tree Blvd.
Cleveland, OH 44131

BFGoodrich Chemical (U.K.) Ltd.
The Lawn
100 Lampton Road
Hounslow, Middlesex TW3 4EB
England

BFGoodrich Chemical (Deutschland)
GmbH
Goerlitzer Str. 1
4040 Neuss 1
Federal Republic of Germany

W.R. Grace & Co,
W.R. Grace & Co./Organic Chem. Div.
55 Hayden Ave.
Lexington, MA 02173

Emerson & Cuming/W.R. Grace & Co.
869 Washington St.
Canton, MA 02021

W.R. Grace Ltd.
Northdale House, North Circular Rd.
London NW10 7UH, UK

C.P. Hall Co.
7300 South Central Ave.
Chicago, IL 60638

Hammond Plastics Div./ Carl Gordon Industries Inc.
1001 Southbridge St.
Worcester, MA 01610

Hardman Inc.
600 Cortland St.
Belleville, NJ 07109

Harwick Chemical Corp./Polymer Application
60 South Seiberling St.
Akron, OH 44305

Hercules Inc.
Hercules Plaza
Wilmington, DE 19894

Hercules Inc./Carbon Fiber Materials
PO Box 98
Magna, UT 84044

Hercules B.V.
8 Veraartlaan, PO Box 5822
2280 HV Rijswijk, The Netherlands

Hercules Ltd., Hercules European Hdqts.
20 Red Lion St.
London WC1R 4PB UK

Hexcel Corp./Chemical Products Div.
215 N. Centennial St.
Zeeland, MI 49464

Hexcel Corp./Rezolin Div.
20701 Nordhoff St.
Chatsworth, CA 91311

Hoechst Celanese Corp.
Route 202-206 North
Somerville, NJ 08876

Hoechst AG
Verhaufkanststoffe, D-6230
Frankfurt (M) 80, West Germany

Hoechst Celanese Plastics Ltd.
78-80 St. Albans Rd.
Watford, Herts
England WD2-4AP

Hoechst Japan
10-33, 4-Chome-Akasaka, Minato-ku
Tokyo, Japan

Huls America Inc.
10 Link Dr.
Rockleigh, NJ 07647

Hüls AG, Chemische Werke
Postfach 1320 D-4370
Marl 1, West Germany

Huls (UK) Ltd.
Cedars House, Farnborough Common
Orpington, Kent BR6 7TE, UK

Hysol Div./The Dexter Corp.
15051 E. Don Julian Road
Industry, CA 91749

Impco Inc.
335 Valley St.
Providence, RI 02908

International Thermoset Inc.
11113 North Dixie Drive
Vandalia, OH 45377

S.C. Johnson & Son, Inc.
3225 North Verdugo Rd.
Glendale, CA 91208

Kama Corp.
668 Dietrich Ave., PO Box 920
Hazelton, PA 18201

Koppers Co., Inc.
1350 Koppers Bldg.
Pittsburgh, PA 15219

LNP Corp.
412 King St.
Malvern, PA 19355

LNP Plastics U.K.
Unit 25 Monkspath Business Park
Solihull, West Midlands
England

M.A. Industries Inc., Polymer Div.
PO Box 2322
Peachtree City, GA 30269

Mitsubishi International Corp./ Chemicals Dept.
277 Park Ave.
New York, NY 10017

Mitsubishi Chemical Industries Ltd.
6-3, Marunouchi 2-chome, Chiyoda-ku
Tokyo, Japan

Mobay Chemical Corp./Organic & Rubber Chem. Div., Plastics Div.
Penn Lincoln Parkway West
Pittsburgh, PA 15205

Mobil Oil Corp./Special Products Dept.
3225 Gallows Rd.
Fairfax, VA 22037

Mobil Polymers U.S. Inc.
591 W. Putnam Ave.
Greenwich, CT 06836

Monsanto Co.
800 N. Lindbergh Blvd.
St. Louis, MO 63167

Monsanto PLC
Monsanto House
Chineham Court, Chineham
Basingstoke, Hants RG24 0UL, UK

Monsanto Europe S.A.
Avenue de Tervuren 270-272
P.O. Box 1
B-1150 Brussels, Belgium

Morton International
1275 Lake Ave.
Woodstock, IL 60098

Morton International/Carstab Div.
2000 West St.
Cincinnati, OH 45215

Morton International/Morton Chem. Div.
101 Carnegie Center
Princeton, NJ 08540, NJ 08540

Morton International/Ventron Div.
150 Andover St.
Danvers, MA 01923

Williams Div./Morton Thiokol Ltd.
Greville House, Hibernia Rd.
Hounslow, Middlesex TW3 3RX, UK

MRC Enterprises Inc.
1716 W. Webster Ave.
Chicago, IL 60614

National Starch and Chemical Corp.
10 Finderne Ave.
Bridgewater, NJ 08807

Represented by:
National Adhesives & Resins Ltd.
Braunston Daventry
Northante NN11 7JL, UK

Neville Chemical Co.
Neville Island
Pittsburgh, PA 15225

Occidental Chemical Corp.
360 Rainbow Blvd, S., POB 728
Niagara Falls, NY 14302

Occidental Chemical Corp./
Durez Resins & Molding Materials
Walck Rd., Box 535
North Tonawanda, NY 14120

Phillips Chemical Co./Subsid. of Phillips Petroleum Co.
14 Phillips Building
Bartlesville, OK 74004

Plastics Engineering Co.
3518 Lakeshore Rd.
Sheboygan, WI 53081

Polymer Applications
3445 River Road
Tonawanda, NY 14150

Polymer Corp., The
2120 Fairmont Ave., PO Box 422
Reading, PA 19603

Polyvinyl Chemical Industries
730 Main St.
Wilmington, MA 01887

Presto Mfg. Co., Inc.
2 Franklin Ave.
Brooklyn, NY 11211

Quantum Chemical Corp.
26250 Euclid Ave.
Euclid, OH 44132

Quantum Chemical Corp., USI Div.
Lange Bunder 7
4854 MB Bavel, The Netherlands

Quantum Chemical Ltd.
365 Evans Avenue, Suite 601
Toronto, Ontario M8Z 1K2

Reichhold Chemicals Inc.
RCI Building
White Plains, NY 10602

Rogers Corp.
One Technology Dr.
Rogers, CT 06263

Rohm & Haas Co.
Independence Mall West
Philadelphia, PA 19105

Rohm & Haas (Australia) Pty. Ltd.
969 Burke Rd., PO Box 115 Camberwell
Victoria 3124 Australia

Rohm and Haas (UK) Ltd.
Lennig House
2 Mason's Avenue
Croydon CR9-3NB
England

Rohm & Haas Co. European Operations
Chesterfield House, Bloomsbury Way
London WC1A 2TP, UK

Rohm and Haas Asia, Ltd.
Kaisei Building
8-10 Azabudai 1-Chome
Minato-ku, Tokyo 106 Japan

TRADENAME PRODUCT MANUFACTURERS

Scott Bader Co. Ltd.
Wollaston, Wellingborough
Northamptonshire NN9 7RL, UK

Solvay & Cie
Rue du Prince Albert 44-
1050 Brussels, Belgium

Soltex Polymer Corp./
Subsid. of Solvay & Cie
3333 Richmond Ave., POB 27328
Houston, TX 77098

Spaulding Fibre Co., Inc.
1 American Dr.
Buffalo, NY 14225

St. Lawrence Resin Products, Ltd.
350 Wentworth St. N.
Hamilton, Ontario, Canada L8L 5W3

Standard Polymers Corp.
45 Legion Dr.
Cresskill, NJ 07626

The Terrell Corp.
820 Woburn St.
Wilmington, MA 01887

Thibaut & Walker
PO Box 296
Newark, NJ 07101

Tra-Con Inc.
55 North St.
Midford, MA 02155

Union Carbide Corp.
39 Old Ridgebury Rd.
Danbury, CT 06817

Union Carbide U.K. Limited
Rickmansworth WD 3 1RB Herts
England

Union Carbide Europe S.A.
15, Chemin Louis-Dunant
1211 Geneve 20
Geneva, Switzerland

Union Carbide Brazil
Rua Dr. Eduardo De Souza Aranha
153, Sao Paulo 04530
Brazil

Union Carbide Japan KK
Toranomon 45 Mori Bldg.
1-5 Toranomon
5-Chome Minato-Ku
Tokyo 105, Japan

USI Chemicals Co.—*See Quantum Chem. Corp./USI Div.*

Washington Penn Plastic Co., Inc.
2080 North Main St.
Washington, PA 15301

Westlake Plastics Co.
PO Box 127
Lenni, PA 19052

Wilmington Chemical Corp.
PO Box 66, Pyles Lane
Wilmington, DE 19899

Xcel Corp.
290 Ferry St.
Newark, NJ 07105

www.ingramcontent.com/pod-product-compliance
Lightning Source LLC
Chambersburg PA
CBHW060754220326
41598CB00022B/2432